I0890532

Redesigning Higher Education Initiatives for Industry 4.0

Arumugam Raman
Universiti Utara Malaysia, Malaysia

Mohan Rathakrishnan
Universiti Utara Malaysia, Malaysia

A volume in the Advances in Higher Education and Professional Development (AHEPD) Book Series

IGI Global
DISSEMINATOR OF KNOWLEDGE

Published in the United States of America by
IGI Global
Information Science Reference (an imprint of IGI Global)
701 E. Chocolate Avenue
Hershey PA, USA 17033
Tel: 717-533-8845
Fax: 717-533-8661
E-mail: cust@igi-global.com
Web site: http://www.igi-global.com

Copyright © 2019 by IGI Global. All rights reserved. No part of this publication may be reproduced, stored or distributed in any form or by any means, electronic or mechanical, including photocopying, without written permission from the publisher. Product or company names used in this set are for identification purposes only. Inclusion of the names of the products or companies does not indicate a claim of ownership by IGI Global of the trademark or registered trademark.

Library of Congress Cataloging-in-Publication Data

Names: Raman, Arumugam, editor. | Rathakrishnan, Mohan, 1965- editor.
Title: Redesigning higher education initiatives for industry 4.0 / Arumugam Raman and Mohan Rathakrishnan, editors.
Description: Hershey, PA : Information Science Reference, [2019] | Includes bibliographical references.
Identifiers: LCCN 2018036353| ISBN 9781522578321 (h/c) | ISBN 9781522578338 (eISBN)
Subjects: LCSH: Higher education--Effect of technological innovations on--Case studies. | Cloud computing--Case studies. | Artificial intelligence--Educational applications--Case studies. | Digial communications--Case studies.
Classification: LCC LB2395.7 .R34 2019 | DDC 378.1/758--dc23 LC record available at https://lccn.loc.gov/2018036353

This book is published in the IGI Global book series Advances in Higher Education and Professional Development (AHEPD) (ISSN: 2327-6983; eISSN: 2327-6991)

British Cataloguing in Publication Data
A Cataloguing in Publication record for this book is available from the British Library.

All work contributed to this book is new, previously-unpublished material. The views expressed in this book are those of the authors, but not necessarily of the publisher.

For electronic access to this publication, please contact: eresources@igi-global.com.

Advances in Higher Education and Professional Development (AHEPD) Book Series

Jared Keengwe
University of North Dakota, USA

ISSN:2327-6983
EISSN:2327-6991

Mission

As world economies continue to shift and change in response to global financial situations, job markets have begun to demand a more highly-skilled workforce. In many industries a college degree is the minimum requirement and further educational development is expected to advance. With these current trends in mind, the **Advances in Higher Education & Professional Development (AHEPD) Book Series** provides an outlet for researchers and academics to publish their research in these areas and to distribute these works to practitioners and other researchers.

AHEPD encompasses all research dealing with higher education pedagogy, development, and curriculum design, as well as all areas of professional development, regardless of focus.

Coverage

- Adult Education
- Assessment in Higher Education
- Career Training
- Coaching and Mentoring
- Continuing Professional Development
- Governance in Higher Education
- Higher Education Policy
- Pedagogy of Teaching Higher Education
- Vocational Education

IGI Global is currently accepting manuscripts for publication within this series. To submit a proposal for a volume in this series, please contact our Acquisition Editors at Acquisitions@igi-global.com or visit: http://www.igi-global.com/publish/.

The Advances in Higher Education and Professional Development (AHEPD) Book Series (ISSN 2327-6983) is published by IGI Global, 701 E. Chocolate Avenue, Hershey, PA 17033-1240, USA, www.igi-global.com. This series is composed of titles available for purchase individually; each title is edited to be contextually exclusive from any other title within the series. For pricing and ordering information please visit http://www.igi-global.com/book-series/advances-higher-education-professional-development/73681. Postmaster: Send all address changes to above address. Copyright © 2019 IGI Global. All rights, including translation in other languages reserved by the publisher. No part of this series may be reproduced or used in any form or by any means – graphics, electronic, or mechanical, including photocopying, recording, taping, or information and retrieval systems – without written permission from the publisher, except for non commercial, educational use, including classroom teaching purposes. The views expressed in this series are those of the authors, but not necessarily of IGI Global.

Titles in this Series

For a list of additional titles in this series, please visit: www.igi-global.com/book-series

Study Abroad Opportunities for Community College Students and Strategies for Global Learning
Gregory F. Malveaux (Montgomery College, USA) and Rosalind Latiner Raby (California State University Northridge, USA)
Information Science Reference • copyright 2019 • 324pp • H/C (ISBN: 9781522562528) • US $195.00 (our price)

Identifying, Describing, and Developing Teachers Who Are Gifted and Talented
Meta L. Van Sickle (College of Charleston, USA) Julie D. Swanson (College of Charleston, USA) Judith A. Bazler (Monmouth University, USA) and Kathryn L. Lubniewski (Monmouth University, USA)
Information Science Reference • copyright 2019 • 302pp • H/C (ISBN: 9781522558798) • US $195.00 (our price)

Challenges and Opportunities for Women in Higher Education Leadership
Heidi L. Schnackenberg (State University of New York at Plattsburgh, USA) and Denise A. Simard (State University of New York at Plattsburgh, USA)
Information Science Reference • copyright 2019 • 353pp • H/C (ISBN: 9781522570561) • US $195.00 (our price)

Examining Student Retention and Engagement Strategies at Historically Black Colleges and Universities
Samuel L. Hinton (Independent Researcher, USA) and Antwon D. Woods (Belhaven University, USA)
Information Science Reference • copyright 2019 • 250pp • H/C (ISBN: 9781522570219) • US $175.00 (our price)

Exploring the Technological, Societal, and Institutional Dimensions of College Student Activism
Michael T. Miller (University of Arkansas, USA) and David V. Tolliver (University of Arkansas, USA)
Information Science Reference • copyright 2019 • 303pp • H/C (ISBN: 9781522572749) • US $185.00 (our price)

Optimizing Instructional Design Methods in Higher Education
Yianna Vovides (Georgetown University, USA) and Linda Rafaela Lemus (Georgetown University, USA)
Information Science Reference • copyright 2019 • 245pp • H/C (ISBN: 9781522549758) • US $175.00 (our price)

Handbook of Research on Field-Based Teacher Education
Thomas E. Hodges (University of South Carolina, USA) and Angela C. Baum (University of South Carolina, USA)
Information Science Reference • copyright 2019 • 749pp • H/C (ISBN: 9781522562498) • US $245.00 (our price)

Student-Centered Virtual Learning Environments in Higher Education
Marius Boboc (Cleveland State University, USA) and Selma Koç (Cleveland State University, USA)
Information Science Reference • copyright 2019 • 281pp • H/C (ISBN: 9781522557692) • US $165.00 (our price)

IGI Global
DISSEMINATOR OF KNOWLEDGE

701 East Chocolate Avenue, Hershey, PA 17033, USA
Tel: 717-533-8845 x100 • Fax: 717-533-8661
E-Mail: cust@igi-global.com • www.igi-global.com

Table of Contents

Section 4
Awareness and Challenges in Redesigning Higher Education 4.0 Initiatives

Detailed Table of Contents

Higher education in the era of Industrial Revolution 4.0 (IR 4.0) has given a brand new drive on Malaysian educational transformation. The impact of a numerous technology innovations in ICT towards Malaysian education is noticeable. Some existing jobs will be eliminated in this era, but more new jobs that require digital literacy are going to be created. Therefore, Education 4.0 must be able to produce graduates with excellent communication and collaborative skills. However, Malaysian Digest states that fresh graduates lack soft skills. Several approaches have been proposed by the Ministry of Higher Education for the purpose of enhancing students' soft skills. Therefore, by having technologies to help overcome this problem is sought after. This chapter describes the utilization of mobile learning and gamification in enhancing learner's soft skills. It elaborates the effectiveness of using gamification in mobile learning technique to develop learners' higher order thinking skills to help them succeed in a global economy.

When a society is undergoing transformational change, it is a challenge for all involved to step outside their immediate context sufficiently to evaluate its implications. In the current digital revolution driving Industry 4.0, the pace of change is rapid, and its scale and complexity can inhibit a proactive, rather than reactive, response. Yet if it were possible to return to the first industrial revolution, armed with twenty-first century knowledge and historical perspective, planning for a healthy society and the future of work could have been very different. This chapter aims to support educational leadership in the development of proactive strategies to respond to the challenges and opportunities of Industry 4.0 to inform the future of work, industry, and society. This is framed through the lens of product design, with its unique position at the nexus of engineering and the humanities, and directly tied to changes affecting manufacturing in the fourth industrial revolution.

The need for students to engage in critical thinking, collaboration, communication, and creativity when utilizing digital technologies is well established given the current push to prepare our students for future work expectations, often referred to as Industry 4.0. In this chapter, 54 Grade 5 students participated in online collaborative mathematical problem solving over a period of 9 weeks. Text-based discussion and software derived artefacts were investigated in an effort to understand how this approach aligns with these expectations. An examination of the frequency/density of technical mathematical vocabulary use and identified examples, Talk Types is used as a means to understand how often students within the online environment are likely to be engaged in work that might be considered productive. The findings of this study suggest that utilizing online synchronous and asynchronous collaborative learning platforms for the purposes of mathematical problem solving is a small but important approach to preparing students for the era of Industry 4.0.

This chapter seeks to take a modest and yet formative approach in proposing computer-based test (CBT) as the inevitable next wave in digital assessment. The major section reports on the practical design of developing and administering CBT in a mathematics competition by HOTS educational innovation development (HEID) in a mathematics competition. This chapter describes the CBT developmental process undertaken by HEID, which focuses on constructing multiple-choice test items, administering CBT, and scoring the test. While CBT offers the possibility of improving assessment, the process of preparing items and conducting online competition have nevertheless encountered several challenges such as getting quality test items, school and teacher readiness, seat time in front of computer, computer compatibility, and internet speed. As with many other educational reforms at infancy, more research in the area of implementing CBT to facilitate international competitions is needed in landscaping the assessment platform of Education 4.0.

In response to increased criticism of the utility of the traditional doctoral dissertation, some institutions have incorporated additional options for students such as articles, portfolios, and industry-specific projects. This trend towards allowing alternatives for doctoral students coincides with the rise internationally of what are variously called professional, applied, or practitioner doctorates in various disciplines. The goal of this review of the relevant research literature and policy documents is to understand the evolution of professional doctoral degrees, and how rigor and quality are evaluated specifically with non-dissertation capstone projects. Findings suggest that universities, accrediting agencies, and professional associations recognize a need for new standards for alternatives to the traditional dissertation in order to make such options rigorous and relevant for the students.

Use of algorithms and data mining approaches are not new to Industry 4.0. However, these may not be common for students and educators in higher education. This chapter compares various classification techniques: classification tree, logistic regression, and artificial neural networks (ANN). The comparison focuses on each method's accuracy, algorithm, and practicality in higher education. This study made use of a dataset from two academic writing courses in a university in Hong Kong with more than 5,000 records. Results suggest that classification trees and logistic regression can be easily used in the higher education context, but ANN may not be applicable in higher educational settings. The research team suggests that higher education administrators take this research forward and design platforms to realize these classification algorithms to predict at-risk students.

Section 4
Awareness and Challenges in Redesigning Higher Education 4.0 Initiatives

Industry Revolution 4.0 (IR 4) is highly rated and greatly establishing itself. The future of our world looks so promising. The revolution is happening, and the impact is already seen in many sectors. Education is one of the sectors that needs to be in line with the growth of the technology to produce a world class work force. In Malaysia, the government had earlier made many changes in education through its National Higher Education Action Plan 2007-2010 in transforming teaching and learning process focusing on technology. Innovative and entrepreneurial skills must be possessed by new graduates and they must deal with complexity with cognitive flexibility. This is because they need to accept robots as their working colleagues apart from humans. The need for better communication and collaborative skills will be far more important than ever. Graduates must acquire self-learning skills to remain relevant in the era of rapid changes.

As programming is one of the core subjects for BSc IT students, it is vital for them to have programming skill. However, students are often terrified of programming since it is a challenging subject. Traditionally, the programming teaching and learning process are limited to lectures and lab exercises. This approach is no longer appropriate in today's learning environment and required skills in the 21st century. Collaborative learning is acknowledged as an approach that fits the programming students. Thus, this chapter has adapted the collaborative approach in Programming 1 class. Three techniques were applied namely problem-based learning, buzz groups, and pair programming. At the end of the semester, an online survey to obtain students' opinions and a focus group discussion were conducted. In addition, the examination results of the repeaters were also compared. Results from the survey and focus group indicate that the collaborative approach is able to increase students' interest and the grade of the repeaters in the Programming 1 subject.

This chapter argues for the importance of equipping academics with the knowledge and impact of IR4.0 and beyond for future graduates. Particularly to prepare for military leaders for Malaysia at the National Defense University of Malaysia (NDUM), a new learning model may be apt and urgently needed. The objectives of the chapter are twofold: firstly, to gauge the readiness and awareness level of academics on

Foreword

Alhamdulillah, all praise and thanks be to Allah, the Lord of the Worlds. It gives me great honor and pleasure to write a few words in the opening pages of this timely published volume that contains important insights to achieving the IR4.0 through the means of education. The editors would have gone through an arduous task of blending such contrasting views into one publication. I commend the editors for their exceptional work and their excellent selection of article contributors. It is indeed a great challenge to bring a new book into the hands of millions, all around the world, especially when the book aims to publish high quality manuscripts on today's very topical issue, the Industry Revolution 4.0 (IR 4.0).

IR 4.0 has become the major focus of all stakeholders from all around the world and many scholars understand its values, but in actual fact only a few spend their time exploring the related issues. This book entitled 'Redesigning Higher Education Initiatives for IR 4.0" aims to provide relevant theoretical frameworks and the latest empirical research findings in this area. It is written by and for professionals, who are paving ways to support and achieve IR 4.0 through education, the core to all revolution and achievement.

From self-driving cars to drone-delivered online shopping, the Fourth Industrial Revolution, is changing and has in fact, changed the way we live, work and communicate. As we are aware, more and more jobs are being taken over by Artificial Intelligence (AI) and students today need to stay relevant for future markets and employment trends. According to The World Economic Forum, an estimated 65% of kids enrolling in primary education today for sure will end up working in jobs that have not been created in the present job market. Gone are the days where students go to college or university to study for a degree that will set them up with a job for life.

With this in mind, two highly valuable scholars from the Universiti Utara Malaysia, have made efforts to serve on this editorial board. I am honored to have Associate Professor Dr. Arumugam Raman and Dr. Mohan Rathakrishnan, who have initiated and worked on collecting high quality articles written by professionals and researchers working in the field of information and knowledge management from various disciplines to provide insights on how education is made to become the path way towards achieving IR 4.0.

This book, majorly provides readers with broad understanding of the impact of Blended Learning, Flipped Classroom, Mobile Learning, Heutagogy, Programming courses, Gamification, Massive Open Online Courses (Moocs), Internet of Things (IoT) and the challenges of IR 4.0 in Higher Education.

Last but not least, my humble hope is that, this book, would be able to support the trends of the future: 2025 and beyond, by making readers understand why it is important that 80% (of Internet users) have unlimited backup space in the cloud; 80% of the world's population have Internet presence; 90% of world's population owns a smart phone; 90% of world's population have Internet access, and what's more, 10% of all vehicles on the roads is driverless! Just by embracing the content of this book: 'Redesigning Higher Education Initiatives for IR 4.0", readers will have better understanding of what makes education rise with the occasion.

The challenges that come with confronting IR4.0 in terms of Education, cry out for the need to be digital natives – which is what Education in the 21st. century is all about! This book which contains collection of thoughts and views by scholars and practitioners have provided us with important clues, if not solutions to meet the challenges of an increasingly complex era. My palms clasped together thanking Associate Professor Dr. Arumugam Raman and Dr. Mohan Rathakrishnan for making this book a reality.

Ahmad Bashawir Hj. Abdul Ghani
Universiti Utara Malaysia, Malaysia

Preface

The Fourth Industrial Revolution (Industry 4.0) involves automation technology giving new challenges to all sectors in the country that require them to make changes in line with the digital transformation to remain competitive. Previously, the industrial revolution was first depended on the use of vapor-powered machines. The second industrial revolution involves electric power followed by a third industrial revolution based on information technology and computer as its pillar.

It is undeniable to say that today's technological developments in education have led to unpredictable changes in the life of human being especially with the emerge of the Industrial Revolution (IR) 4.0. It has lots of influences the way we live in today's society. The Fourth Industrial Revolution (Industry 4.0) provides an education reassess the quality of teaching and learning for learners of the twenty-first century. It is very crucial for providers of education, universities and community colleges as well as schools to work together to provide quality instruction to all learners regardless of ethnic and cultural background.

In line with the democratization to equal access of education, IR 4.0 needs to upgrade and improve education we provide to our graduates or school leavers. Lecturers and teachers have to learn new skills and competencies related to Information Communication Technology (ICT) so that they can be relevant and competitive to serve as future employees in a more challenging environment in the future. Malaysian Ministry of Education (MOE) and education providers around the world have to put in a lot of efforts to integrate ICT into learning and facilitation of various subjects. Universities need to expose lecturers and teachers with new skills, knowledge, and abilities that can maximize ICT into teaching and learning processes.

Higher education according to John Dewey (1916) is a process of experience in learning new things. He mentioned that education helps students to build their inner growth without any age barrier. He indicated that higher education would create skills in processing phase to develop a person. Hacker (1998) said that higher education has a deliberate effort chosen to influence and assist learners to improve science, physiology, and morality so that changes in technology can gradually deliver its purpose to achieve its ultimate goal. Educators should take this opportunity to integrate computers as an instructional medium to facilitate the teaching and learning process.

In general, higher education has never really been a public discourse in many countries. There are many problems in the education system that has to be resolved. Various things can be blamed as the main problem that hampers the progress of education in the world. However, we have to find a solution in improving the conventional learning process that relies on face to face between teachers and students, lecturers and students and coaches with participants training. Ineffectiveness is the most suitable word for this system, because along with the times of development, the information exchange becomes faster and advance, but most of the institution still using traditional way of teaching and learning (at the level of high school, we think of giving information) and not in line with IT development. This conventional system should have been abandoned since the discovery of multimedia communications media. Due to the nature of the internet that can be contacted at all times, it means that students can utilize educational programs provided on the Internet network at any time according to their leisure time so that space and time constraints they face while searching for learning resources can be solved. With the rapid development especially in the Industrial Revolution (IR) 4.0, the field of telecommunications, multimedia, and information, listening to lectures, writing on paper is certainly outdated.

Industry 4.0 covers the discovery of new technologies such as automation, Internet of Things (IoT), analysis and big data, simulations, system integration, robotic and cloud utilization that will accelerate the development of the modern world landscape. This revolution marks the emergence of cyber-physical systems involving a whole new capability for human beings, machines and new technology technologies. In other words, automation technology is seen as a technological capability that does not necessarily involve human beings. Industry 4.0 can also address dependency on energy sources that will significantly change the future of the world of occupation. So, if the complex new technological challenge is not given due weight by a country, it can cause to lag behind in global competition. Recently, World Bank's Chief Expert, Francisco Marmolejo in the Industry Revolution Seminar 4.0 said that the higher education system across the country will also undergo changes in the industrial revolution of 4.0. Imagine the knowledge we have today contributing only one percent by the next 30 years and then the job market will be filled with individuals or highly skilled, creative and critical thinking workforce. "So we need a higher-education institute that is more flexible and ready to face the new challenge so that any group of people will not be left out in the globalization and digital era," he said at a seminar organized by the Higher Education Ministry in the Kuala Lumpur, Malaysia (Ranai, 2017).

Garrison and Archer (2000) indicated that the improvement on the learning system using advance online as a "community of learning" or "community of inquiry model" involving three components that are the educator presence, cognitive presence and social presence. Educator presence in online learning is vital to design, and organize the learning process. Through an adequate educator presence defined educational outcomes will be achieved. However, redesigning higher education to make IR 4. online learning effective and involve serious engagement, cognitive presence is necessary to support the development and growth of critical thinking skills. Cognitive presence is defined by the process of inquiry that moves from problem definition to exploration of relevant content and ideas, integrating those ideas into a meaningful structure, and then directly testing the validity of the outcome of higher education learning.

OBJECTIVES

This book aims to provide relevant theoretical frameworks and the latest empirical research in four different sections namely the important of IR4.0, teaching and learning using up-to-date technology, digital transformation and case study. It is written for professionals or readers who want to seek knowledge, ideas, strategies involved in IR 4.0. The main objectives are to:

1. Provide broad understanding impact of Industrial 4.0 towards higher education
2. Identify current status and readiness of Industrial 4.0 in higher education
3. Reveal the challenges of Industrial Revolution 4.0 in higher education
4. Explore the role of artificial intelligence in higher education

5. Identify the impact of cloud computing in higher education
6. Assess digital transformation in higher education
7. Identify strategies to redesign higher education and Industry 4.0
8. Identify promising delivery of higher education (heutagogy)

TARGET AUDIENCE

This book will be very helpful to professionals and researchers working in the field of information and knowledge management in various disciplines, e.g. library, information and communication sciences, administrative sciences and management, academicians, computer scientists, cloud -computing experts, artificial intelligence specialists, educational technologist, researchers, higher education administrators instructional technologist, planners, strategic managers who are seeking for knowledgeable information . In the context of demanding requirements of IR 4.0, readers of this book would be exposed on how the delivery of soft skills can be made more effective. Efforts should be made on how to make digital or even data literacy more accessible to the overall student regardless of discipline. Hence, the need for human resources in the key technical areas in IR 4.0 such Internet of Things (IoT), big data and analytics, cloud computing, virtual/augmented reality, and robotics should be assessed. Moreover, the book will provide insights and support executives concerned with the higher-educational management expertise.

ORGANIZATION OF THE BOOK

The book is organized into nineteen chapters with four sections. A brief description of each of the chapters follows:

Chapter 1 explains how Industry 4.0 vision entails collaborations between technology specialists and industry experts who are adept at integrating their specialized knowledge and disciplinary expertise. The author of this chapter identifies how graduates prepare for their future involvement in such interdisciplinary collaborations that requires educators to design and implement interdisciplinary teaching in higher education. In order to foster students' engagement in interdisciplinary learning, a meaningful understanding of how interdisciplinary learning actually occurs in the interdisciplinary workplace is essential.

Chapter 2 elaborates how Internet of Things (IoT) for higher education can be adopted. Universities are advised to play their role as proving ground for educating the future generation and innovation. However, the universities should have a close collaboration with industry and stakeholders to implement Education 4.0. In the Malaysian Higher Education System, it is important to look back on how the delivery of soft skills can be made more effective. Efforts should be made on how to make digital or even data literacy more accessible to the overall student regardless of discipline.

Chapter 3 revealed how Heutagogy approach using Trello online learning on the critical thinking skills amongst students with different learning styles has become one of the important strategies for the University Utara Malaysia Management Foundation Programme students. Heutogogy approach teaches other students to think critically in conveying their ideas and become more proactive and creative in critical thinking subject. Learners become highly autonomous and self-determined while learning online. By using Trello online discussion tool, the students could use their capacity and capability to enhancing their Macro-critical thinking among active-reflective learning style.

Chapter 4 discussed an overview of massive open online courses (MOOC) pedagogies and distance education in Malaysia. Massive open online courses (MOOCs) are the instructed course with the various number of students using the internet. MOOCs are not a new tool for educational modules outline and strategy, this tool captivating, expanded thought and examination in advanced education. At the end of this chapter, effective teaching strategies have been proposed.

Chapter 5 shows a brief review of Blended Learning. This chapter covers the definition, design, and implementation of blended learning in higher education. It also discusses blended learning models and assessment tools in the recent education setting around the world. The education system worldwide has to transform for the imminent FIR as the job market will be largely prompted by the advancement of digital economy, robotics, artificial intelligence and automation technology.

Chapter 6 discussed pedagogical orientation topics among educators, parents and other community stakeholders. This is due to the current developments in digital technologies and the onset of the fourth industrial revolution which is set to alter the way people live and work. This book chapter suggests the implementation of the flipped classroom model as an appropriate approach for equipping students with creative abilities, problem-solving skills, thinking skills, and lifelong learning skills

Chapter 7 indicates the implementation of flipped classroom in Malaysia's teacher education institution. The definition of flipped classroom was explained in detail and furthermore flipped classroom activities are based on social and cognitive constructivist theory that is suitable to be implemented in teacher education institution in Malaysia because it is student centered and involved students actively during the learning process.

Chapter 8 describes how Edmodo was used to manage blended learning, with the focus on online activities created in posts, small groups, assignment tool, and polling tool. The activities were planned based on the five ingredients of blended learning by Carman (2002). The result of a poll done in Edmodo showed 74% of the students agreed that they were able to explicate Shakespeare after attending the course and 81% opined that the course was effectively organized in Edmodo.

Chapter 9 explain about Gamification meets Mobile-Learning. The impact of numerous technology innovations in ICT towards Malaysian education is noticeable. Some existing jobs will be eliminated in this era subsequently more new jobs that require digital literacy are going to be created. Therefore, Education 4.0 must be able to produce graduates with excellent communication and collaborative skills. However, Malaysian Digest (2016) states that fresh graduates are lack of soft skills. Several approaches have been proposed by the Ministry of Higher Education for the purpose of enhancing students' soft skills.

Chapter 10 evaluate the future product design education of Industry 4.0. This chapter aims to support educational leadership in the development of proactive strategies to respond to the challenges and opportunities of industry 4.0 to inform the future of work, industry, and society. This is framed through the lens of product design, with its unique position at the nexus of engineering and the humanities, and directly tied to changes affecting manufacturing in the fourth industrial revolution.

Chapter 11 revealed the active use of digital technologies in mathematical problem-solving. An examination of the frequency/ density of technical mathematical vocabulary use and identified examples Talk Types is used as a means to understand how often students within the online environment are likely to be engaged in work that might be considered productive. The findings of this study suggest that utilizing online synchronous and asynchronous collaborative learning platforms for the purposes of mathematical problem solving is a small but important approach to preparing students for the era of Industry 4.0.

Chapter 12 immerse the important of a computer-based test (CBT) as the inevitable next wave in digital assessment. This chapter describes the CBT developmental process undertaken by HEID, which focuses on constructing multiple-choice test items, administering CBT, and scoring the test. While CBT offers the possibility of improving assessment, the process of preparing items and conducting online competition have nevertheless encountered several challenges such as getting quality test items, school and teacher readiness, seat time in front of the computer, computer compatibility, and internet speed.

Chapter 13 elaborate the alternatives to the traditional doctoral dissertation. In response to increased criticism of the utility of the traditional doctoral dissertation, some institutions have incorporated additional options for students such as articles, portfolios, and industry-specific projects. This trend towards allowing alternatives for doctoral students coincides with the rise internationally of what are variously called professional, applied, or practitioner doctorates in various disciplines.

Chapter 14 explained the redesigning prediction algorithms for at-risk students in higher education. Algorithms and data mining approaches are not new to Industry 4.0. However, these may not be common for students and educators in higher education. This study compares various classification techniques: classification tree, logistic regression and artificial neural networks (ANN). The comparison focuses on each method's accuracy, algorithm, and practicality in higher education.

Chapter 15 discussed IR 4.0 and education system in Malaysia: In Malaysia, the government had earlier made many changes in education through its National Higher Education Action Plan 2007-2010 in transforming teaching and learning process focusing on technology. Innovative and entrepreneurial skills must be possessed by new graduates and they must deal with complexity with cognitive flexibility. This is because they need to accept robots as their working colleagues apart from humans. The need for better communication and collaborative skills will be far more important than ever

Chapter 16 presents a case study on collaborative learning for programming course to enhance students' interest and achievement in Universiti Utara Malaysia has been done. As programming is one of the core subjects for BSc. IT students, it is vital for them to have programming skill. However, students are often terrified of programming since it is a challenging subject. Traditionally, the programming teaching and learning process are limited to lectures and lab exercises. Collaborative learning is acknowledged as an approach that fits the programming students.

Chapter 17 explained about the readiness of IR4.0 and it awareness at the National Defense University of Malaysia. The author discussed the importance of equipping academics with the knowledge of IR4.0 and beyond for future graduates. Particularly to prepare for military leaders for Malaysia at the National Defense University of Malaysia (NDUM), a new learning model may be apt, and urgently needed. The objectives of the chapter are twofold; firstly, to gauge the readiness and awareness level of academics on implementing IR4.0 during their classroom teaching, and secondly, to investigate academics' views on their own teaching practices.

Chapter 18 is based on various studies conducted by Malaysian researchers, there are needs for MOHE to redesign higher education via quality in-service Continuing Professional Development Training (CPDT). There are many reasons lecturers fail to integrate ICT into their classrooms. Lecturers should try to implement better ways and approaches to ensure successful ICT integration in classrooms. CPDT should be carefully redesigned to encourage more lecturers to use ICT in order to enhance their student learning outcomes and achievements. Guidelines of redesigning CPDT would surely able to design and develop digital age learning experiences and assessments, facilitate student learning with ICT, increase students' productivity, promote and model digital citizenship and responsibility and engage in professional growth and leadership.

Chapter 19 shows the growing changes brought about by new concepts and new technologies, such as Industry 4.0, have demanded that educational institutions seek new teaching and learning methodologies, as well as new resources that can contribute to the training of the future engineer. This chapter demonstrates some practices adopted in the process of training the future engineer in a University Center in the interior of the state of São Paulo through the use of the CDIO initiative.

In short, the current education system is not competent enough and need changes in order for us to see a better future. With the emergence of new technologies, education needs to be in line with it in order for us to produce world-class talent. Current curriculum system is unable to stand with the sophisticated standard which is required. Many changes need to be done because the future for good achievement looks so bright. We might be living in hi tech world in the future and to make sure it education system need to be flexible and open.

Arumugam Raman
Universiti Utara Malaysia, Malaysia

Mohan Rathakrishnan
Universiti Utara Malaysia, Malaysia

REFERENCES

Dewey, J. (1916). *Democracy and Education. An introduction to the philosophy of education* (1966 ed.). New York: Free Press.

Garrison, D. R., & Archer, W. (2000). *A transactional perspective on teaching-learning: A framework for adult and higher education*. Oxford, UK: Pergamon.

Hacker, D. J. (1998). Definitions and empirical foundations. In D. J. Hacker, J. Dunlosky, & A. C. Graesser (Eds.), *Metacognition in educational theory and practice* (pp. 1–23). Hillsdale, NJ: Lawrence Erlbaum Associates.

Ranai, M. (2017). *World Bank praises Malaysia's educational reforms*. Retrieved from http://news.moe.gov.my/2017/10/25/malaysias-education-reforms-get-thumbs-world-bank/

Acknowledgment

I wish to share my sincere gratefulness to my coworker and also co-editor, Dr. Mohan Rathakrishnan (School of Languages, Civilization & Philosophy, Universiti Utara Malaysia (UUM)). I want to thank all the authors for their chapter contributions. Their valuable contributions have provided provoking insights into the Higher Education 4.0.

Many thanks likewise to all those served as reviewers, without constructive comments and critical suggestions this book never been materialize. Special references hereof need to go to UUM's Vice Chancellor Professor Dr. Ahmad Bashawir Haji Abdul Ghani for his significant foreword for this book, Consistent supports from Deans of School of Education and Modern Languages and School of Languages, UUM, Civilization & Philosophy, UUM.

I am forever indebted to my close family members who remain to inspire me to complete this book. Thanks to the wonderful staff at IGI Global who took part in the total growth and also successful conclusion of this wonderful task. Lastly, I am really thankful to the Editorial Board of Advisers (EAB) group.

Arumugam Raman
Universiti Utara Malaysia, Malaysia

Section 1

Redesigning Drivers of Higher Education Initiatives for Industry 4.0

Chapter 1
Interdisciplinary Higher Education:
Redesigning for Industry 4.0

Mohd Nazri Mahmud
Universiti Sains Malaysia, Malaysia

Thomas William Ridgman
University of Cambridge, UK

ABSTRACT

The Industry 4.0 vision entails collaborations between technology specialists and industrial experts who are adept at integrating their specialized knowledge and disciplinary expertise. Preparing graduates for their future involvement in such interdisciplinary collaborations requires educators to design and implement interdisciplinary teaching in higher education. In order to foster students' engagement in interdisciplinary learning, a meaningful understanding of how interdisciplinary learning actually occurs in interdisciplinary workplace is essential. This chapter provides a useful description of, and explanation for, interdisciplinary learning that are based on a multi-case study of three completed interdisciplinary projects.

INTRODUCTION

Our industrial system is experiencing a transformation that has been known as the Fourth Industrial Revolution (4IR). This industrial revolution has been enabled by an unprecedented confluence of several technological breakthroughs, most notably, by the Internet-of-Things, Machine-to-Machine communications, 3-D printing, autonomous robotics, artificial intelligence, and digital businesses. Many believe that such a technological confluence would lead to the realization of the Industry 4.0 (In4.0) vision. The vision had envisaged a close coupling between the different subsystems that make up our industrial system. The industrial subsystems that are expected to involve includes the physical subsystem of industrial machines and robots, the digital subsystem of e-commerce, the cyber subsystem of communication and control networks, and the biological subsystem of the human and animal species (Lasi et al., 2014; Lee

DOI: 10.4018/978-1-5225-7832-1.ch001

Copyright © 2019, IGI Global. Copying or distributing in print or electronic forms without written permission of IGI Global is prohibited.

et al., 2015; Wollschlaeger et al., 2017). The most likely consequence of this revolutionary confluence and tight coupling would be an increase in the complexity of the future industrial system.

Inevitably, a complex industrial system envisaged by In4.0 brings many opportunities, but it also presents many challenges. As well as offering the opportunity to improve operational efficiencies through the integration of advanced technologies, Industry 4.0 complexities are likely to challenge our technological specialists and industrial experts from different disciplines to collaborate more effectively, especially in synthesizing and integrating their specialized knowledge and skills. Interdisciplinary work is necessary for creating innovative industrial products and services, but also for mitigating several risks, such as the risk of mass unemployment of skilled workers arising from an industry-wide adoption of robotic automation systems, and the risk of exposing our industrial communication and control networks to the threats from cyber-criminals and cyber-terrorists. Since 4IR and In4.0 both rely heavily on interdisciplinary work practices, this chapter is positioned within an interdisciplinary perspective.

More specifically, in the context of higher education for Industry 4.0, the authors' position is that preparing graduates for the 4IR requires the design and implementation of interdisciplinary teaching and learning that is aligned to the In4.0 vision. This position is supported by various studies that have established that employees nowadays are increasingly expected by their employers to be more interdisciplinary in enabling their companies to evolve successfully into In4.0 operations (Dujin et al., 2014). Xing and Marwala (2017) contend that In4.0 demands more interdisciplinary teaching. Hence, this chapter sets out to address the issue of re-designing higher education initiatives for Industry 4.0 by focusing on interdisciplinary teaching and learning as a means to redesign learning for Industry 4.0

Within this focus, the objective of this chapter is to suggest how interdisciplinary learning should be conceptualized so that the design and implementation of interdisciplinary teaching and learning in higher education can be aligned to the In4.0 vision. This chapter fulfils this objective by presenting a multi-case study of interdisciplinary projects in industrial settings. Findings from the study informs the conceptualization of interdisciplinary learning that occurs in interdisciplinary workplaces.

Background

Much has been written over the past few decades about the need for graduates who have the ability to engage in interdisciplinary teamwork as well as about the demand for employees who have interdisciplinary expertise for enabling their companies to produce innovative products and services. However, a proper understanding of the term 'interdisciplinary' is essential for understanding the underlying requirements.

The Organization of Economic Cooperation and Development (OECD) OECD had define the term 'interdisciplinary' in 1972. It refers to the term as "an adjective that describes the interaction between multiple disciplines" (Apostel et al., 1972;p.25-6). However, over the years, the term has been used interchangeably with other terms, most notably, the term 'multidisciplinary'. It is important to clarify the distinction between the two terms so that the requirements of interdisciplinary work could be ascertained and distinguished from other possible forms of interaction between different disciplines. Scholars of interdisciplinary studies have succinctly clarified the distinctions between the adjectives 'interdisciplinary' and 'multidisciplinary' by providing the definitions of their corresponding nouns, 'interdisciplinarity' and 'multidisciplinarity'.

For the noun 'interdisciplinarity', different definitions have converged towards the process-centric definition (Borrego & Newswander, 2010; Lattuca & Knight, 2010), which was proposed as follows:

Interdisciplinarity is a process of answering a question, solving a problem, or addressing a topic that is too broad or complex to be dealt with adequately by a single discipline or profession… [It] draws upon disciplinary perspectives and integrates their insights through construction of a more comprehensive perspective (Newell & Klein, 1998;p.393-4).

This definition of 'interdisciplinarity' has been clearly contrasted from the noun 'multidisciplinarity', where knowledge integration is not required (Lattuca & Knight, 2010).

Thus, from this definition it is clear that interdisciplinary teamwork requires team members from different disciplines to interact productively by integrating their disciplinary knowledge and expertise to address complex issues and problems. Employees with interdisciplinary expertise are adept at interacting across disciplines and at integrating knowledge beyond disciplinary boundaries, and thereby producing innovative and practical solutions to complex problems.

The understanding of these requirements has led to the promotion of the interdisciplinary teaching in higher educational institutions. 'Interdisciplinary teaching' refers to an emerging educational approach whereby educators expose university students to complex issues and problems that demand the latter to draw on two or more disciplines. Unlike the traditional mono-disciplinary teaching, the interdisciplinary approach emphasizes teaching the disciplines by showing how they are dynamically related to other disciplines in the context of societal problems. Students from different disciplines interact among them as well as with people outside their own disciplines, and they are expected to integrate insights from different disciplines to enhance their understanding of the complex issues and problems at hand (Apostel et al., 1972; Holley, 2017; Lattuca et al., 2012; Newstetter, 2011; Repko, 2008; Richter & Paretti, 2009; Stentoft, 2017).

The adoption of the interdisciplinary approach to teaching has been going on for several decades. Since 1972, the Organization of Economic Cooperation and Development (OECD) has been advocating its adoption in higher education as a means for universities to produce graduates who are capable of addressing complex societal problems and challenges, such as problems related to environmental sustainability. This has led to its diffusion across the higher education landscape. Most notably, the last quarter of the previous century saw the proliferation of this approach in the form of interdisciplinary degree programs. For example, Brint et al. (2009) have reported that the number of interdisciplinary degree program in the US had grown from 674 in 1975/1976 to 1663 in 2000/2001.

The rise of interdisciplinary teaching and learning in higher education during the first decade of the 21st century has been driven mostly by policy directives issued by governmental bodies. For example, in the engineering discipline, the National Academy of Engineering (NAE) of the United States has mandated all engineering schools in the US to "introduce interdisciplinary learning in the undergraduate environment", stating that "students would benefit from at least cursory learning about the interplay between disciplines embodied in real-world problems" (NAE, 2005;p.55). Similarly, the Royal Academy of Engineering (RAEng) of the United Kingdom has imposed the requirements for graduates "to have awareness and basic skills beyond their traditional discipline boundaries." (King, 2007;p.13), and has been advocating the need to embed a multidisciplinary approach in UK's undergraduate education. In the recent years, almost all national and international accreditation bodies for degree programs have included interdisciplinary requirements in their accreditation criteria. For example, most engineering schools have been adhering to either the ABET criterion 3(d), which specifies the "ability to function on multidisciplinary teams" (ABET, 2011) or the IEA criterion, which specifies the ability to "function effectively as an individual, and as a member or leader in diverse teams and in multi-disciplinary settings" (IEA, 2015).

Interdisciplinary teaching is believed to have produced widespread benefits for learning. Although the original goal has been mainly to increase students' readiness for interdisciplinary practice (Borrego & Newswander, 2008; Ivanitskaya et al., 2002; Mansilla & Duraising, 2007), the adoption of interdisciplinary teaching has been associated with better student engagement; improved higher-order cognitive skills such as knowledge application, analysis, synthesis and evaluation; greater tolerance for ambiguity; greater sensitivity to ethical issues; reducing disciplinary, political, or religious bias; and more creativity and humility (Holley, 2017; Lattuca et al., 2004; Newell et al., 2001). Indeed, there have been some exemplary implementations of interdisciplinary teaching that have resulted in successful outcomes (Lattuca et al., 2011).

However, a more comprehensive review of the literature on interdisciplinary learning reveals a more complete picture. Figure 1 represents the conceptual framework of interdisciplinary higher education that emerges from the literature review.

The conceptual framework represents the relationships between several factors related to interdisciplinary teaching. It shows how the design and implementation of interdisciplinary teaching is influenced mainly by how educators conceptualize interdisciplinary learning. Consistent with the process-centric definition of interdisciplinarity, the literature reports that most educators conceptualizes interdisciplinary learning as a process that typically begins by exposing students to an interdisciplinary problem or phenomenon and engaging them in the context of interdisciplinary interaction. This process-based conceptualization has been informed mostly by the process-based theory of interdisciplinary studies (Newell & Gagnon, 2013; Newell et al., 2001). According to the theory, the first step in interdisciplinary studies involved exposing academicians to a chosen interdisciplinary problem or phenomenon and engaging them in the context of interdisciplinary interaction. During the interaction, academicians would acquire insights from the different disciplines involved. Then, by integrating those insights, the theory suggests that the interdisciplinary academicians would attain a comprehensive understanding of the problem at hand.

By taking students through this interdisciplinary learning process, as the literature suggests, the students are able to benefits from their attainment of the intended learning outcomes. As well as attaining

Figure 1. Conceptual framework of interdisciplinary higher education

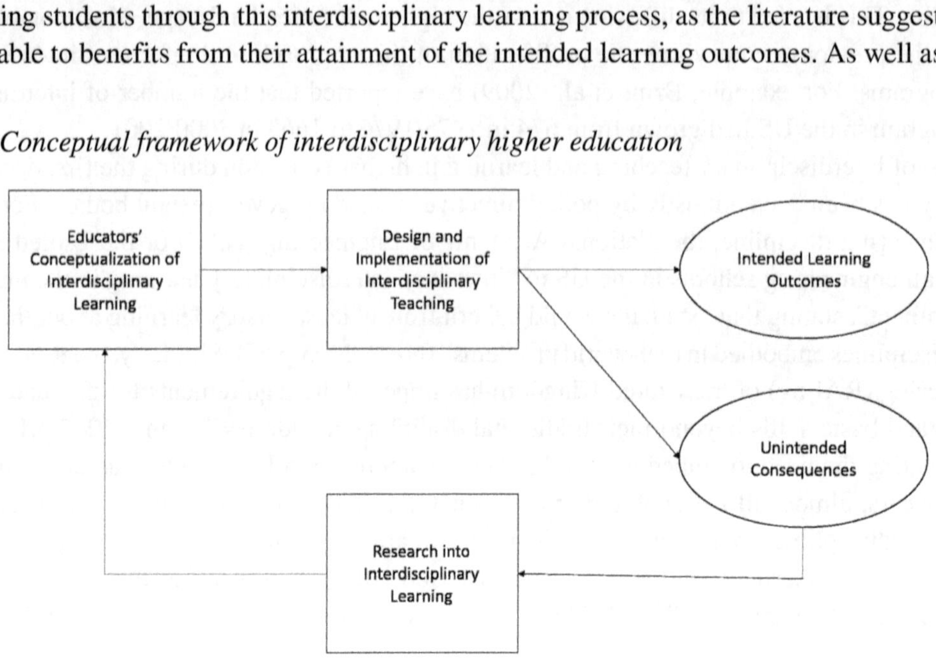

readiness for interdisciplinary practice (Borrego & Newswander, 2008; Ivanitskaya et al., 2002; Mansilla & Duraising, 2007), the students also attain higher-order cognitive skills such as knowledge application, analysis, synthesis and evaluation; greater tolerance for ambiguity; greater sensitivity to ethical issues; reducing disciplinary, political, or religious bias; and more creativity and humility (Holley, 2017; Lattuca et al., 2004; Newell et al., 2001).

However, many studies have revealed how the actual implementations of interdisciplinary teaching have also resulted in outcomes that depart from the intended outcomes. This departure is represented as an arrow pointing towards the 'unintended consequences' outcome of the conceptual framework in Figure 1. There have been many problematic implementations mostly due to learning barriers that confront students during interdisciplinary interactions. For example, in their case study of an interdisciplinary course, Richter and Paretti (2009) had revealed that the barrier to interdisciplinary learning constitutes both the failure to relate between a discipline and an interdisciplinary topic, and the failure to connect between a discipline and other disciplines.

More recently, MacLeod (2016) has raised a concern that interdisciplinarity scholarship has been facing problems in describing the cognitive barriers that tend to impede interdisciplinary learning. Since the process-centric conception of interdisciplinary learning does not specify these barriers, it offers only a limited utility for informing interdisciplinary teaching in higher education. The process-based conceptualization has yet to offer recommendations of useful learning practices that could help students negotiate and overcome barriers to interdisciplinary learning. Newstetter (2011) had argued that a general process model without more detail information of the associated learning practices has not been sufficient for informing how to make classrooms more interdisciplinary.

These problematic implementations have motivated educational researchers to begin conducting empirical research into interdisciplinary learning. As represented in Figure 1, such a research seeks to directly inform the design and implementation of interdisciplinary teaching in higher education. In addition, such a research could also exert an indirect influence. Empirical research on interdisciplinary learning could seek to develop a practice-based theory of interdisciplinary learning, and thereby facilitating the adoption of a practice-based conceptualization of interdisciplinary learning. The recommendation of a practice-based conceptualization of interdisciplinary learning is the main purpose of this chapter. Before making this recommendation, the next section discusses the main focus of this chapter, which is on interdisciplinary learning researches that seek to inform interdisciplinary teaching and learning in higher education.

MAIN FOCUS OF THE CHAPTER: INTERDISCIPLINARY LEARNING RESEARCH

Scholars generally agree that there is a need for a theory that could inform students' engagement in interdisciplinary learning in higher education. However, due to the persistent scarcity of empirical research into interdisciplinary practice to develop such a theory (Lattuca et al., 2017; Nersessian & Newstetter, 2014), many educators and researchers have been relying on the interdisciplinary studies literature instead. Interdisciplinary studies scholars mainly focus on theorizing how academicians engage in interdisciplinary studies, such as the study of the acid rain problem. They have developed a process-based theory of interdisciplinary studies (Newell & Gagnon, 2013; Newell et al., 2001). According to the theory, the first step in interdisciplinary studies involved exposing academicians to a chosen interdisciplinary problem or phenomenon, and engaging them in the context of interdisciplinary interaction. During the interaction, academicians would acquire insights from the different disciplines involved. Then, by integrating those insights, the theory suggests that the interdisciplinary academicians would attain a comprehensive understanding of the problem at hand. In some cases, the process culminates in a solution that integrates disciplinary knowledge (Ivanitskaya et al., 2002; Klein & Newell, 1997; Mansilla & Duraising, 2007; Repko, 2008). Informed by the process-based theory of interdisciplinary studies, the implementation of interdisciplinary teaching in higher education typically begins by exposing students to complex interdisciplinary problems. Students also participate in interdisciplinary teams where they are expected to learn how to draw, acquire and integrate knowledge from team members of different disciplines. However, issues, controversies and problems related to interdisciplinary learning tend to emerge mainly due to barriers that confront students during interdisciplinary interactions.

Barriers to Interdisciplinary Learning

MacLeod (2016) had raised a concern that interdisciplinarity scholarship has been facing problems in describing the cognitive constraints that tend to impede interdisciplinary learning. Since the process-based theory of interdisciplinary studies have yet to identify these barriers, it offers only a limited utility for promoting interdisciplinary learning in higher education. The theory has yet to offer recommendations of useful learning practices that could help students negotiate barriers to interdisciplinary learning. As a result, many students succumb to these barriers. Newstetter (2011) has argued that a general process model without more detailed information on the associated learning practices has not been sufficient for informing how to make classrooms more interdisciplinary. One of the main problems in interdisciplinary learning research is the lack of focus on identifying useful learning practices and the associated barriers, and on explaining how students could sustain their learning practices in the face of those barriers.

Furthermore, research carried out in the academia alone, such as research on interdisciplinary studies, is unlikely to provide enough insights for informing interdisciplinary higher education teaching. This is mainly due to the fact that interdisciplinary higher education typically seeks to prepare graduates for interdisciplinary practices outside of the academia also. Educationists, such as Nancy Nersessian and Wendy Newstetter, have maintained that workplace settings provide more realistic requirements and challenges for interdisciplinary learning than the academic context, and therefore should be the basis on which to inform interdisciplinary teaching and learning in higher education (Nersessian, 2009; Nersessian & Newstetter, 2014).

Moreover, interdisciplinary studies scholars themselves, such as William Newell and Pauline Gagnon, have alerted interdisciplinary studies researchers that the bulk of interdisciplinary activity has shifted from the context of academic studies to the context of real world applications (Newell & Gagnon, 2013). They had observed that interdisciplinary collaborations increasingly involve non-academics and professionals, and had prompted scholars to study them. Newell and Gagnon observes that interdisciplinary collaborations outside the academia tend to pursue outcomes that are different from the interdisciplinary academics. Whereas academic collaborators in interdisciplinary studies tend to focus on achieving a comprehensive and integrative understanding of the interdisciplinary problems at hand, non-academics and professionals tend to focus also on achieving an integrative solutions to complex problems. Indeed, Industry 4.0 demands interdisciplinary work practices that results in the development of innovative solutions, not merely in the understanding of the issues surrounding the fourth industrial revolution. Thus, the current reliance on interdisciplinary studies to formulate the learning outcomes represents a controversial issue that is worth discussing in more detailed.

Speculative Formulation of the Learning Outcomes

In the era of outcome-based education, the design and implementation of interdisciplinary teaching requires clear specifications of the learning outcomes. Typically, the outcomes are expressed in terms of a set of abilities that students must possess upon graduation. However, due to the scarcity of empirical research in interdisciplinary classrooms and workplace settings, many educators have been relying on 'speculative' learning outcomes derived from their review of interdisciplinary studies literature in the humanities (Mansilla & Duraising, 2007; Newell et al., 2001; Repko, 2008). Since the interdisciplinary studies literature have yet to focus on interdisciplinary practices outside the academia, educational researchers have so far been cautious in prescribing these speculative learning outcomes.

Nevertheless, many studies have focused on the deployment of various pedagogical approaches – such as problem-based learning, project-based learning and active learning – in interdisciplinary curricula, some of which have resulted in successful outcomes (Lattuca et al., 2011). However, there has been a lack of understanding of how and why learning outcomes had been achieved or not. Some hypothetical explanations have been proposed (see for example, Lattuca et al. (2004))) based on the literature on cognition, but there is a lack of systematic study that seek to explain the achievements of the outcomes.

Despite the scarcity of research in interdisciplinary practice to inform interdisciplinary curricular design, there has been an emerging research approach being promoted in recent years. The so-called translational approach seeks to study interdisciplinary learning practices in real interdisciplinary settings in order to translate them into transformative learning practices inside the classrooms. This approach has been used to inform the design and implementation of a biomedical course based on an investigation of interdisciplinary practices in bioengineering research laboratories. Proponents of this approach have maintained that workplace settings provide more realistic requirements and challenges for interdisciplinary learning than the academic context, and therefore should be the basis for informing interdisciplinary learning in higher education (Nersessian, 2009; Nersessian & Newstetter, 2014).

Translational approaches to curricular design seek to align between the workplace learning and higher education learning. This alignment espouses the socio-material view of learning that underpins recent studies of workplace learning (Fenwick et al., 2014; McMurtry et al., 2016; Reich et al., 2015). The socio-material view recognizes that in today's society, knowledge is not only generated by, and embodied in, practitioners within communities, but knowledge is also generated by, and embedded in material things, such as plants, animals, artefacts as well as their representations (i.e. drawings, models etc.). Furthermore, knowledge, in its various forms and representatives, is also becoming widely circulated around the world and across professional boundaries (Nerland & Jensen, 2014a).

The result of such a wide circulation of knowledge is the increase in availability and accessibility of knowledge. However, possible differences in the veracity of knowledge claimed by different sources can also result in ambiguity. Therefore, to learn from encounters with different knowledge from different sources, a learner has to perform some learning actions, known collectively as 'epistemic practices', for processing the knowledge that s/he encounter in practice. The socio-material perspective focuses on identifying the epistemic practices, and proposes the relevance of learning through epistemic practices (Karseth & Nerland, 2007; Nerland & Jensen, 2012, 2014b).

The next sections recommends and describes a research study that integrates the translational approach and the socio-material perspective of learning and draws some lessons that could address the ongoing issues.

SOLUTIONS AND RECOMMENDATIONS

A multi-case study of interdisciplinary projects is recommended as a solution for addressing the issues and controversies surrounding interdisciplinary teaching and learning in higher education. The study offers a conceptualization of how interdisciplinary learning is enacted in interdisciplinary projects in industrial settings. The objectives of the study were twofold. First, it sought to describe interdisciplinary learning in terms of the concept of 'epistemic practice' and the corresponding learning outcomes. Second, it sought to explain the emergence of epistemic practices and the corresponding learning outcomes.

The study was informed by the theoretical framework depicted in Figure 2.

The theoretical framework relates the learning barriers confronted by the interdisciplinary practitioners to the learning practices they employ for achieving the outcomes. The framework helps the study to focus on the identification of the different epistemic practices and the corresponding learning outcomes as well as on the identification of learning barriers that provide explanations for the emergence of the different epistemic practices.

The study investigates the different aspects of interdisciplinary learning that occur in one sub-class of interdisciplinary collaborations that involves engineering and the life sciences disciplines, known as engineering for the life sciences sub-class (Niemeyer, 2017). In this sub-class, practicing engineers in one or more engineering disciplines – such as robotic engineering for industrial automation – collaborate with specialists from the life sciences disciplines – such as cell-culture biology for bio-medicines production – for addressing opportunities and challenges in the life science domain, such as the problem of low productivity in bio-medicines production. This sub-class is of relevance to the fourth industrial revolution as the confluence between advances in engineering and the life sciences in this sub-class have resulted in the emergence of many new industries, for instances Synthetic Biology, Regenerative Medicines and Bio-nanotechnology industries that partly constitute the Industry 4.0 vision. A suitable research design inspired by the translational approach to educational design facilitates the transferability of the findings from the workplace context to the educational context.

Study Design

The multi-case study was designed according to the 'building-block' case study design offered by George and Bennett (2005). In this approach, a small number of cases that together represent a particular sub-class of the phenomenon of interest are studied in series, and the findings are then used to develop a theory. In this study, three interdisciplinary projects were chosen to develop the conceptualization of interdisciplinary learning.

Figure 2. The theoretical framework of the case study

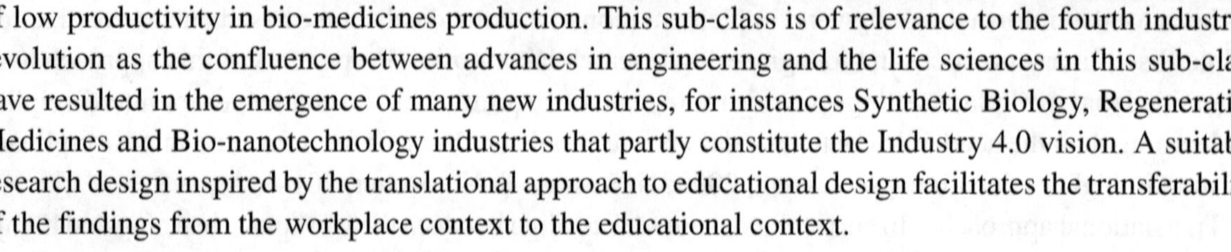

The first case, studied in 2014, had involved an integration between robotic automation engineering and cell-culturing biology for addressing the problem of scaling up a manual lab-based cell culturing method into an automated industry-scale production of a type of bio-medicine, called Erythropoeitin (EPO). Following the 'building-block' theory-building approach, this first case was used as a heuristic case to develop a tentative theory. Then, the second and the third cases were studied to support comparative cross-case analyses and validations that have helped to evolve the tentative theory into a generalizable one.

The second case, studied in early 2015, had involved the integration of chemical engineering and biological lab experimentation for designing and developing an automated bioreactor, which is used in an early stage bioprocess development. The third case, studied in late 2015, had involved the fusion between an advanced image processing technology and a pulmonary function diagnostic method for developing a non-invasive device to diagnose lung diseases.

Twenty three participants from engineering and the life sciences disciplines were interviewed, and materials related to their projects, such as archived documents and artifacts, were gathered for analysis. The researcher uses these retrospective interviews and archived materials as 'empirical traces' to develop, test and validate a set of plausible descriptions and explanations. The analyses of these 'empirical traces' and the validation of the findings adhere to the principles of data analysis offered by Wynn and Williams (2012).

Cases Studied

The first case is a project that involves an interdisciplinary collaboration that sought to automate a manual cell culturing method for scaling-up the production of bio-medicines. The manual cell-culturing method was called a roller-bottle cell-culturing method, whereby genetically-modified mammalian cells are manipulated manually in roller bottles to generate bio-medicines. The manual method had traditionally been a lab-scale method, used by a leading UK biotechnology company to produce a small amount of a therapeutic hormone called Erythropoietin (EPO). However, the company had won a contract from a leading pharmaceutical company, which demanded a sudden increase in the production rate. This demand was made known to a Cambridge-based engineering consultancy company, whose team of four engineers and a biochemist had proposed to automate the manual method using robotics technology. However, the manual cell culture practices were once considered as a form of 'art-and-craft' that requires 'green-fingers' and intuition, and therefore were not initially thought to be amenable to automation (Archer & Wood, 1992; Stacey, 2012). In addition, there had been no precedent automation of this lab-scale method. Nevertheless, the bio-technology company agreed to form an interdisciplinary team consisting of engineers, biochemists and cell-culturing practitioners to solve the problem of scaling-up the production of its EPO.

The second case is a project that had involved the development of an automated experimental tool for an early-stage bioprocess development. Early stage bioprocess development refers to a set of research practices whereby life scientists conduct bioprocess experimentations in laboratories using conventional tools, such as shake flasks and micro-titre plates, in order to identify cells and operational conditions that could optimize the large-scale production of biomedicines. Early stage bioprocess development is especially challenging and time-consuming due to the intensive multi-factor statistical experimentations that are involved in the process (Bareither & Pollard, 2011). Yet, there is no guarantee that such experimentations would result in an optimal production since the conventional scale-down tools are not representatives of the industrial-scale bioreactors.

A team of engineers and bio-process development scientists had collaborated to develop a novel scale-down tool that automates the laborious experimental practices. The tool also faithfully represents the industrial-scale bioreactors in terms of its functional properties. The tool is popularly known as an automated micro-scale bioreactor.

The third case is a project that had involved the development of a non-invasive diagnostic device for lung functions. The non-invasive device differs from the conventional diagnostic devices in that the latter require patients to be physically attached to the diagnostic devices using wires.

Descriptive Findings

The study had identified four categories of epistemic practices and four categories of learning outcomes, the combination of which constitutes the proposed description of interdisciplinary learning in practice.

Categories of Epistemic Practices

The four categories of epistemic practice have been conceptualized as follows:

1. **Consultational Epistemic Practice (CEP)**

The CEP is conceived as a set of related learning actions that interdisciplinary practitioners perform to gain further understanding on the knowledge of other disciplines. The actions largely involve consultations with people who are perceived as knowledgeable about the knowledge. The CEP is likely to occur when the interdisciplinary practitioners perceive that their disciplinary knowledge is insufficient for assessing the relevance or usefulness of the knowledge suggested by team members from different disciplines.

2. **Translational Epistemic Practice (TEP)**

The study conceives the TEP as a set of related learning actions taken on the contents and forms of knowledge from other disciplines in order to produce the corresponding knowledge contents and forms that are more useful for developing solutions to the problem at hand. The study found that interdisciplinary practitioners often encounter knowledge elements in forms that are not directly useful for developing solutions. For example, the description of biological aspects are often provided in qualitative descriptions laden with disciplinary jargons or are tacitly embodied in everyday practices of the practitioners of other disciplines. Therefore, solution developers often find it necessary to translate them into something quantitative and explicit in order to design and develop solutions.

3. **Evidential Epistemic Practice (EEP)**

EEP refers to a set of related learning actions taken to gain and show confirmation on the usefulness of the different contents and forms of knowledge to be integrated into the solutions under development. Interdisciplinary collaborators often face situations whereby evidence of knowledge usefulness or relevance is required to be produced and shown before deciding whether to integrate the knowledge into solutions under development.

4. **Comparative Epistemic Practice (CPEP)**

The CPEP refers to the practice of comparing knowledge suggested or practised by others from different disciplines against a set of criteria that are underpinned by prior knowledge of the learner. Actions that constitute this practice include judging the extent to which the knowledge suggestions would satisfy the criteria, and predicting how the solutions would fare if the knowledge suggestion are adopted.

In addition to these epistemic practices, the study also describes the outcomes that are likely to be achieved from engaging in those practices.

Categories of Learning Outcomes

Four learning outcomes were identified and had been conceptualized as follows:

1. **Knowledge Adoption**

This study has conceptualized the 'knowledge adoption' outcome as the attainment of understanding, appreciating, and reusing the relevant knowledge of the other disciplines while retaining its original contents and meanings. To the extent that solutions work as expected, the practitioners often found the necessity to maintain knowledge in its original form, such as reusing disciplinary jargon in knowledge description and replicating the existing procedural knowledge of the other disciplines.

2. **Knowledge Translation**

The 'knowledge translation' outcome has been conceptualized as encompassing a broad range of achievements that result from developing knowledge through changing original knowledge contents, or/and forms into more useful ones. Interdisciplinary practitioners gain and use knowledge elements that are distinct from the knowledge of the other disciplines from which those elements were derived.

3. **Knowledge Avoidance**

Interdisciplinary practitioners often find themselves in a difficult situation whereby their collaborators from other disciplines insist on the re-use of existing knowledge related to the other disciplines. When the practitioners perceive that such a re-use would result in suboptimal solutions, they tend to avoid reusing it. Most of the time, the knowledge is embedded in the existing tools, processes, or procedures that the users of the other disciplines are already familiar with, but are hindrances to the improvement in the effectiveness and efficiency of their practices. As a result, achieving the knowledge avoidance outcome usually entails significant efforts by the interdisciplinary practitioners to challenge the status quo.

4. **Knowledge Addition**

The study had conceived the concept of knowledge addition outcome as the ability to contribute knowledge that is new to the collaborators of other disciplines who then adopt it as their domain knowledge. It involves the diffusion of knowledge from a practice domain of one discipline to that of the others.

It was found that the interdisciplinary practitioners acknowledge the relevance and usefulness of the knowledge contributed by their counterparts from other disciplines; however, they also perceive that unless the knowledge is complemented by other knowledge elements, its usage would be insufficient for solving the problems at hand. They are likely to experience much difficulty in arguing for the acceptance of it by their counterparts who lack familiarity with the knowledge.

The identifications of those epistemic practices and learning outcomes accomplish the first objective of the study. For the second objective, the study had produced explanations for the engagement in those practices and the achievements of those outcomes.

Explanatory Findings

The engagements in the four epistemic practices were found to be necessitated by the perception of four types of barriers that tend to emerge during interdisciplinary interactions. However, successful engagement in those practices are made more likely when interdisciplinary practitioners orientate into four modes of epistemic engagement. Successful engagements in the four practice and orientations into the modes of epistemic engagements increase the likelihood of achieving the four learning outcomes.

Barriers to Interdisciplinary Learning

The study identified four types of barriers to interdisciplinary learning that explain the necessity for engaging in the four epistemic practices. The barriers are situational perceptions that constrain interdisciplinary learning, and thereby creating the need for engaging in specific practices to overcome those barriers.

1. **Communication barriers**

Communication barriers have been conceived by this study as a form of predicament in getting the knowledge that interdisciplinary practitioners think they need for developing solutions. It was found that the perception of communication barriers is likely to cause engagements in the TEP, and subsequently to the achievement on knowledge translation outcome. In addition, the study also found that the perception of communication barriers is likely to relate to another category of outcome - the knowledge avoidance outcome. This means that overcoming communications barriers through engagement in the TEP also contributes to the identification of knowledge elements that can hinder successful development of solutions. This is especially the case when interdisciplinary practitioners create representations of the knowledge of other disciplines for eliciting the implicit form of knowledge or for converting its form. The removal of the communication barrier thus allows interdisciplinary practitioners and their counterparts to jointly evaluate, or test, the usefulness of the knowledge. This facilitates their joint agreement to avoid knowledge that is not evidently useful for solving problems.

2. **Contradictory barriers**

Contradictory barriers are more likely to arise when there is no perceived communication barrier to communicating and understanding the knowledge of other disciplines but the knowledge itself is perceived as discordant since the actual practices of the practitioners from other disciplines appear to contradict the knowledge they espouse. It is conceived by this study as a form of hesitation or reluctance to learning the knowledge of other disciplines due to the perception that it contradicts the observed practices. It was found that contradictory barriers relate causally to engagements in EEP, and is likely to lead either to the knowledge adoption outcome or to the knowledge avoidance outcome.

3. **Contributory barriers**

The study conceptualizes the difficulty in contributing knowledge to other disciplines without first providing the relevant evidence as a form of contributory barriers. Such barriers are likely to arise due to the difficulty of the interdisciplinary practitioners in foreseeing the results of complementing their knowledge with the knowledge proposed by the other disciplines. Although the knowledge itself may be understood, the value of accepting the knowledge contribution may not be appreciated easily without going through tests that generate evidence of the expected results. It was found that contributory barriers relate causally to engagements in EEP and is likely to lead to the knowledge addition outcome.

4. **Implicational barriers**

Implicational barriers are likely to arise when there is an absence of knowledge from all the collaborating disciplines about the implications of the suggested knowledge to the solutions that are being developed. It is conceived by this study as a form of difficulty in deciding whether or not one should learn the knowledge due to the uncertainty of its implications for the solutions under development. Such barriers are likely to arise during the engagement in the CEP due to the uncertainty surrounding the implications of the knowledge suggestion for the solutions to be developed or under development. It was found that implicational barriers relate causally to engagements in the EEP, and is likely to lead either to the knowledge adoption outcome or to the knowledge avoidance outcome. Although the collaborating disciplines are able to communicate the knowledge well, they often do not know enough about the solutions proposed by others from different disciplines. Therefore, they often do not know the possible implications of the knowledge to the solutions. Consequently, the development of an adequate understanding of the knowledge is impeded by the uncertainty about the possible implications (i.e. the implicational barrier). As a result, practitioners often resort to engage in the EEP in order to learn specifically about the implications. Depending on the evidence, overcoming the implicational barrier through engagement in the EEP is likely to cause either the adoption of knowledge, if it is shown to affect the development of solutions, or the avoidance of it, if the evidence shows otherwise.

Although the existence of the four barriers contingently explains the necessity for engaging in those practices, there is no guarantee that interdisciplinary learners could successfully overcome the barriers, engage in those practices, and achieve the learning outcomes. This study has established that successful engagements in the four epistemic practices and the achievements of the corresponding learning outcomes are contingent upon orientations into four modes of epistemic engagements. When faced with barriers to interdisciplinary learning, engagements in the corresponding epistemic practices and achievements of the outcomes become more likely when interdisciplinary practitioners orientate into the specific modes of epistemic engagement.

Modes of Epistemic Engagement

There are four modes of epistemic engagement that relate in specific ways to the four barriers in that the modes are more likely to enable the interdisciplinary practitioners to engage in the epistemic practices and achieve the outcomes.

1. **Perspectival mode**

The engagement in this mode is likely to enable interdisciplinary practitioners to overcome, rather than succumb to, communication barriers and subsequently engage in the TEP. In this mode, the practitioners tend to engage with knowledge from other disciplines by viewing it from multiple perspectives, such as from the users' perspective as well as from the perspective of their own disciplines.

These perspectives help the practitioners to harness their knowledge and skills in producing representations. Their orientation into this mode enables them to produce knowledge representations that address the needs of others for participating in the process of translating knowledge into a more useful form. Since the participation of others is also necessary to the development and the adoption of the solutions, the orientation into the perspectival mode could increase the propensity of achieving the desired outcomes.

2. **Justificational mode**

This mode is likely to enable interdisciplinary practitioners to overcome, rather than to succumb to, contradictory barriers and subsequently engage in the EEP. In this mode, the practitioners tend to become more tolerant to ambiguity in knowledge claims or contradictions in practice, and see the contradictory claims and practices as opportunities to interrogate the different justifications to those claims and practices. Their orientation into this mode enables them to identify the kinds of tests and evidence that would be required for interrogating and testing the different justifications. This facilitates a collaborative rationalization on what should be done to the knowledge. As a result, it guides their engagement in the EEP that could lead to the appropriate outcomes.

3. **Complementary mode**

This mode is likely to enable the interdisciplinary practitioners to overcome, rather than to succumb to, contributory barriers and subsequently engage in the EEP. In this mode, they tend to become more visionary in finding ways in which they could position their prior knowledge as complementary to the knowledge of other disciplines suggested to them. They are more likely to formulate and articulate a vision of the realization of the benefits of the complementary combination of knowledge. The responses of their counterparts to the formulated vision can then be used to guide the choice of tests and evidences on the envisioned benefits of the knowledge combination. As a result, it guides their engagement in the EEP that leads to the knowledge addition outcome.

4. **Implicational mode**

This mode is likely to enable the interdisciplinary practitioners to overcome, rather than to succumb to, implicational barriers and subsequently engage in the EEP. In this mode, the practitioners tend to direct their investigation towards determining whether the knowledge they encounter has implications for the solutions under development. They would be more likely to design suitable tests of the possible impact of the knowledge on the solutions. As a result, it guides their engagement in the EEP that could inform the appropriate outcomes.

Recommendations for Interdisciplinary Learning Practice and Research

The descriptive and explanatory findings of this study contribute towards a practice-based theory of interdisciplinary learning that could have profound implications for interdisciplinary learning in higher education.

Firstly, the description of interdisciplinary learning in terms of epistemic practices implicates the need to facilitate repeated engagements in knowledge practices, or the so-called 'epistemic practices', rather than only to expose students to interdisciplinary problems and contexts. Educators need to intervene with suggestions of the appropriate epistemic practices when students encounter knowledge in interdisciplinary interactions.

Secondly, the explanation for the necessity of those epistemic practices shows that engagements in epistemic practices are contingent on a number of situational perceptions. This implicates that instructors should not simply intervene by dictating engagements into a specific knowledge practice without first knowing their students' perceptions of the situations. Instead, instructors should ensure that engagement in any knowledge practice should result from students' efforts to analyze and perceive different situations. More generally, students should be encouraged to engage in situational judgement for choosing knowledge practices that could help them overcome learning barriers and sustain interdisciplinary learning. Thus, it is recommended that situational analysis and judgement on the most appropriate knowledge practice should form a necessary part of interdisciplinary teaching and learning in higher education.

Thirdly, the perceptions of various barriers to interdisciplinary learning are prevalent in interdisciplinary interactions. It is therefore important to create awareness of their occurrences in interdisciplinary teamwork. More importantly, students should be trained to become more adept at overcoming those barriers by repeatedly orientating themselves into the enabling modes of epistemic engagement. However, since the barriers depend on the individual perceptions of situations, it is recommended that instructors should first seek to interpret and validate the type of barriers faced by students before intervening with suggestion for orientation into the appropriate mode of epistemic engagement.

Fourthly, the identification of the four learning outcomes implicates the need for promoting and assessing the attainment of the outcomes in educational settings. These outcomes need to be expressed in a form that is amenable to summative assessments of students' readiness for interdisciplinary practices. Table 1 below translates the four outcomes identified by this study into the corresponding abilities that can be assessed summatively at the end of a course or a program.

The dependence of the four outcomes on prior orientations into modes of epistemic engagement and engagements in epistemic practices has a profound implication for the summative assessment of students' abilities. It implicates the need to complement summative assessments with formative assessments that foster and assess the gradual development of abilities. Thus, the findings can be translated into 'intermediate learning outcomes', a set of leading indicators of student progress towards the final learning outcomes. For example, one of the modes of epistemic engagement, the 'justificational mode of epistemic engagement' can be translated into an assessable statement of 'ability to inquire into the different justifications to knowledge suggestions'. By promoting and assessing their achievement of the

Table 1. The four learning outcomes and the corresponding statements of abilities

Outcomes of Interdisciplinary Learning	Assessable Abilities
Knowledge Adoption.	Ability to understand, appreciate the importance of, and reuse the relevant knowledge of other disciplines while retaining its original contents and meanings.
Knowledge Translation.	Ability to develop and use knowledge whose terms and forms are usefully different from, but correspond to that which is used in, or provided by, the other disciplines.
Knowledge Avoidance	Ability to avoid pursuing the learning and use of suggested knowledge contents and forms that do not contribute to the successful development of the solution.
Knowledge Addition	Ability to add knowledge that is new to the collaborators from other disciplines and evidently useful for improving their practices.

intermediate outcomes prior to the summative assessment of the outcomes, instructors can track progress in learning and deploy interventions in a timely manner. This would progressively increase the likelihood of achieving the final outcomes.

Finally, the decisions on what, and how, knowledge could be acquired, translated, avoided or added are largely undertaken emergently by the interdisciplinary learners themselves. This implicates the need to give autonomy to students during interdisciplinary learning and to tolerate changes in their decision-making. Therefore, interdisciplinary approaches to learning in higher education needs to embrace some form of self-determined approach to teaching and learning, also known as heutagogy. By embracing heutagogy within the interdisciplinary approach, instructors can relinquish ownership of the learning process to the learner, who autonomously determines what and how knowledge will be learnt as well as when and why the determinations should change.

FUTURE RESEARCH DIRECTIONS

Since this study did not seek to exhaustively describe and explain interdisciplinary learning in interdisciplinary workplace, it is important to pursue future research into interdisciplinary learning. Future research should be directed towards a number of areas that could increase the comprehensiveness of the descriptions and explanations offered by this study.

Studying Ongoing Cases of Interdisciplinary Learning in 4IR-Related Projects

The most obvious direction for future research is to study cases that are still ongoing rather than those that had been completed in the past. Researchers should engage with current interdisciplinary projects that directly address the Industry 4.0 requirements. Such an engagement should prioritize on observing the learning practices of professionals at the actual sites of interdisciplinary collaborations. Preferably, a study of ongoing cases of Industry 4.0 projects should be carried out as one large-scale longitudinal research covering all the whole process of interdisciplinary collaboration. As well as conducting observations and interviews, the data collection could also include focus group so that alternative descriptions and explanations could be entertained and adjudicated. It is recommended that future research adheres to the practice-based approach. Their results could cumulatively increase, if not exhaust, the range of learning practices and outcomes identified.

Replicating the Study in Educational Settings

Another high priority area of research would be to replicate the study in educational settings. Preferably, the study of interdisciplinary learning in educational settings should engage ongoing cases of interdisciplinary student projects related to Industry 4.0, such as interdisciplinary engineering capstone design projects. As well as describing and explaining learning practices that result in both the desired and undesired outcomes, such as study should also seek to identify other personal, curricular and institutional contingencies that could increase, if not ensure, student engagements in the learning practices and the achievements of the learning outcomes.

With the larger scope of factors related to those personal, curricular and institutional contingencies, researchers can increase the applicability of the findings in interdisciplinary workplace setting to the educational setting. These descriptive and explanatory studies could be usefully followed by action-research type of studies. Researchers can implement and then evolve the description and explanation into those that could be used to further support and sustain learning in educational settings.

Studying Less Successful Cases of Interdisciplinary Collaborations

Another theoretically useful research direction is to study past interdisciplinary collaborations that had failed to create, or sustain, solutions to problems that they had attempted to solve. These can be called less successful, or 'negative cases'. Although they cannot be prioritized highly due to the overreliance on retrospective views of the informants and the possible lack of data for validating inferences, they could still be theoretically useful. If the concepts that describe and explain the successful cases are absent from the 'negative cases', then our confidence in the proposed description and explanation could be increased. Since access to data on such cases would be difficult for non-participant researchers, researcher should try to become 'participants' in interdisciplinary collaborations and keep documentary evidences easily accessible for comparing between cases of less successful and more successful ones.

CONCLUSION

This chapter has shown how a multi-case study of interdisciplinary projects in the industrial workplace setting could generate a more useful conceptualization of interdisciplinary learning that is aligned to the demand of Industry 4.0. The recommendations that have been derived from findings could have profound implications for interdisciplinary higher education. First, the concept of epistemic practice, which conceptualizes what practitioners actually do with the knowledge from different disciplines that they encounter in interdisciplinary projects, provides a practice-based perspective to redesigning higher education for Industry 4.0. In contrast to the process-based conceptualization that emphasizes repeated exposure to interdisciplinary problems, the practice-based conceptualization offered by this chapter implicates the need to emphasize repeated engagements in the different categories of epistemic practices as well as deliberate orientations into the different modes of epistemic engagement. Future research in industrial and education settings is required to identify more epistemic practices and modes of epistemic engagement that could further inform the design and implementation of interdisciplinary teaching in higher education.

The necessity for engaging in epistemic practices tend to emerge from the perception of various barriers to interdisciplinary learning. Since the perception of various barriers depends on individuals' interpretations of the situations they face during interdisciplinary interactions, educators need to design interdisciplinary learning activities that foster situational analysis and judgement in order to develop the ability to choose the appropriate epistemic practices to be deployed. Repeated practice of analyzing interdisciplinary situations in terms of learning barriers could help students understand the difficulties they experience in interdisciplinary learning as well as help them understand that there are ways of sustaining their learning towards achieving the intended outcomes.

The identifications of the learning outcomes has been shown to facilitate the formulation of assessable abilities which guide the assessment of graduates' readiness for interdisciplinary practice. These are the abilities that are aligned to the demand of Industry 4.0, whereby employees are expected to be able to develop solutions that address opportunities and challenges arising from the fourth industrial revolution.

This chapter concludes by reinforcing that there are many valuable lessons to be learnt from studying interdisciplinary learning practices in the industrial workplace. Educational researchers are encouraged to pursue further engagements with interdisciplinary projects that are related to Industry 4.0. These engagements could produce more valuable insights for redesigning higher education initiatives for Industry 4.0.

REFERENCES

ABET. (2011). *Criteria for Accrediting Engineering Programs.* Accreditation Board for Engineering and Technology Inc.

Apostel, L., Berger, G., & Michaud, G. (1972). *Interdisciplinarity: problems of teaching and research in universities*. Paris: OECD.

Archer, R., & Wood, L. (1992). Production tissue culture by robots. In R. Spier, J. Griffiths, & C. MacDonald (Eds.), *Animal Cell Technology: Developments, Processes and Products* (pp. 403–408). Oxford, UK: Butterworth-Heinemann. doi:10.1016/B978-0-7506-0421-5.50094-5

Bareither, R., & Pollard, D. (2011). A review of advanced small-scale parallel bioreactor technology for accelerated process development: Current state and future need. *Biotechnology Progress*, *27*(1), 2–14. doi:10.1002/btpr.522 PMID:21312350

Borrego, M., & Newswander, L. K. (2008). Characteristics of Successful Cross-disciplinary Engineering Education Collaborations. *Journal of Engineering Education*, *97*(2), 123–134. doi:10.1002/j.2168-9830.2008.tb00962.x

Borrego, M., & Newswander, L. K. (2010). Definitions of interdisciplinary research: Toward graduate-level interdisciplinary learning outcomes. *The Review of Higher Education*, *34*(1), 61–84. doi:10.1353/rhe.2010.0006

Brint, S. G., Turk-Bicakci, L., Proctor, K., & Murphy, S. P. (2009). Expanding the social frame of knowledge: Interdisciplinary, degree-granting fields in American colleges and universities, 1975–2000. *The Review of Higher Education, 32*(2), 155–183. doi:10.1353/rhe.0.0042

Dujin, A., Geissler, C., & Horstkötter, D. (2014). *Industry 4.0.* The new industrial revolution-How europe will succeed. Munich: Roland Berger Strategy Consultants.

Fenwick, T., Nerland, M., & Jensen, K. (2014). *Professional Learning in Changing Contexts.* Oxon, UK: Routledge.

George, A. L., & Bennett, A. (2005). *Case Studies and Theory Development in the Social Sciences.* MIT Press.

Holley, K. (2017). Interdisciplinary Curriculum and Learning in Higher Education. *Oxford Research Encyclopedia of Education.* Retrieved from http://education.oxfordre.com/view/10.1093/acrefore/9780190264093.001.0001/acrefore-9780190264093-e-138

IEA. (2015). *25 Years > Washington Accord 1989-2014 Celebrating International Engineering Education Standards and Recognition.* International Engineering Alliance.

Ivanitskaya, L., Clark, D., Montgomery, G., & Primeau, R. (2002). Interdisciplinary learning: Process and outcomes. *Innovative Higher Education, 27*(2), 95–111. doi:10.1023/A:1021105309984

Karseth, B., & Nerland, M. (2007). Building professionalism in a knowledge society: Examining discourses of knowledge in four professional associations. *Journal of Education and Work, 20*(4), 335–355. doi:10.1080/13639080701650172

King, J. (2007). *Educating Engineers for the 21st Century.* London: Royal Academy of Engineering.

Klein, J. T., & Newell, W. H. (1997). Advancing interdisciplinary studies. Handbook of the undergraduate curriculum: A comprehensive guide to purposes, structures, practices, and change, 393-415.

Lasi, H., Fettke, P., Kemper, H. G., Feld, T., & Hoffmann, M. (2014). Industry 4.0. *Business & Information Systems Engineering, 6*(4), 239–242. doi:10.100712599-014-0334-4

Lattuca, L. R., & Knight, D. (2010). *In the eye of the beholder: Defining and studying interdisciplinarity in engineering education.* Paper presented at the American Society for Engineering Education, Louisville, KY.

Lattuca, L. R., Knight, D. B., & Bergom, I. M. (2012). *Developing a measure of interdisciplinary competence for engineers.* Paper presented at the American Society for Engineering Education.

Lattuca, L. R., Knight, D. B., Ro, H. K., & Novoselich, B. J. (2017). Supporting the development of engineers' interdisciplinary competence. *Journal of Engineering Education, 106*(1), 71–97. doi:10.1002/jee.20155

Lattuca, L. R., Trautvetter, L., Codd, S., & Knight, D. (2011). Working as a team: Enhancing interdisciplinarity for the engineer of 2020. *Proceedings of the 118th Annual Conference of the American Society for Engineering Education Vancouver.*

Lattuca, L. R., Voigt, L. J., & Fath, K. Q. (2004). Does interdisciplinarity promote learning? Theoretical support and researchable questions. *The Review of Higher Education*, *28*(1), 23–48. doi:10.1353/rhe.2004.0028

Lee, J., Bagheri, B., & Kao, H. A. (2015). A Cyber-Physical Systems architecture for Industry 4.0-based manufacturing systems. *Manufacturing Letters*, *3*, 18–23. doi:10.1016/j.mfglet.2014.12.001

MacLeod, M. (2016). What makes interdisciplinarity difficult? Some consequences of domain specificity in interdisciplinary practice. *Synthese*, 1–24. doi:10.100711229-016-1236-4

Mansilla, V. B., & Duraising, E. D. (2007). Targeted assessment of students' interdisciplinary work: An empirically grounded framework proposed. *The Journal of Higher Education*, *78*(2), 215–237. doi:10.1080/00221546.2007.11780874

McMurtry, A., Rohse, S., & Kilgour, K. N. (2016). Socio-material perspectives on interprofessional team and collaborative learning. *Medical Education*, *50*(2), 169–180. doi:10.1111/medu.12833 PMID:26812996

NAE. (2005). *Educating the Engineer of 2020: Adapting Engineering Education to the New Century*. Washington, DC: The National Academies Press.

Nerland, M., & Jensen, K. (2012). Epistemic practices and object relations in professional work. *Journal of Education and Work*, *25*(1), 101–120. doi:10.1080/13639080.2012.644909

Nerland, M., & Jensen, K. (2014a). Changing cultures of knowledge and professional learning. In S. Billett, C. Harteis, & H. Gruber (Eds.), *International Handbook of Research in Professional and Practice-based Learning* (pp. 611–640). London: Springer. doi:10.1007/978-94-017-8902-8_23

Nerland, M., & Jensen, K. (2014b). Learning through epistemic practices in professional work. In T. Fenwick & M. Nerland (Eds.), *Reconceptualising Professional Learning: Sociomaterial Knowledges, Practices and Responsibilities* (pp. 25–37). London: Routledge.

Nersessian, N. J. (2009). How Do Engineering Scientists Think? Model-Based Simulation in Biomedical Engineering Research Laboratories. *Topics in Cognitive Science*, *1*(4), 730–757. doi:10.1111/j.1756-8765.2009.01032.x PMID:25163455

Nersessian, N. J., & Newstetter, W. C. (2014). Interdisciplinarity in engineering research and learning. In A. Johri & B. M. Olds (Eds.), *Cambridge Handbook of Engineering Education Research* (pp. 713–730). Cambridge, UK: Cambridge University Press.

Newell, W. H., & Gagnon, P. (2013). The State of the Field: Interdisciplinary Theory. *Issues in Interdisciplinary Studies,* (31), 21.

Newell, W. H., & Klein, J. T. (1998). *Interdisciplinarity: Essays from the Literature*. College Board.

Newell, W. H., Wentworth, J., & Sebberson, D. (2001). A theory of interdisciplinary studies. *Issues in Interdisciplinary Studies*, *19*, 1–25.

Newstetter, W. C. (2011). *Unpacking the interdisciplinary mind: Implications for teaching and learning The challenge.* Paper presented at the American Society for Engineering Education, Vancouver, Canada.

Niemeyer, C. M. (2017). Engineering for life sciences: A fruitful collaboration enabled by chemistry. *Angewandte Chemie International Edition*, *56*(8), 1934–1935. doi:10.1002/anie.201611865 PMID:28105666

Reich, A., Rooney, D., Gardner, A., Willey, K., Boud, D., & Fitzgerald, T. (2015). Engineers' professional learning: A practice-theory perspective. *European Journal of Engineering Education*, *40*(4), 366–379. doi:10.1080/03043797.2014.967181

Repko, A. F. (2008). Assessing interdisciplinary learning outcomes. *Academic Exchange Quarterly*, *12*(3), 171.

Richter, D. M., & Paretti, M. C. (2009). Identifying barriers to and outcomes of interdisciplinarity in the engineering classroom. *European Journal of Engineering Education*, *34*(1), 29–45. doi:10.1080/03043790802710185

Stacey, G. (2012). Current developments in cell culture technology. In M. Balls, R. Combes, & N. Bhogal (Eds.), *New Technologies for Toxicity Testing* (pp. 1–13). Springer. doi:10.1007/978-1-4614-3055-1_1

Stentoft, D. (2017). From saying to doing interdisciplinary learning: Is problem-based learning the answer? *Active Learning in Higher Education*, *18*(1), 51–61. doi:10.1177/1469787417693510

Wollschlaeger, M., Sauter, T., & Jasperneite, J. (2017). The future of industrial communication: Automation networks in the era of the internet of things and industry 4.0. *IEEE Industrial Electronics Magazine*, *11*(1), 17–27. doi:10.1109/MIE.2017.2649104

Wynn, D., & Williams, C. K. (2012). Principles for conducting critical realist case study research in information systems. *Management Information Systems Quarterly*, *36*(3), 787–810. doi:10.2307/41703481

Xing, B., & Marwala, T. (2017). *Implications of the Fourth Industrial Age on Higher Education.* arXiv preprint arXiv:1703.09643

ADDITIONAL READING

Acton, R. (2017). Place-people-practice-process: Using sociomateriality in university physical spaces research. *Educational Philosophy and Theory*, (March): 1–11.

Dall'Alba, G., & Barnacle, R. (2015). Exploring knowing/being through discordant professional practice. *Educational Philosophy and Theory*, *47*(13-14), 1452–1464. doi:10.1080/00131857.2014.947562

Decuypere, M., & Simons, M. (2016). On the critical potential of sociomaterial approaches in education. Teoría de la Educación. *Revista Interuniversitaria*, *28*(1), 25–44. doi:10.14201/teoredu20162812544

Fenwick, T., Edwards, R., & Sawchuk, P. (2011). *Emerging Approaches to Educational Research: Tracing the Socio-material.* Oxon: Routledge.

Fenwick, T., Nerland, M., & Jensen, K. (2012). Sociomaterial approaches to conceptualising professional learning and practice. *Journal of Education and Work*, *25*(1), 1–13. doi:10.1080/13639080.2012.644901

Gherardi, S., & Nicolini, D. (2002). Learning in a constellation of interconnected practices: Canon or dissonance? *Journal of Management Studies*, *39*(4), 419–436. doi:10.1111/1467-6486.t01-1-00298

Klein, J. T. (1990). *Interdisciplinarity: History, Theory, and Practice*. Detroit: Wayne State University Press.

Knight, D. B., Lattuca, L. R., Kimball, E. W., & Reason, R. D. (2013). Understanding interdisciplinarity: Curricular and organizational features of undergraduate interdisciplinary programs. *Innovative Higher Education*, *38*(2), 143–158. doi:10.100710755-012-9232-1

McMurtry, A., Rohse, S., & Kilgour, K. N. (2016). Socio-material perspectives on interprofessional team and collaborative learning. *Medical Education*, *50*(2), 169–180. doi:10.1111/medu.12833 PMID:26812996

KEY TERMS AND DEFINITIONS

Epistemic Practices: Actions taken by practitioners on the knowledge they encounter in their work practices.

Heutagogy: Self-determined learning approach whereby learners autonomously decide what, why, and how to learn.

Interdisciplinary Learning: Learning the knowledge of other disciplines other than one's won in the context of interaction with others from different disciplines.

Interdisciplinary Teaching: An emerging educational approach whereby students are presented with complex issues and problems that demand them to draw on two or more disciplines in order to advance their understanding of those issues and problems.

Mode of Epistemic Engagement: Orientations of the learners when they encounter knowledge that is new to them.

Socio-Material Perspective of Learning: A view of learning as interactions that involve interactions among human beings and material items with which they work.

Chapter 2
Adopting Internet of Things for Higher Education

Ruuhina Mohd Sani
Universiti Utara Malaysia, Malaysia

ABSTRACT

To respond to the needs of IR 4.0, universities must continue to play their role as proving ground for educating the future generation and innovation. However, close collaboration with industry and stakeholders will be even more important to implement Education 4.0. Many of the components of IR 4.0 already exist in the Malaysian higher education system, but perhaps in separation such as soft skills that IR 4.0 demands have already been introduced to higher education in the country. In the context of demanding requirements of IR 4.0, it is important to look back on how the delivery of soft skills can be made more effective. Efforts should be made on how to make digital or even data literacy more accessible to the overall student regardless of discipline. Hence, the need for human resources in the key technical areas in IR 4.0 such internet of things (IoT), big data and analytics, cloud computing, virtual/augmented reality, and robotics should be assessed. This chapter will focus on integrating IoT in higher education.

INTRODUCTION

In effort to support the needs of IR 4.0, universities and college must continue to lead and become the proving ground for educating the next generation and innovation. Hence, close collaboration between industry and stakeholders is important to enable the implementation of Education 4.0 will be success. So, it is important for national educational system to have not just technology blueprint for the major economic but also for industrial sectors of the country. Such technology blueprint will provide the right path that may include two important aspects which are to improve education across disciplines in terms of content, delivery, management; and also create special education programs for technical workers development.

Many of the aspects related to IR 4.0 already applied in the Malaysian higher education system, but perhaps in separation for example soft skills that becoming demanding for IR 4.0 which have already been introduced to higher education. In the context of understanding the needs of IR 4.0 in today's

DOI: 10.4018/978-1-5225-7832-1.ch002

Copyright © 2019, IGI Global. Copying or distributing in print or electronic forms without written permission of IGI Global is prohibited.

higher educational system, it is important to look back on how the conveyance of soft skills can be made more effective. The efforts on how to make digital or even data literacy more accessible to the overall student regardless of their field of study should be priority. Hence, the need for human resources in the key technical areas in IR 4.0 such Internet of Things (IoT), big data and analytics, cloud computing, virtual/augmented reality and robotics should be assessed. This chapter will focus on integrating IoT in higher education.

ICT educators in higher education institutions suggest some incentives for teaching IoT concepts, such as:

1. IoT winds up critical pattern in educational system and the present students will require understanding on the best way to create and keep up these systems.
2. IoT may present processing in a way that widens cooperation in figuring since it is integrative.
3. IoT systems extent significant part of modern computing, including embedded systems, networking, client–server architectures, cloud services, service-oriented architectures, and human-computer interaction.
4. IoT provides greater chance to connect with the students by demonstrating an application domain that is part of their daily life and may indicate them more extensive uses of computing.
5. IoT help to broaden students' knowledge areas as they can deliver, analyze, and manage data to maintain systems or even build solutions in ethical and secure ways.

CONCEPT OF INTERNET OF THINGS

The Internet of Things (IoT), also known as Internet of Everything or the Internet of Objects is a latest advancement featured as a global network of machines and devices into a network that enable them to interact with each other. The idea of IoT is to integrate all devices such mobile devices, vehicles, appliances and additional into the network, which may be managed from net and within the end, it will offer data and knowledge in real time conjointly permitting the interaction with those who use it.

In 2014, Gartner forecasts that by 2020 the IoT will reach 26 billion units, up from 0.9 billion in 2009 and then will impact the information available to be access anytime and anywhere. This notion is making progress due to advances in nanotechnology that permits the creations of devices capable of connecting to the net efficiently. The new prospect of the net, that allows devices to integrate and communicate with one another and share information which may then be processed for decision-making and later becomes 'Smarts'. This whole conception is called as 'Internet of Things'.

Since the start of *Internet of Things* (IoT), many researchers have tried to outline IoT in many ways like *Internet of Everything, Internet of Anything, Internet of People, Internet of Signs, Internet of Services, Internet of Data* or *Internet of Processes* (Cornel, 2015). According to Mark Weiser, "The most profound technologies are those that disappear. They weave themselves into the fabric of everyday life until they are indistinguishable from it" (Weiser, 1999). However in 1999, Kevin Ashton a British technology pioneer, who is known for coining the term Internet of Things to describe a system where the Internet is connected to the physical world via ubiquitous sensors. According to Oriwoh and Conrad

(2015), IoT represents as *'anything at all, depending on requirements'* whereas Cisco defines IoT as a network of connected physical objects. Besides, Cisco additionally uses the term *Internet of Everything* for both physical and virtual objects. Cisco also states that "*IoE brings together people, process, data, and things to make networked connections more relevant and valuable than ever before* (Barakat, 2016).

Generally, the IoT organize interfaces diverse sorts of devices like personal computers, laptops, tablets, smartphones and PDAs with one another. Alternative examples are devices for blood pressure measures, heart rate, devices to contact emergency services, robots, autonomous vehicles, and home appliances. These devices accumulate useful information needed with a variety of sensors and data collection technology, and then transmit it to other processing devices for translation and decision-making. According to Marques, Garcia and Pombo (2017), IoT based systems ordinarily communicate through wireless technologies such as RFID (Radio-Frequency Identification), NFC (Near Field Communication), WSN (Wireless Sensor Network), WLAN (Wireless Local Area Network), DSL (Digital Subscriber Line), UMTS (Universal Mobile Telecommunications System), WiMax (Worldwide Interoperability for Microwave Access), ZigBee, GPRS (General Packet Radio Service), or LTE (Long-Term Evolution).

In any case, the same number of the ideas basic IoT, for example embedded programming and distributed system that in fact propelled, the inquiry remains is the manner by which to bring down the section level for finding out about and investigating IoT advances, particularly for young audiences. In addition, the key challenges to IoT would be much same as others new technologies including security and privacy, availability, mobility, reliability, performance, interoperability, scalability, trust and management (Al-fuqaha, Guizani, Mohammadi, Aledhari & Ayyash, 2015). This paper discusses the role of IoT particularly in the field of higher education. In this chapter, the recent research, challenges and future impact of IoT in education will be presented.

Figure 1. Concept of internet of things

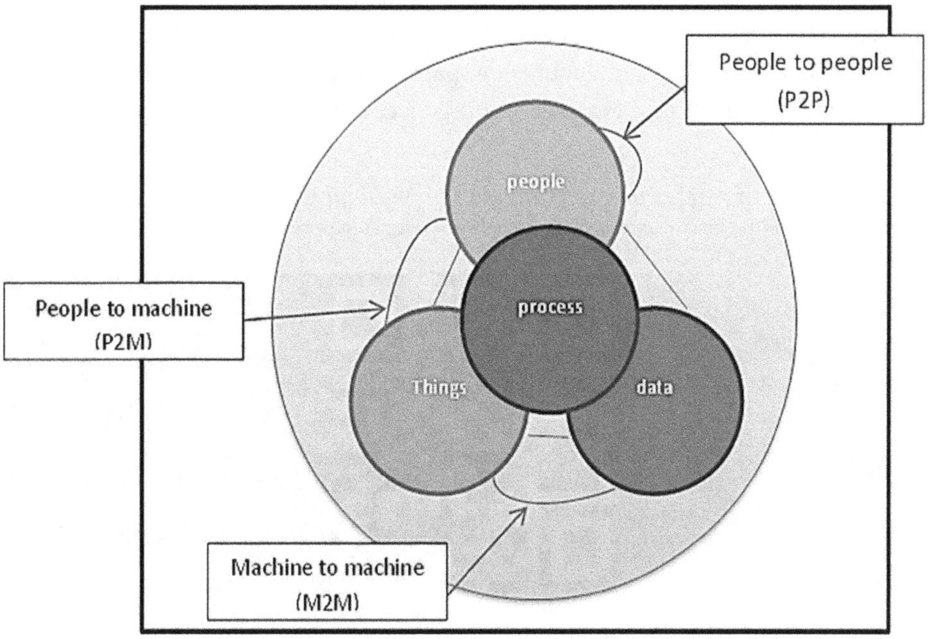

ENABLING TECHNOLOGIES FOR THE IoT

As below figures shows, tagging an object to reference a proxy Web service can be achieved through a variety of technologies such as NFC tags, QR code or BLE tag, however there were some issues with the early primary contenders that hindered their adoption. Next section will explain details some of these technologies (NFC tags, QR code and BLE tag) in IoT.

RFID and Near-Field Communication

Since early 2000s, RFID was viewed as a standout amongst the no doubt innovations to quicken the development of the IoT. EPC Global built up another UHF RFID tag standard aimed further automating retail transactions and supplanting barcodes with a tag that was machine readable at a distance of up to 10 feet. However, after a few preliminaries with couple of enormous companies such as Walmart and Tesco, the EPC standard indicated impediment. It was discovered that practically speaking, a significant number of tags were imperceptible due to factors in retail environments such as poor product/tag orientation and the occurrence of materials caused interference with the wireless identification process. Hence, the most recent open door for RFID innovations can be found as near-field communication (NFC) as a provision for electronic payments such as in iPhone 6. Despite the fact that not all of smartphone products incorporate NFC transceivers, the potential for this capability to be utilized for altogether future smartphones is high.

Optical Tags and Quick Response (QR) Codes

Another technology for affordable and popular tagging is the optical or printed tag and the most popular 2D optical standard would be the quick response (QR) code. The high-quality of the QR code standard is at once associated to the ubiquity of the reader, regarded as an application of the high-resolution digital camera found in all cutting-edge smartphones. A QR code features when it is extracted and decoded from a scene the use of image-processing methods yielding a number, text, or URI. QR codes can be discovered printed on many products, including newspapers, magazines, billboards, and coupons or even appear on prime-time television ads. However, in practice, many QR advertising campaigns result in

Figure 2. Different forms of electronic tags support the physical web
(a) Near-field communication (b) Quick response (QR) code (c) Bluetooth loe energy (BLE) (NFC) tag

a poor customer response because of the requirement that a preinstalled application is required to read QR codes which can be a barrier for some users. Additionally, problem in positioning the smartphone so that the digicam can center the attention and precisely decode the image also one of the barriers; while some advertisers also experience that a visible QR code spoils the aesthetics of their campaign.

Bluetooth Low Energy (BLE)

Bluetooth low energy (BLE) is part of the Bluetooth v4.0 standard (Bluetooth Smart), adopted by the Bluetooth Special Interest Group in 2010. Bluetooth silicon vendors have covered BLE in their brand new chipsets, and all smartphones released in the last few years with more than a few ranges of functionality relying on operating system support included in those smartphones. BLE is specifically designed for low power, low facts functions and is very well suited to many interesting IoT applications, such as wearables and battery operated sensor networks due to bluetooth silicon can be pared down to solely encompass the BLE aspects of the standard, removing the need for compatibility with classic Bluetooth. (Burd, Divitini, Barker & Perez, 2017). This new technology know-how standard, along with the availability of affordable BLE tag hardware and tag readers already built-in with smartphone hardware and regarded as catalyst in these days communication world. Since all modern smartphones aid BLE and provide APIs to have interaction with BLE devices, it's fantastically handy to build apps for these devices. Many well-known computer companies, and some wide variety of startups, are experimenting with products and enterprise opportunities related with these tags. As with the other tagging technology, this is also an enabler for the IoT, but with extra opportunities for ubiquitous deployment, higher-accuracy tag reads, and the capacity to blend in invisibly with a product.

INTERNET OF THINGS IN HIGHER EDUCATION

Regarding the educational field, the goal is to make contribution to improve teaching and learning through Internet of Things. It can be determined in an interactive English instructing conducted by Wang (2011) which integrates voice and visual sensors to obtain the pronunciation of students. Besides, students also learn at their personal capacity in accordance to their capabilities, so they are now not constrained by an application that constructed with certain functions.

In 2013, Gómez, Huete, Hoyos, Perez and Grigori (2013) conducted an experiment to find out the effect of adopting Internet of Things a tool to support the teaching process in instruction and the results shown student academic performance improves. Additionally, they ascertain by using real objects and partner them as a learning resource through the Internet of Things allows significant learning, as it allows linking specific knowledge to an actual context. Regarding the use of the system Internet of Things in the experimental group, the result confirmed that students elevated their learning, which can be considered by using the result of measuring academic outcomes compared to the control group (Gómez et al., 2013). In this case, Bo Xing and Tshilidzi Marwala (2017) mentioned the goals of higher education include:

1. To ensure value of learning through teaching; and one of the primary duties of every university is to educate the youth. Therefore, it is imperative to implement appropriate **teaching** strategies and to organize work in a way that nurtures the learning so that it will assist facilitate learning programs create better learning experience, and lifelong learning attitude.

2. To allows students to get the latest knowledge through exploratory **research**. Based on the recent development of technology, the challenges toward global competition in higher education require institutions to put a massive amount of effort into research and development (R&D). Experts consider these forces range from new technology deployment to global cooperation and collaboration.
3. To withstand the improvement of societies without depending on other reasons. To enable sustainability of the competitive function among world higher education system, Bo Xing and Tshilidzi Marwala consider that the higher education system need to radically improve educational **services** by organizing more innovation and competition into education surrounding.

Based on the increasing usage devices by students for educational purpose, the universities should provide much better service networks. This is important because based on the study by Educause Center for Analysis and Research (ECAR) on 2014, more students own mobile devices and it becomes increasing, with 86% of under-graduates own a smartphone in 2014 (up from 76% in 2013), and almost half of them (47%) own a tablet (up from 31% in 2013). In term of the rapid growth of Internet of Things especially in educational purpose, the use of three or more devices to the service network at the same time become common. 54% undergraduates mentioned that they typically connect at least two devices to the college/university network at the (54%) say they usually connect at least two devices to the university network at the same time. Furthermore, the student with the age younger (25 years old and below) identified as the power users of university networks almost twice as many students which connect two or more devices at a time to the service network (61% compared with 35% for students age more than 25 years old). Additionally, the study shown about 8% of the university students will connect three or more devices to the service network at the same time, and based on this trend it can be predicted that it will likely increase as portable devices and the adoption of Internet of Things grow up into everyday devices that students can use and afford. So, it is crucial for every university to enhance their service networks to accommodate not just diverse devices but also operative systems.

Figure 3. Students rate on their network experience as good or excellent
Source: Dahlstrom, Walker & Dziuban, (2013)

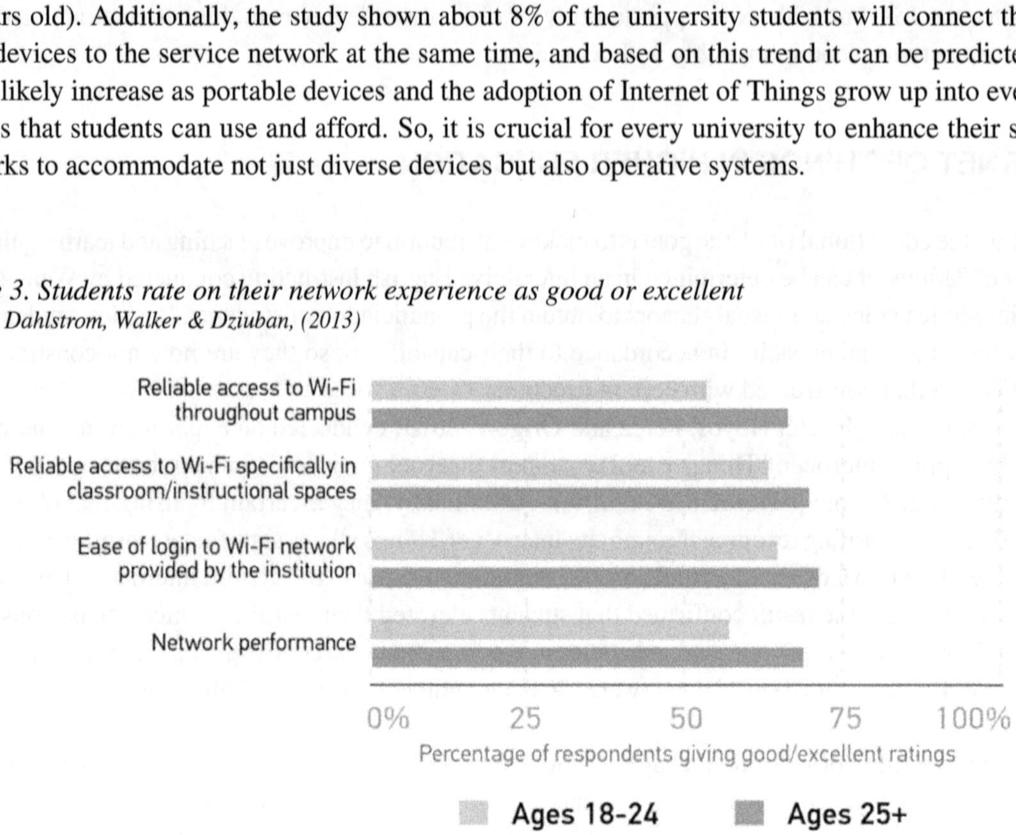

Another study done by the Open University in United Kingdom on October 2011 which they have redesigned their introductory computer science curriculum by adopting the Internet of Things at the center of students' experience in a course they called My Digital Life. The objective of the new course is to place the IoT at the core of the first year students to prime students from beginning for the coming evolution in society and technology. Besides, My Digital Life also designed to improve students view with IoT as a medium to understand and interrogate the students' world, and recognize their own part in realizing the potential of IoT could offer. My Digital Life is a completely new IoT teaching infrastructure that consists of the SenseBoard, Sense and a cloud infrastructure that connects the SenseBoard of all students together. Overall, they found that the approach of teaching computer science with the adoption of IoT worked. The result showed those students able to develop the skills in developing SenseBoard application faster and able to developed network application.

CURRENT MEGATRENDS RELATED TO IoT IN HIGHER EDUCATION

Currently, all graduates face a world changed by using modern day technology. This means that the Internet, cloud computing, and social media prepares different opportunities and challenges for formal education systems. As students reflect on consideration on life after graduation, universities management are dealing with inquiries about their own purpose especially employment because these technologies

Figure 4. The sense programming environment based on my digital life course
Source: Kortuem, Bandara, Smith, Richards & Petre (2013)

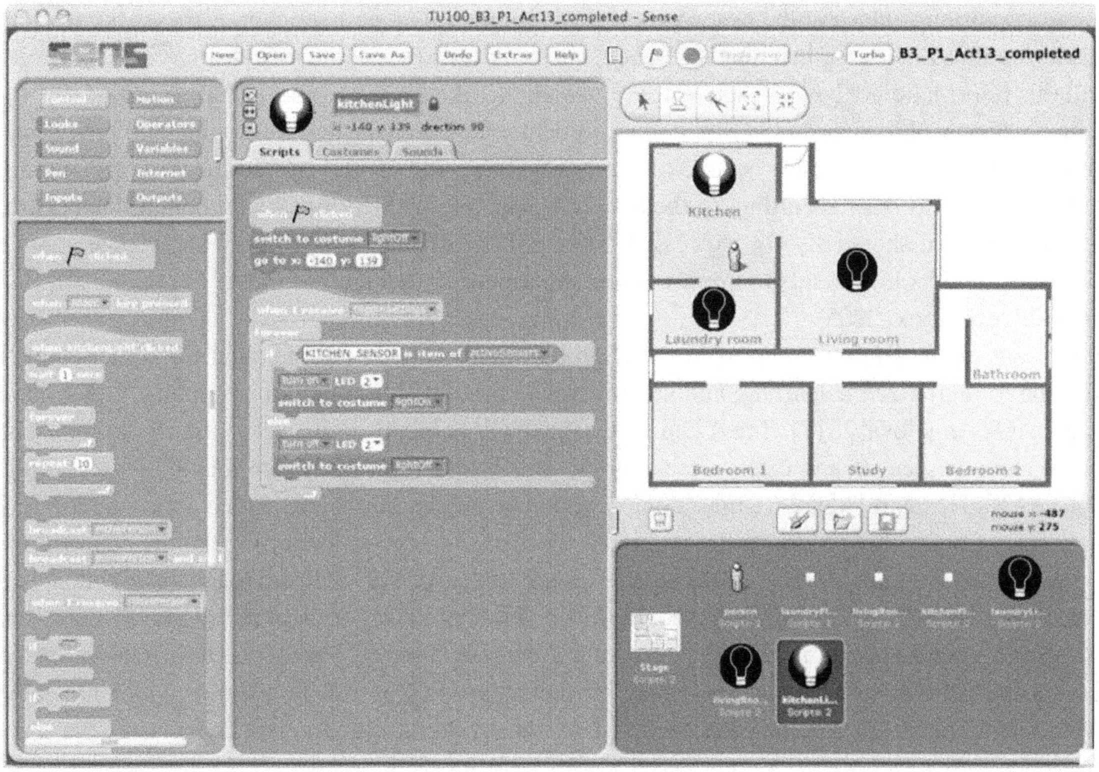

powered by artificial intelligence are so much changing the world that social concepts such as "post-work" are more and more defining the present period. The world today requires particular skills that are more complex than ever before where information technology was the only fundamental. However, these skills should come with critical thinking, people management, emotional intelligence, judgement, negotiation, cognitive flexibility and also knowledge production and management. Based on those skills it will be important for not only universities management but also the education ministry to commence investigating some of the current megatrends as well as their consequences that already becomes wide-spread such as IoT-based Smart Environment, IoT-based Smart Campus, IoT-based Smart Classroom and IoT-based Smart Lab.

One of the current megatrends that widely recognize is *IoT-based Smart Environment.* Mark Weiser's was a chief scientist at Xerox PARC in the United States, and he is known as the father of ubiquitous computing, a term he coined in 1988. His idea about smart environments is that smart devices and smart environments will be available everywhere, for anyone to perform their hobbies or even routine tasks also acknowledge as ubiquitous computers to help overcome the problem of information overload (Weiser, 1991). Additionally, smart environments can also be smart homes, smart offices, smart classrooms and other smart places (Lucke, Constantinescu, & Westkämper, 2008). There are three main goals in Smart environment need to be understood which are to learn, to reason and to predict. It means, smart environments need to examine how their environment works and thinks and to be aware of exactly how to react when some motion or signal occurs. A smart environment (also recognized as SE) can be described "as one that is capable to gather and apply knowledge about the environment and its inhabitants in order to improve their journey in that environment" (Fernandez-Montes, Ortega, Alvarez & Gonzalez-Abril, 2009). Moreover, the primary purpose of IoT-based smart environments is to facilitate everyday routine tasks. For example, when people are in their holidays outside country, they would want to know about the road conditions, which is the best route, if there is any traffic jams or even if they want to change the radio stations they being listening to. Hence, via using sensors, actuators and smart devices, one can get all this information only with their voice (Husnjak, Perakovic & Jovovic, 2014). Additionally, there are three primary objectives in the Smart environment which are learning, reasoning and predicting. It means smart environments must be able to learn or understand how the environment works and thinks finally must able to react according to the action or situation needed for certain conditions occurs. A smart environment should also be expressed "as one that can acquire and apply knowledge about the environment and its inhabitants to improve their experience in that environment" (Youngblood, Heierman, Holder & Cook, 2005).

IoT based smart Digital Campus: Smart Digital Campus becomes demanding for higher education institution to digitalize the learning and administrative process in the digital environment (Kiryakova, Yordanova & Angelova, 2017). Smart Digital Campus will help university reduce the operational costs, improve campus security and offers a technology tools not just for the student but staff. According to Aldowah (2017), smart Digital Campus comprises in two main components which are connectivity and IoT devices. Connectivity requires the use of IT service delivery platform to make the connection all over the digital campus while the IoT devices will support the service platform to enhance student learning and management. According to Cisco, there are five main categories in the digital campus which is Building Control and Management, Security and Access Control, Video and Information Systems, Location and Attendance Systems, and Energy Monitoring and Control (Sivakumar, Muhammad Fariz & Aina Nornidya, 2017). Implementation of IoT in a smart digital campus will connects all the sensors

Figure 5. Illustration of system architecture of smart digital campus
Source: Sivakumar, Muhammad Fariz & Aina Nornidya (2017)

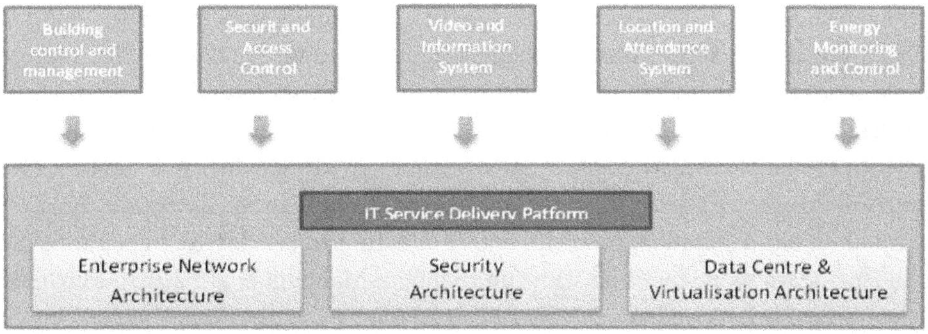

to these five main categories of system. In a digital campus, sensors will alert the system and transfer data collected to the IT Service Delivery Platform.

A Smart Campus can be a series of more than one smart object in a single system. Since most of all university campuses these days are connected to the Internet, the Smart Campus finally is implemented when physical objects like projectors, printers, labs, parking, and building provided by university management connects with each other. This is possible by using sensors, RFID, NFC, QR tags and such different IoT innovations that enable these objects to be converted to Smart objects (Caţă, 2015). A smart campus may include Smart E-learning Application with IoT, Smart IoT-based Classroom, Smart IoT-based LAB Room, IoT Sensors for Notes Sharing, IoT Sensors for Mobiles Devices; IoT- enabled Hotspot for Campus which enables interaction with physical spaces for learning purposes or communication (Veeramanickam & Mohanapriya, 2017). Besides, in an e-learning platform application, these IoT technologies will help educators to observe the teaching activity and let them to communicate with students and share it via connected data recording. Then, without too much troubles, the traditional classroom instruction will effectively transform into a smart classroom efficiently at the same time synchronize it to e-learning application. In addition to above, a smart campus also offers many other smart features in helping the current needed in university management like smart parking, smart inventory, smart lighting, and smart tracking of students, goods and equipment using RFID technology (Caţă, 2015).

IoT based smart Classroom: Modern technologies, information and communication systems allow learning to take place wherever and whenever the learner desires. With the help of new technologies will allow students to learn faster and improve their knowledge and help educators to teach easily (Kubiatko & Haláková, 2009). The concepts of smart classroom means an intelligent environment fortified with advanced learning aids based on latest technology or smart things. These smart things can be either cameras or microphones or even others that can be used to measure student learning satisfaction or many other related things. The best offers about smart classrooms is that they allow educators to observe how students truly want to learn and allow educators to give students the information they want it without having to worry about time and location restrictions. It benefits both educators and students. Smart classroom gathered information, and store it as digital data in an e-learning platform memory (Veeramanickam & Mohanapriya, 2017). Educators and students can choose learning approach based on their curiosity. It affects their memory while learning to remember some of it. Another important thing about smart classrooms is that they allow students to see a real purpose for using technology and everyone learns better when they see some technology's actual purpose (Firmin & Genesi, 2013).

With the help of modern sensors and actuators, educators can get information about the students before they click the button and type how they feel so that they can perform in smart classrooms (Santana-Mancilla, Echeverría, Santos, Castellanos, & Díaz, 2013). In addition, with the help of smart thing, smart classroom also allow educators to measure whether students are satisfied with what they are learning and, finally, classroom management becomes very simple and easier to understand than without smart devices. The term "classroom management" represents the way in which an educator achieves order in teaching and with the help of smart devices; the educator can know when it is the best time to take a break, when to talk louder and when the concentration of students drop (Rytivaara, 2012). With some cheap sensors and lot of imagination, powerful and stimulating smart classroom could be built that can really help students learn more easy and with fun (Santana-Mancilla et al., 2013). Furthermore, some daily routine activities such as student attendance can be track using the connected devices (Sivakumar et al., 2017). The smart object gives ease and comfort with the use of IoT. IoT can help to improve the learning and teaching environment. The data collected must be kept to analyze the student report and to draw up a future school or university plan.

1. **Smart Classroom Management:** The term "classroom management" means a way or approach a educator uses to control/manage his/her classroom with the help of smart devices (Gul, Asif, Ahmad, Yasir, Majid & Arshad, 2017). Smart devices have made it possible for an educator to decide whether he should speak louder when students lose interest, or if their concentration level is decreasing in classroom (Rytivaara, 2012). The use of IoT devices for teaching and learning purposes is becoming popular among institutions around the world they provide a new and advanced approach to better education and classroom management. According to Gul et al., some of the common used IoT devices in the smart classroom management in educational system are:
    a. Interactive Whiteboards
    b. Tablets and Mobile devices
    c. 3-D Printers
    d. eBooks
    e. Student ID Cards
    f. Temperature Sensors
    g. Wireless door locks1
    h. Attendance Tracking Systems
    i. Security Cameras and Video
    j. Smart HVAC systems
    k. Electric Lighting and Maintenance
    l. Room Temperature Sensors

Moreover, smart classrooms allow educators to discern what students need to learn and how they want to learn and help students to understand the real purpose of using technology that benefit both educators and students (Chang, 2011). The technological innovation in education has made it easy for educators to design classrooms which are creative, valuable, and collaborative that managed through IoT. The majority of recent studies propose different models for smart classrooms which demonstrate that many advanced and innovative concepts are being proposed or introduced in education. One of them is that IoT technology with crowdsourcing in e-learning can be useful for improving learning and teaching processes (Gul et al., 2017).

2. **Smart Classroom Attendance System:** Taking attendance of a class manually is a time-consuming task. However, with the help of IoT technologies such as Smart classroom roll caller system (SCRCS) that using IoT architecture to collect or record student attendance after every period would be really helpful and can save time and effort not just for educators but also students (Gul et al., 2017). This technology requires RFID tags attached to the Students' ID cards and the SCRCS will be installed in each classroom. It will then read the students' identity card collectively based on RFID tags. It shows not only the total of LED display at the beginning of any class, but also the identity of the card owner on multiple SCRCS slots. The record of a student's attendance will also be kept at the academic office (Chang, 2011).
3. **Real-Time Feedback on Lecture Quality:** One of the main reasons for understanding and achieving students in class is the quality of the lecture. Student feedback on lecture contents plays a vital role improving the quality of the lecture. A study conducted by Chew, Mahinderjit-Singh, Wei, Sheng, Husin & Malim (2015) proposes a creative environment that allows educators to monitor and observe lecture reactions using sensing and monitoring technology. This IoT-based smart classroom provides real-time feedback on lecture quality, helping to improve the lecture quality (Chew et al., 2015).

IMPACT AND FUTURE OF IoT IN EDUCATION

There will be significantly impact in many ways with the innovation in technology especially with the Internet of Things (IoT), which will end up considerably vital in the context of Information and Communication Technologies and the improvement of society. With the guide of IoT, educational institutions can enrich learning outcomes by providing more contented learning experiences, extended operational efficiency, prepares real-time learning, expand the opportunity while the student could benefit from these by increasing their performance in classroom. It can be anticipated that IoT tools will offer a more interesting, flexible, enticing and quantifiable system of education that fulfills the ever-changing needs of a great variety of students (Gul et al., 2017). Besides, higher education institutions can work across disciplines and lead the progress of the IoT technologies, business models, ethics and leaders of the IoT economy in the future (Aldowah, 2017). IoT is not just a technology advance and change within the industry, but it can lead to expand the change to the whole society including higher education institutions (Aldowah, 2017). For example, the university instructors are directing IoT labs for the expansion of the IoT technologies, create and prepare the IoT courses for students' business model, also develop and discover solutions for RFID and cloud computing through IoT technologies.

The future application in higher education depends in the three phases based on the development of IoT which are students' progressive evaluation, integration of current teaching platforms and improvement of educational middleware (Zhiqiang & Junming, 2011). This root then amplified convenience and interest for students, hence makes the teaching process more meaningful for instructors. In the end, instructors can focus on the actual learning that is more valuable to the students rather than just completing assignment given when somehow the connected devices and technology actually able to support them in class.

Moreover, IoT will benefit especially for instructors to increase the learning experience by providing for real-time and actionable insights into student performance (Aldowah, 2017). Nowadays, universities in Malaysia are gradually moving away from using textbooks in lectures to new technologies such as tablets and laptops. This enables students to learn at their own capability and let them to have an identical learning experience not just in classrooms but also at homes thanks to the advanced in e-learning applications (Aldowah, Ghazal & Muniandy, 2015). This resulted in increases on gratification for students as well as helping the instructors to deliver individual instruction and persistent student assessments (Ghazal, Samsudin & Aldowah, 2015). In addition, instructors also able to gather information on the students' performance and finally decide which ones need more attention besides helps them improve lesson plans and methods for future classes.

Furthermore, IoT also enables educators to teach crucial concepts of Programming language to students and also another system uses objects with tags and Learning Management System to gather data and analyze students' learning method using learning analytics techniques (Chin & Callaghan, 2013; Chen & Liao, 2012). In addition, the advanced of mobile technology and the IoT allow universities to improve the security of campuses, increase the access to information and applications whenever and wherever while keep track of core resources (Gubbi, Buyya, Marusic & Palaniswami (2013). According to the Citrix 2020 Technology Landscape Report (2015), in the next five years, IoT technology will improve the learning experience not like before because IoT is now gradually changing the student learning experience and universities facilities management by connecting individual, data and things.

The potential of technology to facilitate teaching, learning, and assessment becomes dialogue for some time and will last in the future. In enabling students to be confident for the real working work is complex because it is not just requires them to have strong academic leadership but also access to a high quality syllabus and the experience them to the effective use of new technology. With the development of IoT nowadays, many institution of higher education have commenced to focus on the related technology and application of the IoT (Aldowah, 2017). This effort is also used in university (Qi & Shen, 2011). As Internet becomes embedded into colleges and universities, e-learning has become common practice in most universities systems (Aldowah et al., 2015) even it is not largely being utilized in all the universities; the applications of the IoT are various, and the implications for this are immense (Aldowah, 2017). Furthermore, IoT will improve teaching and learning process in future (Gul et al., 2017) because it will facilitate both students and educators in instruction and help the students learn better, and educators will be able to perform their responsibilities more proficiently. It can be expected that IoT tools will offer a more interesting, adaptable, engaging and quantifiable system of education that fulfills the different requirements of a vast number of students.

Additionally, IoT enable for higher operational efficiency in all learning environments; as it can sustenance classroom instruction by refining learning setting, improve learning resources, improve learning approaches and methods, nurture management efficiency and save management expenses (Aldowah, 2017). Even there are resources available for learning via the traditional devices, like e-books, which are more engaging and more fun; however, the need for new technologies based on IoT for learning process, for example, high-speed wireless networks with the bandwidth for streaming audio and video lessons is still become priority. An important aspect of today's technological advancement is the development of open source hardware and software platforms for unrestricted prototyping and experimentation. One of the examples is the Arduino open hardware platform that has been used in higher education (Kortuem, Bandara, Smith, Richards & Petre, 2013). In the end, educators must be able to identify properly integrate the right technology in the classroom for learning to grow.

IoT allows educators and students in the IoT topic stack to expand the subjects included in computer science or even information science at many levels. In addition, smart devices can be considered in the curriculum; hardware, network connectivity, human-computer interaction, and ethics play increased roles for future research. Although the main directions of IoT technologies still far to be predicted, the point is certainly that many contents are the result of new development phase. Given the need for more professional research, setting the IoT major is relatively easy and applicable for graduate students; but for undergraduate students, wide range of basic courses are still needed (Aldowah, 2017). Therefore, new training for undergraduate students will be explored in the future (Tan, 2015). In addition, different universities must explore the appropriate approach based on their own background, established and improved the system approach and courses content and finally generate many new training methods and cross-cutting in the future (Ning & Hu, 2012).

CHALLENGES OF IoT IN HIGHER EDUCATION

IoT brings tremendous challenges and opportunities to higher education. The transformation of learning method by integrating IoT in educational can present new challenges and problems which are associated with the changes in the educators' cultural background and broadening the scope of training for educators and students (Aldowah, 2017). Successfully in integration of IoT in educational especially in classroom environment may have to face some difficulties such as network bandwidth, reliable Wi-Fi connection, cloud computing, instructional technologies, mobility application, security and privacy, quality and ethics, financing management, and research computing (Aldowah, 2017; Gul et al, 2017). The distinctive growth of ubiquitous computing and IoT technologies development such as cloud computing, big data and analytics help to improve the core values of teaching research quality in current higher educational system especially in Malaysia. Despite of improving the core values in teaching, it also helps in developing an IoT society and foster a new digital culture. However, no matter how advance a technology could be, there must be some challenges that need to be addressed. Some of the IoT challenges in higher education sector include:

- **Capability to integrating IoT in classroom:** IoT offers many benefits to educators and students; but it also leaves some concerns about its implementation. To ensure the successfully integrated into the classroom, educators should be able to provide necessary equipment such as robust network bandwidth, professional training for educators and student devices (Sivakumar, 2017). Continuous need for new education technologies, such as high-speed wireless networks will always be demanding, providing bandwidth for audio and video streaming lessons. A reliable Wi-Fi connection is therefore crucial to ensure that lessons can be carried without any problems.
- **Cloud Computing:** Cloud computing offers many advantages for educators and students such as centrally controlled computer systems that enable processes huge amounts of data (e.g: graphics and audio); manage complex tasks; or even store enormous amounts of keywords searchable information, while allow users to share this information while interact with other users through low-cost devices (e.g: smartphones, tablets and notebooks) (Weber, 2013). The clouds also provide continuous connections and services to IT services with ubiquitous computing. Business architecture in many higher education institutions depends on hybrid cloud infrastructures with computer platforms on private clouds, while enterprise and training applications move gradually to public clouds (Aldowah, 2017). Hence, business architecture in these institutions needs to reduce response time due to the demand for content, such as the increased audio and video required for instructions.

- **Instructional Technologies:** The demands use of learning management systems (LMS) such as Moodle and Blackboard creates a great deal of structured and unstructured data such as audio and video content. Moodle is a course management system (CMS) which is an open source software package designed with sound pedagogical principles, to help educators create effective online learning communities while Blackboard is software packages used by North American and European universities. Blackboard divided by three key areas which are instruction, communication and assessment. Modern electronic classroom equipped with not only lecture capture systems but also web streaming provide the opportunity to access information they needed at any time (Aldowah, 2017).
- **Mobility Applications:** IoT applications nowadays are increasingly being used to integrate mobile learning applications and systems for evaluation and grading in higher learning (Aldowah, 2017). IoT's ideal application can help students use learning resources, manage tasks, and allow educators to teach highly specialized concepts, complex physical, scientific simulation, or even social topics in the classroom. This would help to make learning interactive and much and much easier for educators and students.
- **Security and Privacy:** The concerns about the broad adoption of the IoT todays are at risk due to privacy and security. To counter this, sound technological solutions must be developed, with national and supra-national legislators (Aldowah, 2017). Since the IoT used to measure and gather data from students, security and privacy issues are at risk because educators or administrator maximize the storage of sensitive data in Internet network of connected devices prone to cyberattacks (Sivakumar, 2017). Moreover, education providers are concerned about consequences of security breaches that could risk a student ID linked to the health record of the individual and other classified information. Hence, tackling these challenges security concerns in IoT devices and services should be a basic priority before it can be used for educational purpose. The essential element an IoT platform is the need to include efficient and reliable privacy and security mechanisms (Mineraud, Mazhelis, Su & Tarkoma, 2016). Higher education is vulnerable to the security and privacy. Further efforts have been made to address the security of the IoT infrastructure in higher education; however there is still no approach to identify business risks related to the data breaches (Aldowah, 2017).
- **Legal/accountability:** The adoption of IoT in educational environment will create new legal challenges for consideration and care. IoT administration should be dictated to the necessary management by a broad stakeholder approach. A common structure for the IoT including all relevant stakeholders is therefore needed. (Weber, 2009). Global responsibility and regulation are also required. Responsibility tends to improve governance efficiency by threatening sanctions (Weber 2011).
- **Quality and Ethics:** The common discussion discussed todays is the quality of online and campus learning and the rising costs of higher education (Aldowah, 2017). It is known that the adoption of IoT in educations offers unique opportunities to develop digital courses; however, it also raises the issues to enhance the value of education and evaluation of students' work. Besides, ethical issues become one of the fundamental aspect that need to be look into thoroughly. IoT educational applications need new technologies not only for educators, but also the science community to improve the research value and address issues of ethics within higher education.

- **Financing:** One of the fundamental challenges with adoption of IoT in higher education would be financing because the cost of information technologies continues to rise every year on the basis of increasing demands on educational technologies, research computing and business technologies. In addition to information technology and laboratory fees, most universities do not have a cost sharing strategy and the identification of total ownership cost for an IoT infrastructure (Aldowah, 2017) which can be seen with the recent economic growth.
- **Management:** IoT program could be hard to manage by novice users. Some devices and applications sometimes are not compatible with each other and impede the organization's ability to make an IT environment that is accessible to all users. Since IoT applications require that objects and devices remotely be managed it will be a complex task, and requires a lot of various actions such as switching device off/on, device/network configuration, firmware/software updating, errors mitigation, monitoring, data gathering and connectivity statistics and others (Aldowah, 2017). For effective adoption of IoT, an educational institution must make sure that both its IT equipment and teaching approaches will support the use of IoT in the classroom.

CONCLUSION

Despite the challenges, education system in Malaysia must enhance the quality of learning environment and management process with the use of IoT. Government needs to provide more funds to the universities and schools to become a Smart Campus or Smart School so that it will give a huge impact on education industry to Malaysia. Higher education could attract students to learn with the integration of new technology in education through Smart campus or Smart School. Besides, university management process will be much easier with the use of IoT which makes administrative cost cheaper and more efficiently. With the adoption of IoT in higher education, students will be able to connect their creativity and create more significant learning spaces. The adoption of IoT in higher education is just beginning and will only continue to increase and this means educators must be prepared to face new challenges in the future.

REFERENCES

Al-Fuqaha, A., Guizani, M., Mohammadi, M., Aledhari, M., & Ayyash, M. (2015). Internet of things: A survey on enabling technologies, protocols, and applications. *IEEE Communications Surveys and Tutorials*, *17*(4), 2347–2376. doi:10.1109/COMST.2015.2444095

Aldowah, H., Ghazal, S., & Muniandy, B. (2015). Issues and challenges of using E-learning in a Yemeni Public University. *Indian Journal of Science and Technology*, *8*(32). doi:10.17485/ijst/2015/v8i32/92160

Aldowah, H., Rehman, S. U., Ghazal, S., & Umar, I. N. (2017). Internet of Things in Higher Education: A Study on Future Learning. *Journal of Physics: Conference Series*, *892*(1), 012017. doi:10.1088/1742-6596/892/1/012017

Barakat, S. M. (2016). Internet of things: Ecosystem and applications. *Journal of Current Research in Science*, *4*(1), 32.

Burd, B., Elahi, A., Russell, I., Barker, L., Pérez, A. F., Siever, B., ... Guerra, J. G. (2017). The Internet of Things in CS education: current challenges and future potential. In *Proceedings of the 2017 ACM Conference on Innovation and Technology in Computer Science Education* (pp. 392-392). ACM. 10.1145/3059009.3081331

Caţă, M. (2015). Smart university, a new concept in the Internet of Things. In *RoEduNet International Conference-Networking in Education and Research (RoEduNet NER), 2015 14th* (pp. 195–197). IEEE. doi:10.1109/RoEduNet.2015.7311993

Chang, C. H. (2011). Smart classroom roll caller system with IOT architecture. In *Innovations in Bio-inspired Computing and Applications (IBICA), 2011 Second International Conference on* (pp. 356-360). IEEE. 10.1109/IBICA.2011.94

Cheng, H. C., & Liao, W. W. (2012). Establishing an lifelong learning environment using IOT and learning analytics. In *Advanced Communication Technology (ICACT), 2012 14th International Conference on* (pp. 1178-1183). IEEE.

Chew, C. B., Mahinderjit-Singh, M., Wei, K. C., Sheng, T. W., Husin, M. H., & Malim, N. H. A. H. (2015). Sensors-enabled smart attendance systems using NFC and RFID technologies. *Int. J. New Comput. Archit. Appl, 5*, 19–29.

Chin, J., & Callaghan, V. (2013). Educational living labs: a novel internet-of-things based approach to teaching and research. In *Intelligent Environments (IE), 2013 9th International Conference on* (pp. 92-99). IEEE. 10.1109/IE.2013.48

Dahlstrom, E., Walker, J. D., & Dziuban, C. (2013). ECAR study of undergraduate students and information technology. Academic Press.

Fernández-Montes, A., Ortega, J. A., Alvarez, J. A., & Gonzalez-Abril, L. (2009). Smart environment software reference architecture. In *2009 Fifth International Joint Conference on INC, IMS and IDC* (pp. 397-403). IEEE. 10.1109/NCM.2009.115

Firmin, M. W., & Genesi, D. J. (2013). History and implementation of classroom technology. *Procedia: Social and Behavioral Sciences, 93*(2), 1603–1617. doi:10.1016/j.sbspro.2013.10.089

Gardner, M., & Elliott, J. (2014). The Immersive Education Laboratory: understanding affordances, structuring experiences, and creating constructivist, collaborative processes, in mixed-reality smart environments. *EAI Endorsed Transactions on Future Intelligent Educational Environments, 14*(1).

Gershenfeld, N., Krikorian, R., & Cohen, D. (2004). The internet of things. *Scientific American, 291*(4), 76–81. doi:10.1038cientificamerican1004-76 PMID:15487673

Ghazal, S., Samsudin, Z., & Aldowah, H. (2015). Students' perception of synchronous courses using skype-based video conferencing. *Indian Journal of Science and Technology, 8*(30). doi:10.17485/ijst/2015/v8i1/84021

Gubbi, J., Buyya, R., Marusic, S., & Palaniswami, M. (2013). Internet of Things (IoT): A vision, architectural elements, and future directions. *Future Generation Computer Systems, 29*(7), 1645–1660. doi:10.1016/j.future.2013.01.010

Gul, S., Asif, M., Ahmad, S., Yasir, M., Majid, M., & Arshad, M. S. (2017). A survey on role of internet of things in education. *IJCSNS*, *17*(5), 159.

Husnjak, S., Perakovic, D., & Jovovic, I. (2014). Possibilities of using speech recognition systems of smart terminal devices in traffic environment. *Procedia Engineering*, *69*, 778–787. doi:10.1016/j.proeng.2014.03.054

Kiryakova, G., Yordanova, L., & Angelova, N. (2017). Can we make Schools and Universities smarter with the Internet of Things? *TEM Journal*, *6*(1), 80.

Kortuem, G., Bandara, A. K., Smith, N., Richards, M., & Petre, M. (2013). Educating the Internet-of-Things generation. *Computer*, *46*(2), 53–61. doi:10.1109/MC.2012.390

Kubiatko, M., & Haláková, Z. (2009). Slovak high school students' attitudes to ICT using in biology lesson. *Computers in Human Behavior*, *25*(3), 743–748. doi:10.1016/j.chb.2009.02.002

Lucke, D., Constantinescu, C., & Westkämper, E. (2008). Smart factory-a step towards the next generation of manufacturing. In *Manufacturing systems and technologies for the new frontier* (pp. 115–118). London: Springer. doi:10.1007/978-1-84800-267-8_23

Marques, G., Garcia, N., & Pombo, N. (2017). A survey on IoT: architectures, elements, applications, QoS, platforms and security concepts. In *Advances in Mobile Cloud Computing and Big Data in the 5G Era* (pp. 115–130). Cham: Springer. doi:10.1007/978-3-319-45145-9_5

Mineraud, J., Mazhelis, O., Su, X., & Tarkoma, S. (2016). A gap analysis of Internet-of-Things platforms. *Computer Communications*, *89*, 5–16. doi:10.1016/j.comcom.2016.03.015

Ning, H., & Hu, S. (2012). Technology classification, industry, and education for Future Internet of Things. *International Journal of Communication Systems*, *25*(9), 1230–1241. doi:10.1002/dac.2373

Oriwoh, E., & Conrad, M. (2015). 'Things' in the Internet of Things: Towards a definition. *International Journal of Internet of Things*, *4*(1), 1–5.

Qi, A. Q., & Shen, Y. J. (2011). The Application of Internet of Things in Teaching Management System. In *Information Technology, Computer Engineering and Management Sciences (ICM), 2011 International Conference on* (Vol. 2, pp. 239-241). IEEE. 10.1109/ICM.2011.238

Rytivaara, A. (2012). Collaborative classroom management in a co-taught primary school classroom. *International Journal of Educational Research*, *53*, 182–191. doi:10.1016/j.ijer.2012.03.008

Santana-Mancilla, P. C., Echeverría, M. A. M., Santos, J. C. R., Castellanos, J. A. N., & Díaz, A. P. S. (2013). Towards smart education: Ambient intelligence in the mexican classrooms. *Procedia: Social and Behavioral Sciences*, *106*, 3141–3148. doi:10.1016/j.sbspro.2013.12.363

Sivakumar, D., Jusman, M. F. B., & Mastan, A. N. B. M. (n.d.). *A Case Study Review: Future of Internet of Things (IoT) in Malaysia*. Academic Press.

Tan, D. (2015). *Engineering Technology, Engineering Education and Engineering Management: Proceedings of the 2014 International Conference on Engineering Technology, Engineering Education and Engineering Management (ETEEEM 2014), Hong Kong, 15-16 November 2014*. CRC Press.

Veeramanickam, M. R. M., & Mohanapriya, M. (2017). IOT enabled Futurus Smart Campus with effective E-Learning: i-Campus. *GSTF Journal of Engineering Technology*, *3*(4).

Weber, A. S. (2013). Cloud computing in education. In *Ubiquitous and mobile learning in the digital age* (pp. 19–36). New York, NY: Springer. doi:10.1007/978-1-4614-3329-3_2

Weber, R. H. (2009). Internet of things–Need for a new legal environment? *Computer Law & Security Review*, *25*(6), 522–527. doi:10.1016/j.clsr.2009.09.002

Weber, R. H. (2011). Accountability in the Internet of Things. *Computer Law & Security Review*, *27*(2), 133–138. doi:10.1016/j.clsr.2011.01.005

Weiser, M. (1991). The Computer for the 21 st Century. *Scientific American*, *265*(3), 94–105. doi:10.1038cientificamerican0991-94 PMID:1675486

Xing, B., & Marwala, T. (2017). *Implications of the Fourth Industrial Age on Higher Education.* arXiv preprint arXiv:1703.09643

Youngblood, G. M., Heierman, E. O., Holder, L. B., & Cook, D. J. (2005). Automation intelligence for the smart environment. *International Joint Conference on Artificial Intelligence*, *19*, 1513.

Zhiqiang, H., & Junming, Z. (2011). The Application of Internet of Things in Education and Its Trend of Development. *Modern Distance Education Research*, *2*, 19.

Chapter 3
Heutagogy Approach Using Trello Online Learning on the Critical Thinking Skills Amongst Students With Different Learning Styles

Mohan Rathakrishnan
Universiti Utara Malaysia, Malaysia

Arumugam Raman
https://orcid.org/0000-0001-5351-8944
Universiti Utara Malaysia, Malaysia

ABSTRACT

Trello online discussion has become one of the important strategies for the University Utara Malaysia Management Foundation Programme student to teach other students to think critically in conveying their ideas and become more proactive and creative in critical thinking subject. In a heutagogical approach, learners become highly autonomous and self-determined while learning online. They use Trello online discussion tool as self-determination learning. Trello enable the learners to use their capacity and capability with the goal of producing learners. Trello online learning was conducted to examine its effectiveness in enhancing macro critical thinking among active-reflective learning style. The students discuss and write their ideas in Trello. Ideas that are posted in Trello will be displayed in front of the class so that the entire learners in the class could see the given ideas. Paul's model was used to analyze learners' critical thinking in Trello online discussion.

DOI: 10.4018/978-1-5225-7832-1.ch003

Copyright © 2019, IGI Global. Copying or distributing in print or electronic forms without written permission of IGI Global is prohibited.

INTRODUCTION

Learning using online learning is a very interesting way to upgrade students learning style. Traditionally, teachers present their teaching lesson to the students using linear model. If the students didn't pay any attention to the teaching, they will be blamed for not concentrating in their studies. The lack of interest in teaching and learning among the teachers and students makes them feel easily bored while they are learning. Nowadays students are changing radically in receiving more advance in learning and expecting interesting design of teaching with the rise of technology (Prensky, 2001). Steward.M (2005) has mentioned that teachers should use new approach in giving student to develop their ideas in learning new technology with open mind and connect students to the digital natives. That is why it has been an ambition to the Education Ministry of Malaysia to encourage teachers to enhance our education to a schooling culture that encourages critical thinking, and become creative in using technology as teaching and learning tool.

Under the Malaysian Education Blueprint (2015-2025) Higher Education, the Malaysian Education Minister emphasizes on creating long life learners who can survive and thrive in world knowledge and apply information technology to produce integrated thinking skills to help students to apply cognitive strategies in problem solving. The government's target to create intellectual human capital by the year 2020 depends on the premise of developing the thinking domain of students today. This is called as heutagogy learning environment. It is not easy to produce students to be good self-determine, independent and thinkers if the education system does not allow teacher or students to develop and create using of relevant skills for educational purposes such as the skills of enquiry, reasoning and information organizing (Mohan, 2016). In heutagogy learning environment, new technologies have created a need for considering new pedagogical approaches.

In Universiti Utara Malaysia, Management Foundation Programme emphasizes the application of critical thinking skill subject (CTSS) to be incorporate into the curriculum to help students' development a well-balanced individual, sensitive people, ability to communicate and make the student to think critically. It is important to accommodate student to thinking maturely, subject that listed in Management Foundation Program course is "Critical Thinking Skill' (AD0043). Among first year university students' Critical thinking skill is pertinently important. Particularly, the critical thinking subject is essential for first year students who undertake the General Studies Malaysian Nationhood (GSMN) subject. The course syllabus indicates that students should think independently and apply language proficient. In addition, GSMN aims to broaden student's global outlook while enabling them to remain mindful of the countries administration and effective communication. In sum, the ability to master critical thinking is the key for first year student in order to achieve the course objectives of GSMN.

Critical thinking skill subject (CTSS) has been documented as an important subject for foundation students. The capability to become critical thinker and give well-reasoned opinions on some issues using effective communication skills are crucial to obtain good grades in CTSS. The abstract and complex nature of the knowledge and subject-matter dealt with in CTSS and the critical thinking skills needed to understand are said to have major influences on students' learning.

Student should learn how to analyze information and evaluate issues on humanities, social sciences, politics, economics, religion, government policies and law as well as legal procedures and law. CTSS students need to understand that there are many ways for them to discover and process information and apply these skills when the teacher gives them assignments. Bloom (1956) referred this as a way of classifying educational goals in terms of complexity.

To reach well-reasoned conclusions on specific issues, CTSS help the students to use their ability and desire to think critically. This is a heutogogy framework that deals with productive thinking where the students' ideas are not confined to the analysis of existing argument but concerned with generating ideas and has consequences for action (Baron, 1985). The student will be able to engage in the intellectual activities, which require critical thinking if the students become skillful in evaluating and analyzing information without depending to the instructor (Mertes, 1991).

If aspects such as thinking, knowledge and expertise were to be stressed in the ill-structured subject such as CTSS, the lectures have to construct their students' understanding to remedy learning deficiencies related to domain complexity in learning process (Spiro, R. J.Jacobson, M. J., & Coulson, R. L. Feltovich, P. J. (1991). The way a lecturer teaches can influence the way the students learn. To encourage critical thinking in the students, the lecturer has to adopt various strategies that enable them to think more deeply while studying or doing assignment. It is not easy to produce students to be good thinkers if the education system does not allow the lecturer or students to adopt the use of relevant skills for educational purposes such as the skills of enquiry, reasoning, information organizing and translation (Lipman, 1991: Mohan 2016).

From the researchers' observation, foundation students are facing problem to identify, communicate and to integrate the syllabus content of CTSS far from what they have learnt from their previous school subject which is more exam orientated. Usually, students like to study with the intention only to pass their exam. Besides, it is observed that students are totally depending on the course facilitator to prepare all the materials, ideas, notes and discourse models of specific academic areas.

In addition to the lack of interest among students and too much relying on course facilitator, the typical university classroom setting would be seen as a major cause. In university, lecture hall has a fix setting which would lead course facilitator to execution learning techniques. For example, sometimes the class is dominated by vocal students, who can consume the class time while some students feel insecure to express their opinion while others ask their question at the end of the class, compromising their learning process.

INITIATE LEARNING PROBLEM

It is important for the university to initiate solution to above mentioned problem. Obviously, it is vital for the institution to design suitable instruction to help the students use their ability and interest to think critically about specific issues. Students would become skillful in evaluating and analyzing information and will try to engage in intellectual activities that require need to think critically.

Adopting back channel communication in the class is one of the ways to tackle these problems is to Back channel is a complementary channel for interactions and collaborations among students and instructors, usually in the form of synchronous chat like Trello, TodaysMeet, Trello and other related tools. Some of these tools allow learner to be anonymous when posting and discussing their ideas. Research by Best Colleges (2015) shows that learning through online or backchannel communications benefit students in classroom or attending lectures. Among the positive benefits documented are the conveniences of asking questions and getting responses from the instructors learn new information from peers and having more fun learning (Du et al., 2013).

Online teachers will face problem as there will be some students who are passive in giving their opinions or to engage in online learning (Harris, R. N., Dwyer, W. O., & Leeming, F. C., 2003). There are students who could not justify the topic given in online learning and sometime reluctant to take part in collaborative discussion. They often rely on their personal opinions, facilitator's idea or anecdotes. The online teachers have to see from the learner-centered perspective the differences among the students' learning styles. Every student has different strategies and capabilities for learning which are functional of prior experience and heredity. In face to face learning, there are four dimensions' of students' learning style (Felder, R. M., & Silverman, L. K., 1988) namely (1) sensory-intuitive, (2) visual-auditory, (3) active- reflective and (4) sequential-global. Meanwhile, sensory learning style involves gathering data through the senses and observing; intuition involves indirect perception by way of the unconscious speculation, imagination, hunches.

Every student uses both the learning style, but most of the students tend to favor one over the other. Visual students remember best what they see: pictures, diagrams, flow charts, time lines, demonstrations and films. If something is simply said to them they will probably forget it. Auditory students remember more of what they hear then what they say. Felder (1998) mentioned that students get a lot out of discussion and prefer verbal explanations to visual demonstration. Active students prefer to do physical experiments, learn by trying and manipulate objects. The students are enjoying working in groups to figure out problems. Reflective students prefer to evaluate options, learn by analysis think and things through.

Active and reflective students work very well when they are offered open-ended problems with some degree of choice, can be electronically apprenticed with timely instructional support and need feedback when they are struggling to reach consensus or make decisions. Sequential learners on the other hand prefer to have information presented linearly in an orderly way. They put together the details in order to understand the big picture that emerges. Global learners prefer a holistic and systematic approach. They see the big picture first and then fill in the details.

Paulsen (1995) indicated that the online teachers have to be more concerned with encouraging students with suitable learning style to participate in certain online learning setting. The teachers who teach online should identify their preferred forms of moderation and pedagogical styles. The online teachers should also know how to move between different roles of instruction to get the students frequently involved in the online learning process.

Therefore, guidelines and directions need to be provided to avoid students become confused and thus complain about online learning. The online teachers have to use simple tools such as the web to display the students' work samples online to give a clear picture about what they are studying to avoid confusion. (Peffers, 1999). The web page such as trello offers opportunities to online learners to development shared document workspaces and idea exchange for self-determine students. The selected tool will help the online teachers to encourage students to participation and interaction that might be expected in traditional settings. With the systematic change in education through heutogogy, UUM foundation students would be more proactive, analytical and creative. They are encouraged to play active role in the classroom inculcating Malaysian values within themselves.

Trello Online Tool

Learning using Trello environment in teaching and learning of Critical Thinking Skill will make the student to be more efficient among university students. Thus, there is a need for a study in Malaysia to know the effects of Trello online learning among students of difference learning style. There are very limited research has been done on examining whether trello can promote critical thinking among undergraduate students. It is therefore, important for the lecturers to understand how undergraduate students can use trello to engage deeply with the instructors, peers and the subject matter or just merely discussing informally about class topic or sometimes "off-topics" that will not enhance their learning process.

Trello is a task management app (Figure 1) that helps giving user an overall idea of client collaboratively. It is related to Kanban System in Japan by Toyota Company. Toyota has developed this system to keep their product in level high and flexible. Trello represent as a board to allow individual or group (team) to track the growth of their products. Trello was used as JIT (just-in-time) to increase manufacturing under the Kanban system. Kanban shows tremendous increase in production involving workers who gives ideas from all level.

Trello are fully editable websites where users can visit, read, edit and update the structure and content (Smallbiztrends, 2018). It is a collaboration tool that gives user a visual overview of what is being worked on, who is working on it, and how far they've gotten. Not like many other software tools that have a steep learning style, requiring training to use and involving few features, trello is a combination of a Web site and a Word document. Trello can be used as an education web that eases online learning. The trello is gaining traction in education as a tool for the increasing amount of self-learning by students after facilitated by instructor. The system of manufacturing JIT has been modified to education system in the form of Trello to upgrade teaching and learning technique in universities and education institution.

Trello is used in the "real human life and world" by people collaborating on projects or sharing issues online like education programme, technical information from users of products, travel journals from abroad, data from research and development projects or clubs. In K-12 education, and cloud learning, educators conduct professional development workshops using trello and they found that their students are very keen to participating in the ongoing creation and evolution of the online discussion.

Trello is also different from any other web blog. The web blog shares writing and multimedia content in the form of "posts" (using point entries) and "comments" (responding to the post). Members of the blog can open the blog using a password given by the host. However, no one is able to change a comment or post made by others unless allowed by administrator. The usual customized work that can be done in the blog by users is to post comments - add and comments without limit.

There are several challenges of trello implementation that are not well understood. There are researchers argue that trello online learning may become a source of distraction among students (Schwartz, 2003; Phalen, 2003). However, Mohan Rathakrishnan, Rahayu Ahmad, Lee Jun Choi (2017) in their research indicated that students are motivated to use backchannel; trello communication and the underlying reasons that are being less examined. Thus, the purpose of this research is to identify the importance on trello online learning. Specifically, this study wants to investigate the effect of heutogogy approach on the students' learning adoption, the role of online learning in facilitating critical thinking and the benefits of the tools in classrooms. The outcome of this research will contribute to Malaysia Education Blueprint 2015-2025 (Higher Education), specifically through globalized online learning and information technology to produce integrated thinking skills to help undergraduate students to understand cognitive strategies and appropriate communication in problem solving. Figure 2 shows how the students in their group convey their ideas in Trello. The facilitator upload questions on critical thinking subject about nationhood.

Figure 1. Trello task management app

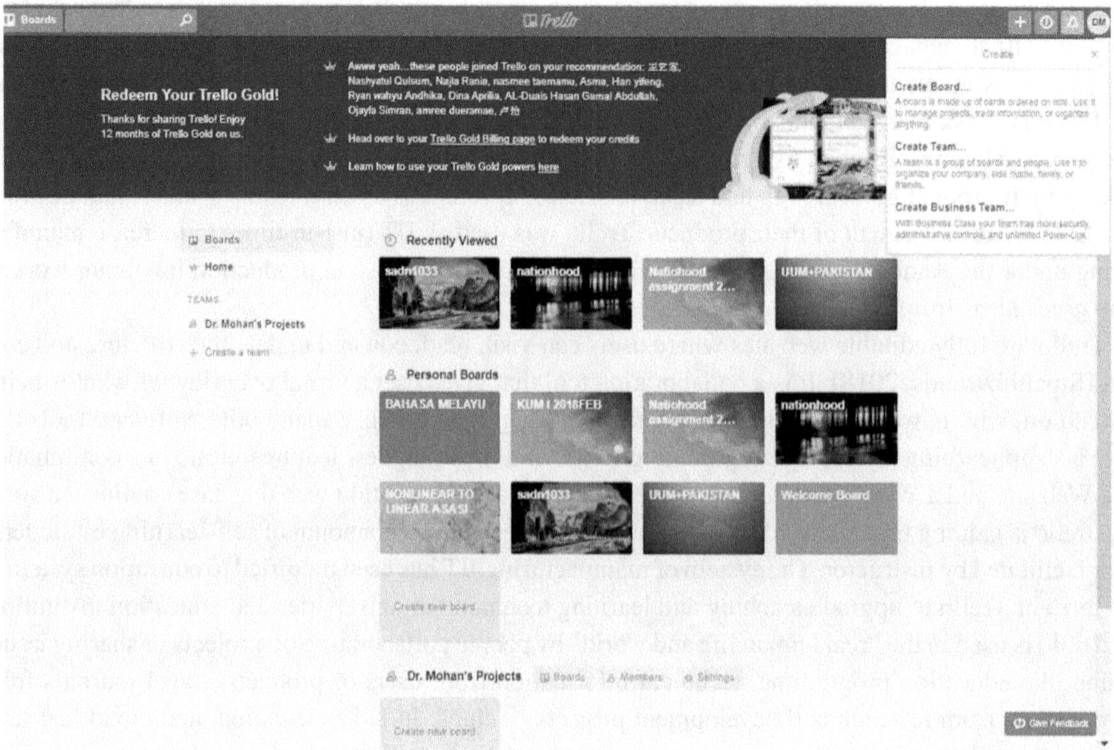

Figure 2. Student's involvement in Trello

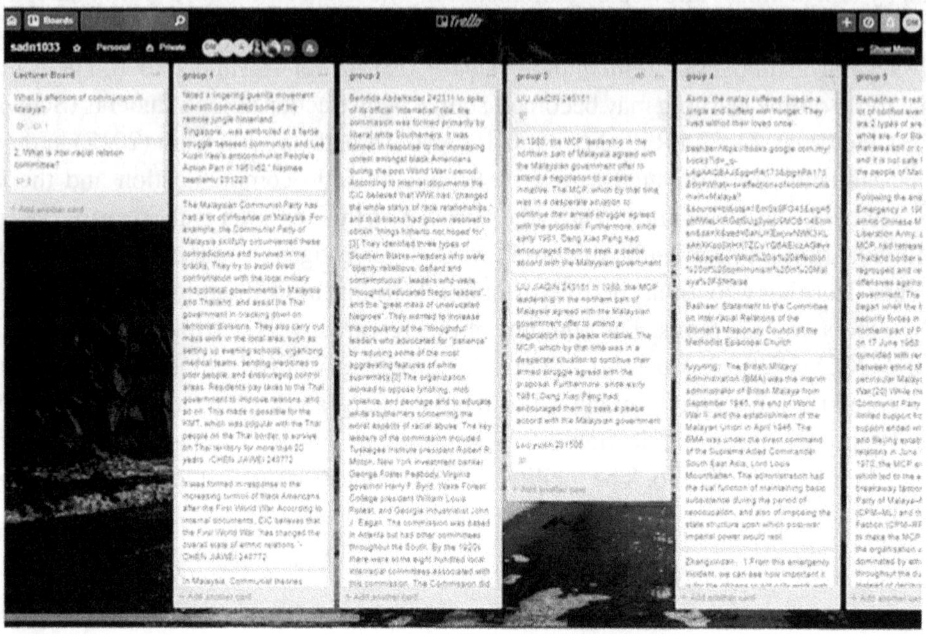

HEUTOGOGY APPROACH

Learning through online has proved enormously useful in the education system. The usage of online tool has provided a common experience and eased for the student to be self-determination. Many researchers such as Reiser and Dempsey (2002) as well as Wang and Gearhart (2006) urged that online learning tool is not simply another way of teaching to replace face-to-face instruction. It provides a learning facility and a suitable environment where institutions and teachers could assist students to understand the features of various instructional technologies (Anderson & Elloumi, 2005). Hu. (2007) mentions that currently there are many requests for online courses and training offered through higher education and government schools to make learning and teaching process effective in self-determination approach or heutogogy approach. Heutagogy expands the role of human abilities in the learning. Hase and Kenyon (2007), mentioned that learner can become their own learning tutor if they are guided in the correct way of learning. Learning using online tool helps the instructional design team to provide information that students need to help them initiate the instructional design process without leaving any important questions unanswered.

A study by Arumugam Raman et al. (2013) found that learning using heutagogy approach has helped students of distance learning courses to get good grades in their examinations. Even though not many students participated in online tutorials at the beginning of his research, after some persuasion, undergraduate and graduate students become confident enough to ask question and actively participate in the online tutorial. He mentioned that using an appropriate online tool would enormously help students to study and use new learning methods and thereby eliminate face-to-face tutorials.

Blaschke (2012) indicated that one of the differences between heutagogy and andragogy. Heutagogy concentrated on learning process which result students' personal experiences. The learner and teacher, work as partners. Heutagogy approach emphasis on developing capability, metacognition and self-reflection of students' own learning process, nonlinear learning and teaching processes.

Students are capable of performing at higher intellectual levels when asked to work in collaborative situations than when asked to work individually (Vygotsky 1978). He also mentions that technology can be learned and change the organizational infrastructure to increase the emphasis on teamwork within the workforce. Students are able to make decisions, think creatively, problem solver and work as a team. Therefore, the development and enhancement of critical-thinking skills through collaborative learning is one of the primary goals of online learning.

Figure 3, describes the basic principles of online learning heutagogy approach among foundation students. Constructivist and humanistic visions of learning, heutagogy approach shows that student could think critically once the facilitator allow students to use online learning trello in their learning. The student become be self-centered; explore how trello would help them in critical thinking subject. Students collaborate and create new idea to post without depending to the facilitator. Heutogogy also shows more recent advances in neuroscience as how people should learn at a cellular level. A simple explanation on online research and its relation to heutagogy can be found in Hase and Kenyon (2013) and Blaschke and Hase (2014). This online research support learner-centered approaches in education and casts doubt on much of the current orthodoxy way of teaching methods. Furthermore, these approaches gives an alternative to students to be more creative and understand how the brain functions seem to have a strong association with the ways in which people learn naturally when they were young and even could adapt their learning style in educational settings. It is not surprising why it is said online learning can help students to improve their ability to think and enhance their knowledge. Self-learning

Figure 3. Online Learning Heutagogy Approach: [Mohan 2018]

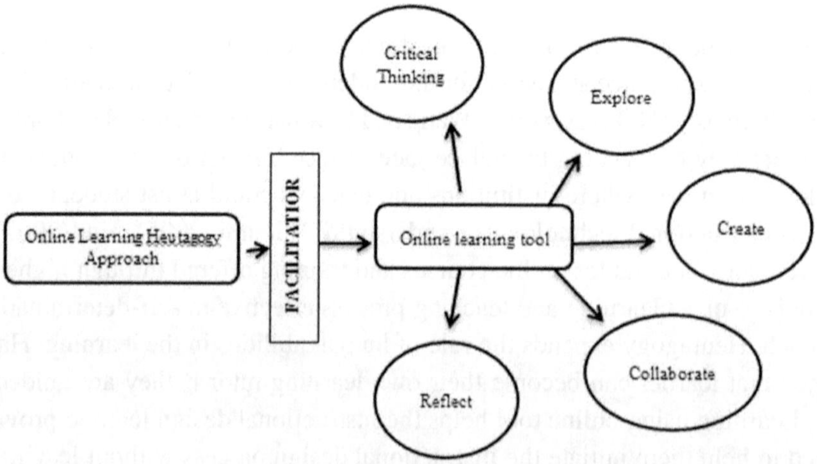

allows the student to be independent without relying on instructors. This situation is in accordance with the online learning because they can learn on the go regardless of the time. The heutogogy approach is very important to ensure student to be high-level thinking skill students to solve individual or group problems in their studies

LEARNING STYLES

A learning style is a biologically and developmentally determined set of personal characteristics that make the identical instruction effective for some students and ineffective for others. (Dunn & Griggs, 2000). Wintergersta, DeCapuab and Itzenc (2001) said that learning styles "…. *are innate preferences of individuals as to how they prefer to go about the process of learning*" (pg.98).

The teacher's ability to provide a suitable learning environment will engage the students' interest in the subject. According to Bloom's taxonomy, learning begins with the reception of information (Anderson et al. 2000). Sarasin (1998, pg 25) noted that learning style as "the preference or predisposition of an individual to perceive and process information". A student's predisposition to a preferred and unique learning style is an undisputed fact about how students learn.

Kolb (1984) published his learning style model setting with four distinct learning styles based on the 4 stages of a learning cycle. In his experimental theory, he mentions that the 'cycle of learning' i.e. the cycle of reflective observation, concrete experience, active experimentation and conceptualization and will help teachers to understand students' different learning styles. Kolb (1984) said, this cycle to provide a basis for 'observation and reflection'. These bases are assimilated into abstract concepts producing new implications on students' learning styles. He defined three stages of a students' development that integrates with four different learning styles (experiencing, reflecting, thinking and acting).

The three stages are:

1. Acquisition- developing basic abilities and cognitive structure

2. Specialization – development of a particular educational, social environment and organizational socialization
3. Integration – experience of non–dominant learning style in work and personal life

Felder (1988) in his research said that learning styles basically involves how the student listens, sees, reacts, hears, reflects, gives reasons, thinks logically, uses intuition, visualizes and analyzes. He indicates that different students have different ways of learning. After Felder did his experiment on engineering and general science education students, he created a learning style model using a web-based instrument called the *Index of Learning Styles*. He found that learning in a structured educational setting involves reception and processing of information. In the reception step, two types of information: external (observable through the senses) and internal (arising introspectively) become available to students. Felder (1988) divided the learning process into four dimensions: (1) sensory-intuitive (2) visual-auditory (3) active- reflective (4) sequential-global.

Felder (1988) also indicate that a students' learning style could be vary and it depends on the teacher to use a suitable learning style according to the students' ability, interest and suitable learning environment. Some students prefer to work with concrete information, facts and experimental data while others think abstractly, innovatively and about the underlying meaning. There are students who prefer visual representations of present material such as pictures, diagrams and flow charts and some students learn verbally using preferred written and spoken explanations.

Active students learn by trying things out and enjoy working in-groups but reflective students learn by thinking things through, prefer working alone or with a single familiar partner. A sequential student is linear in the thinking process and learns in small incremental steps whereas a global learner progresses toward holistic thinking process and learns in large leaps.

CRITICAL THINKING (CT) ON ONLINE LEARNING

The development of CT among students stresses the objective of various disciplines of study and goals that every school aspires to achieve. The Malaysia Education Departments hopes that teachers will apply CT skills in their teaching and include problems solving question in exams that test students' critical thinking skills (National Education Blueprint 2006-2010, 2007). Schafersman (1991) indicated that the purpose of teaching critical thinking in any discipline is to improve the thinking skills of students and thus better prepare them to survive and succeed in the world. Therefore, students should learn to ask appropriate questions, gather relevant information, efficiently and creatively sort out and sieve through this information, reason logically from this information, and come to reliable and trustworthy conclusions.

Paul (1993) defined CT as disciplined self-directed thinking, which exemplifies a particular domain of thinking. Paul identified two forms of critical thinking: sophistic or a weak sense of CT (which serves the interest of particular individuals) and a strong sense of critical thinking (the interest of diverse people or groups) or fair minded. Paul (1993) divided the student's critical thinking into four parts: element of reasoning, standard of critical thinking, intellectual abilities and intellectual traits. There are 12 standards of critical thinking encompassed in Paul's model such as precision, specificity, clarity, accuracy, relevance, consistency, logic, and fairness. All these standards of critical thinking support students' group work in performing common tasks and provide a shared interface for groups to work with.

Lipman (1991) in his research portrayed high-order thinking as critical thinking which is guided by ideas and creative meaning. Lipman (1991) resembles Bloom's analysis and creative thinking as critical thinking to show that students will think deeply and critically when analyzing a given topic. He mentioned that a critical thinking student would have sensitivity to the context of learning, reliance on criteria and indulge in self- correction. Similar to Lipman and Ennis (1996), Tama (1989) believes that the critical thinking student focuses on belief and action, and is involved in high-order thinking. However, they agreed that their taxonomy does not place emphasis in subject matter area that might be either exclusive or overlapping.

TRELLO ONLINE DISCUSSIONS: STUDIES TO PROMOTE CRITICAL THINKING

For this study, an experimental research design applied in this research. A purposive sample were chosen from Universiti Utara Malaysia Management Foundation Programme students who are studying Critical Thinking subject. There are seventy students from two classes involved as sample in this research from the population of 360 students. Each class consists of thirty-five students and has the same ability and qualification as they are selected students after their Sijil Pelajaran Malaysia (SPM) government exam. A study has carried out a few strategies to promote critical thinking by using online writing and discussion among foundation students in UUM.

A small number of groups learning were choosen to learn online and analyze a case study through role-playing. First, a topic is presented for online discussion and each student develops questions and answers based on the topic discussed. They discuss in group without the help of facilitator. Next, the facilitator who monitors the learning process posts comments on the given online answer. After looking through the given comment, each student represents a role in different points of views on specific issues and presents them by suggesting, analyzing, evaluating and giving explanations. The learning process helps the learners to compare their reasoning with their peers critically. The finding and ideas in the online learning will be summarized and presented for feedback. The instructor inspects the students' ideas posted in Trello without giving any comments or help the students to expand the ideas, so that the students will be self-determined.

RESEARCH PROCEDURES

Class X was selected as the experimental group (using trello online discussion) and Class Z became the control group (classroom discussion) using heutogogy approach while learning Critical Thinking subject. Trello online discussion is a tool used to plot discussion in a group on the topic given by the facilitator.

Even though there are four dimensions of learning styles as introduced Felder and Solomon, for the purpose of this research, the active-reflective dimension was used in this study on trello. The reason of using this dimension is because trello enable the students to discuss and give their own ideas or support their ideas with data, facts and creativity. The participants worked collaboratively or individually to share their ideas. An active student was expected to convey more ideas, comment and input engagement compared to a reflective student.

In this active-reflective dimension, there are four divisions involved, as indicated in Figure 4. The related items are indicated in the parentheses. The first box in each question resembles the active learning style and the second box resembles the reflective learning style.

The instrument uses a multiple-choice format in presenting two options. Even though there are 44 items to answer, only 11 items on active-reflective learning style questionnaires are evaluated. The students responded to the learning style that they prefer during the learning. The students were rewarded one point for each item in option "a" and "b".

The treatment students were exposed on how to use trello online discussion on the first week (8 hours). In the treatment class, the students are divided into seven groups with five members in a group (second week). Every group discuss the topic given using trello online discussion to All the group members discuss and write their ideas in trello. Ideas that are posted in trello will be displayed in front of the class so that the entire group in the treatment class could see the given ideas.

The research procedure begins with a topic given in an online discussion and each student in a group develops questions and answers based on the topic discussed without facilitator interruption. The peers in every group then posts comments on the given ideas online. After looking through the given comment, each student takes on roles representing different points of views on specific issues and presents them by suggesting, analyzing, evaluating, and giving explanations.

The learning process helps the learners to compare their reasoning's with their peers critically. The outcome of this research is similar that has shown in McCarthy, J. & Boyd, D (2004), which indicated that student become more critical in their thinking when they try to defend their ideas in discussion. All the finding and ideas in the online learning will be summarized and presented for feedback. The macro critical thinking (MaCT) (Table 1) assessment instrument was developed for this study to measure the student's macro critical thinking. MaCT is adopted from Paul's model of critical thinking who considers critical thinking as a fundamental structure of humans thought. In this study, the MaCT will extract student's writing quality and thinking skills from ideas writing on trello online discussion.

Figure 4. Four types of component in active and reflective learning style

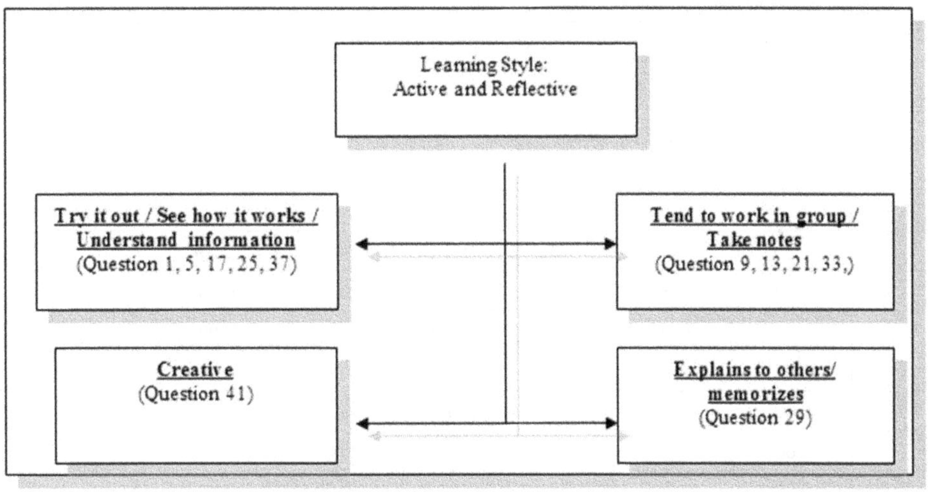

The ideas and point discuss based on topic given in trello will be print out. Four evaluators were appointed to analyzed students' idea and sentences after they have posted in trello using critical thinking rubric (MaCT)) as shown in Table 1.

Members of other group are allowed to give comments on the idea posted and at the same time, they give their own ideas on the topic discussed. After one hour of discussion, the respective teacher asked one of the member of each group to come in front of the class to explain what they have discuss about the topic (based on trello).

Table 2 shows the calculation on how students' critical thinking scores after they post their ideas in trello. Student 1 contributed two inputs, one input on Ma1 and one input on Ma2. Student 2 contributed only one input on Ma3. Student 3 contributed one input on Ma4. Student 4 did not contribute any input on macro critical thinking. Student 5, and Student 6 did not contribute any input that related to macro critical thinking.

DISCUSSION

In heutagogical approach, facilitators help their students by providing them the context of subject online and creating opportunities for them to explore the subjects fully on trello. They encourage students to find their own problems and questions to answer around the context area. The foundation students seek

Table 1. Critical thinking rubric

Level	MaCT	Score
M_{a6}	Evaluating Arguments	6
M_{a5}	Analyzing Arguments	5
M_{a4}	Making interdisciplinary connection (giving logical sequence)	4
M_{a3}	Clarifying Issues (elaborate issues discussed)	3
M_{a2}	Generating Solutions	2
M_{a1}	Refining Generalizations (remove defects / identify mistakes)	1

Table 2. Calculation for macro critical thinking

STUDENT	CRITICAL THINKING						Mark	SCORE
	MA1	MA2	MA3	MA4	MA5	MA6		
1	1.00	1.00	0.00	0.00	0.00	0.00	3.00	1.18
2	0.00	0.00	1.00	0.00	0.00	0.00	3.00	1.18
3	0.00	0.00	0.00	1.00	0.00	0.00	4.00	1.57
4	0.00	0.00	0.00	0.00	0.00	0.00	0.00	0.00
5	0.00	0.00	0.00	0.00	0.00	0.00	0.00	0.00
6	0.00	0.00	0.00	0.00	0.00	0.00	0.00	0.00
Total	**1.00**	**2.00**	**3.00**	**0.00**	**0.00**	**0.00**	**10.00**	**3.93**

out areas of uncertainty and complexity in the subjects they study by giving ideas critically instead of simply completing the tasks teachers assign, there are a few strategies that have been used to encourage students to get involved in online writing (Wilson, 2002). 'Crafting Questions' an online learning coordinator in United State of America have carried out strategies such as the use of convergent questions for Online Discussions (2002).

Under this strategy, convergent questions were used to ask learners to analyze information by breaking down parts, recognizing patterns, forming assumptions and inserting relevant ideas. Sometimes convergent questions were used to check the content information necessary to form a good essay.

Promoting active learning among the foundation students have become an important agenda in educating new intake students in the Malaysian university. Hence, to accommodate the importance of thinking maturely, one of subject listed is "Critical Thinking Skill'. Therefore, critical thinking skill is pertinently important among first year university students. The course syllabus indicates that students should be independent in thinking and have proficient in the use of language.

Blanchette (2001) in his research on online discussion mentioned that divergent questions can be used to explore different possibilities, variations and alternative answers. Wilson (2002) said divergent questions usually stimulate creative and critical thinking and often challenge learners to synthesize information. He also pointed out that divergent questions in online discussions could provide opportunities to expose learners to alternative possibilities, and new solutions are presented by different learners.

Clark (2004) indicates that evaluative questions promote critical thinking in online discussion by giving opportunities to reflect deeply on the topic. Evaluative questions require higher levels of cognitive thinking and well-prepared condition to answer the questions. He mentions that in evaluative questions, learners evaluate issues by giving comments, assessing, analyzing, arguing according to a set of criteria, beliefs and collection of resources to support their opinion. Discussions can become intense and emotional, and facilitation is critical to prevent argumentative interactions (Crafting Questions for On-line Discussions, 2002).

There are several studies demonstrating the benefits of having backchannel communications in classroom or lectures. Among the positive benefits documented are the conveniences of asking questions and getting responses from the instructors, learn new information from peers and having more fun learning

Walker (2005) explains that the best strategy to promote critical thinking is Socratic-questioning. In Socratic-questioning, students have to think deeply for matured answers, appropriate ideas with concrete evidence, various perspectives, examine reasons, vary concepts, and indulge in consequences. Through Socratic-questions, learners will identify causes, analyze assumptions and probe by asking, "Why do…", "How do..." and "Can you..." (The Foundation for Critical Thinking, 2002). Examples of Socratic of questioning methods are shown below:

1. What do you mean by…?
2. How did you come to that conclusion?
3. Why do you believe that you are right?
4. What is the source of your information?
5. What assumption has led you to that conclusion?

CONCLUSION

The implications for practice by higher education instructors addressing critical thinking skills were the findings from this study. The results of this study shows heutagogy approach in online learning could improve students' critical thinking. It is highly recommended that the higher education curriculum should adjust the course content to reduce the amount of replicated information taught in courses. Based on problem-based learning, active learning, and a constructivist approach is recommended to facilitate critical thinking skills and learning that is meaningful and lasting.

Besides building community interaction and enabling self-determined learning, trello as an online tool has helped the students to perform better in their writing and developing their CT skills. Trello has influenced the way UUM foundation students structure their knowledge to write the text with relevant information and in a matured manner.

This study has revealed that the heutagogy approach plays an effective role in enhancing students' performance and critical thinking skills. This approach has helped the students to participate and interact online. The students have become self-motivated, focused on goal setting and learn from feedback given on the topic discussed online. This gives an opportunity to their peers to collaborate and be innovative in giving ideas. Even though, the students' use online learning as a platform, they gained maturity of thought and understanding through the inputs posted in trello, the heutagogy approach enabled the students to analyze, evaluate issues across disciplines and express understanding as well as apply critical and creative thinking skills through informed personal responses as required in fulfillment foundation students' assessment objectives and skills.

REFERENCES

Blanchette, J. (2001). Questions in the online learning environment. *Journal of Distance Education, 16*(2), 13–49. Retrieved from http://cade.athabascau.ca/vol16.2/blanchette.html

Bloom, B. S. (1956). *Taxonomy of educational objectives: The classification of educational goals.* New York: McKay.

Clark, D. (2004). *Gains in critical thinking using online discussions.* International Forum of Educational Technology & Society. Retrieved Jun 16, 2018, from https://mail.fit.fraunhofer.de/pipermail/ifets/2004q1/000553.html

Du, H., Xu, H., Rosson, M. B., & Carroll, J. M. (2013). Effects of Fear Appeals and Point of Reference on the Persuasiveness of IT Security Communications. In IEEE Intelligence and Security Informatics 2013. IEEE Digital Library. doi:10.1109/ISI.2013.6578791

Ennis, R. H. (1996). *Critical thinking.* Upper Saddle River, NJ: Prentice Hall.

Felder, R. M., & Silverman, L. K. (1988). Learning and teaching styles in engineering education. Engr. *Education, 78*(7), 674–681.

Harris, R. N., Dwyer, W. O., & Leeming, F. C. (2003). Are learning styles relevant in Web-based instruction? *Journal of Educational Computing Research, 29*(1), 10–27. doi:10.2190/YHL4-UP7P-K0GD-N5LJ

Lipman, M. (1991). *Thinking in education.* Cambridge, UK: Cambridge University Press.

Malaysian Education Blueprint (2015-2025) Higher Education. (2018). Retrieved November 18, 2008, from http://www.mohe.gov.my

McCarthy, J. F., & boyd, d. (2004). Digital Backchannels in Shared Physical Spaces: Attention, Intention and Contention. *Proc. of the 2004 ACM Conf. on Computer Supported Cooperative Work (CSCW 2004)*, 550-553. 10.1145/1031607.1031700

Mertes, L. (1991). Thinking and writing. *Middle School Journal, 22*(1), 24–25. doi:10.1080/00940771.1991.11496002

National Education Blueprint. (2006-2010). Putrajaya: Kementerian Pelajaran Malaysia.

National Education Blueprint. (2015-2025). Putrajaya: Kementerian Pelajaran Malaysia.

Arumugam. (2012). The usage of technology among education students in University Utara Malaysia: An application of extended Technology Acceptance Model. *IJEDICT, 7*(3).

Carrol. (1989). Critical thinking has a place in every classroom. *Journal of Reading, 33*(1). Retrieved from https://smallbiztrends.com

Paulsen, M. P. (1995). Moderating educational computer conferences. In Z. L. Berge & M. P. Collins (Eds.), *Computer-mediated communication and the on-line classroom in distance education* (pp. 113–258). Cresskill, NJ: Hampton Press.

Peffers, K., & Bloom, S. (1999). Internet-based innovations for teaching IS courses: The state of adoption, 1998-2000. *Journal of Information Technology Theory and Application, 1*(1), 10–41.

Prensky, M. (2001). *Digital natives, digital immigrants.* Available at: www.marcprensky.com/writing

Rathakrishnan, Ahmad, & Choi. (2017). Padlet Online Discussion in Enhancing Students' macro Critical Thinking Skills. *International Journal of Latest Research in Humanities and Social Science, 1*(3), 121-126.

Rathakrishnan, M., & Pbwiki, S. (2016). On-Line Learning: The Effectiveness of Pedagogical Role Online Teachers' On Students with Difference Learning Style In Their Critical Thinking. *Proceeding of the First International Conference on ICT for Transformation*, 98-101.

Schwartz, L., Clark, S., Cossarin, M., & Rudolph, J. (2004). Educational wikis: Features and selection criteria. *The International Journal of Research in Open and Distance Learning, 5*(1), 162–691.

Spiro, R. J., Feltovich, P. J., Jacobson, M. J., & Coulson, R. L. (1991). *Cognitive Flexibility, Constructivism and Hypertext.* Retrieved July 6, 2018, from http://www.ilt.columbia.edu/ilt/papers/Spiro et al.html

Stewart, M. (2001). Towards a global definition of patient centred care. *British Medical Journal, 322*(7284), 444–445. doi:10.1136/bmj.322.7284.444 PMID:11222407

Walker, G. (2005). *Critical thinking in asynchronous discussions.* Retrieved April 5, 2018, from http://www.itdl.org/Journal/Jun_05/article02.htm

Chapter 4
An Overview of Massive Open Online Courses (MOOCs), Pedagogies, and Distance Education

Mohd Faiz Mohd Yaakob
Universiti Utara Malaysia, Malaysia

Mohd Muslim Md Zalli
Universiti Pendidikan Sultan Idris, Malaysia

ABSTRACT

This chapter discussed an overview of massive open online courses (MOOC) pedagogies and distance education in Malaysia. Meanwhile, topics such as MOOC in Malaysia, distance education, theory, and concept and other sub-topics are discussed. Massive open online courses (MOOCs) are the instructed course with the various number of students using the internet. Indeed, MOOCs are not a new tool to educational modules outline and strategy. At the end of this chapter, effective teaching strategies are proposed.

INTRODUCTION

Massive open online courses (MOOCs) are the instructed course with various number of students using the internet. Indeed, MOOCs are not new tool to educational modules outline and strategy, this tool captivating, expanded thought and examination in advanced education. Universities in the United States and around the world started to offer MOOCs either through corporations with MOOC providers or on their own websites (Pappano 2012; Vardi 2012).

In addition, distance education refers to learning interventions during which geographically separated instructor and learners interact and communicate via telecommunication (Moore 1973; Simonson et al. 2012). Accordingly, distance education has several benefits and advantages over traditional courses, such as accessibility and time flexibility (Buckley 2003; Chen et al. 2009; Washer 2001).

DOI: 10.4018/978-1-5225-7832-1.ch004

Copyright © 2019, IGI Global. Copying or distributing in print or electronic forms without written permission of IGI Global is prohibited.

Then, it presents educators and students with various kinds of difficulties, including the communication amongst students and teachers. In distance education, students may never get to a physical grounds or build up associations with their teachers (Bolliger & Martindale 2004). But, arbaugh (2000) argued that learners' interactions with others in online courses increase their satisfaction with these courses.

In the meantime mid-1970s, a creating of imperically study has appropriated with hypothesis structure concerning study on the distance education. Despite the fact that MOOCs are viewed as a type of distance education (Romiszowski 2013), they are unique in relation to online courses that colleges have been putting forth for quite a long time utilizing a few instructional advances (Kim 2013; "MOOCs: Online Education 2012"). According to Allen and Seaman (2015), MOOCs contrast from ordinary online courses, because MOOC members are not enlisted students at the colleges offering them, MOOCs are intended for unlimited cooperation and free access through the internet, and the MOOC student, by and large, gets no acknowledgment for the fulfillment of the MOOC.

Meanwhile, by understanding several distance education from theories can help to understand the learner experience in terms of interaction in and satisfaction with MOOCs. This is supported by the theory of independent learning and teaching and the three types of interaction model introduced by Moore (1973, 1989). Furthermore, Moore (1973,1989) defined interactions in distance education and identify three significant types of interaction; learner-content interaction, learner-instructor interaction, and learner-learner interaction. By understanding the theory of independent learning and teaching, MOOC, pedagogies, and distance education will be successfully implemented.

MOOC IN MALAYSIA

Presently, MOOCs Malaysia has accessible 63 interesting courses with 137,946 students from 80 countries that follow MOOCs Malaysia. MOOCs has been emphasized according to agenda 9: Globalized Online Learning under the Malaysian Education Development Plan (Higher Education) 2015-2025. MOOC in Malaysia are an extremely contemporary improvement. Besides, these reasons can be understood as potential benefits and they have compelled Malaysia's own initial exploration of MOOCs in this country. Now, MOOCs aplication in Malaysia is growth with important national planning such as the 11th Malaysia Plan (2016-2020); National Economic Model, Economic Transformation Programme and the anticipated Malaysian Education Blueprint for Higher Education. Besides, online learning, as an essential component of the delivery mechanism in MOOCs, is also clearly addressed in the blueprint.

In another word, the ubiquity of MOOC is expanding with the association of different establishments and scholastics, particularly from field experts. Besides, the components encapsulated in MOOC, for example, online courses, adaptable outline, and intuitiveness improvement likewise add to the fervor of MOOC utilization and also advantage the understudies. The enlightening ideas connected in the MOOC contrast from other web based learning and make it an advantage when the receptiveness highlights draw in more individuals to enlist and utilize MOOC as an extra learning medium. The principle trademark accentuated in this online course is the substance of its learning assets that suit the necessities of its understudies and additionally the desires for the course.

One of the keys to making online courses is an interaction. In the meantime, the course learners likewise need to effectively include themselves to help understudies over the span of their investigations and constantly mindful of the requirements of their understudies. Understudies utilizing the MOOC can understand leeway of having the capacity to trade data and thoughts with associates far and wide and pick up information that isn't educated in the classroom. Getting the hang of utilizing MOOC is all the more difficult and adaptable. This is on the grounds that the understudy will learn without anyone else and this will prepare the basic and basic reasoning aptitudes, the method for correspondence, the time administration and train of an understudy. Nonetheless, additionally investigate regarding the acknowledgment and utilization of MOOC, MOOC's usage by the focal point of study or speaker who conducts courses and view of MOOC's engineers, clients, and course coaches should at present be examined particularly with regards to advanced open education movement in Malaysia.

Open Education Movement: MOOCS

Movement online learning and focused in higher education has been investigated during the most recent more than 20 years and very huge group of literature in, open education, E-Learning, and independent learning environments (Adams, Yin, Vargas-Madriz & Mullen, 2014). However, MOOCs due to their open and high enrollment characteristics seem to be a different phenomenon on its own. Study is still investigating and exploring different existing online learning models and MOOC environments seem to be better investigated and explored as courses some way or another related with customary web based adapting, yet in addition demonstrating novel particularities (Breslow, Pritchard, DeBoer, Stump, Ho & Seaton, 2013). Accordint to Kennedy (2014), three major aspects of MOOC characteristics have been identified: openness, barriers to persistence and design models.

Firstly, "openness" identifies with the utilization of open advancements and programming, open substance, open access and open condition where members share learning. Secondly, "barriers to persistence" refer to the issues students experience while taking the courses. These boundaries are identified with innovation mastery, English mastery if MOOC is offered in English, course structure, and time constraints (Rodriguez, 2012). Different student agec, backgrounds and profiles that lead to different learning objectives and motivations to take the MOOC are also imperative factors that influence course culmination rates and persistence. More research inside these areas is encouraged (Fidalgo-Blanco, Garcia-Peñalvo, & Sein-Enchaluce, 2013). These authors hypothesize that the lack of homogeneity within students' backgrounds, and level of education is a major element of the low completion rates.

Designing activities and course content for altogether different shoppers is a test for instructional fashioners and field specialists that may be better investigated. On the other side, there are designer that inquiry the handiness of estimating understudy accomplishment by taking a gander at fruition rates, for the most part as a result of the underlying transparency and the distinction in introductory inspiration levels (Alraimi, Zo and Ciganek, 2014). This findings opens the door to other ways of tracking and measuring student success.

Finally, the third from MOOC characteristic is design models, there seems to be two main models within MOOC's: from connectivist model targeting lifelong learners (cMOOC) and another model applied to the post secondary education similar to the traditional online courses (xMOOC).

DISTANCE EDUCATION, THEORY, AND CONCEPT

Since the 1970s and 1980s a few academic study were suggested that went for being complete and including the entire region of distance learning. These scholars defined distance education, theory, principals concepts, and primary focus on distance education. Peters (1983) stated principal concept distance education are from industrial and post-industrial. This concept focused on industrialized education. Meanwhile, Moore (1973) developed transactional distance (dialogue and structure); learner autonomy as important thing for distance education. Holmberg (1983) introduced motivation; empathy; non-contiguous communication; learner autonomy; interpersonal communication as principal concept of distance education. Keegan (1986) studied on reintergration; intersubjectivity; two-way communication as a principal concept of distance education using theory of reintergration of the teaching and learning acts. In addition, Garrison (1985;1987) stated that inseparability of technology-collaborative; educational transaction; self directed learning as a principal concept of distance education.

From an investigation study of the normal components of these fundamental scholars in the field of distance learning, two ideas developed as focal: distance and communication. Furthermore, especially because of the impact of the most seasoned of these scholars, to be specific Peters (1983), there is likewise a third hidden idea which they share: the marvel of industrialized training.

These three wide ideas - or classifications - give a beginning stage to a more profound returning to of the establishments of separation training. Peters (1983) investigated creation procedures of modern and post-mechanical society which affect education. Intending to modern culture, remove is dealt with from the point of view of new educating organizations. Under this model, the transcendence of creation forms supersedes the interests of people who are liable to these procedures. The dehumanized relationship among educators, students, and substance is an element of industrialized society investigated by Peters. The indifferent and institutionalized connections are featured. With regards to a mechanical society, instructive procedures are near creation and sequential construction system forms, portrayed by a Ford-ist producing model. In this model, land remove is definitely not a key driver.

The determinants which oversee the new authoritative practices are spoken to by the method of generation in modern culture. The separation amongst teachers and students is viewed as a business open door for the training establishment to accomplish an expansive volume of students who are geologically scattered. In this unique situation, serial generation forms take into account the qualities spoke to by an institutionalized and vast scale sort of instruction.

Concerning the post-industrial perspective, Peters (1983) also viewed distance as a boundary which can be surmounted by more finely custom fitted innovations and decentralized basic leadership structures. Procedures among instructive specialists can be united in order to render relations more acculturated. This procedure of moving operators nearer together using advances is likewise respected by Keegan (1993) as a re-mix of the instructing and learning acts. In other words, the geographic separation between students, teachers, and substance in post-mechanical society tends, to be risen above by progressively refined innovative procedures. This jumps out at a point where the thought of spatial separation starts to vanish through and through from investigations by later scholars (Garrison, 2000). The transcendence of educating and learning connections dominating separation between them renders the physical separation between specialists immaterial.

New innovations allow the connection amongst teaching and learning, reflected in face-to-face models, to be re-established since the operators can communicate, work together and share with each other. The thought of distance education fundamentally is progressively truant in remove instruction contemplates.

Different terms become the dominant focal point, for example, "intervened instruction" or "innovation interceded training" (Kanuka and Conrad, 2003). The rubric "instruction" measurement conveys more weight than the expression "distance education". Scholars stretch the significance of educating and realizing which might be improved by the utilization of two-way communication assets, as revealed by Keegan (1993) and Garrison (1989).

The scholars analyzed in the present examination, two merits exceptional consideration in connection to the investigation they performed in view of the idea of distance: Holmberg and Moore. Holmberg (1995) authored the term " non-contigous communication" to portray correspondence which happens in the connection between instructive specialists who are physically isolated by time and place. The communication procedure is of two sorts: (1) one-path activity, as pre-delivered course materials sent from the supporting association and including students in cooperation with writings, chronicles and information bases, which can be portrayed as mimicked correspondence, and (2) two-way movement, i.e., genuine communication amongst understudies and their supporting association, in composing, on the phone, by fax or email (Holmberg, 1995).

Holmberg's theory centers on the fact that teachers and students are physically separated and seeks a means of remedying this fundamental gap between them. Holmberg beliefs it probable to reproduce the teaching and learning condition of classrooms by utilization of different methodologies. He chooses various theories in light of a few factors, for example, relational correspondence, which should fill in as a methods for modifying discourse amongst teacher and student, whereby correspondence procedures would likewise be available at the teaching structure level or correspondence materials and techniques: printed materials, composed remarks by the instructor and so forth. It is an offer to reestablish the connection amongst teachers and students in the teaching and learning process. Learning process is connecting through the learning pedagogies.

DISTANCE LEARNING PEDAGOGIES

Connectivist Pedagogy

Connectivist pedagogy in distance learning pedagogies introduced that learning occurs when autonomy, diversity, openness and interactivity happen. The learners are able to remix and repurpose concepts towards the design of new knowledge, while moving forward within the context of the course (Anderson and Dron, 2011).

This pedagogy approach has all the earmarks of being a decent method for outlining an open online condition where understudies can repurpose a portion of the substance and make significant methods for learning. On the opposite side, it represents a testing circumstance for teachers to oversee and organize because of the gatherings and the connections between understudies, educator, and substance. It is additionally testing from a planning point of view, making assignments that permit this kind of opportunity and proprietorship, represents a few difficulties (Rodriguez, 2012). It is another method for learning for a few understudies and they have to combine and utilize diverse innovations that may diminish their attention on content. The course structure can likewise be free and requires an additional exertion frame the facilitator to manage understudies through these kind of methodologies.

Cognitivist-Behaviorist Pedagogy

Cognitivist-behaviorist pedagogy imperically developed at the end of the 20th century, and it defined the action of learning as a change in behavior of an individual when responding to a stimuli (Anderson and Dron, 2011). This instructional design methodology was broadly connected toward the start of separation instruction history when video chatting was the best way to help remove learning. This system has its esteem when connected to circumstances of one-to-numerous or coordinated correspondence, and it is generally an instructor or instructional planner.

Teachers, or the course design, control students communications incorporating how they relate with each other and the substance. This teaching method connected to new conditions, for example, MOOCs, builds up that learning destinations must be obviously disclosed to enable students to explore through the substance and achieve a reasonable comprehension of the materials. In this sense, the learning happens exclusively and it is understudy centered. This teaching method is as yet being utilized today in xMOOC situations where content is the primary focal point of the plan.

Social-Constructivist Pedagogy

Social–constructivist pedagogy derives from the work of Dewey (1916) and Vygotsky (1978) and views learning as a social job. Each learner constructs his/her understanding at the same time that interacts with instructor and other learners in the group (Anderson and Dron, 2011). Understudies increase new konwledge and expand upon their previous one to make new encounters; learning is gained up separately however inside a social domain. Meanwhile, strengths and weaknesses of social-constructivist pedagogy are online learning is based in communication, both synchronous and asynchronous. This pedagogy offers open learning for students and access to other means of content and social interaction, but it is very similar to traditional classrooms. Emphasizing communication adds pressure on technology and sometimes technology accessibility can be an issue.

A summary of the previously described distance learning pedagogies is provided in table 1

Table 1. Distance learning pedagogies

Pedagogies	Type of MOOC	Strengths	Weaknesses
Connectivist	cMOOC	Different approaches to content and ways of delivering information. Exposure to different ideas, points of view and resources.	This way of delivering information can be difficult for some type of students. Loose structure, needs structure Requires high instructor engagement.
Cognitivist-behaviorist	xMOOC	Courses can adapt to very large number of students Adaptation to scalability and low prices	Teaching and social presences are low
Social-constructivist	xMOOC with an emphasis on community knowledge building	Similar to traditional face to face environtments High emphasis on social presence	Technology accessibility is a key element

Communication and Interaction

Communication and interaction through transactional distance can be displayed a role in a face-to-face instructional. However, transactional distance takes merge and center stage in the distance education setting where physical distance between students is better. The communication and interaction spaces which permeate the relationship between the agents, and which characterize transactional distance, actually gain a new implementation in the processes of instructional approach.

Communication

The word of communication has an extensive scope of implications which renders the term uncertain and reliant on specific utilization and applications. Correspondence is a field of concentrates in its own privilege in which numerous hypotheses, scholars, and schools exist. In this area, how this idea fits inside the individual hypothetical structures in regards to separate instruction is explored with a view towards meeting educating and learning goals.

Table 2 summarizes the dominant concepts and focus by which five theorists deal with communication. Those theories with closer links to classic concepts of pedagogy and education, particularly those of Holmberg and Keegan, set forward the improvement of data and correspondence innovations, setting more prominent accentuation on correspondence forms as instruments of educational intercession amongst instructors and understudies. Since instructors and understudies are isolated by spatial separation and utilize supporting analogical materials (e.g., ponder aides, messages, and printed content, sound accounts), the focal point of these scholars is on instructive setting – its constraints and potential – and on proposing systems to advance the teaching and learning relationship.

Peters (1983) viewed interactive and communicative forms of teaching are the focus on communication. Meanwhile, he pointed out that self-learning, telelearning, social intercourse are the central concepts of communication. Moore (1973) and Holmberg (1986) agreed that psychological and communication space is the focus on communication. In addition, Moore stated that dialogue and structure are the central concepts while Holmberg (1983) viewed didactic, empathy and motivation are the central concepts of communication. Futhermore, Keegan (1986) viewed interpersonal communication and two-way communication are the focus on communication and learning materials, the variety of techniques are the central concepts of communication. Finally, Garrison (1985,1987) pointed out that two-way communication is the focus on communication and dialogue and debate are the central concepts. For the conclusion, communication has their own focus and central concepts according to the scholar.

Table 2. Communication-related concepts emerging from literature

Authors	Communication Focus	Communication Central Concepts
Otto Peters (1983)	Interactive and communicative forms of teaching	Self-learning; tele-learning; social intercourse
Michael Graham Moore (1973)	Psychological and communication space	Dialogue and structure
Börje Holmberg (1983)	Psychological and communication space	Didactic; empathy; motivation
Desmond Keegan (1986)	Interpersonal communication; two-way communication	Learning materials; variety of techniques
Randy Garrison (1985; 1987)	Two-way communication	Dialogue and debate

Interaction

According to the theory of independent learning and teaching, distance education is an instructive framework in which the student and teacher convey by print, electronic, or another nonhuman medium. Moore (1973) noticed that these autonomous learning and encouraging instructive frameworks comprise the accompanying subsystems: a student, an educator, and methods of correspondence. These three subsystems have fundamental attributes that separate them from getting the hang of, instructing, and correspondence in other training frameworks (Moore 1973). An unmistakable comprehension of the idea of collaboration to remove instruction and how to encourage association through intelligent media transmission is basic (Moore and Kearsley 2011). Moore (1989) identified the following as three significant types of interactions: learner-content interaction, learner-instructor interaction, and learner-learner interaction.

Learner-Content Interaction

Moore (1989) viewed out that the teacher in distance learning must to encourage the communication of understudies with the topic that is introduced in the course. Student's association with content is viewed as a basic normal for training since it changes and enhances the student's understanding and capacity (Moore 1989). Halasek et al. (2014) contended that in MOOCs, students' interests and individual inspiration in taking MOOCs to decide if and how they connect with the course materials. The teacher's part in separate teaching isn't just to help every understudy in associating with the substance yet additionally to change over the substance into the client's close to home learning system (Moore 1989).

Learner-Instructor Interaction

Interaction between the student and teacher is imperative for most students and wanted by general educators. In the event that self-coordinated students interfaced alone with the substance introduced in the course, given a restricted information of the topic, they may require help for the purpose of utilization (Moore 1989). Ponti (2014) contended that despite the fact that expanded access to open instructive assets and advanced media gives students more chances to access and increment information, students still need help from teachers to appreciate diverse portrayals of ideas and calculated comprehension of a train.

Learner-Learner Interaction

Interaction between the learner and other learners is valued in distance education. It enables students to consider thoughts, examine them with each other, and test displayed content. Long range interpersonal communication advances encourage student cooperation by sharing perspectives and additionally encounters (Moore 1989; Moore and Kearsley 2011). Blau, Mor, and Neuthal (2013) found that in the web-based learning condition in a college course, student connection, for example, sharing encounters and asking for criticism increments when students request it unequivocally and when they need dynamic assistance by educators. In the MOOC, for example, it is normal that the learning happened not just because of the restricted exchange of substance (from teachers to students) however more basically through arranged and crowdsourced synergistic cooperation among students (Porter 2014; Ravenscroft 2011; Siemens 2005). In conclusion, communication and Interaction are the important keys to facilitate distance education on MOOC course. But, effective teaching strategy plays an important role in blending teaching and learning.

Effective Teaching Strategy

A large and growing body of literature has investigated on effective teaching strategy. Laurillard (1993) emphasized that teaching and learning activities should be constantly updated and improved in line with the development of current teaching and learning techniques. He advised lecturers equip themselves by learning and mastering contemporary teaching techniques that are more geared towards media usage and technology. This approach is supported by Bates (1984) stating that the use of technology in the teaching and learning process can extend the scope of teaching, improve quality of teaching as well as reduce costs. It also gives students the opportunity to master their learning and encourage interaction and feedback from students.

Rashidi Azizan and Abdul Razak (1996) outlined some of the key criteria in determining the effectiveness of teaching and learning processes in distance education. Among the criteria are:

(a) A teacher needs to identify the teaching objectives and ensure that the objectives are suitable. The teaching objective will affect the teaching and learning process.
(b) The teacher should deliver the contents of the lesson in order and smoothly. As well as starting from the basic concepts, the use of precise and appropriate examples, clear explanations and adjust the simple and easy language of the students.
(c) The teacher should ensure that the student can hear clearly in addition to the teaching's speech must be clear and fluent. It's a lot depending on the intonation or correct sound suppression. For reinforcing the message delivered, the lecturer can using face-to-face changes as well as using gestures.
(d) Teachers should also optimize the use of teaching aids. Teaching aids is in addition to delivering teaching verbally and helping students understand it.
(e) Teachers should use the teaching method centered students because students should be guided using and develop their potential maximum for acquiring knowledge and skills.
(f) Teachers need to know that evaluation for the teaching process and learning is a continuous process as well bilateral process.

Based on these criteria of effective teaching and learning, teaching are consciously aware of or not, should always be present positive attitudes and traits. Negative qualities will only affect perceptions and expectations of students towards their teachers. The study conducted by Saroyan and Snell (1997) showed the teaching Effectively depends on many factors, including the philosophy and education background of a teacher, knowledge about pedagogy, existing teaching materials and circumstances a conducive environment for teaching and learning. Miller and Tilstone (1998) states that in the distance education program, the system of Flexible teaching is necessary to ease the burden of time and the stress experienced by these students. Kimber and Wang (2000) argues that the effectiveness of the teaching is closely related by way of learning the student itself.

The study conducted by Saroyan and Snell (1997) showed the teaching Effectively depends on many factors, including the philosophy and education background of a teacher, knowledge about pedagogy, existing teaching materials and circumstances a conducive environment for teaching and learning. Miller and Tilstone (1998) states that in the distance education program, the system of Flexible teaching is necessary to ease the workload of time and the stress experienced by these students. Kimber and Wang (2000) argues that the effectiveness of the teaching is closely related by way of learning the student itself.

CONCLUSION

MOOC was instinctive from distance learning. Communication and interaction is an important element in the success of this course. Malaysia has participated in distance learning by introducing MOOC. Hence, the understanding and ability to implement this agenda is important. This section has revealed MOOC, distance learning, communication and interaction as the key to success in the course. The possibly huge part MOOCs can play should to be recognized in the bigger setting of long-lasting learning. This is significant not exclusively to decide the best push ahead where MOOCs are concerned yet, in addition, to use on MOOCs as a development that can support the deep-rooted learning cause in this nation. Malaysia has pronounced long-lasting learning as the third column in human capital advancement: an instructive plan similarly essential to the school and advanced education frameworks and one that underlines the formation of learning open doors for the whole the populace, regardless of whether for formal, non-formal, or casual learning premiums. Education is for all, no matter who, when, where and how to manage. Accordingly, it is the responsibility of each higher education institution to giving a fair and just opportunity to the individual who wants to continue the education to a higher level.

REFERENCES

Adams, C., Yin, Y., Vargas Madriz, L., & Mullen, C. (2014). A phenomenology of learning large: The tutorial sphere of xMOOC video lectures. *Distance Education, 35*(2), 202–216. doi:10.1080/01587919.2014.917701

Allen, I., & Seaman, J. (2015). *Grade level: Tracking online education in the United States*. Babson Survey Research Group.

Alraimi, K., Zo, H., & Ciganek, A. (2015). Understanding the MOOCs continuance: The role of openness and reputation. *Computers & Education, 80*, 28–38. doi:10.1016/j.compedu.2014.08.006

Anderson, T., & Dron, J. (2011). Three generations of distance education pedagogy. *The International Review of Research in Open and Distributed Learning, 12*(3), 80. doi:10.19173/irrodl.v12i3.890

Arbaugh, J. B. (2002). Managing the on-line classroom: A study of technological and behavioral characteristics of web-based MBA courses. *The Journal of High Technology Management Research, 13*(2), 203–223. doi:10.1016/S1047-8310(02)00049-4

Azizan & Habib. (1995). *Pengajaran dalam bilik darjah: Kaedah dan strategi*. Masa Enterprise.

Bates, A. W. (1984). Putting it together: Now and the future. In A. W. Bates (Ed.), *TheRole of Technology in Distance Education*. London: Croom Helm.

Blau, I., Mor, N., & Neuthal, T. (2013). Interacting for learning: Digital portfolios for a learning community in a university course. *Learning, Media and Technology, 38*(3), 241–255. doi:10.1080/17439884.2012.709864

Bolliger, D. U., & Martindale, T. (2004). Key factors for determining student satisfaction in online courses. *International Journal on E-Learning, 3*(1), 61–67.

Breslow, L. B., Pritchard, D. E., DeBoer, J., Stump, G. S., Ho, A. D., & Seaton, D. T. (2013). Studying learning in the worldwide classroom: Research into edX's first MOOC. *Research & Practice in Assessment, 8*, 13–25.

Buckley, K. M. (2003). Evaluation of classroom-based, web-enhanced, and web-based distance learning nutrition courses for undergraduate nursing. *The Journal of Nursing Education, 42*(8), 367–370. PMID:12938900

Chen, C.-Y., Chang, R.-E., Hung, M.-C., & Lin, M.-H. (2009). Assessing the quality of a web-based learning system for nurses. *Journal of Medical Systems, 33*(4), 317–325. doi:10.100710916-008-9193-1 PMID:19697698

Cormier, D., & Siemens, G. (2010). The open course through the open door: open courses and research, learning and engagement. *Educase Review,* 31-39.

Dewey, J. (1916). *Democracy and education: an introduction to the philosophy of education*. New York, NY: MacMillan Company.

Fidalgo-Blanco, Á., Sein-Echaluce, M. L., & García-Peñalvo, F. J. (2013). MOOC cooperativo. Una integración entre cMOOC y xMOOC. *Proceedings from II Congreso Internacional sobre Aprendizaje, Innovación y Competitividad, CINAIC 2013.*

Garrison, D. R. (1985). Three generations of technological innovations in distance education. *Distance Education, 6*(2), 235–241. doi:10.1080/0158791850060208

Garrison, D. R., & Baynton, M. (1987). Beyond independence in distance education: The concept of control. *American Journal of Distance Education, 1*(3), 3–15. doi:10.1080/08923648709526593

Garrison, R. (2000). Theoretical challenges for distance education in the 21st century: A shift from structural to transactional issues. *International Review of Research in Open and Distance Learning, 1*(1), 1–17. doi:10.19173/irrodl.v1i1.2

Halasek, K., McCorkle, B., Selfe, C. L., DeWitt, S. L., Delagrange, S., Michaels, J., & Clinnin, K. (2014). A MOOC with a view: How MOOCs encourage us to reexamine pedagogical doxa. *Invasion of the MOOCs,* 156. Retrieved from http://www.parlorpress.com/pdf/invasion_of_the_moocs.pdf

Holmberg, B. (1983). Guided didactic conversation in distance education. In Distance education: International perspectives (pp. 114-122). London: Croom Helm.

Holmberg, B. (1995). *The sphere of distance education: Theory revisited.* Retrieved from www.eric.ed.gov/ERICDocs/data/ericdocs2sql/content_storage_01/0000019b/80/14/25/f8.pdf

Kanuka, H., & Conrad, D. (2003). The name of the game: Why "distance education" says it all. *Quarterly Review of Distance Education, 4*(4), 285–394.

Keegan, D. (1986). *The foundations of distance education*. Londres: Croom Helm.

Keegan, D. (1993). Reintegration of the teaching acts. In Theoretical principles of distance education (pp. 113-134). London: Routledge.

Kember, D., & Wong, A. (2000). Implications for evaluation from a study of students' perceptions of good and poor teaching. *Higher Education Journal., 40*(1), 69–97. doi:10.1023/A:1004068500314

Kennedy, E. (2014). The flipside of blended learning. *Journal of Lifelong Learning in Europe*, 4.

Kim, J. (2013). *On bubbles, online education, and confused reporting*. Retrieved from https://www.insidehighered. com/blogs/technology-and-learning/bubbles-online-education-and-confused-reporting

Laurillard, D. (1993). *Rethinking university teaching: A framework for the effective use of educational technology*. London: Routledge.

Miller, C., Smith, C., & Tilstone, C. (1998). Professional development by Distance Ediucation: Does distance lend enhancement? *Cambridge Journal of Education, 28*(2), 221–230. doi:10.1080/0305764980280207

MOOCs. (2002). *Online education for the masses!* Retrieved from http://www.distancelearningportal.com/articles/ 241/moocs-online-education-for-the-masses.htm

Moore, M. (1973). Toward a theory of independent learning and teaching. *The Journal of Higher Education, 44*(12), 661–679. doi:10.2307/1980599

Moore, M. (1973). Toward a theory of independent learning and teaching. *The Journal of Higher Education, 44*(9), 661–679. doi:10.2307/1980599

Moore, M. (1989). Three types of interaction. *American Journal of Distance Education, 3*(2), 1–6. doi:10.1080/08923648909526659

Moore, M. (1990). Distance education theory. *American Journal of Distance Education, 5*(3), 10–15.

Moore, M., & Kearsley, G. (1996). *Distance education. A system view*. Belmont, CA: Wadsworth Publishing Company.

Moore, M. G., & Kearsley, G. (2011). *Distance education: A systems view of online learning* (3rd ed.). Belmont, CA: Wadsworth.

Pappano, L. (2012). The Year of the MOOC. *The New York Times, 2*(12).

Peters, O. (1983). Distance teaching and industrial production: A comparative interpretation in outline. In D. Sewart, D. Keegan, & B. Holmberg (Eds.), *Distance education: International perspectives*. London: Croom Helm.

Ponti, M. (2014). Self-directed learning and guidance in non-formal open courses. *Learning, Media and Technology, 39*(2), 154–168. doi:10.1080/17439884.2013.799073

Porter, J. E. (2014). Framing questions about MOOCs and writing courses. *Invasion of the MOOCs,* 14. Retrieved from http://www.parlorpress.com/pdf/invasion_of_the_moocs.pdf

Ravenscroft, A. (2011). Dialogue and connectivism: A new approach to understanding and promoting dialogue-rich networked learning. *International Review of Research in Open and Distance Learning, 12*(3), 3. doi:10.19173/irrodl.v12i3.934

Rodriguez, C. O. (2012). MOOCs and the AI-Stanford like courses: Two successful and distinct course formats for massive open online courses. *European Journal of Open, Distance and E-Learning.*

Romiszowski, A. J. (2013). Topics for debate: What's really new about MOOCs? *Educational Technology, 53*(4), 48–51.

Saroyan, A., & dan Snell, L.D. (1997). Professional development by Distance Education: Does distance lend enhancement? *Cambridge Journal of Education, 28*(2).

Siemens, G. (2005). Connectivism: A learning theory for the digital age. *International Journal of Instructional Technology and Distance Learning, 2*(1), 3–10.

Siemens, G., & Downes, S. (2009). *Connectivism & Connective Knowledge*. Retrieved from http://ltc.umanitoba. ca/connectivism/

Simonson, M., Smaldino, S., Albright, M., & Zvacek, S. (2012). *Teaching and learning at a distance: Foundations of distance education* (5th ed.). Boston, MA: Pearson.

Vardi, M. Y. (2012). Will MOOCs destroy academia? *Communications of the ACM, 55*(11), 5. doi:10.1145/2366316.2366317

Vygotsky, L. (1978). *Mind and society*. Cambridge, MA: Harvard University Press.

Washer, P. (2001). Barriers to the use of web-based learning in nurse education. *Nurse Education Today, 21*(6), 455–460. doi:10.1054/nedt.2001.0583 PMID:11466008

Section 2

Pedagogical Strategies in Higher Education 4.0

Chapter 5
Blended Learning in Higher Education 4.0:
A Brief Review

Arumugam Raman
https://orcid.org/0000-0001-5351-8944
Universiti Utara Malaysia, Malaysia

Mohan Rathakrishnan
Universiti Utara Malaysia, Malaysia

ABSTRACT

Rapid technology advancement in the Fourth Industrial Revolution (FIR) has undoubtedly posed a great challenge to the education system in higher education institutes and changed every aspect of our lives including education. Higher education institutions across the world are amalgamating technology in modern learning and teaching approaches such as blended learning so that future teachers are equipped with the latest 21st century knowledge for further innovations and creations. This chapter covers definition, design, and implementation of blended learning in higher education. Further it also discusses blended learning models and assessment tools in recent education setting around the world. The education system worldwide has to transform for the imminent FIR as the job market will be largely prompted by the advancement of digital economy, robotics, artificial intelligence, and automation technology.

INTRODUCTION

The landscapes of education and training in Higher Education have been transformed in the past decade as a result of drives such as rapid developments in Information and Communication Technologies (ICTs). These developments generate great optimism among instructors and students due to changes in lifestyles, culture, globalization, changing patterns of work, expectation of stakeholders and the persistent demand of effective and efficient in output. Companies want to train their employees anywhere, anytime, while Universities aiming on featuring simplest teaching and learning via net from any part of the world.

DOI: 10.4018/978-1-5225-7832-1.ch005

Copyright © 2019, IGI Global. Copying or distributing in print or electronic forms without written permission of IGI Global is prohibited.

Gradually, the mixture of insight classroom learning and online learning is actually generating fresh resourceful and inventive ways to accentuate and make the learning enjoyable, thrilling as well as more effective. Considering technology progressed thus did Blended Learning (BL) instruction approaches and also software applications. BL offers a powerful keep of track history delivering conventional lecture rooms within the tech-friendly twenty first century.

REVIEW OF BLENDED LEARNING

There is no single definition of "blended". However, Graham, Woodfield and Harrison (2013) outlined Blended Learning (BL) as a mixture of Face-to- Face and on-line instructional models. A more detailed analysis of BL is provided by Sharpe, Benfield, Roberts, and Francis (2006) and identifies features such as synchronous or asynchronous learning activities and communications, the place where the learning takes place (campus, home, workplace), the usage of ICT tools (web 2.0 and 3.0), context, pedagogy, types of learners, learning style and etc. According to Cassidy, Colmenares, Jones, Manolovitz, Shen, and Vieira (2014) BL arising as one of the instructional models used in Higher Education Institutions (HEIs) due to tremendous development in net technology. In this sense, BL provides, for instance, benefits such as: 1) greater active communications between lecturers and students; 2) increased collaboration between students; 3) encourage student-centred assignment and information; 4) increased versatile teaching and learning; and 5) apposite usage of technologies in teaching and learning process (Garrison and Kanuka 2004; Garrison and Vaughan 2008; Chew 2009; Köse 2010; Graham, Woodfield & Harrison 2013).

Blended courses increased access and convenience. Student engages online courses no longer attend class to take the course. It provides more flexibility and freedom compared to face to face courses by transforming significant amount of face to face sessions online. US Department of Education report (2009) suggests blended courses are more effective compared to onsite classes. The report also distinguishes blended course with fully online courses and found that "instruction combining online and face-to-face elements had a larger advantage…than di purely online instruction". Stein and Graham (2014) listed why blended is more effective than face to face courses, include:

1. **Improve instructional design:** Blended courses intentionally designed compare to face to face course which are designed by instructional designers from the faculty.
2. **Increased guidance and triggers:** Within a blended study course, the course atmosphere delivers a transparent path via methods, things to do, and assessments with precise assistance each phase from the way.
3. **Simpler access to learning functions:** Placing materials and activities on-line lets many students to participate by having their personal time table which could guide to help much more comprehensive discovering.
4. **Individualized learning opportunities:** Given that online digital resources may be accessible according to students' personal preferences, and examined following requirement, the provision of electronic digital resources enables learners' self-direct particular learning pursuits to fulfill their knowledge gaps. Computerized evaluations frequently used in on-line training conditions might also offer instant, helpful responses in which redirects learners to review resources.

5. **Greater involvement via societal connections:** Learners during a face-to-face setting probably have constrained chances to interact with every fellow student, and face-to-face conditions alone might prevent some learners from engaging. The online settings that assist course interactions, cooperation and collaboration, and so on could raise the volume of student-to-student communication. This will likely, in turn, enhance their concentration together with the subject matter and provide inspirational benefits from the increased societal interaction.
6. **Time on task:** Blended and internet-based courses have tendency to accentuate learner's concentration. This can be due to elevated guidance and access, and enhanced instructional design.

In addition, Blended sessions could lessen expenditures to lecturers, learners, and universities. Instructor and learners can benefit from much less journey hours, transportation savings, and fewer parking costs. From an institutional standpoint, use of physical campus means can be diminished. Each time a blended program cuts its onsite time by at the least fifty percent this reduction can offer sizeable useful resource price savings to establishments challenged with maximizing physical classroom space. Using the illustration of a Tuesday/Thursday course referenced higher than (Figure 1), we can see this opens the Thursday classroom slot for an additional blended training course, primarily doubling the classroom's scheduling capability (Figure 2).

Drivers for Successful Implementation of BL in Higher Education

BL is now a notable approach associated with content material distribution within Higher Education (HI). Experts have discovered that enthusiasm, interaction, and study course structure tend to be three elements that will make difference towards the entire accomplishment involving BL programs along with learners' gratification along with BL programs. Aside from this, Graham, Woodfield and Harrison (2013) investigated six cases of institutional of BL to identify and elaborate on core issues related to institutional strategies, structure, and support, spanning the adoption stages. They argued many Higher Education Institutions differentiate between online and traditional courses, but do not clearly categorize what lies between as shown in Figure 3.

Figure 1. Two blended learning courses maximizing a single classroom
Source: Essentials for blended learning: a standards-based guide/ by Jared Stein and Charles Graham.

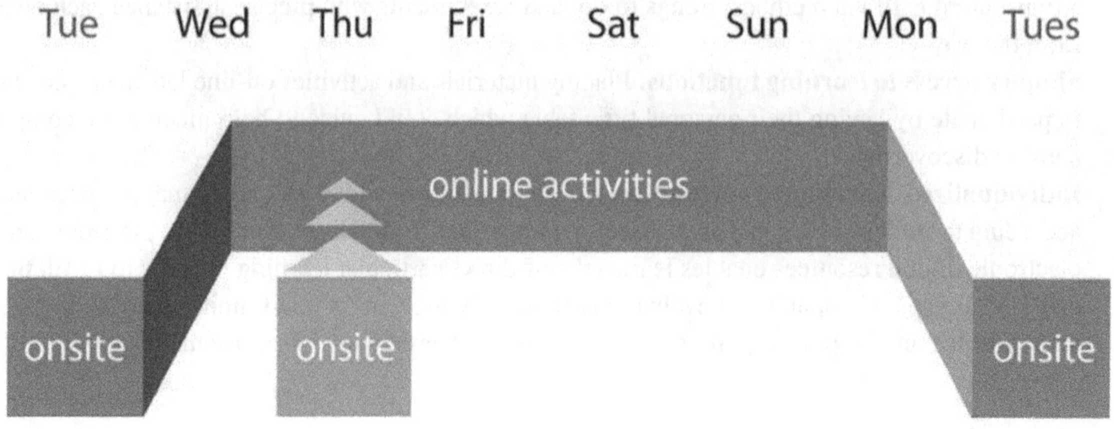

Figure 2. Moving learning experiences online
Source: Essentials for blended learning: a standards-based guide/ by Jared Stein and Charles Graham.

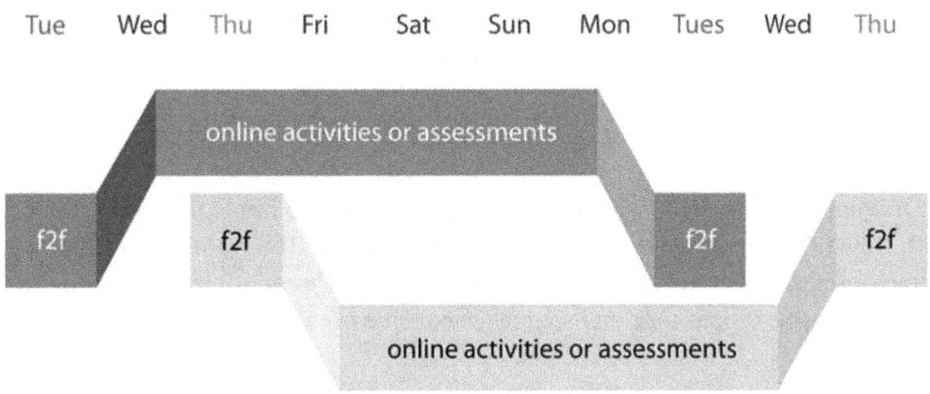

Figure 3. Spectrum of course-delivery modalities in higher education
Sourse: Graham, Woodfield and Harrison (2012)

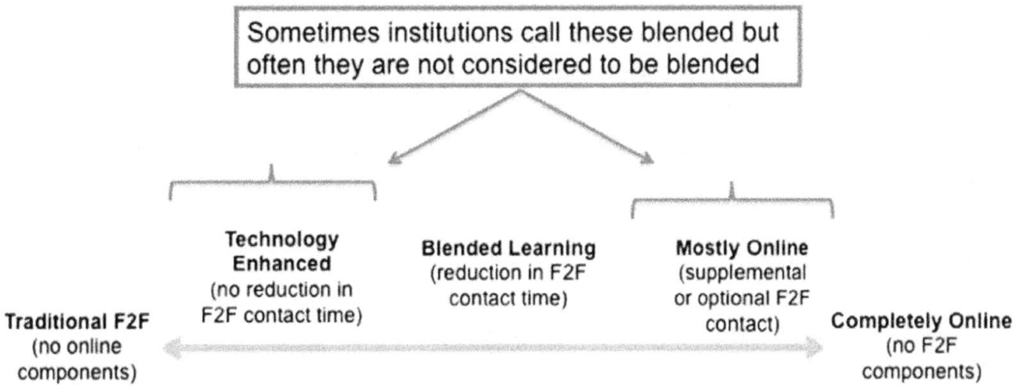

Even more, Graham, Woodfield and Harrison (2013) detailed policy and technological construction that affect adoption of BL among the Greater Institutions specifically technological innovation, possession, definitions and seat time, incentives and Analysis. Garrison and Kanuka (2004) noted that a transparent institutional course and plan are critical to correctly adopting a BL initiative. They also recommended that institutions could desire to convene a process force to strategically handle concerns, challenges, and chances that arise through implementation. Niemiec and Otte (2009) mentioned the need to obviously identify how BL can help the Higher Education Institution (HEI) fulfill its mission and goals, providing many examples of how an institution may well use BL to overcome obstacles to its main ambitions. The literature has also emphasized the importance of enough funding and time for you to achieve the desired results of the BL initiative (Piper, 2010; Rossett, Douglis, & Frazee, 2003).

Physical and Technological Infrastructure

Kanuka and Garrison (2004) noted Higher Institutions (HI) must provide physical and technological infrastructure to adopt BL. These infrastructure could include computer peripherals, network facilities (servers, routers, cables and etc.), and essential software. This is in line Powell's (2011) case study of New Zealand reported lack of technological infrastructure impede the online learning implementation. Approved in these terms, Harris, Connolly and Feeney stated that technology should be compliant to the function of teaching and the desired learning outcomes.

Skills Training

Insufficient training is discovered to be a significant threat for BL implementation in HI (beadle and Santy, 2008). New talents required to mastering new strategies in BL. Training for trainers encompassing ICT literacy, e-learning strategies, multi-tasking during synchronous virtual learning setting should be considered by faculty. Proper time and sources need to be used to provide training if skill-gap arise.

Pedagogy

According to Harris, Connolly and Feeney (2009) learning process, learning outcomes and the learning condition or environment must be considered in BL. Philosophical and Psychological factors must take into account when designing BL for Higher Institutions. Further the researchers suggested evidence-based instructional design should underpin the learning. The learning should be adult-oriented. BL course designers must recognize that the roles of instructor and college student in BL environment equally congruent with, and different from, their roles in a traditional face-to-face classroom setting. Garrison and Anderson (2003) suggested specific guidelines to be considered before and during e-learning experience are:

1. Establishing curriculum;
2. Identifying resources;
3. Defining clear expectations and goals (process and content);
4. Addressing technological concerns;
5. Structuring activities (collaborative and individual);
6. Setting time-frames;
7. Devising assessment processes and instruments; and
8. Selecting media.

Incentives

Providing Incentives for staff has been proven increase the chances of successful implementation (Shea, 2007; Martin, 2003; Garrison & Kanuka, 2004). According to Incentive Theory (Cherry, 2018), incentives can be used to get people to engage in certain behaviors. In the BL setting staff get motivated if there exists a possibility of material incentive (Shea, 2007). Approved in these terms Martin (2003) presented several tips for incentives, which includes financial compensation, release time, and tools. Put bluntly, broader incentives such as funding allocations will help BL implementation in Higher Institutions (Garrison & Kanuka, 2004; Watson et at., 2010).

Evaluation

Pombo and Moreira (2012), concludes four elements to evaluate BL courses namely a) purpose of evaluation -to enhance students' engagement resources, or overall program quality; b) who must be evaluated -Instructors, learners, course leaders; c) How and when evaluation should be conducted - Approaches of data collection during the BL or at the end; 4) What ought to evaluated - resources, teaching and learning, assessment. In this sense, Bowyer and Chambers (2017) analyzed few BL and e-learning evaluation framework and introduced new conceptual framework for evaluating BL.

The outer sphere is **situation** encompassing two variables namely **context** (socio-economic, ethical, legal, cost, accessibility, cultural, geographical and support) **and institution** (support and administration). The mid-sphere is **course organization** that includes four variables such as **Design and Planning** (Curriculum management, Organization BL teaching Flexibility and support), **Content** (relevance and scope, quality, breadth of content, breadth of methods of presentation and activates, Validity, Accuracy and balance, Interactivity, Accessibility, Organization, Currency (up-to-dateness) and support), **Assessment** (Diversity, Fit/relevance, support), **Technology** (Interface design, Ease of use, Security, Reliability, Usability, Maintenance, Accessibility, Organization, Availability, Personalization, Interactivity, Currency (up-to-dateness), Support). The inner sphere is about individual perspectives which focusing on teacher (Attitude towards computers and technology, Attitude towards learners, Technological experience, Teach-

Figure 4. Conceptual framework for evaluating blended learning
Sources: Jessica Bowyer Exeter University and Lucy Chambers Research Division (2017) Research Matters: A Cambridge Assessment publication. http://www.cambridgeassessment.org.uk/research-matters/ © UCLES 2017

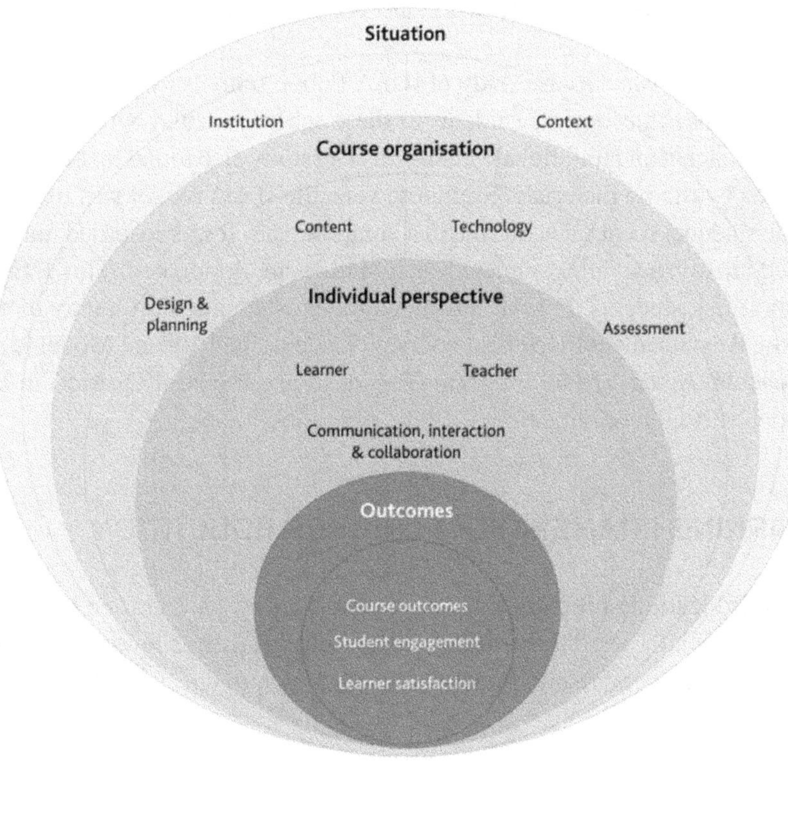

ing experience, Subject knowledge, Response time, Feedback, Support and Provision of information) and learner (Attitude towards computers and technology, Attitude towards learning, Attitude towards teaching staff, Motivation to take the course, study habits, technological experience, prior knowledge and learning experiences, Convenience, autonomy, perceived usefulness, perceived enjoyment, peer interaction, group working and collaboration) elements. There is also the core of the sphere names **outcomes** that contains three variables namely **learner satisfaction** (with course (overall), with learning, with teaching, and utility of course for future plans), **student engagement** (psychological and cognitive engagement, Behavioral and Emotional engagement) and **course outcomes** (Grades and marks, online activity, attendance, dropout rates). These spheres and elements are important for a successful blended learning course. The researchers expected this framework can be used beyond BL.

BLENDED LEARNING MODELS

Researchers can find four standard types of blended learning that have verified to fulfill leaner needs and provide overall flexibility with instructional configurations. The four blended learning (Figure 5) approaches consist of: a) The rotation design, where HI students rotate between face-to- face and online learning; b) Self- Blend model, in which students' experience most learning online, while instructor provide a personalized contact; c) The self-blend product, in which learners choose their classes from a menu to supplement their regular, non-technical school choices; d) The enriched-virtual model, a complete school knowledge by which learners divide their time among attending a brick-and mortar campus and understanding remotely, employing on-line supply of content and instruction. Fortunately, it does not require a huge expenditure to consider actions inside the route of BL. The following resources are free of charge, relatively effortless to work with, and very advantageous when it comes to escalating choices for learner learning.

DreamBox Learning published a case study of IDEA Public schools in Texas claiming that the rotational BL increase student achievement. Students of these schools rotated between intelligent software to learn mathematics concept and traditional classroom. The study also found that students become more active and challenged by the by materials. Similarly, versatile (flex) model was used by AdvancePath Academy where the students spent most of time in a computer lab. It was recorded ninety percent (90%) of students enrolled at AdvancePath completed their studies and graduated. This reflects this method quite promising for at-risk students. For the self-blend model students must highly motivated and suitable for those eyeing Advancement Placement courses. The enriched Virtual Model is an alternative to full-time online schools. Aiken Virtual Program (Rapides Parish School District) in Louisinia is one example of an Enrich Virtual program.

BLENDED LEARNING FRAMEWORK IN HIGHER EDUCATION

There are many blended learning framework have emerged for past 10 years and help us to enhance our knowledge in a lots of crucial ways. The subsequent overview discusses a few of most influential frameworks, which discussion delivers assorted lenses through which to check out the differences amongst these frameworks.

Figure 5. Four types of blended learning
Source: Christensen Institute

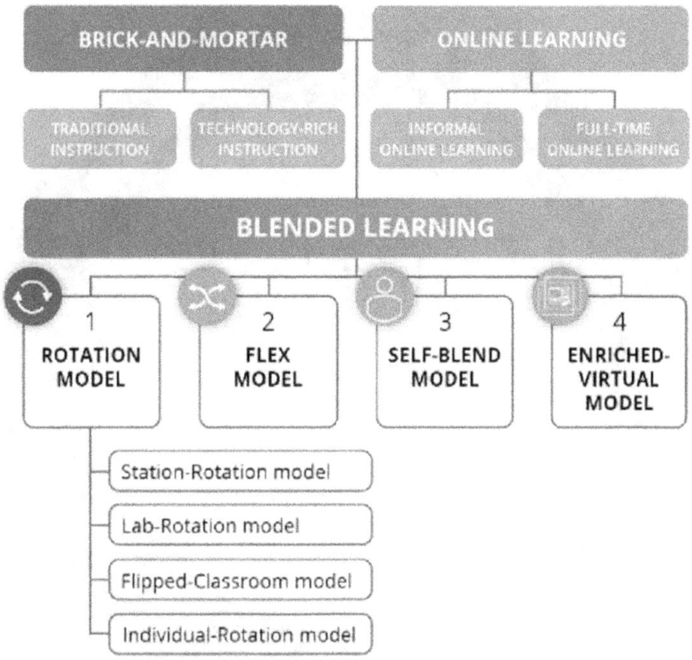

Khan's (2005) octagonal framework introduced eight dimensions (Figure 6) in relation to the design, delivery, implementation and evaluation of e-learning: pedagogical, technological, interface design, evaluation, management, resource support, ethical and institutional. In his framework, Pedagogy relates to analysis of audience, content analysis, media analysis and suitable strategies methods of delivery to help learners to accomplish the training objectives. Technology focuses technological infrastructure planning, hardware and software that enable e-learning delivery. Interface design includes elements such as page and site design, content design, navigation, and usability testing. Evaluation is assessment of learners and evaluation of the teaching and learning process.

Management is about maintenance of the learning environment involving instructors, students, course designers and others. Resource support concerns online and offline support and resources required to foster meaningful learning settings. Ethics includes issues such as social and political influence, cultural diversity, bias, and legal issues. Institution encompasses academic affairs, policy, administrative affairs, student services, organizational change, staff support related to e-learning.

The e-Learning Planning Framework (eLPF) is a tool to aid educational institutions and instructors to evaluate their e-learning capacity. It could possibly help common self-review and subsequent advancement of e-learning skills and knowledge. It provides a road map and practical steps faculty can take to improve their practices. There are five faces in this framework namely Pre-emerging, Emerging, Engaging, extending and Empowering. The first phase is about the lack of awareness about e-learning in teaching and learning, inaction to explore e-learning, ad hoc technology usage and no strategic plan to adopt technology. The second phase concerns about the faculty may possibly be concentrating on investigating, planning to include technology within faculty's vision and mission. The third phase is regarding developing and connecting planning throughout the school as well as trialing ways to utilize

Figure 6. E-Learning framework (Khan, n.d.)
Source: Adapted from Khan (n.d.) under the Fair use policy

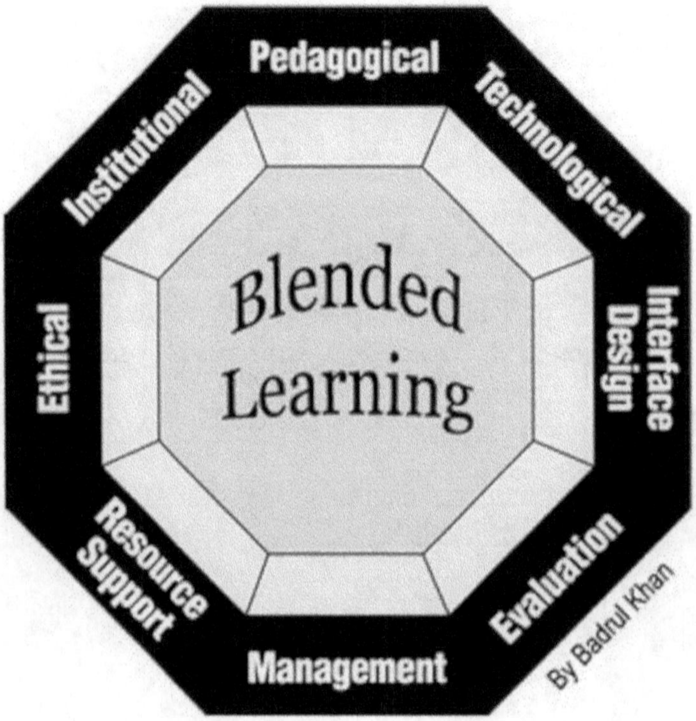

technology properly to satisfy employees, neighborhood, and students' demand. The technology starts to boosts the learning experience. In the fourth phase the university may have successfully aligned processes and procedures throughout the college and community. The use of technologies is acceptable and enables significant adaptation of understanding encounters to meet all learners' needs. In the classroom, instructors and students may work to gather to use technologies as part of reliable source of information. The last phase (fifth) the faculty expected regularly plan, review and evaluate together. In this phase technology considered ubiquitous and embedded in real world.

McSporran and King's (2005) simple framework pertaining to blended learning, recommends the selection of delivery strategies in line with learning requirements and accessible resources. Again, this model accommodates only to one element of blended learning, subject material delivery, which is useful in directing the delivery of blended learning at a program level rather than leading the execution at an institutional level.

In line with this, the Community of Inquiry (CoI) framework (Figure 7) provides three elements namely cognitive presence, social presence, and teaching presence.

Each of the presences reflects categories and indicators widely operationalized in examining interaction in both synchronous and asynchronous modes as a part of online learning. Social presence concerns open communication, group cohesion and affective that enabling risk-free expression, encouraging collaboration and expressing emotions. The cognitive presence relates triggering event, exploration, integration and resolution indicating having sense of puzzlement, exchanging information, connecting ideas and applying new ideas. Finally, teaching presence categorized as design and organization, facilitation of discourse and direct instruction (Garrison & Vaughan, 2008).

Figure 7. Community of inquiry framework

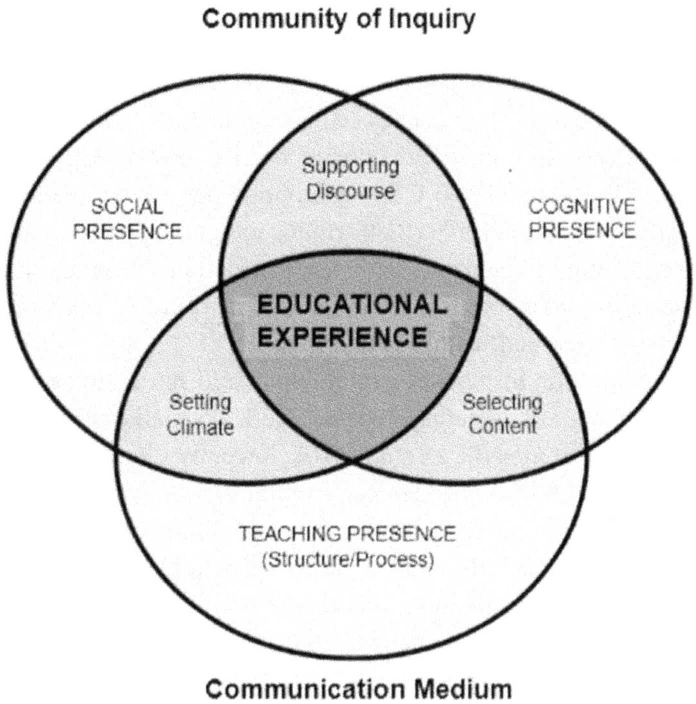

As demonstrated from the over discussion, each and every of these types has its personal issues and focuses and examines blended mastering from various perspectives. Despite the fact that they have got all contributed to our understanding of blended learning in one way or a different, none continues to be able to offer a complete capture of blended learning as none has explored blended finding out utilizing a posh adaptive programs approach. As a result, blended learning still is apparently an enormous puzzle, consisting of intertwined disjointed components, all attempting to link.

It has been challenging to begin to see the entire picture of blended finding out since each factor, in isolation, only provides component of its landscape without having interconnection. The above discussion recognizes this hole and makes an attempt to discover what constitutes blended learning and how its constituting things function individually and alongside one another, and attempts to examine the impetus that drives blended studying forward.

BLENDED LEARNING TOOLS

This section explores some of the popular applications and strategies to developing a BL including course management systems, learning management systems, blogging, micro-blogging, wikis, virtual reality, augmented reality, digital storytelling, and flipped classrooms.

Course Management Systems (CMS) and Learning Management Systems (LMS)

Glassman (2016) distinguishes Course Management Systems (CMS) and Learning Management Systems (LMS). Course Management Systems are prepackaged collection of education-based applications to reach distributed sets of students and assist instructors conduct tasks such as testing, grading and response. Most well-known CMS is Blackboard/Web CT and limited open source platform Moodle. Learning Management Systems (LMS) are designed as platform for integrating more instruction applications to create holistic educational setting for students tied with instructional theories. It is more to intelligent tutoring system which incorporates theoretical and practical underpinning that makes the course efficient or technology user friendly for the staff and learners.

Almost every Higher Education Institutions around the world started to use Learning Management System (LMS) in their teaching and learning process. The LMS market is expected to be worth over $15.72 billion in 2012 (marketsandmarkets.com, 2016). Recently cloud-based LMS controlled huge proportion in supporting Distance, Online and e-learning industries. For instance, Docebo is backed by respected firms for example Thomson Reuters, Bloomberg and Sharp. Docebo is passionate promoter of the significance regarding gamification throughout student involvement. Badges, accolades and ratings usually are used in order to boost participation in addition to study course achievement. This LMS might be employed collectively by using various online video webinar applications namely Adobe Connect, Big Blue Button, Blue Jeans and SkyMeeting. Docebo's API permits end users to help conveniently incorporate third-party software directly into the LMS without any further trouble. Its user-friendly interface permits program designers to easily control written content for certification testing and retraining. The Docebo LMS Online Instruction Program could be utilized for marketing courses on line. But, Higher Education Institutions simply cannot adopt this commercial based LMS due different philosophy, culture, faculty, and different type of learners. LMS main features are:

- **Administration:** Including monitoring and reporting, material administration, notifications, single sign-on, API, plus more.
- **Monitoring and Reporting:** Using these tools, guide scholar progress by means of conditions and studying paths, export consumer data and create real-time reviews, and produce timed programs and inactivity timeouts.
- **Evaluation & Testing Applications:** Some kind of on-line test software along with numerous assessment instruments. Several attributes that could be reward are usually query combining, question randomization, expected responses, and automated self-grading.
- **Compatibility and Supported Equipment:** User interface that adjusts to various components equipment, like desktops, laptops, and tablets. Specific platforms which may be supported are PC, Mac, Linux, iOS (iPads/iPhones), Andriod, Blackberry, and more.
- **Course Interactivity:** Interactive features that may critical are High Definition streaming online video, audio, visulas, SCORM presentations, assignments, and assessments.
- **Analytics & Reporting:** Participant records will all be gathered in a type of electronic diary, from where you can access them any time, and pull off important stats to examine the health of your teaching/learning practices.

Blogging in Classroom

Oravec (2003) emphasized weblogs are a versatile medium that may be utilized in strategies that give educational contributors that has a 'middle space' of alternatives concerning ways to integrate face-to-face and on the net modes. Weblog building encourages the development of particular person, significant voices inside the broader context of classroom interactions. In a review articles reporting on attempts to integrate blogging activity in curricula Kim (2008) found blogging to be capable of increase student self-motivation though shared commentary. Blogs were also found to increase student autonomy. Deng and Yuen (2011) claimed blogs offer high levels of connectivity but more individual expression than constructing a shared knowledge base. Ellison and Wu (2008) found student seem to enjoy reading other students' post. Further they concluded students did not get that much out of direct comments to their own posts and that comments were often repetitive. Studies suggest the importance of interest in content for successful blogging. Teachers can use extrinsic motivation such as grades and badges to establish regular blogging however this should not hinder students' autonomy that leads to a successful online learning (Rovai and Jordan, 2004).

Micro-Blogging

Micro-blogging including Twitter streams individual posting about their happenings in lives at the time. Twitter started using Short messaging services (SMS) into shared system so every person in the network could see others thoughts. Dunlop and Lowenthal (2009) argue how Twitter might be used to increase social presence. Students should be encouraged to tweet immediate experiences to eliminate boundaries. Ebner et al (2010) tracked thousands of students' posts and found a continuous flow of information between students. Consequently, the ideas could be remixed, reframed then learning become less of isolating experience. According to Junco et al (2013) micro-blogging can be used to draw students into course. Students' engagement increased after Twitter posts were required. Micro-blogging is much less likely to have social and emotional triggers of effective blog posts.

Virtual Reality and Augmented Reality

The major difference between virtual reality (VR) and augmented reality (AR) is that virtual reality is the use of computer technology to create a simulated environment such as animated scene, artificial or actual place photographed and included in VR application (Google earth) whereas, AR defined as an enhanced version of reality by use of technology to add digital information in smartphones and tablets. Researchers, medical and engineering students are making great use of immersive technology. There are two types of AR namely a) devices can reflect a three-dimensional image; and b) Video mixing technologies can be used to merge with objects (Glassmann, 2016).

SUMMARY

Initially BL adopters contented that now we have numerous options presented simply by the technologies higher instruction (Garrison, Anderson, & Archer, 1999). Generally there are a number of good reasons powering the particular drive to integrate ICT within the actual educational framework. First, demand to employ ICT at Higher Institutions happens out of developments throughout university student demography. According to Concannon, Flynn and Campbell (2005) the tremendous increase in "full time and part time students is a phenomenon of recent years, where school leavers take part-time jobs whilst attending university" (p.502). Regarding part time students' versatile design and style caters to their own hectic daily activities. Without this mobility, the learners may hardly become ready to engage in their degrees. BL settings accommodate college students who seem to favor insight classroom interaction within Blended studying environments suit students who prefer face-to-face interaction in addition to students who prefer online learning.

REFERENCES

Beadle, M., & Santy, J. (2008). The early benefits of a problem-based approach to teaching social inclusion using an online virtual town. *Nurse Education in Practice*, *8*(3), 190–196. doi:10.1016/j.nepr.2007.07.004 PMID:17855168

Bowyer, J., & Chambers, L. (2017). Evaluating blended learning: Bringing the elements together. *Research Matters, 23*, 17-26.

Cassidy, E. D., Colmenares, A., Jones, G., Manolovitz, T., Shen, L., & Vieira, S. (2014). Higher education and emerging technologies: Shifting trends in student usage. *Journal of Academic Librarianship*, *40*(2), 124–133. doi:10.1016/j.acalib.2014.02.003

Chew, E. (2009). *A blended learning model in higher education: A comparative study of blended learning in UK and Malaysia* (Doctoral dissertation). University of Glamorgan.

Deng, L., & Yuen, A. H. (2011). Towards a framework for educational affordances of blogs. *Computers & Education*, *56*(2), 441–451. doi:10.1016/j.compedu.2010.09.005

Dunlop, J. C., & Lowenthal, P. R. (2009). Horton Hears a Tweet. *EDUCAUSE Quarterly Magazine*, *32*(4). Retrieved 6th July 2018, from http://goo.gl/P9Wig

Ebner, M., Lienhardt, C., Rohs, M., & Meyer, I. (2010). Microblogs in Higher Education–A chance to facilitate informal and process-oriented learning? *Computers & Education*, *55*(1), 92–100. doi:10.1016/j.compedu.2009.12.006

Ellison, N., & Wu, Y. (2008). Blogging in the classroom: A preliminary exploration of student attitudes and impact on comprehension. *Journal of Educational Multimedia and Hypermedia*, *17*(1), 99–122.

Garrison, D. R., Anderson, T., & Archer, W. (2003). A theory of critical inquiry in online distance education. Handbook of distance education, 1, 113-127.

Garrison, D. R., & Kanuka, H. (2004). Blended learning: Uncovering its transformative potential in higher education. *The Internet and Higher Education*, *7*(2), 95–105. doi:10.1016/j.iheduc.2004.02.001

Garrison, D. R., & Vaughan, N. D. (2008). *Blended learning in higher education: Framework, principles, and guidelines*. John Wiley & Sons.

Glassman, M. (2016). *Educational psychology and the internet*. Cambridge University Press. doi:10.1017/CBO9781316155134

Graham, C. R., Woodfeld, W., & Harrison, J. B. (2013). A framework for institutional adoption and implementation of blended learning in higher education. *The Internet and Higher Education, 18*, 4–14. doi:10.1016/j.iheduc.2012.09.003

Harris, P., Connolly, J., & Feeney, L. (2009). Blended learning: Overview and recommendations for successful implementation. *Industrial and Commercial Training, 41*(3), 155–163. doi:10.1108/00197850910950961

Junco, R. (2014). *Engaging students through social media: Evidence-based practices for use in student affairs*. John Wiley & Sons.

Khan, B. H. (Ed.). (2005). *Managing e-learning: Design, delivery, implementation, and evaluation*. IGI Global. doi:10.4018/978-1-59140-634-1

Köse, U. (2010). A blended learning model supported with Web 2.0 technologies. *Procedia: Social and Behavioral Sciences, 2*(2), 2794–2802. doi:10.1016/j.sbspro.2010.03.417

Lothridge, K., Fox, J., & Fynan, E. (2013). Blended learning: Efficient, timely and cost effective. *The Australian Journal of Forensic Sciences, 45*(4), 407–416. doi:10.1080/00450618.2013.767375

Martin, M. H. (2003). *Factors influencing faculty adoption of Web-based courses in teacher education programs within the State University of New York* (Doctoral Dissertation).

McSporran, M., & King, C. (2005, June). Blended is better: Choosing educational delivery methods. In *EdMedia: World Conference on Educational Media and Technology* (pp. 4932-4939). Association for the Advancement of Computing in Education (AACE).

Niemiec, M., & Otte, G. (2009). An Administrator's Guide to the Whys and Hows of Blended Learning. *Journal of Asynchronous Learning Networks, 13*(1), 19–30.

Oravec, J. A. (2003). Blending by blogging: Weblogs in blended learning initiatives. *Journal of Educational Media, 28*(2-3), 225–233. doi:10.1080/1358165032000165671

Piper, T. (2010). *What policy changes do experts recommend K-12 instructional leaders enact to support the implementation of online instruction and learning?* (Doctoral Dissertation). Available from ProQuest Dissertations and Theses database. (UMI No. 3430711)

Pombo, L., & Moreira, A. (2012). Evaluation framework for blended learning courses: A puzzle piece for the evaluation process. *Contemporary Educational Technology, 3*(3), 201–211.

Powell, A. (2011). *A case study of E-Learning initiatives in New Zealand's secondary schools*. Pepperdine University.

Rossett, A., Douglis, F., & Frazee, R. V. (2003). Strategies for building blended learning. *Learning Circuits, 4*(7), 1-8.

Rovai, A. P., & Jordan, H. (2004). Blended learning and sense of community: A comparative analysis with traditional and fully online graduate courses. *The International Review of Research in Open and Distributed Learning*, *5*(2). doi:10.19173/irrodl.v5i2.192

Sharpe, R., Benfield, G., Roberts, G., & Francis, R. (2006). *The undergraduate experience of blended e-learning: A review of UK literature and practice*. The Higher Education Academy.

Shea, P. (2007). Bridges and barriers to teaching online college courses: A study of experienced online faculty in thirty-six colleges. *Journal of Asynchronous Learning Networks*, *11*(2), 73–128.

Stein, J., & Graham, C. R. (2014). *Essentials for blended learning: A standards-based guide*. Routledge. doi:10.4324/9780203075258

Watson, J., Murin, A., Vashaw, L., Gemin, B., & Rapp, C. (2010). *Keeping pace with K-12 online learning: An annual review of policy and practice*. Academic Press.

Chapter 6
Pedagogical Orientation in the Fourth Industrial Revolution:
Flipped Classroom Model

Byabazaire Yusuf
https://orcid.org/0000-0002-8043-1112
Universiti Utara Malaysia, Malaysia

Abdul Hamid Busthami Nur
Universiti Utara Malaysia, Malaysia

ABSTRACT

Redesigning higher education is one of the most widely discussed topics among educators, parents, and other community stakeholders. This is due to the current developments in digital technologies and onset of the fourth industrial revolution which is set to alter the way people live and work. This chapter suggests the implementation of the flipped classroom model as an appropriate approach for equipping students with creative abilities, problem solving skills, thinking skills, and lifelong learning skills. Eight in-service teachers pursuing a postgraduate course on a part-time basis participated in a qualitative study. The study was conducted using asynchronous virtual focus group dialogue sessions. The study addressed three main questions: 1) why they adopted a flipped classroom model, 2) how they implemented it, and 3) what fourth industrial revolution skills were developed using this model. On the whole, this approach created an active and collaborative environment which enabled students to demonstrate their creativity and problem solving skills needed for future careers.

INTRODUCTION

Redesigning higher education has been identified as one of the important elements for preparing teachers for their career success in the fourth industrial revolution. Therefore, higher education and teacher education in particular have entered a critical phase of transformation (Marks, 2015). Rani (2018) emphasized that the process of teaching and learning needed to be transformed to suit current student needs and their

DOI: 10.4018/978-1-5225-7832-1.ch006

Copyright © 2019, IGI Global. Copying or distributing in print or electronic forms without written permission of IGI Global is prohibited.

learning environment. Teaching strategies are one main concern and need to be examined and probably transformed. Despite the current demands for the 21st-century skills and approaching fourth industrial revolution, teaching approaches have not changed in light of the new educational challenges. Although the learning and teaching landscape is changing, the lecture method remains a common way of delivering instruction to learners (Cashin, 1995) and a preferred approach in many countries around the world (Saavedra & Opfer, 2012). Yet, many resources widely available online have become a main source of reference for students. Besides, borderless communication has made peer interaction and peer-to-peer learning possible. As for lecturers, their new role as a facilitator in student learning process will require a new teaching approach different from that of the main provider of knowledge. Included among the important elements to be prioritised in the teaching process in the fourth industrial revolution are learning spaces, pedagogies, the curriculum, and teaching technologies (Rani, 2018).

Therefore, teacher educators not only face the challenge of preparing teachers for new ways of teaching but also must embrace and model best practices for the new ideas (Marks, 2015). However, educators have always looked for alternative ways of engaging their students in the classroom (Bacsich et al. 2013). For some teacher educators, the flipped classroom model is considered a better approach because the focus of higher education is in a transition from teaching to learning (Marks, 2015). This means that students are indirectly being prepared for the fourth industrial revolution for which they will need to engage in learning to acquire skills for jobs that do not existed and that may continue for the rest of their lives (Study Malaysia, 2018). Thus the flipped classroom strategy not only prepares students for lifelong learning (Bland, 2006), but also responds to the teaching and learning challenges of the present time.

Teacher educators have considered the extensive research dedicated to understanding how people really learn and the best possible approaches for teaching them to think creatively (Ambrose et al., 2010). This is very important to understand because students from different cultures and from different backgrounds as well as age groups learn in different ways (Choy et al., 2015).

As we have entered the fourth industrial revolution (4IR), teacher educators will need to impart knowledge in ways that will nurture the current skill sets that are in demand such as problem solving, critical thinking and creativity. That only happens when teacher educators realize that they must train students to possess adaptive and flexible minds. This would equip students with the cognitive agility required for the fast-paced knowledge and digital challenges of future learning environments (Gleason et al., 2011). As such, a need exists for pedagogical transformation from the lecture format with a professor transferring knowledge to large number of students to a different approach. This is because this approach to student learning is outdated and does not deliver the skills required in the era of the fourth industrial revolution (4IR). As such, teacher educators need to employ some new pedagogical approaches.

This chapter focuses on the aspects of redesigning pedagogies and their likely effects on student acquisition of skills, especially higher-order thinking skills (HOTS), which are relevant for the fourth industrial revolution. The skills required include: discussion, analysis, evaluation, creativity and authentic application of course learning objectives (Marks, 2015). Some of the current pedagogies for honing student skills include: heutagogy (self-determined learning), paragogy (peer-oriented learning), and cybergogy (virtual-based learning). There are many educators who have valued traditional pedagogical approaches in imparting knowledge to their learners. However, time has come to make some tough decisions based on new developments in the teaching profession. As Marks (2015) stated, "Changing times mean that teacher educators need to examine teaching from new perspectives and make pedagogical choices based on pre-service teachers' needs and those needs of the profession" (p. 241).

One of the pedagogical ways through which university lecturers and teacher educators can provide personalized learning environments and student engagement to maximize learning is through the flipped classroom model. Current practices of lecturers in higher education institutions based on the flipped learning model are limited (Jensen, Kummer & Godoy 2015); however, one of its common forms is a blended learning approach known for raising the learners' responsibility in line with their individual learning styles and performances (Staker & Horn, 2012).

THE FLIPPED CLASSROOM MODEL

It may be considered relatively new, but the flipped classroom model takes many forms and is currently popular among teachers at different levels of education (Hamon, 2014). Active learning is at the heart of the flipped class teaching approach (Auster & Wylie, 2006). When employed properly, this approach creates a learning environment in which instructors facilitate students for peer collaborations in the learning process to attain higher-order thinking skills. Several studies have proven that when learning becomes interactive, working with peers on in-class activities promotes confidence and improves overall learning (Gehringer & Peddycord, 2013). The concept basically refers to the use of teacher-created video clips or any interactive content so that students get exposed to new learning materials outside of class, which was a reversal of traditional teaching strategy. Hence, class time is reserved for activities, tasks and discussions aimed at solving problems through which knowledge is assimilated (Brame 2013) by engaging in an active learning process.

The Flipped Classroom Strategy Maximizes Bloom's Taxonomy

See Figure 1.

MOBILE LEARNING

An important element that is very crucial for the success of the flipped classroom strategy is mobile learning, which could be used to share instructional content as part of the preparatory work. This gives learners an opportunity to study the material before turning up for in-class activities. The fourth industrial revolution is characterized by the pace of technological development that requires the fastest and cheapest means of generating and sharing knowledge. Hence, the use of mobile phone in teaching and learning has become a viable instructional strategy in education in this era. The increased ownership and usage of mobile devices among the younger generation has made this technology potentially potent for improving teaching-learning process particularly in Malaysia. It was reported that the penetration rate of mobile devices in Malaysia for the year 2010 was 106%, suggesting multiple subscriptions among users (MCMC). More than 30% of mobile subscribers in Malaysia were from the youth mobile subscribers category (Ng, Voges, & Goi, 2010). Because of this penetration, it has become imperative for educators to understand the concept of using mobile-learning in teaching and learning delivery approach.

Figure 1. Model of Bloom's taxonomy in the flipped classroom. Source: Bloom (1956)

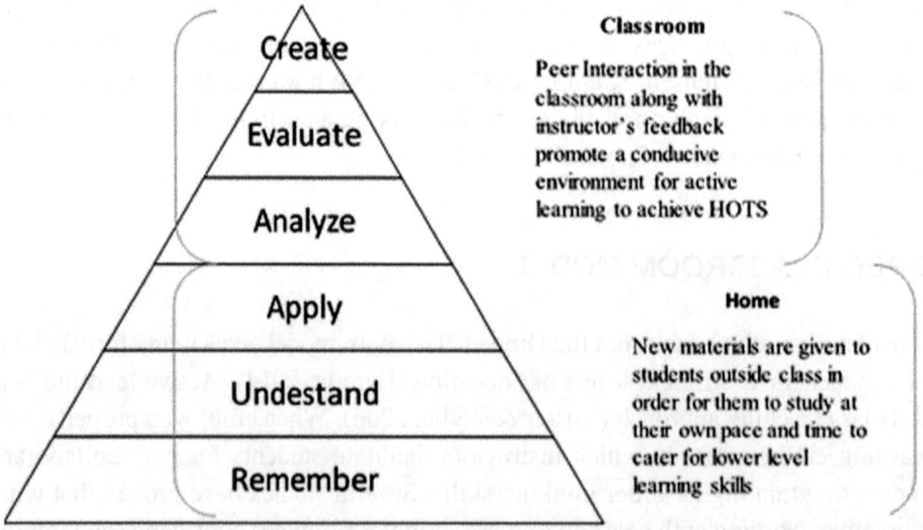

Many studies regarding mobile learning, which is also called m-learning, have examined various approaches and ways in which the idea of learning mobility should be implemented. The concept of m-learning has been around for some time now with several studies reporting their findings on respondents' readiness for mobile learning in terms of their ownership and user knowledge of mobile devices (Sharples, 2002). M-learning is a distinctive and useful way of learning as higher education students are always on the move nowadays. They could be working on part-time basis, doing internships, or even involved in various games or sports tournaments. They can simply access learning material, get instructions and other course content related applications anytime and anywhere. Students want to access resources that are open where they can learn and study anytime and anywhere and without any barriers to accessibility or interactivity. They would definately prefer to access information which is free, copyable and remix-able (Johnson et al., 2013).

Issham, Rozhan and Thenmolli (2010) suggested that students enjoyed using mobile phones as communication tools and could check such information as class schedules, class delays or postponements and other current information or university updates while on the go. This reflected that, to keep up with the demands of a student's daily life, mobile phones are a very important tool not only for communication but for training and learning. Therefore m-learning can be a very useful and complementary strategy for the flipped classroom approach. The instructional content in the form of teacher-created or curated videos, textbook materials, assigned articles or online quizzes would be shared with the students together with assigned tasks prior to attending the next class. This would give learners ample preparation time for discussions and activities to be undertaken during in-class activities. As the flipped classroom strategy encourages a student-centred learning approach, students could even study the contents collaboratively in more detail by accessing resources beyond the instructional content shared by the teacher. They also could post their questions to the teacher regarding concepts that they might find confusing or difficult to understand. In turn, this would help a teacher to identify and prepare the appropriate responses for areas of concept difficulty or misconceptions during in-class activities sessions.

The Flipped Learning Strategy

A flipped classroom model that is based on a strong foundation is among the most widespread blended learning models in recent history (Bergmann & Sams, 2012; Khan, 2011). It comprises two main phases. First, students engage in individual preparatory work by undertaking low-level cognitive activities outside of the designated school environment. Second, students attend in-class activities where concepts are analyzed collaboratively in an active learning format to promote higher-order thinking skills (HOTS). This book chapter deliberates a pedagogical design based on a flipped learning strategy, how it benefits students in the fourth industrial revolution and how it will assist teachers in advancing higher-order thinking skills (HOTS) such as analysis, evaluation, creativity and problem solving.

A flipped learning model is known to addresses challenges such as absenteeism (Bergmann & Sams 2012) poor motivation, the inability to complete homework, a lack of concentration during class-time, poor teacher knowledge about the background knowledge of students on topics of study and a lack of communication skills among students. The approaches employed by a flipped learning model to deal with such matters were demonstrated in qualitative research based on a focus group discussions strategy. The participants in the study were in-service teachers pursuing a post-graduate program on a part-time basis. Their participation helped the researcher to understand the flipped learning strategy better. They were able to relate their personal experiences to that of their own students as the researcher had employed a flipped classroom strategy with them in a previous course. Among other skills, the flipped learning model can help to prepare learners for the fourth industrial revolution environment by acquiring skills through critical direct instruction thus encouraging critical thinking skills through reasoning, reflection and sound judgement. Such skills are necessary when it comes to problem solving in our private or professional lives. The flipped learning strategy also has the potential to promote knowledge as well as skill retention and knowledge transfer through discussion and participation in collaborative and active learning activities. Finally, Anderson and Krathwohl (2001) asserted that problem solving is one of the main propositions for "meaningful learning". Therefore, problem solving is a broad goal in an education process and skills related to it could be developed in a flipped learning environment (Nitko & Brookhart 2007). This way a teacher who plans and pursues a flipped learning approach for teaching can equip students with appropriate 21st-century skills, which are relevant for the fourth industrial revolution. (Keane 2014).

This chapter is based on a pilot study in which postgraduate in-service teachers implemented a flipped classroom model for five months. Among other reasons for implementing this model were: a desire for teachers to maximize the learning potential of students, creating an active learning environment, and promoting the development of higher-order thinking skills (HOTS) among the students. The book chapter highlights some crucial pedagogical elements that every teacher should know about a flipped learning model. It also shares the experiences of in-service teachers in flipping their classrooms to enhance student engagement for a better teaching and learning process suitable for equipping students with fourth industrial revolution skills.

How the Model Works

On the whole, no particular model exists for the flipped classroom, although it has its roots in blended learning models (Bergmann & Sams, 2012). Hence, the basis of its operation is the blended learning pedagogy comprising online and face-to-face interactions. This arrangement enhances group and social interaction opportunities while promoting individual efficiency. The term is generally employed to describe almost any class structure that provides preparatory instructional content to be studied outside the classroom environment followed by in-class activities. For as Mazur (1997) points out, the learning processes should start out of school and proceed in-depth while at school with peer interactions. Sharing pre-lecture resources with students helps to reduce in-class cognitive load as students study the new materials at their own pace and at a time suitable to them (Clark, Nguyen, & Sweller, 2005). In one familiar model, students might view several lectures of five-to-seven minutes each aimed at raising their curiosity as well as questions on difficult or confusing aspects. Meanwhile, online quizzes or activities can be included to test what students have learned to enable the lecturer to identify areas of emphasis during discussions for in-class activities. The provision of immediate quiz feedback and the ability to replay lecture segments may help to clarify unclear content while motivating students to engage with learning content. Teachers might lead in-class discussions or turn the classroom into a studio in which students may individually or collaboratively engage in active learning processes using the content to analyse authentic issues relating to real life (King, 1993). In this approach, teachers act mainly as facilitators or on-site experts, suggesting various approaches in a discussion session, clarifying unclear content, and evaluating student performance. Whenever necessary, teachers might organize students into an impromptu workgroup to solve a difficult problem that several others may have struggled to understand. This approach introduces a comprehensive change in the class dynamics and may bring abstract concepts to life. As such, some teachers may choose either to employ only a few elements of the flipped model to make the best out of the face-to-face sessions or might decide to flip only a few selected topics or class sessions highly deemed crucial for advancing problem solving skills, analysis and knowledge creativity among learners during a term or pre-determined learning session (FLN 2012). Furthermore, students can extend newly acquired skills such as collaboration, creativity, problem solving and critical thinking to respond to challenges of the fourth industrial revolution because these are the most demanded skills (Gray, 2016).

Regulating Cognitive Load and Learning Motivation

See Figure 2.

Figure 2. Concept of regulating Cognitive Load Theory (CLT) and promoting motivation in the flipped classroom

Sources: Bergmann and Sams (2012) and Abeysekera and Dawson (2015).

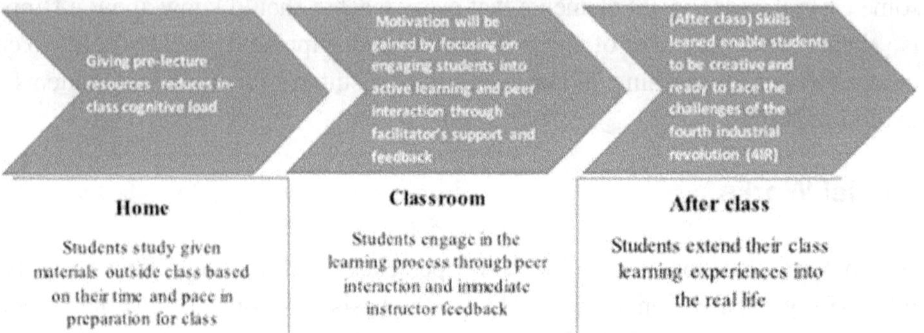

Significance of the Flipped Learning Model

When students attend a traditional lecture, their focus is centred on capturing what is being delivered at the moment when the instructor says it. Usually students do not have time for reflecting upon what is being said during the lecture. They do not want to miss some significant points while reflecting as they would unable to write down the instructor's main points. On the other hand, the use of instructional content such as video and other pre-recorded media gives students complete control over their learning process.

They can watch, rewind, replay and fast-forward as the situation may require them to do. Smith (2013) emphasized that the course videos play an essential role in a flipped classroom model, and based on his study findings, 97% of the participant students found the course videos useful. Meanwhile, Herreid and Schiller (2013) suggested that course videos play an important role in the attitude, behaviour, and performance of students. The recommendation, however, is that such videos should be kept short and manageable so as to retain student attention (approximately 10 minutes) for them to remain effective (Alvarez, 2012; Forsey, Low, & Glance, 2013; See & Conry, 2014).

When implementing the flipped classroom model, students are assured of sufficient support from their classmates and effective feedback from their instructors via digital applications at any time they want (Hamdan et al., 2013b). Therefore, a flipped classroom model, when implemented properly, can help to establish a healthy communication network between instructors and students (Cockrum, 2014). When a healthy communication network is established in this process;, this creates an active interaction process, making it easier for them to learn from one another and for those of varying skill levels to support their peers in the learning process (FLN 2012). Hence a flipped learning environment has potential to promote active learning for the mastery of critical thinking, creativity and problem solving skills (Braxton et al. 2008)

Promoting Active Learning

A flipped learning model supports active learning techniques that are essential for nurturing 21st-century skills for the fourth industrial revolution. This approach presents students with an opportunity to become real active learners rather than objects of instruction (Hamdan, McKnight & Artfstrom, 2013b). Moreover, active learning techniques are designed to enhance learning the content at the deepest levels of Bloom's Revised Taxonomy (Anderson & Krathwohl 2001) by prompting learning curiosity and eagerness (Foust, 2012). Furthermore, Hamdan et al. (2013b) observed that teachers recorded "increased student engagement, critical thinking, and better attitudes towards learning" when active learning techniques were employed. Among other things, active learning provides opportunities for instant interactions between educator and students, and between students and their peers both within and out of school settings (Cockrum, 2014; Hamdan et al., 2013b)

Active learning environments provide opportunities for effective and efficient use of learning time (Johnson, 2013) while encouraging the development of cooperative learning skills (Abeysekera & Dawson, 2015; Bergmann & Sams, 2012; Hamdan et al., 2013b; Strayer, 2012) thus exposing students to a variety of problem solving strategies (Bland, 2006). Moreover, when students are engaged in active learning, the environment encourages peer support in the learning process (Gojak, 2012) leading to better conceptual understanding of issues through fostering critical thinking skills (Berrett, 2012). Because active learning is mainly student centred; each student assumes his/her own learning responsibility (Bergmann &

Sams, 2012; Hamdan et al., 2013b), hence, reducing disciplinary problems (Cockrum, 2014) and class absenteeism (Alvarez, 2012).

Wlodkowski (2011) emphasized that motivation is key in student learning success and is one of the outcomes of an active learning environment provided by a well-planned flipped classroom model. To sustain motivation among the students, educators should play their roles by providing appropriate learning techniques and prompt feedback to support knowledge scaffolding as this could influence both the individual performance of students as well as group work (Pritchard, 2015).

DISADVANTAGES OF THE MODEL

Although the flipped classroom model offers benefits and the concept is straightforward, it requires careful planning and implementation. The preparation process, which includes recording video lectures, is tedious and requires meticulous planning, effort and time on the part of instructors (Davies et al., 2013; Herreid & Schiller, 2013). Meanwhile, out-of-class and in-class elements must be carefully synchronized and integrated for students to understand the model and be motivated to prepare for in-class activities instead of feeling anxious (Strayer, 2009). Consequently, introducing a flipped classroom model requires additional efforts and may demand new skills on the part of the instructor, although this learning curve could be mitigated by introducing the model gradually (FLN, 2012) or through collaborative ventures with other educators.

The flipped classroom model should be explained to students and applied properly right from the beginning. Otherwise, some might complain about the loss of face-to-face lectures especially if they prefer traditional approaches because of their passive role in the learning process (Missildine et al., 2013). This could make it difficult for students and educators who view constructivist approaches as a waste of time and, hence, fail to appreciate the value of the in-class hands-on activities of the model (Mangan, 2013). Moreover, students who prefer attending class to listen to lectures only may feel it safe to skip a class that focuses on activities. This could lead to missing the real intended value of the flipped classroom model. Finally, even when students decide to embrace the model, sometimes the facilities may be inadequate and access might not always support fast delivery of video content.

However, when an ideal learning environment is in place with all the required infrastructure and resources, teachers can do a lot of good things with their students when this model is implemented correctly.

Data Collection for This Book Chapter

This section of the book chapter is based on data collected for a pilot study of the flipped classroom model. This study addressed three main questions. They were: 1) Why did teachers adopt a flipped classroom model? 2) How did teachers employ a flipped classroom model?, and lastly 3) What fourth industrial revolution skills were developed using a flipped classroom model?

The research questions were addressed by employing a qualitative research design approach which utilized a focus group strategy. Four participants were initially contacted by the researcher via face-to-face interactions at a public university campus in Malaysia. By way of snowball technique, each of the initial four participants introduced and recommended other teachers who employed the flipped classroom model in their teaching practices. After a list of participants was finalized, a follow up email message was sent to each one of them. The researcher sought to confirm their commitment to participate in the study. In total, twelve (12) participants replied and affirmed their commitment and were scheduled to participate in focus group dialogue sessions. However, only eight (8) participants were able to participate in the study as the other four (4) participants opted out at the last minute due to unknown reasons.

At the initial stage, care and consideration of the influences of environmental factors, individual differences and interpersonal factors were addressed (Stewart & Shamdasani, 2015). According to Kruger (1988), focus groups are usually conceived in three phases, namely, 1) conceptualization, 2) interview sessions, and 3) analysis and reporting. During the conceptualization phase, the purpose of the study was established, a sample of study determined, and a plan of study developed. In the interview phase, an interview guide with appropriate questions was developed collaboratively with teachers who had implemented the flipped classroom model in their teaching practices since 2013 (with a good mixture of what, how and why) mainly for probing participants for more in-depth understanding of ideas and opinions on the concept.

In this study, before any asynchronous virtual focus group dialogue sessions (text-based communication postings that did not occur in real time using the Google Groups platform) were conducted, a Wiki survey tool located at www.allourideas.org was used to collaboratively develop the interview guide with the best ideas contributed by the participants. During the data collection process, the researcher acted as the moderator by starting a discussion thread and shared new questions with the participants at intervals previously agreed upon. He also engaged participants by probing for comments from them to obtain a consensus of ideas for every issue or topic. The participants responded at a time of their own convenience but within a stipulated deadline. The text-based focus group dialogue sessions were downloaded for the analysis phase. During the analysis and reporting phase, the researcher was engaged in the verification and analysis exercise, which resulted in a descriptive report of evolving issues and themes from the study.

Because the interview guide questions were developed collaboratively using the Wiki survey tool at the free and open-source website (www.allourideas.org), it was unnecessary to conduct a pilot group to ensure their reliability for the study process. Finally, the scissor-and-sort technique was applied to the data analysis process. The report summary from the data analysis was shared in the virtual focus dialogue groups for review and scrutiny by the participants for the purpose of validation.

Discussion of Results From the Study

Three asynchronous virtual focus group dialogue sessions (comprising 3-4 in-service teacher participants per group) were carried out. A total of eight (8) in-service teachers participated in the study. All participants were secondary school teachers teaching various subjects including maths, language, art and geography. Of the eight (8) participants; two (2) were males and six (6) were females. The study participants had varied teaching experiences ranging from 5-10 years (four participants), 10-15 years (two participants) and more than 15 years (two participants).

Reasons for Employing a Flipped Classroom Model

The participants shared various reasons for employing a flipped classroom model. Each of the participants valued this pedagogical approach in a different way as it was used to address unique challenges in their teaching and learning process. Despite their differences, there were some common reasons for employing a flipped classroom strategy among the participants as discussed below.

Maximizing Student Learning Potential

Generally, most participants aimed at maximizing the learning potential of students through the application of the flipped learning strategy. In regard to this endeavour, it was understood that the participants needed to decrease or overcome classroom anxiety among some students as this was one of main concerns of the teachers in the teaching-learning process. When students interacted with the learning content in advance, this interaction helped them to overcome anxiety and prepare for the in-class activities. In this regard, some teachers encouraged their students to submit their questions about the topic before the next class. This helped the teachers not only to prepare the answers in advance, but also to assess the knowledge of the learners on the topic. This observation was echoed by one participant.

The flipped classroom model helps students to assess their understanding of the topic and allows them time to formulate the right questions on the subject matter. It also gives me adequate time which allows me to prepare exhaustive answers to students' questions.

Another participant remarked:

The flipped classroom model has reduced classroom anxiety among my students. Some students were always nervous whenever a new topic was introduced; as a result they did not have confidence to ask questions on the new topic.

In reference to one student who always missed class and was unable to catch up with the rest of the classmates, one participant observed:

Previously a boy who often missed class due to unknown reasons was always struggling to catch up with the subject content and could never complete his homework. This pedagogical strategy provided him with a supportive and interactive learning environment which made him realize his potential.

Moreover, the flipped classroom model moves direct instruction from the group learning space to the individual learning spaces of students. This implies that an educator can categorize learners and apply a differentiation strategy in the teaching-learning process. This improves teacher delivery as learners with the same instructional needs are grouped together for efficient instructional management. In this particular case, one participant stated:

The flipped classroom model enabled me to quickly differentiate between learners in terms of knowledge levels, suitable instructional approaches and learning materials. When some students requested for audio content instead of printed copies I was able to determine their learning style. Hence, I did not waste time preparing materials that did not suit their instructional preference.

Creating an Engaging Learning Environment

As the flipped classroom model reverses the traditional instructional approach, class time is transformed into a supportive and interactive learning environment. A teacher devotes his time and energy towards guiding students to ensure that they apply the new concepts while engaging creatively in the subject content. This approach stimulates discussion among students and encourages learners to concentrate on hands-on activities while in the classroom. One participant reflected on her classroom situation before she had implemented a flipped model:

Previously there was no time to engage students in hands-on activities as all the time was spent on delivering the new content. Students neither interacted with their peers nor engaged in fruitful academic conversations with each other.

Another one added:

A new learning culture based on a learner-centred approach has evolved in my classroom. Most of the in-class time is now being devoted to exploring content in a more in-depth approach and hence creating rich learning opportunities for everyone in the classroom.

Another participant interjected:

Previously, whenever I gave homework some students would complain that it was too difficult for them to complete at home as there was no one to help them out when they faced difficulties with some mathematical concepts. Now classwork has become a fun activity done in collaborative groups with teacher on hand to provide the required assistance.

Another one added:

Although I teach English language, there was less talking before I introduced a flipped classroom model. A dynamic and interactive learning environment has resulted into spontaneous conversations among learners. The students have acquired more vocabulary and their pronunciations have improved significantly.

Promoting and Teaching Higher-Order Thinking Skills (HOTS)

Generally, most participants were of the view that a flipped classroom model was most suited for teaching higher-order thinking skills (HOTS). The reversal of the traditional instructional approach could easily be converted into a dynamic and interactive learning environment for teaching such skills. In this regard, Brookhart (2010) identified three types of higher-order thinking skills which include those that define thinking in terms of: 1) transfer, 2) critical thinking, and 3) problem solving. In today's learning environment, such skills are regarded as critical in preparing learners to work and compete favourably in the fourth industrial revolution environment. In this regard, one participant said:

A flipped classroom model provides ample opportunities for engaging learners in problem-based learning scenarios. The knowledge and skills acquired can be transferred and applied to new real life situations outside of school.

Another one added:

A flipped classroom model provides an environment for teachers to engage their learners by bringing real-life problems into the classroom for students to provide solutions. That way, the learning content becomes authentic and relevant for stimulating fourth industrial revolution skills. Students have to think critically and creatively in order to come up with suitable solutions to problems.

In reference to the use of a flipped learning approach to advance critical thinking skills among the learners, one participant observed:

The interactive learning environment provided by a flipped classroom model can be used for nurturing critical thinking skills among the learners to enable them to employ wise judgement in different situations. Such skills as the ability to judge the credibility of a source of information, comprehending the purpose of a written text, and identifying the right audience are essential fourth industrial revolution skills.

Ways of Employing a Flipped Learning Model

Generally, the most popular approach of employing a flipped classroom model among the participants was through the creation and use of video content. Most of the teachers had access to the Frog Virtual Learning Environment (VLE) for Malaysian schools. Hence, it was relatively easy for them to post or upload a video for their students to watch the content at home before coming to school the next day, as one participant explained:

The process starts with a teacher explaining a concept on video. It is done by explaining a concept on camera, or using screen capture tools or presentation software with a voiceover capability.

Another one added:

The video content helps to establish the connection and humor, and also set the pace for great teaching. Video technology provides plenty of scope for teachers to unleash their creativity. There are plenty of online guidelines detailing how teachers create video learning content to engage their students in powerful classroom learning experiences.

Once video learning content is created, the teacher shares it with the students through the Frog Virtual Learning Environment (VLE) or other instructional content sharing resources such as a classroom Facebook page. One participant explained how learners can engage with the learning content.

Students watch the shared video content at home before the next class. They choose the most suitable way to engage with the video content. They can watch it on a smart phone, tablet, laptop or desktop. They can re-watch, rewind, pause, fast-forward and read about the topic from other sources as well.

Students turn up in the classroom when they are well prepared with substantial knowledge and ideas about the topic. They may even send their questions to the teacher for in-depth understanding of the topic. The teacher may use the questions to determine the instructional approach to the topic such as personalizing learning experiences for individuals or groups of students depending on their needs, as one participant observed.

Students spend quality time in the classroom applying knowledge and skills to understand new concepts by engaging in hands-on and collaborative activities. The teacher spends time going around the classroom assisting students individually or in small learning groups. This way, every student has a chance to get the deserved attention.

On the other hand, another participant was of the view that:

The flipped classroom model encourages a student-centred learning environment. It makes learning experiences more engaging, enjoyable and productive for students. Students engage in active learning; analysing, synthesizing and evaluating the subject content. Teacher's role is to facilitate the learning process and provide immediate feedback wherever necessary.

Apart from using video content as the main tool for implementation of the flipped classroom model, other resources such as PowerPoint presentation slides and printed hand-outs of topic contents or book chapters are also used for the same purpose.

Fourth Industrial Revolution Skills Developed in a Flipped Learning Environment

Generally, the most common fourth industrial revolution skills reported amongst the participants of this study as observed among students after employing the flipped classroom model were as discussed below. For example, one participant explained that:

Most of the students in my class have acquired the skill of solving problems. They discuss with their peers and even conduct research until they get a solution for their learning problem. Sometimes they have more than one solution for a problem. They do not depend on me for answers any more.

Furthermore, a participant added that:

The problem based approach gives them an opportunity to look at their learning content in a more authentic way. It challenges their creativity and helps to nurture their critical thinking skills. The ability to solve problems in school prepares them for the challenging life in their communities or workplace after graduation.

A more elaborate explanation was given by one participant who felt that this model enabled students to acquire more than one set of fourth industrial revolution skills at the same time.

The model promotes collaboration and team spirit among students. They use their critical thinking skills to judge and decide the most appropriate solutions to their learning problems. In most cases, the process requires some creativity and imagination. This way, they are able to feel a sense of accomplishment and become independent learners. They are more likely to transfer this knowledge to their daily real lives outside the classroom as it is part of their personal experiences.

Moreover, one of the requirements for one to succeed in a knowledge based society is to have the competence to analyze and synthesize information or even conduct research. This may be a pre-requisite for certain academic, personal or professional engagements, as one of the participants reported regarding his students who were involved in a flipped classroom model.

There is a remarkable improvement as far as my students' analytical skills are concerned. They are more logical in their thinking and have sound reasons for whatever they do in the class. Their imaginative and creative skills seem to have improved over time as is noticeable in their written assignments and completed group projects.

Another participant who had carefully observed the gradual development of fourth industrial revolution skills among his students added that:

When discussing with peers, you hear one of them asking for the logic behind the choice of their answers. They are more independent in their thinking now as I do not have to point out each and every step when they perform their group or individual academic activities.

CONCLUSION

A flipped classroom model can be a transformative approach for promoting active learning among students. It enables teachers to make better use of the teaching space by creating a highly motivating and engaging environment in which students excel regardless of their differences in individual abilities. The teacher has the opportunity to provide feedback in real time while providing the necessary assistance to individuals or groups of students. When implemented properly this approach has the potential to equip students with the necessary fourth industrial revolution skills such as problem-solving, critical thinking, synthesizing information and research skills that are essential for living and participating in a competitive fourth industrial revolution environment.

LIMITATION OF THE STUDY

As is normally the case with the use of a focus group strategy, the findings of this study might not be generalised; however, this study should provide a baseline for future researchers and teacher education providers regarding the development of fourth industrial revolution skills.

IMPLICATIONS AND WAY FORWARD

The study encouraged participants to reflect on their technology skills for flipped classroom model strategies in addition to knowledge sharing and collaborative activities. This was a good way of nurturing teacher professional knowledge and developing unique technology skills. School leaders could encourage, support or even reward teacher autonomy in employing specific pedagogical approaches which result in effective teaching and learning opportunities. To implement a successful flipped classroom model, educators should familiarize themselves with the necessary technologies and current pedagogical strategies for planning and managing a flipped learning environment.

For flipped classroom model approaches to succeed teachers should endeavor to acquire technological proficiency in related fields to realize other fourth industrial revolution professional development objectives. These include information and communications technology (ICT), information literacy and digital media literacy. This would enable them to implement flipped classroom model strategies suitable for specific subject domains and learning environments. Hence, institutional support can go a long way towards making these fourth industrial revolution professional development goals achievable.

Furthermore, educators should have an open mind to assume new instructional roles such as taking a back seat while enabling students to take ownership of the learning process. This way, students will "do" and "talk" more while educators should continuously observe their students and provide them with instant feedback relevant at every essential moment. The feedback provided in real time is crucial for sustaining student engagement and learning motivation.

The implementation of flipped classroom model approaches without the necessary preparations including training can lead to adverse consequences for an institution. Therefore, institutional leaders should be at the forefront of planning for innovative pedagogical training for teachers. Moreover, they should draw up professional development practice guidelines governing the implementation of flipped classroom model initiatives in schools. Efforts to stimulate teachers' creativity by infusing digital technologies in flipped classroom model strategies should be the focus of pre-service teacher training programs.

Research orientation, web-based inquiry, reflective practice and teachers networking with each other are essential elements for improving the instructional process. Such elements should be encouraged to ensure teachers' success in their creative and innovative pedagogical approaches (Fullan, 1992). Moreover, teachers should be receptive to constructive criticisms as a way of embracing innovative ideas to transform their professional practices.

Technology skills suitable for teacher pedagogical strategies should result from creative processes of imagination and originality, reflective practices, as well as features of digital technologies for multimodal presentation and communication to achieve success (Loveless, Burton, & Turvey, 2006).

REFERENCES

Abeysekera, L., & Dawson, P. (2015). Motivation and cognitive load in the flipped classroom: Definition, rationale and a call for research. *Higher Education Research & Development*, *34*(1), 1–14. doi:10.1080/07294360.2014.934336

Alvarez, B. (2012). Flipping the classroom: Homework in class, lessons at home. *Education Digest: Essential Readings Condensed for Quick Review*, *77*(8), 18–21.

Ambrose, S., Bridges, M. W., Lovett, M. C., DiPietro, M., & Norman, M. K. (2010). *7 Research-based principles for smart teaching*. San Francisco, CA: Jossey-Bass. Retrieved from https://firstliteracy.org/wp-content/uploads/2015/07/How-Learning-Works.pdf

Anderson, L., Krathwohl, D., & Airasian, P. (2001). *A taxonomy for learning, teaching, and assessing: A revision of Bloom's Taxonomy of educational objectives*. New York, NY: Pearson, Allyn & Bacon.

Anderson, L. W., & Krathwohl, D. R. (Eds.). (2001). *A taxonomy for learning, teaching, and assessing: A revision of Bloom's taxonomy of educational objectives*. New York, NY: Longman.

Auster, E. R., & Wylie, K. K. (2006). Creating active learning in the classroom: A systematic approach. *Journal of Management Education, 30*(2), 333–353. doi:10.1177/1052562905283346

Bacsich, P., Bristow, S. F., Camilleri, A., Op de Beeck, I., Pepler, G., & Phillips, B. (2013). *Virtual schools and colleges. Providing alternatives for successful learning, 2*. Roosbeek, Belgium: ATiT bvba. Retrieved from https://www.pedocs.de/volltexte/2013/8286/pdf/VISCED_Handbook_2013_2.pdf

Bergmann, J., & Sams, A. (2012). *Flip your classroom: Reach every student in every class every day*. Eugene, OR: International Society for Technology in Education.

Bergmann, J., & Sams, A. (2013). Flip your students' learning. *Educational Leadership, 70*(6), 16-20. Retrieved from ASCD website: http://www.ascd.org/publications/educational-leadership/mar13/vol70/num06/Flip-Your-Students%27-Learning.aspx

Berrett, D. (2012). How "flipping" the classroom can improve the traditional lecture. *Education Digest, 78*(1), 36–41.

Bishop, J., & Verleger, M. (2013). The flipped classroom: A survey of the research. In *ASEE Conference Proceedings*. Available at http://www.asee.org/public/conferences/20/papers/6219/download

Bland, L. (2006). *Applying flip/inverted classroom model in electrical engineering to establish lifelong learning*. Presented at the 2006 ASEE Annual Conference & Exposition, Chicago, IL. Retrieved from https://peer.asee.org/491

Bloom, B. S. (1956). *Taxonomy of educational objectives, Handbook I: The dognitive domain*. New York, NY: David McKay Co Inc.

Brame, C. (2013). *Flipping the classroom*. Vanderbilt University Center for Teaching. Retrieved from http://cft.vanderbilt.edu/guides-sub-pages/flipping-the-classroom/

Braxton, J., Jones, W., Hirschy, A., & Hartley, H. III. (2008). The role of active learning in college student persistence. *New Directions for Teaching and Learning, 2008*(115), 71–83. doi:10.1002/tl.326

Brookhart, S. (2010). *How to assess higher-order thinking skills in your classroom*. Alexandria, VA: ASCD. Retrieved from http://www.ascd.org/Publications/Books/Overview/How-to-Assess-Higher-Order-Thinking-Skills-in-Your-Classroom.aspx

Cashin, W. E. (1995). *Student ratings of teaching: The research revisited. IDEA Paper No. 32. Center for Faculty Evaluation and Development in Higher Education*. Manhattan, KS: Kansas State University. Available from https://eric.ed.gov/?id=ED402338

Choy, C. S., Sedhu, D. S., Liew, Y. L., Lee, M. Y., Malanee, A., & Anua, N. (2015). Influence of culture on students' awareness of how and why they learn. *Malaysian Journal of Learning and Instruction, 12*, 49–67.

Clark, R. C., Nguyen, F., & Sweller, J. (2005). *Efficiency in learning: Evidence based guidelines to manage cognitive load*. San Francisco, CA: Pfeiffer & Co.

Cockrum, T. (2014). *Flipping your English class to reach all learners: Strategies and lesson plans. Abingdon-on-Thames*. Routledge.

Collins, R. (2014). Skills for the 21st Century: Teaching higher-order thinking. *Curriculum & Leadership Journal, 12*(14). Retrieved from http://www.curriculum.edu.au/leader/teaching_higher_order_thinking,37431.html?issueID=12910

Davies, R. S., Dean, D. L., & Ball, N. (2013). Flipping the classroom and instructional technology integration in a college-level information systems spreadsheet course. *Educational Technology Research and Development, 61*(4), 563–580. doi:10.100711423-013-9305-6

Davis, C. (2013). Flipped or inverted learning: Strategies for course design. In E. G. Smyth & J. X. Volker (Eds.), *Enhancing instruction with visual media: Utilizing video and lecture capture* (pp. 241–261). Hershey, PA: Information Science. doi:10.4018/978-1-4666-3962-1.ch017

EDUCAUSE Learning Initiative. (2012). *7 things you should know about flipped classrooms*. Retrieved from http://net.educause.edu/ir/library/pdf/ELI7081.pdf

Flipped Learning Network (FLN). (2012). *7 things you should know about flipped classrooms*. Retrieved from https://library.educause.edu/resources/2012/2/7-things-you-should-know-about-flipped-classrooms

Flipped Learning Network (FLN). (2014). *The four pillars of F-L-I-P*. Retrieved from https://flippedlearning.org/definition-of-flipped-learning/

Forsey, M., Low, M., & Glance, D. (2013). Flipping the sociology classroom: Towards a practice of online pedagogy. *Journal of Sociology (Melbourne, Vic.), 49*(4), 471–485. doi:10.1177/1440783313504059

Foust, T. (2012). Special guest article: A tip of the hat to the flip of the class. *Illinois Music Educator, 73*(2), 100.

Fullan, M. (1992). *Successful school improvement: The implementation perspective and beyond*. Birmingham, UK: Open University Press.

Gehringer, E. F., & Peddycord, B. W. III. (2013). Experience with online and open-web exams. *Journal of Institutional Research, 2*, 10–18. doi:10.9743/JIR.2013.2.12

Gleason, B. L., Peeters, M. J., Resman-Targoff, B. H., Karr, S., McBane, S., Kelley, K., ... Denetclaw, T. H. (2011). An active-learning strategies primer for achieving ability-based educational outcomes. *American Journal of Pharmaceutical Education, 75*(9), 1–12. doi:10.5688/ajpe759186 PMID:22171114

Gojak, L. (2012). *To flip or not to flip: That is not the question*. National Council of Teachers of Mathematics. Retrieved from http://www.nctm.org/about/content.aspx?id=34585

Gray, A. (2016). *The 10 skills you need to thrive in the Fourth Industrial Revolution*. New York, NY: World Economic Forum. Retrieved from https://www.weforum.org/agenda/2016/01/the-10-skills-you-need-to-thrive-in-the-fourth-industrial-revolution/

Hamdan, N., McKnight, P. E., McKnight, K., & Artfstrom, K. M. (2013a). *The flipped learning model: A white paper based on the literature review*. Retrieved from http://flippedlearning.org/cms

Hamdan, N., McKnight, P. E., McKnight, K., & Artfstrom, K. M. (2013b). A review of flipped learning. *Flipped Learning Network*. Retrieved from https://flippedlearning.org/wp-content/uploads/2016/07/WhitePaper_FlippedLearning.pdf

Hamon, K. (2014). *The benefits of flipped classes on student workloads*. Retrieved from: https://www.fractuslearning.com/2014/07/22/flipped-classes-student-workloads/

Herreid, C., & Schiller, N. (2013). Case studies and the flipped classroom. *Journal of College Science Teaching*, *42*, 62–66.

Issham, I., Rozhan, M. I., & Thenmolli, G. (2010). Motivation, psychology and language effect on mobile learning in University Sains Malaysia. *International Journal of Interactive Mobile Technologies*, *4*(4), 31–36. doi:10.3991.ijim.v4i4.1408

Jensen, J. L., Kummar, T. A., & Godoy, P. D. (2015). Improvements from a flipped classroom may simply be the fruits of active learning. *CBE Life Sciences Education*, *14*(1), 1–12. doi:10.1187/cbe.14-08-0129 PMID:25699543

Johnson, G. B. (2013). *Student perceptions of the flipped classroom* (Unpublished Master's Thesis). University of British Columbia. Retrieved from: https://open.library.ubc.ca/cIRcle/collections/24/items/1.0073641

Johnson, L., Adams, S., Cummins, M., Estrada, V., Freeman, A., & Ludgate, H. (2013). *The NMC Horizon Report: 2013 higher education addition*. Austin, TX: The New Media Consortium. Retrieved from http://www.nmc.org/pdf/2013-horizon-report-HE.pdf

Keane, L. (2014). *The 21st century skills of the flipped Classroom*. Retrieved from https://www.linkedin.com/pulse/20141117162612-162064004-the-21st-century-skills-of-the-flipped-classroom

Khan, S. (2011). *Let's use video to reinvent education, Speech presented at TED 2011*. Retrieved from https://www.ted.com/talks/salman_khan_let_s_use_video_to_reinvent_education

King, A. (1993). From sage on the stage to guide on the side. *College Teaching*, *41*(1), 30–35. doi:10.1080/87567555.1993.9926781

Krueger, R. A. (1988). *Focus groups: A practical guide for applied research*. Newbury Park, CA: Sage Publications.

Loveless, A., Burton, J., & Turvey, K. (2006). Developing conceptual frameworks for creativity, ICT and teacher education. *Thinking Skills and Creativity*, *1*(1), 3–13. doi:10.1016/j.tsc.2005.07.001

Mangan, K. (2013). Inside the flipped classroom. *The Chronicle of Higher Education*. Retrieved from https://www.chronicle.com/article/Inside-the-Flipped-Classroom/141891

Marks, D. B. (2015). Flipping the classroom: Turning an instructional methods course upside down. *Journal of College Teaching and Learning*, *12*(4), 241–247. doi:10.19030/tlc.v12i4.9461

Mazur, E. (1997). *Peer instruction: A user's manual*. Upper Saddle River, NJ: Prentice Hall.

Missildine, K., Fountain, R., Summers, L., & Gosselin, K. (2013). Flipping the classroom to improve student performance and satisfaction. *The Journal of Nursing Education*, *52*(10), 597–599. doi:10.3928/01484834-20130919-03 PMID:24044386

Moore, A. J., Gillett, M. R., & Steele, M. D. (2014). Fostering student engagement with the flip. *Mathematics Teacher*, *107*(6), 420–425. doi:10.5951/mathteacher.107.6.0420

Ng, P. Y., Voges, K., & Goi, C. L. (2010). The importance of mobile phone applications to young consumers: An exploratory study in Malaysia. Proceedings of Australian & New Zealand Marketing Academy.

Ntiko, A. J., & Brookhart, S. M. (2007). *Educational assessment of students* (5th ed.). Upper Saddle River, NJ: Pearson.

Poore, R. (2013). Flipping the classroom. *Focus on Boise State, 34*(1), 21-22. Retrieved from http://news.boisestate.edu/update/files /2013/08/BSU_FOCUS_FALL2013_web.pdf

Pritchard, A. (2015). *Öğrenme yolları, Sınıfta öğrenme teorileri ve öğrenme stilleri* [Learning pathways, learning theories in classroom and learning styles] (M. Çevikbaş & S. Çevikbaş, Trans. & Eds.). Ankara: Nobel Akademik Yayıncılık.

Rani, M. (2018). *MANDATE: Embracing Industry 4.0.* Retrieved from http://news.moe.gov.my/2018/01/27/2018-mandate-embracing-industry-4-0/

Saavedra, A. R., & Opfer, V. D. (2012). *Teaching and learning 21st century skills: Lessons from the learning sciences.* Presented at the Joint AARE/APERA Conference, Sydney, Australia. Retrieved from https://www.aare.edu.au/data/publications/2012/ Saavedra12.pdf

See, S., & Conry, J. M. (2014). Flip my class! A faculty development demonstration of a flipped-classroom. *Currents in Pharmacy Teaching and Learning*, *6*(4), 585–588. doi:10.1016/j.cptl.2014.03.003

Sharples, M., Corlett, D., & Westmancott, O. (2002). The design and implementation of a mobile learning resource. *Journal of Personal and Ubiquitous Computing*, *6*(3), 220–234. doi:10.1007007790200021

Smith, J. D. (2013). Student attitudes toward flipping the general chemistry classroom. *Chemistry Education Research*. Available from https://pubs.rsc.org/en/content/article landing/2013/rp/c3rp00083d#!divAbstract

Smith, J. D. (2013). Student attitudes toward flipping the general chemistry classroom. *Chemistry Education Research and Practice*, *14*(4), 607–614. doi:10.1039/C3RP00083D

Stakyer, H., & Horn, M. B. (2012). *Classifying K–12 blended learning*. Mountain View, CA: Innosight Institute, Inc.

Sterling, S. (2012). *The flipped classroom: A pedagogy for differentiating instruction and teaching essential skills*. Retrieved from http://www.scilearn.com/blog/flipped-classroom-pedagogy-differentiating-instruction

Stewart, D. W., & Shamdasani, P. N. (2015). *Focus groups - Theory and practice. Applied Social Science Research Series* (3rd ed.). Thousand Oaks, CA: Sage Publications, Inc.

Strayer, J. (2012). How learning in an inverted classroom influences cooperation, innovation and task orientation. *Learning Environments Research*, *15*(2), 171–193. doi:10.100710984-012-9108-4

Strayer, J. F. (2009). *Inverting the classroom: A study of the learning environment when an intelligent tutoring system is used to help students learn*. Saarbrücken, Germany: VDM.

Study Malaysia. (2018, May 2). *The Fourth Industrial Revolution (IR 4.0) and what it means for students like you*. Retrieved from https://www.studymalaysia.com/education/top-stories/the-fourth-industrial-revolution-ir-4.0-and-what-it-means-for-students-like-you

Sweller, J., Ayres, P. L., & Kalyuga, S. (2011). *Cognitive load theory*. New York: Springer. doi:10.1007/978-1-4419-8126-4

Wlodkowski, J. R. (2011). *Enhancing adult education to learn. A comprehensive guide for teaching all adults* (3rd ed.). Hoboken, NJ: John Wiley & Sons.

Chapter 7
Implementation of Flipped Classroom at Malaysia's Teacher Education Institute

Malar Muthiah
Institut Pendidikan Guru Sultan Abdul Halim, Malaysia

Raamani Thannimalai
SMK Kepala Batas, Malaysia

ABSTRACT

In a teacher education institute in Malaysia, blended learning is considered as a strategic and effective learning instruction with unique features to help improve student-teacher's discussion skills, improve their communities of practice, and achieve their course purposes. One of the tools of blended learning is flipped classroom. Flipped classroom should be implemented especially among student teachers. The definition of flipped classroom was explained in detail, and furthermore, flipped classroom activities are based on social and cognitive constructivist theory that is suitable to be implemented in teacher education institutions in Malaysia because it is student-centered and involved students actively during the learning process. A research done in one of the teacher education institutions was discussed to find out the effect of flipped classroom in achievement history test and motivation among the 20 student teachers who enrolled for Early European History until 21st Century as one of their major papers.

INTRODUCTION

The role of teachers is increasingly challenging and they need to prepare for the Fourth Industrial Revolution (IR4.0) or risk being left behind. The rapid development of technology requires a paradigm shift among teachers. Are teachers ready for a generation of students who are digital natives? We need teachers who are committed to forming new pedagogical tools and adapting to new teaching methods (Goh, 2017). Thus, it is expected that our teacher education institute should focus on teacher training programs that need to be in line with current developments and be updated with new technology to equip student teachers with tools to face this new age.The role of teachers as the main source of knowledge and infor-

DOI: 10.4018/978-1-5225-7832-1.ch007

Copyright © 2019, IGI Global. Copying or distributing in print or electronic forms without written permission of IGI Global is prohibited.

mation may be irrelevant today as teaching and learning now emphasise on creativity and innovation. Teaching and learning for student teachers should focus more on active learning, project-based learning, problem-solving and inquiry with opportunities to engage with the real world should be a practice and to improve their overall lack of engagement, in class. As future teachers, student teachers should be given the freedom to customise their learning process through blended learning, flipped classroom and Bring Your Own Device. It is to make them realise the importance of to adapt they are teaching to students' interest and abilities. Learning is not only concentrated in the classroom, but can also happen anywhere, at home, workplace and even at the bus stop. Baeten et al. (2010) strongly believes that student intrinsic motivation increases in a constructivist learning environment, but have not indicated if student motivation mediates achievement in a learning environment such as the flipped classroom.

The Fourth Industrial Revolution powered by the Internet of Things will have an effect on every aspect of our lives including learning and teaching methods at Malaysia's Teacher Education Institute. . The education system worldwide has to transform for the imminent FIR as the job market will be largely prompted by the advancement of digital economy, robotics, artificial intelligences and automation technology. According to Gleason (2018), institutes of Higher Education must work to reinforce the higher order thinking that is needed in the FIR. Lewis (2018) posited that the ability to learn new skills, accept new approaches, and cope with continual social change will be essential in the FIR.

Blended learning is actually one of the techniques that is assisted through mixeds media modern technology and also can easily produce student teachers of the 21st century. Hybrid mix-learning, blended learning, and discovering are likewise inclusive in blended discovering. The principle of blended knowing combines the traditional learning model with interactive media devices to create a combined setting finding out where students learn with direct active discovering in the classroom while it is actually incorporated with various other components including on the internet understanding outside the class. Students are going to be much more proactively entailed throughout the finding out procedure so they may seek much better methods of discovering that matches them thus that they can participate in the dual-way method to the greatest of their capacities. The task of a instructor is merely as a facilitator and negotiator who produce a favourable learning setting for the learning procedure.

Interaction between the teacher and the student remains in a two-way direction whereby the student and also the teacher can share knowledge to allow an efficient understanding method. The Blended Learning concept will include the typical discovering method through its own usage. Poon (2014) pressured that combined discovering is a mix in between classroom activities as well as internet learning adventures outside the course. The principal concentration of blended learning is also the integration in between learning in person and on-line learning as well as it has its own benefit which is actually to create the understanding process much more affordable.

Blended learning is a fundamental redesign of the instructional model with a shift from lecture based to active student- centred instruction that involves different types of designed learning activities and multilevel interaction between classmates and their instructors during classroom. Educators worldwide have been advocating for a more student-centred teaching approaches to promote active learning among students. Blended learning involves active learning and is a term for teaching practices which places emphasis on student activity and their involvement in the process of learning (Prince, 2004).

According to Singh and Reed (2001), blended learning as a learning program uses more than one delivery model to increase students' academic achievement. It is a combination of technology and classroom instruction in a flexible approach for learning that recognizes the benefits of delivering some training and assessment online. In fact, blended learning also uses other modes to make up a complete training program which can improve learning outcomes and save costs. Blended learning becomes one of the best alternatives for the usage of e-learning that helps to settle issues among students who felt alone and believed they lacked of any skills (Tayebinik, 2012; Zhang & Han, 2012). It is strongly believed that blended learning helps to increase student engagement and the restructuring of the traditional class hours. Tayebinik (2012) and Zhang and Han (2012) added that blended learning blends the best of traditional and Web-based learning experiences.

The blended learning model can be implemented dynamically and rigorously in institutions of higher education. Blended learning has emerged as a potential platform to transform higher education institutions to supplement the traditional classroom setting and increase the quality of student learning in classroom (Garrison &Hanuka, 2004). These include allowing classroom time to be utilised more effectively, higher rates of student achievement, the engagement, flexible use of technology to enhance students' learning experiences and meeting their needs and expectations have been main the priority in higher education institutions over three decades (Demirer&Sahin, 2013; Garrison &Hanuka, 2004).

The Ministry of Education (MOE) and the Ministry of Higher Education (MOHE) are conducting teacher education and training in Malaysia for primary and secondary school teachers by training primary school teachers at Institutes of Teacher Education (ITE) (previously known as Teacher Training Colleges) and training of secondary school teachers is carried out by government-funded universities. The Teacher Education Division (TED) is a section in the Malaysian Ministry of Education, which supervises teacher education programs in Malaysia. The TED has several departments that aids in its operations. Based on the recent changes in the Malaysian education scenario, universities have changed their approach according to the teacher education and training programs. In Malaysia, there are many institutes and universities that provide both pre-service and in-service teachers with teacher education programs. Recently in Malaysian higher education institutions, blended teaching and learning approaches have attracted great interest and support (Puteh&Hussin, 2007; Siew-Eng, Ariffin, &Rahman, 2010, Embi, 2011). Blended learning plays a major role especially in higher education institutions because it involves a wide usage of the internet and if it is implemented it can widen access, improve the quality of lessons, increase cost effectiveness and promote Malaysia globally. Recently, Malaysian institutions offer a wide variety of blended courses for students. For example, at the time of this study, 17 universities and colleges in Malaysia offer teacher education courses and some of these universities and colleges are delivering these courses for teachers in the form of blended learning (Azizan, 2010).

TEACHER EDUCATION INSTITUTE

In Malaysia, teacher education programs are offered via methods such as Problem-based learning (PBL), Work based learning (WBL), Project based learning (PBL) and blended learning. Blended learning can also bring a combined meaning of various media, methodology, activity, teaching aid material and also a combination between learning with face-to-face interaction and e-learning. There are a few features in blended learning such as transitioning from lecturer centered learning to student centered learning; interaction among students; student and content; and student and other external information from other sources. Blended learning also provides formative and summative evaluation that could be integrated for lecturers and students. Additionally, the students also can afford to determine their own learning speed. This factor will definitely reduce the challenges for lecturers during teaching sessions. Students are also encouraged to join discussions in the form of forums by contributing ideas during the learning process.

In teacher education programs, blended learning is considered as a strategic and effective instructional method with many unique features to help improve student teacher's discussion skills; improve their communities of practice, and achieve their course purposes (Means, Toyama, Murphy, Bakia, & Jones, 2009). Blended learning in lecture rooms at higher education institutions was encouraged to be used widely as it helps promote critical thinking among students and makes them more responsible for their own learning process. During lecture time in institutions of higher learning, using different types of blended learning models will encourage instructors/lecturers to focus on students' different learning styles rather than relying entirely on their lecture content. Blended learning directly helps to form behavioural engagement among the students such as on-task attention; effort; persistence and positive conduct; and also to build emotional engagement that relates to student's interest, belonging, value, and positive emotions.

Flipped Classroom as a Blended Learning Tool

The flipped classroom is the reverse method of the traditional classroom and is also considered as a blended learning tool that is used to create a learning environment that optimizes teaching and learning. According to Bergmann and Sam (2012), teachers can use video lectures outside the classroom and in-class time for doing assignments or active learning activities such as problem-based learning. However, the online component of blended classrooms usually happens during on-going class time, along with student-teacher interaction. The concept is gaining popularity, perhaps due to the ubiquitous nature of the tools that teachers use to accomplish the flip such as iTunes, YouTube, and Learning Management Systems (LMS). Furthermore, twenty-first century students are very comfortable with electronic devices and they also thrive on learning in social situations. Students are also accustomed to widely interacting with audio and video using electronic devices.

The Goal of Flipped Classroom

The goal of flipped classroom were to focus on instructors role from sage to guide, focus more on understanding, application and analysing than on recall facts. It also enables students to have more control on the ability to learn and feel responsible for their space of learning. Lastly it creates platform for the students to learn from their peers and focus more involvement on active learning strategies. The instructor is also able to focus on individual learning compared to the whole class. Instructors will make adjustment to find a suitable pace for the average student to learn. By making continuous adjustment, instructors are able to motivate them to succeed.

Benefits of Using Flipped Classroom

The flipped class appeals to instructors to for several reasons. The main focus is on the most rewarding part of teaching and active interactions with students during the lesson. The hands-on activities in class engage the instructors as much as the student and a way to bring in experts to the classroom by applying podcasts and internet video of content area. The flipped classroom is a method of instruction for a course delivery method. Flipping classroom also helps to cover more content for students during class time and they have more free time for collaborative learning projects. Furthermore flipped classroom activities allow students to own their knowledge gain when they learn a new content or topic. Students also can use online resources found via Google or on YouTube and they will be more engaged in looking for help on the Internet, if their teacher uses videos as part of their teaching for entire class time. Lastly as instructors at school. Lecturers do not only deliver content of the subject but, also encourage, motivate and be supportive for students to achieve their vision. So flipped classroom help instructors build better relationship with students.

OBJECTIVE OF THE RESEARCH

Firstly, the objective of the research was to ascertain the effectiveness of flipped classroom method on student teachers' achievement in Early European History until 20th century Civilization. Besides that, the study also aimed to find out student teachers learning motivation at Teacher Education Institute in Sungai Petani.

Research Questions

The questions which will be answered by this study are:

1. Is there any difference in the achievement among student teachers at the Teacher Education Institute after using flipped classroom?
2. Is there any improvement among student teacher's motivation at the Teacher Education Institute after using flipped classroom?

Conceptual framework is the researcher's understanding of how the particular variables in his study connect with each other. Thus, it identifies the variables required in the research investigation.

In this research (Figure 1), the independent variable was flipped classroom, a tool of blended learning. Dependable variable is achievement among student teachers after implementation of flipped classroom for five weeks.

Flipped Classroom and Constructivist Theory

This study was done based on the Piaget (1967) and Vygotsky (1978) theory. Flipped classroom reflects many characteristics of learning based on cognitive and social constructivism. It is also supported by Strayer (2012) that the flipped classroom as a student- centered instructional approach is grounded in the constructivist theory of learning. The constructivist theory lie in the fact that students are given the opportunity to build knowledge through and active process of interrelationship between prior learning

Figure 1. Conceptual framework of study

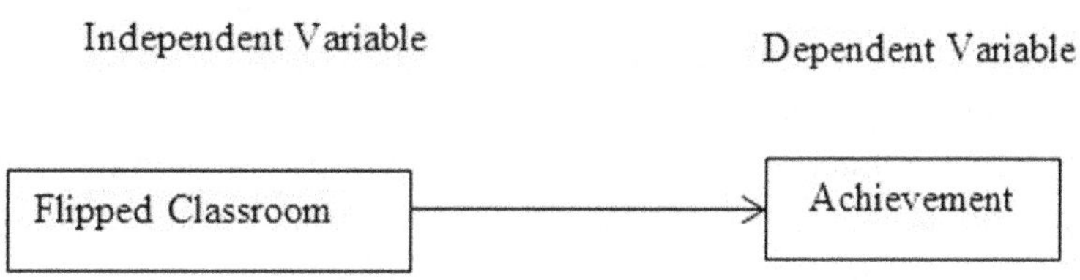

and present learning. In constructivist classrooms, students to form meaning based on existing knowledge. Furthermore, the constructivist theory also stimulates students to build their understanding of concepts through a combination of their existing knowledge and new experiences by actively participate in their learning process and not passively receiving from the outside.

In constructivist classroom teacher's learning process is more important than just imparting knowledge to the students. Therefore, it's the instructor's role in the classroom as the organizer, facilitator and motivator by utilizing settings, cooperation, and dialogue to motivate students' interests, activities, critical thinking and creativeness. Before the flipped classroom, students knew the learning tasks by watching the videos or studying the materials made by the instructors/teachers to construct new knowledge and in class, the instructors organize and guide the students to collaborate with each other to discuss the task given and what they have learned, so that students can gain help of their peers and instructors/teachers. Constructivist theory focuses learning process through collaboration, interaction and engagement of learners. Most research on the flipped classroom employ group-based interactive learning activities inside the classroom, citing student- centered learning theories based on the works of Piaget (1967) and Vygotsky (1978).

The flipped classroom approach aims to create a student-centered learning environment in which students take care of their own learning and become more active and interactive in class. In the flipped classroom, students work through problems, advance concepts, and engage in collaborative learning (Tucker, 2012). Such collaborative activities are effective in supporting students' higher level of understanding (Yang & Wu, 2012).

The exact nature of these activities varies widely between studies especially being assigned *homework*. The flipped classroom label is most often assigned to courses that use activities made up of web-based video lectures and closed-ended problems or quizzes. Thus, the flipped classroom actually represents an expansion of different type of curriculum, rather than a mere re-arrangement of activities.

The flipped classroom creates a structure that is very complementary to the constructivist learning environment. Students are able to shape and take responsibility for their own learning experiences while viewing, listening, or revise lecture notes at home. During their learning process the next day, students will collaborate with their peers and process the information on higher critical thinking and apply the knowledge they have constructed at home. The instructors/teachers will guide students through this process, reaffirming or correcting their interpretations of information or content. Collaborative learning task in the classroom allows students to do work together such as cooperative learning method encourages students to work in group work and emphasis on individual contributions together. It is important for instructors/teachers to understand the idea about student-centered learning that give priority to curiosity, interest and experience before implementing flipped classroom in their teaching.

Flipped Classroom at the Teacher Education Institute in Malaysia

Teacher education programs need to be designed according to two main purposes. First, these programs should educate student teachers in subject matter content. Second, these programs should enable these student teachers to transfer their knowledge into practice and prepare themselves to carry out simultaneously. So according to Dewey (1924) and Robinson (2011) the core idea for teacher education is to educate the student teachers in order to meet the needs of the society. This can become a reality if student teachers are not only enriched with knowledge but are also able to use the knowledge in future when they are posted to schools. So, it is important that student teachers learn technology use in classrooms and various instructional approaches. However, being a relatively new trend, most implementations of the flipped classroom are reported in blogs, online magazines and newspapers instead of academic papers and conferences. There seems to be little rigorous research done to measure the effects of this pedagogy (Goodwin & Miller, 2013), and what has been published so far seems far from conclusive. In the Malaysian context, research on the implementation of flipped classroom in Teacher Education Institute has not been widely carried out. Flipped learning helps to change this and our students get an opportunity to not just learn the theoretical content but also practice their knowledge in class. They can experience many Web 2.0 tools which they could use for different types of content. They can discuss 21st century student skills, discover various Web tools they could use with different teaching methods, design technology integration projects for schools, the usage of mobile technologies in teaching and learning, experience virtual and augmented reality and discuss how to use it in learning, embedding games in their teaching and learning processes and learn different formative and summative assessment techniques.

Flipped classroom learning enables students in teacher education programmes to gain lots of experience and explore how to use their knowledge in the classroom. Student teachers will be more motivated, satisfied, productive and active than before. Flipped classroom in teacher education will help to promote critical thinking among student teachers and make them more responsible for their own learning process. As lectures in Teacher Education Institutions are usually limited to one or two hours at a time, flipped learning will allow lecturers to focus on students' different learning styles instead of relying entirely on lecture content (McLaughlin et al., 2014).

As flipped classroom is implemented to give more face-to-face interaction time for hands-on activities, this method has been considered suitable for any kind of writing assignment or project. Flipped classroom encourages students to have more time to improve their writing with the guidance from the lecturer in class after they have read and watched videos about the lecture before attending the class. Students can get prompt feedback and guidance from the lecturer if they have any problems in the process of writing assignments. Furthermore, student teacher's engagement is closely related to their involvement in classroom activities to enhance learning process. Many researchers always focus on studying educational constructs through active participation and involvement of students in the classroom activities by students. The term engagement in classroom has been defined by several scholars and researchers rather differently, based on different conceptualisations and dimensions. It shows that the lecturer plays a major role in integrating the delivery content with the instructional approaches in the classroom. Students will feel engaged when they are able to apply the skills and information learnt in the writing assignment process from the materials given outside of the lecture room.

Herreid and Schiller (2013) strongly believe that implementation of flipped classroom by using interactive videos will encourage engagement among students as it generates a new conception of incorporating active, student- centred learning with content. Additionally, project-based learning activities in class will also help to increase students' understanding of the content and at the same time encourages them to be more engaged in their activities. Using the flipped classroom approach, student teachers are encouraged to do their own research for their writing assignments or projects. Thus, they are exposed to independent learning and are given the latitude to express their opinions or ideas and conduct peer evaluation through online mediums created by their instructor.

Flipped classroom as an educational teaching method in Institute of Teacher Education consists of two parts: interactive and active learning groups activities inside the classroom/lecture hall, and direct computer-based individual instruction outside the classroom/lecture hall. Implementation of new pedagogical methods such as flipped classroom needs a brief explanation of the implementation procedure to the student teachers in Teacher Education Institute. Student teachers should be given an explanation in detail about flipped classroom before implementation starts in order to avoid any unpleasant feelings, worries and emotions among student teachers. In addition, explanation about flipped classroom before the implementation process can also prepare students to accept a new approach. The lecturer will conduct a workshop for the student teachers two weeks before the implementation of the flipped classroom.

Model of Flipped Classroom

In this research, a model of flipped classroom by Reeve (2013) was adapted. The adapted model is shown in Figure 2.

Reeve (2013) explained that flipped classroom is based on mixed learning and is upheld by constructivist hypothesis (intellectual and social constructivism) that incorporates web-based learning and vis-à-vis collaboration. The characteristics of the model are as explained below:

1. By uploading videos online, instructions are carried out from the school setting and it also permits more time to implement effective teaching in the classroom. Besides that, it also offers more learning opportunities that indirectly increase the quality of face-to-face learning.
2. Online and face-to-face learning environments are utilized in the flipped classroom model; therefore, indirectly conducive learning environments are created during the learning processes in classrooms.
3. The flipped classroom model develops 21st century learning and emphasizes on learning activities which are assessed in real life context. It also stimulates active and meaningful learning.

Figure 2. Flipped classroom model adapted from Reeve (2013)

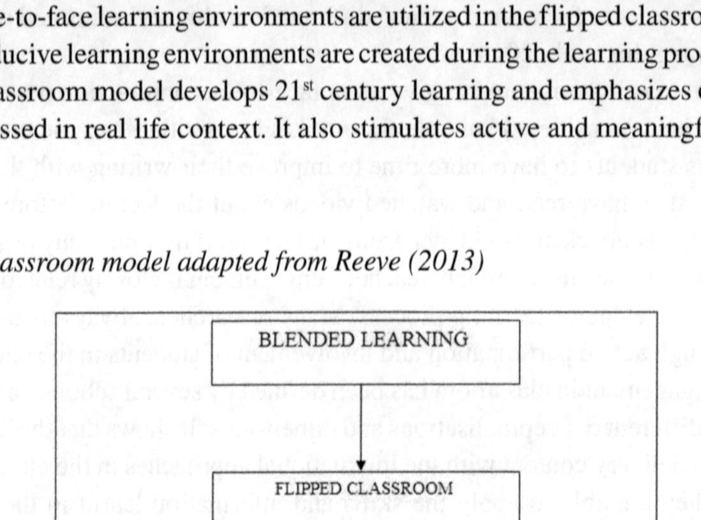

RESEARCH METHODOLOGY

Due to flipped classroom pedagogy raising popularity in this digital era, a study on flipped classroom in Teacher Education Institute which situated in Sungai Petani was carried out. The following section goes on to highlight how the flipped classroom pedagogy was implemented at different stages of pre-lesson, lesson and post lesson follow up. The process of learning the different types of pedagogy among student teachers is an important activity to build efficiency and competency of their training especially in terms of intellectual development and character as future teachers. Furthermore, pedagogy teaching plays an important role in fulfilling the needs of education and organizational education to progress and develop better in terms of knowledge and skills which are suitable with demands of the education institution. Education and teaching training allocated for the student teachers are parallel to the vision of the National Education Philosophy (Ministry of Education, 2000).

Research Sampling

The sample for this research comprised a class of 20 students who enrolled for Early European History until 20th century as a major paper. The student teachers were between the ages of 21 to 22, were highly motivated, academically average and hardworking. The rationale for the choice of sample is because they comprised of three races (Malay, Chinese and Indians). Other reasons are samples comprised of male and female, similar family background and their academic achievement in foundation are similar.

Research Design

This was a quantitative research using the quasi experimental design (Creswell, 2014). The research was carried out for five weeks. In the first week the student teachers were administered a pre-test. A day after the pre-test the student teachers were given the questionnaire to measure their motivation towards learning the Early European History until 20th century. The Early European History until 20th century is one of the important major papers that student teachers find it difficult to understand and remember the facts. From the second week to the fourth week (over a period of three weeks) the student teachers were taught using flipped classroom during their lectures. The student teachers were taught three topics using flipped classroom for three weeks with each lecture lasting two hours and one hour of tutorials. The lecturer planned the implementation of flipped classroom in two phases as below:

Pre-Implementation of Flipped Classroom

The first step towards the implementation of the flipped classroom was the selection of topics which are important and focuses on their course content. Firstly, the selection of lecture videos that suitable according to the topics and mostly referred to Khan Academy videos and also from You Tube. The selected content of the videos should be coherent, concise and clear, free of background noise and be professionally edited. The lecturer made sure the immediate priority should be to get the initial batch of videos ready; improved versions can be prepared for the next round if time permits. According to the pre survey done by the lecturer, none of the students in this class had heard of "flipped classroom" and only a small number were familiar with the word flipped classroom. Before introducing flipped classroom during the lecture, the student teachers are given briefing during a workshop conducted by the lecturer. This is to reduce the impact of the biggest obstacle of all is psychological in nature: converting a conventional class to a flipped class is a major change during lecturers. Student teachers expectations need to be appropriately set during the first lecture, and some amount of buying-in needs or persuasive approach to be done to convince student teachers that flipping is beneficial and suitable to be implemented during the lectures.

The Implementation of the Flipped Classroom Is as Below

The videos were usually uploaded two days before the flipped lesson was conducted to give student teachers ample time to watch the recordings as well as read the supplementary readings or refer to online resources that were distributed before teaching each week by the lecturer.

1. Once the video was uploaded in their class Blog Spot, student teachers were informed to view the recorded video that consists of lectures and video clips at home while simultaneously doing a task associated with the content of the topics.
2. The lecturer also prepared task as part of the course pack, came in a variety of forms – from study notes with blank gaps in information, short quizzes, short structure questions (according to different level of Bloom Taxonomy, 2001) to complete a mind map.
3. Lecturer also prepared study notes and quizzes to engage students in meaningful ways and enables them to create a set of revision notes for future reference at the end of the exercise
4. Student teachers were also told to note down any question that arose when watching the videos.
5. During the lecture, student teachers will be evaluated on their understanding of the content by asking a sequence of the questioning by the lecturer
6. The lecturer's role during the lecture room as facilitator in learning process and prepared different activities to engage the student teachers in learning process such as Student Cooperative Team Achievement Division, Think Pair and Share, Hot Seat etc. Apart from that, student teachers also have to enrich their understanding of the content through different level of questions of different stages according to the Revised Bloom Taxonomy (2001).
7. The lecturer also gives spontaneous quiz or tasks individually to evaluate the understanding of each student teacher.
8. At the end of each lesson, the links to the video lectures and self-check quizzes for the next session were put online. Students were expected to prepare for the next class by watching all the assigned videos and attempting the corresponding self-check quizzes until they got a perfect score.

To conduct this study the lecturer had uploaded three types of videos that are suitable with the content of the topic and task according to the topics in the course Blog Spot. Table 1 shows the implementation of the study .

Implementation of Flipped Classroom

During the first lesson, the lecturer explained how the flipped classroom and the importance of each member in group activities, set expectations about attendance and pre-class preparation. As a general rule of thumb, the one-hour in-lecture sessions were exclusively used for student teachers to critically apply what they learned from the videotaped lectures. The lecturer, however, spent some time clarifying the content of the lectures at the start of each lesson. For example, during lesson one, The Greek

Table 1. Implementation of the study

Week	Learning Title	Research Handling
First week		Pre Test of Achievement History Test Pre-questionnaire of Motivated Learning
Second week	Early European Civilization - Greek and Roman Civilizations	Source:Video Khan Academy-Early Civilization (Flipped Classroom)
Third week	European Middle Ages	Source: Video Middle Ages in Europe (Flipped Classroom)
Fourth week	Dark Ages	Source:Suction of You Tube: Dark Ages (Flipped Classroom)
Fifth week		Post-test of History achievement Test Post Questionnaire of Motivated Learning Strategies Questionnaire

civilization, the lecturer explained about the location and the rise of Athens and Sparta as one of the early township (provided in the video recording) in order to correct misconceptions and fill in gaps in the student teachers' understanding. It will emphasize on their prior knowledge of Greek history on this topic which they learned from secondary school History subject. The lecturer regularly used activities such as a quick Think-Pair-Share, Gallery Walk, Hot Seat or a pop quiz. Next, student teachers applied what they learnt to the main task of the lesson. With knowledge of the factors that differentiate the growth and contribution of Athens and Sparta student teachers were asked to evaluate which was the most important aspect to mankind. According to Hyland (2004) this could be done either in pairs or groups so that students could benefit from sharing their perspectives, ideas and then presenting their findings to the class. However, in order to check for individual understanding, the lecturer will questioned a number of modified questions according to higher level of Revised Taxonomy Bloom (2001). During the lecture an individual task will be given to monitor the progress of individual student teachers. Table 2 shows the questions and checks for learning and understanding in the lesson describe

Research Instruments

Instruments used for this study is history test and Motivated Strategies for Learning Questionnaire (MSLQ) by Pintrich et al. (1991). The history test consisted of 4 structural question and 2 essay questions. This structured question consisted of four main questions and was followed by a question which required the student teacher to fill in a blank with a short word or phrase. The lecturer also developed two essay questions and student teachers were required to answer all essay questions. Student teachers were given 2 hours and 30 minutes to answer 4 structural questions and 2 essay questions. The history test questions were based on the Revised Taxonomy Bloom (2001). History test questions measured the understanding and achievement of student teachers in three topics that were thought using flipped classroom method. This history test is used to find out the performance of students in European History before and after the lecture using flipped classroom method.

Pre-test and post-test questions were the same that is 4 structural questions and 2 essay questions which were prepared by the lecturer. To ensure the validity of the history test, the lecturer made sure that the topics chosen is suited the ability levels of the student teachers. Two experienced lecturers with more

Table 2. List of activities, questions, and checking for learning in the flipped classroom

Activities Prepared by the Lecturer	Possible Questions	Checking for Learning
Prior to Class (Students to watch recorded lectures) Induction of the topic- video clips about Greek History in brief- short checklist questions	What were the factors that led to development of Greek civilization?	Recap in the form of quiz or other methods during the introduction at the beginning of the lesson.
During Class (Students to apply what they know) Questioning, Hot Seat, Gallery Walk, Think Pair Share	Which was the aspect that differentiated Athens and Sparta?	Can be carried out throughout the class, preferably done in pairs / groups
After Class (Students to demonstrate what they know in individual tasks) Questions for Individual Tasks	"Historians believe that the political ideology was the most important contribution of Athens and Sparta to mankind." To what extent do you agree with this statement?	To be completed as individual work to measure individual learning.

than fifteen years of teaching experience checked the essay and structural questions suited the ability levels of the student teachers. The answers and the marking scheme for all the questions in the history test were prepared by the lecturer after discussions with two experienced senior lecturers from History department from Teacher Education Institute.

The researcher had also conducted pilot studies to test the reliability and validity of the instrument used in the study. According to Thomas and Nelson (2001), pilot study is important to avoid the methodological problems in the actual study and therefore it is important to run the pilot study to obtain the reliability of the research instrument before the actual study is conducted. The reliability value for the history test was 0.8895. Thus, the test is considered to be of high reliability and acceptable for research purposes.

The questionnaire on Motivated Strategies for Learning Questionnaire(MSLQ) comprised 20 items on Likert Scale. The lecturer adapted 20 items from a modified questionnaire of MSLQ by Pintrich et al. (1991). The student teachers responses were measured on a four point scale, that as a range value of 1 to 4, where 4 indicated total agreement and a 1 indicated total disagreement for the positive items. As for the negative items, the scores were reversed. There were 20 items, 16 positive items and 4 negative items. Table 3 shows the score for the Motivated Strategies Learning Questionnaire (MSLQ).

Table 3. Likert Scale for motivated strategies learning questionnaire

Statement	Positive Items 1,2,3,4,6,7,8,9,10,11,12,13,14,16,17,19	Negative Items 5,15,18 and 20
Total Agreement	4	1
Agreement	3	2
Disagreement	2	3
Total Disagreement	1	4

Pilot Study

A pilot study was conducted for both instruments (history test and Questionnaire) used in this research to ensure their validity. The purpose of the pilot study was to gauge the suitability of the test in terms of content, time given for answering the questions and clarity of instructions. The pilot was study among senior student teachers who has enrolled for this course previous year. The findings of the pilot study revealed that the students understood all the instructions given in the history test and the time period of two hours allocated to answer the questions are sufficient. The lecturer also distributed the questionnaire to the same senior student teachers to test its validity. The validity of the questionnaire was tested using Statistical Packages for the Social Sciences Version 21.0. The Coefficient Reliability used was "Cronbach's Alpha". The Alpha value for the questionnaire was 0.883

The first section contained several relevant data on the respondent's demographic information such as gender, age, specialization, and the semester, which was needed to provide information of the respondents. The other sections contained a total of 20 items measuring motivation among students which originally developed by MQSL Pintrich (1991) with an alpha level of 0.88.

FINDINGS AND DISCUSSION

Descriptive statistic (frequencies, percentages, means and standard deviations) were used to describe demographic information of the respondents as well as the levels of motivation among student teachers after using the flipped classroom method. The data was then analysed with an alpha level of .05 for all significance tests in the study. The Likert-type scales were considered to be interval scales.

Achievement Among Student Teachers at The Teacher Education Institute in Sungai Petani After Using Flipped Classroom

Table 4 shows the increase in the mean for student teachers after teaching using flipped classroom that is 37.000. This increase of score mean is significant ($t = 36.739$, $df = 19$, $p = 0.000$). The findings of this study was line with the findings of Dickenson (2014); Prodoehl (2015); Keenge, Onchwari&Oigara (2014); Finkell (2012) and Marlowe (2012) who strongly believed that flipped classroom has successfully transformed the traditional classroom atmosphere into a fun classroom environment. The many types of activities enabled students to actively participate in inquiry learning, project and group discussions that incorporated student's attitude of responsibility. They worked together and positively relied on complementing the tasks or activities prepared by lecturers in the classroom. Subsequently, the video's which were uploaded earlier initially helped the student in forming new knowledge before entering the classroom hence more confident of their ability to tackle different types of task given during the lecture

Table 4. T-Test showing effect of flipped classroom on achievement

Pre Test	Post Test	Mean Difference	df	t	p-Value
32.500	69.500	37.000	19	36.739	0.000

sessions. This time around the student teachers had to do the questions in class, and as a consequence, benefited greatly because they actually spent much more time completing their task. The atmosphere in the classroom helped them reinforce new knowledge by linking the available knowledge and adapting the ideas or concept in their daily life. Students also felt confident to discuss or seek help from their friends if they did not understand the contents of the lesson during their preparation at home. Therefore, the flipped classroom method gave a time frame for student teachers themselves to master the contents of their lessons according to their level of intelligence. This finding also showed that student teachers ranked high the statements that described the understanding of the content course as well as the ability of improving classroom practices and student centred lessons from this teacher education program. The findings showed that there has been a significant increase in the achievement among the student teachers before and after the use of flipped classroom during lectures. They worked together and positively worked on completing the tasks or activities provided by lecturer during the lecture time. Subsequently, the uploaded video's initially helped the student teachers in forming new ideas and knowledge during their active learning in the classroom, understanding the historical concepts before entering the class and encouraging the student teachers' to be more think critically . The flipped classroom also helped the student teachers also involved actively especially during group work discussion such as Student Cooperative Teams Achievement Division (STAD), Think Pair and Share and Gallery Walk. The research findings was also supported by Bowers &Keisler (2011) who strongly believed that the activities planned during the flipped classroom had to be augmented with other teaching strategies (pair/group work, presentations, initial didactic teaching, and individual homework) in order to be highly effective. A tailored approach that consisted of teaching strategies /methods/ approaches were used in different combinations that engage or appeal to the students should be an effort undertaken by educators.

Improvement of Motivation Among Student Teachers After Using the Flipped Classroom Method

Total mean score of student teachers before teaching using flipped classroom methods was 45.894 and the score mean after application of flipped classroom method was 68.894. Mean difference =23.00, t=20.404, p=0.000. The findings showed that there was a significant increase in motivated strategies for learning among student teachers after the use of flipped classroom method. The results of the study indicated that high levels of motivation and satisfaction among student teachers in teacher education institute in Sungai Petani. The findings of this study are in line with the findings of the study by Noor Erma et al., (2014) andZulzana et al., (2013) which proved that there is a positive correlation between the tendency of motivation towards learning and the quality of teaching in the classroom. These findings corresponded with the results of a previous work in this field of Subadrah Nair and Malar Muthiah

Table 5. Improvement of motivation

Pre Test Mean	Post Test Mean	Mean Difference	df	t	p-Value
45.894	68.894	23.000	19	20.404	0.000

(2005) that proved that the appropriate method that focused on hands-on activities can help students appreciate historical events and the importance of historical facts. In addition, it could also eliminate the boredom, lack of interest and attitude of the students on history. The finding is consistent with the study byBliuc*et al.*, (2011) that increase of high levels of motivation among student teachers in teacher education program due to a change in the learning culture among the student teachers were more inclined to take ownership of their learning because the availability of the video lectures empowered them to do so. In previous cases, the student teachers were usually unable to be involved totally in the activities and complete the hands on task that followed by re-watching the videos until they were convinced that they had understood the contents .Themain aim of adding the online dimension to the face-to-face discussion was to provide the best context for a broader range of students to participate in discussion. Thus, when instructors/lecturers used methods that allowed student teachers interact with materials and historical events, then it would increase the understanding of events and help them to remember historical facts. They experienced increased collaborative activities and interaction between student and student as well as between student and instructor through tasks, and activities. t.

IMPLICATION OF RESEARCH

The data analyses show that student teachers from Teacher Education Institute have accepted readiness in flipped classroom. However, appropriate training on the usefulness and effectiveness of the technology can play an important role to concretize the adoption of flipped classroom among the lecturers/ teachers and students. Hence, the research implication is that different steps need to be taken by Ministry of Higher Education and stakeholders to train student teachers and novice teachers in the use of flipped classroom. Also, student teachers should be encouraged to embrace this new method.

CONCLUSION

Flipped Learning can have numerous advantages for the higher education.Normally, students learning in such environment may initiate to more effectively and efficiently educate each other, and embark on taking the responsibility for their learning. Many studies in the area of flipped classroom also reveal that when this method is used, the result will be more learning. This study also has shown that student teachers in Teacher Education Institute during this course who were exposed to the flipped classroom had enjoyed the experience with a significant number believing that it was an effective pedagogy method. The repeatability of the videos at any time and place allowed student teachers to prepare thoroughly for class, and the displacement of classroom lectures meant additional contact time for more useful and engaging learning activities that also helped them build up their confidence and enjoyment of the subject matter. Depending on the active learning activities chosen for classroom time, there could also be more opportunities for students to interact and learn from one another. The flipped classroom benefited the student teachers as they are able to repeat the video clips as many times as needed and viewing the videos at home "saved time" so that more could be done in class. In fact, some student teachers were glad that they were "forced" to come to class prepared. The flipping classroom had brought about a positive change to student teachers' ownership and responsibility toward learning. Much higher engagement during class time was observed because of ability to easily and effortlessly produce and use multimedia on PCs, smart phones, tablets and personal media players, and the enhanced accessibility of Web-based tools for the purpose of collaboration and communication are considered as two factors that have a great contribution to the popularity of the flipped classroom setting. It is recommended that the flipped classroom be incrementally introduced to other courses in teacher education Institute in Malaysia that are likely to benefit from this flipped classroom pedagogy so as to prepare future teachers to be ready to face the challenges or IR 4.0.

REFERENCES

Almacen, B. (2010). *Teacher training program: The Malaysian perspective*. Retrieved on 3 July 2012 from http://independent.academia.edu/DrBayaniAlmacen/Books/408843/Teacher_Training_Program_The_Malaysian_Perspective

Azizan, F. Z. (2010). Blended Learning in Higher Education Institution in Malaysia. *Proceedings of Regional Conference on Knowledge Integration in ICT.*

Baker, J. W. (2000). The "classroom flip": Using web course man- agement tools to become the guide on the side. *11th International Conference on College Teaching and Learning.*

Baker, L. M., & Settle, Q. (2013, September). Flipping the classroom and furthering our careers. *NACTA Journal, 57*(3), 75.

Barak, M., & Shakhman, L. (2008). Teachers and Teaching. *Theory and Practice, 14*(3), 191-208.

Bergmann, J., & Sams, A. (2012). *Flip your classroom: Reach every student in every class every day.* Arlington, VA: International Society for Technology in Education.

Bloom, B. S., Engelhart, M. D., Furst, E. J., Hill, W. H., & Krathwohl, D. R. (1956). *Taxonomy of Educational Objectives: The Classification of Educational Goals. Handbook I: Cognitive Domain.* New York: Longmans.

Demirer & Sahin. (2013). *Effect of blended learning environment on transfer of learning: an experimental study*. doi:10.1111/jcal.12009

Dickenson, P. (2014). Flipping the Classroom in a Teacher Education Course. In J. Keengwe, G. Onchwari & J. Oigara (Eds.), Promoting Active Learning through the Flipped Classroom Model (pp. 145-162). Hershey, PA: Information Science Reference. doi:10.4018/978-1-4666-4987-3.ch008

Embi, M. A. (2010). e-Learning in Malaysian Higher Education Institutions. Klang: Ministry of Higher Education, Malaysia.

Eunmi, Y. U. (2017). Influence of Flipped Classroom Model on Korean Translation Teaching. *2017 International Conference on Education Innovation and Economic Management (EIEM 2017).*

Finkel, E. (2012). Flipping the script in k12. *District Administration, 48*(10), 28–30.

Fleck, J. (2012). Alternative Models Blended learning and communities: Opportunities and challenges. *Journal of Management Development*, *31*(4), 398–441. doi:10.1108/02621711211219059

Garrison, D. R., & Vaughan, N. (2008). *Blended Learning in Higher Education: Framework, Principles, and Guidelines.* Jossey-Bass.

Garrison, R., & Kanuka, H. (2004). Blended learning Uncovering its transformative potential in higher education. *The Internet and Higher Education*, *7*(2), 95–105. doi:10.1016/j.iheduc.2004.02.001

Gleason, N. W. (2018). *Higher education in the era of the fourth industrial revolution*. Retrieved from https://link.springer.com/content/pdf/bfm%3A978-981-13-0194-0%2F1.pdf

Goodwin, B., & Miller, K. (2013). Research says evidence on flipped classrooms is still coming in. *Educational Journal of Information Systems Education, 25*(1).

Greason N. W. (2018). *Higher education in the era of the fourth industrial revolution.* PELLGRAVE.

Hativa, N. (2001). *Teaching for effective learning in higher education.* Kluwer Academic Publishers.

Herreid, C., & Schiller, N. (2013). Caswe Studies and The Flipped Classroom. *Journal of College Science Teaching*, *42*, 62–66.

Hyland, K. (2004). *Genre and second language writing.* University of Michigan Press.

Järvelä, S., Volet, S., & Järvenoja, H. (2010). Research on motivation in collaborative learning: Moving beyond the cognitive-situative divide and combining individual and social processes. *Educational Psychologist*, *45*(1), 15–27.

Keengwe, J., Onchwari, G., & Oigara, J. (2014). *Promoting Active Learning through the Flipped Classroom Model.* Hershey, PA: Information Science Reference. doi:10.4018/978-1-4666-4987-3

Khan Academy. (2012). *Watch. practice. learn almost anything for free.* Retrieved from http://www.khanacademy.org/about

Kim, M. K., Kim, S. M., Khera, O., & Getman, J. (2014). The experience of three flipped classrooms in an urban university: an exploration of design principles. *The Internet and Higher Education*, *22*, 37-50.

Kurt, G. (2017). Implementing the flipped classroom in teacher education: evidence from Turkey. *Educational Technology & Society*, (99), 211-221.

Lewis, K. E. (2018). *How to ensure the Fourth Industrial Revolution is "Made in the USA"*. Retrieved from http://theconversation.com/how-to-ensure-the-fourth-industrial-revolution-is-made-in-the-usa-81385

Marlowe, C. A. (2012). *The Effect of the Flipped Classroom on Student Achievement and Stress* (Unpublished Master of Science Thesis). Montana University.

McLaughlin, J. E., Roth, M. T., Glatt, D. M., Gharkholonarehe, N., Davidson, C. A., Griffin, L. T. M., ... Mumper, R. J. (2014). The flipped classroom: A course redesign to foster learning and engagement in a health professions school. *Academic Medicine*, *89*(2), 236–243. doi:10.1097/ACM.0000000000000086 PMID:24270916

Means, B., Toyama, Y., Murphy, R., Bakia, M., & Jones, K. (2009). *Evaluation of Evidence-based Practices in Online Learning: A Meta-analysis and Review of Online-learning Studies*. Washington, DC: U.S. Department of Education.

Mohamad, M. M., Saud, M. S., & Ahmad, A. (2011). The need in training and retraining for TVET teachers in Malaysia. *Journal of Technical Education and Training*, *1*(1), 51–57.

Pacansky-Brock, M. (2013). *Best practices for teaching with emerging technologies*. New York: Routledge.

Pintrich, P. R., Smith, D. A. F., Garcia, T., & McKeachie, W. J. (1991). *A manual for the use of the Motivated Strategies for Learning Questionnaire (MSLQ)* (Tech. Report No. 91-B-004). Boards of Regents, University of Michigan.

Poon, J. (2013). Blended learning: An institutional approach for enhancing students' learning experiences. *Journal of Online Learning and Teaching*, *9*(2). Available at http://jolt.merlot.org/vol9no2/poon_0613.htm

Prodoehl, D. (2015). Flipping First-Year English: Strengthening Teacher-Student Conferencing through Online Modules. In A. Abigail (Eds.), Implementation and Critical Assessment of the Flipped Classroom Experience (pp. 1-24). Hershey, PA: Information Science.

Puteh, M., & Hussin, S. (2007). *The role of Malaysian government it policies & e-learning in the tertiary sector.* Paper presented at the National E-Community Seminar on Bridging the Digital Divide: Malaysia's Initiative, Kuala Lumpur, Malaysia.

Shen, Y. W., Reynolds, T. H., Bonk, C. J., & Brush, T. A. (2013). A Case Study of Applying Blended Learning in an Accelerated Post-Baccalaureate Teacher Education Program. *Journal of Educational Technology Development and Exchange*, *6*(1), 6. doi:10.18785/jetde.0601.05

Siew-Eng, L., Ariffin, S. R. B., & Rahman, S. B. (2010). Diversity in education using blended learning in Sarawak. *US-China Education Review*, *7*(2), 83–88.

Singh, H., & Reed, C. (2001). *A White Paper: Achieving Success with Blended Learning*. Retrieved July 15,2015 from www.leerbeleving.nl/wbts/wbt2014

Strayer, J. F. (2007). *The effects of the classroom flip on the learning environment: A comparison of learning activity in a traditional classroom and a flip classroom that used an intelligent tutoring system* (Unpublished doctoral dissertation). Ohio State University.

Strayer, J. F. (2012). How learning in an inverted classroom influences cooperation, innovation and task orientation. *Learning Environments Research*, *15*(2), 171–193. doi:10.100710984-012-9108-4

Teaching and Teacher in Blended Learning Model. (n.d.). Retrieved from http://www.google.com

Thomas, J. R., Nelson, J. K., & Silverman, S. J. (2011). *Research methods in physical activity*. Champaign, IL: Human Kinetics.

Tucker, B. (2012). The Flipped Classroom. *Education Next*, *12*(1).

Vygotsky, L. S. (1986). *Thought and language*. Cambridge, MA: MIT Press. Wainer.

Chapter 8
Edmodo in an Institute of Teacher Education:
A Platform for Blended Learning

Quinie Ong Kooi Loo
https://orcid.org/0000-0002-9201-8624
Institut Pendidikan Guru Kampus Sultan Abdul Halim, Malaysia

Kathleen Dass
Institut Pendidikan Guru Kampus Sultan Abdul Halim, Malaysia

ABSTRACT

Edmodo is a free online learning platform for teachers to communicate and collaborate with students. It is a digital communication tool that has been designed to work across a range of devices, suitable for blended learning. To maximize learning and meet the challenges of the current digitalized classroom, supplemental blended learning was adopted in an English literature course in an institute of teacher education in Kedah, Malaysia. This chapter describes how Edmodo was used to manage blended learning, with the focus on online activities created in posts, small groups, assignment tool, and polling tool. The activities were planned based on the five ingredients on blended learning by Carman. The result of a poll done in Edmodo showed 74% of the students agreed that they were able to explicate Shakespeare after attending the course, and 81% opined that the course was effectively organized in Edmodo. Based on the positive feedback, Edmodo is recommended for subsequent blended learning courses.

INTRODUCTION

Blended learning has received much attention in recent years due to its ability to accommodate the needs of 21st century learning where learning goes beyond the classroom walls. Blended learning refers to the mixing of traditional face-to-face instruction with online learning, and it includes the use of digital content, resources, and tools to enhance, extend and transform the learning process (Pape, Sheehan & Worrell, 2012). Blended learning can be defined as a combination of any online and traditional brick-and-

DOI: 10.4018/978-1-5225-7832-1.ch008

Copyright © 2019, IGI Global. Copying or distributing in print or electronic forms without written permission of IGI Global is prohibited.

mortar learning environments which also utilizes other modes to improve learning outcomes and save cost. (Banados, 2006). Therefore, blended learning can be defined as the integration of complementary face-to-face and online strategies (Garrison & Vaughan, 2008).

Implementing blended learning programs is one alternative for English learners who do not get enough exposure to the language in class or opportunities to communicate, as well as make explorations regarding the language (Tomlinson & Whittaker, 2013). Educational systems such as Moodle as well as Edmodo allow learners to begin interaction and share information outside the class. Particularly if instructors come across an absence of time with their students, it is much better to review a topic with them by means of social media in educational ways (Yagci, 2015). As a result, students can execute the tasks outside the usual course duration by using online educational platforms. (Tomlinson & Whittaker, 2013).

There are many learning management systems (LMS) available and one prominent international education network which offers its users communication, cooperation and also training techniques is Edmodo (Cauley, n.d.; Edmodo, 2013; Renard, 2016). Edmodo is a secure environment and devoid of advertisements, video games, as well as other diversions that might interrupt student learning. Teachers can post announcements and also tasks for their students on Edmodo. Students use the network to communicate with their teachers to share suggestions, problems, and useful tips; ask questions concerning lessons and research; and also team up with various other classmates on tasks and group work (Arroyo, 2011; BusinessWire, 2014; Edmodo, 2014; Mills & Chandra, 2011; Renard, 2016).

In this chapter, Edmodo was chosen by the authors to be the platform to facilitate blended learning because the institutes of teacher education in Malaysia do not provide nor prescribe any official online platform for use in the institutions. In the literature course taught by the authors, *Drama in English,* Edmodo was used by four lecturers teaching the course to manage the virtual classroom, share materials, communicate and collaborate among themselves; and with two classes of students which added up to 27 students. Four features of Edmodo were utilized to create online tasks. They were ***posts, small groups, assignment tool, and polling tool***. Other digital applications were also used such as Kahoot, Padlet, Word Search Maker and Instant Online Crossword Puzzle Maker. With Edmodo, online tasks were organized and recorded in a more systematic way. Hence, the objective of this chapter is to share how Edmodo was used as a platform to maximize learning in a literature course through blended learning.

BACKGROUND

The ways of teaching and learning have changed in this fast-growing world of technology. Learning is no longer within the confines of a school compound. Virtual classrooms are becoming more popular with the presence of the World Wide Web. To support online learning, many learning management systems have been created. An LMS is a system that enables users to “create, administer, manage, and analyze eLearning courses and trainings. It is the engine that powers eLearning” (CAE Computer Aided USA Corp., 2018, para. 1). Besides, it is also a massive depot for the storing and tracking of information. With a login and password, one can access online resources at anytime and anywhere (Pappas, 2017).

There are many LMSs available; some are more suited for the academia while others are for the corporate and business world. Although the LMSs have basic features, each has its own distinct core elements, which makes it more suitable for different audiences. Some basic features that qualify a software system to be an LMS are platforms for managing learning content, managing user roles, and delivering the learning content. They are the digital platforms that move, track and enhance learning (Quigley, 2018).

One of the LMSs widely used in academia is Edmodo. Unlike most LMSs, Edmodo was actually created by teachers for teachers. Two teachers, Nic Borg and Jeff O'Hara, launched Edmodo in 2008 (Edmodo, 2017). Edmodo is a free online learning platform for teachers to manage a virtual classroom, share materials, communicate and collaborate with students and other teachers in real time. In recent years, Edmodo has been applied in language classrooms by countless educational institutions around the world due to its appealing features (Delacruz, 2013); user-friendliness (Capterra, 2018b; Witherspoon, 2011); and free and secure online environment (Kongchan, 2013). Edmodo, undoubtedly aids students' learning experience in various ways (Capterra, 2018b; Purnawarman, Susilawati, & Sundayana, 2016) and enables educators to set up, run and manage their online classes easily (Witherspoon, 2011). In the context of teacher education programs, Ekici (2017) noted that practitioners can utilize Edmodo to form their own online community.

Edmodo is a digital communication tool which is device agnostic. It operates across various types of devices, including desktop computers, laptops, tablets and smartphones. In addition, it has countless resources which are created and shared by fellow teachers, making them relevant for the students. It also allows digital assignments and auto-graded quizzes to be conducted efficiently (Edmodo, 2018). In 2018, Edmodo was ranked the most popular LMS by Capterra (2018a) with 350,000 customers and 58 000 000 users. However, it was not because of its popularity that Edmodo was the chosen LMS for the course *Drama in English*. Having used Edmodo before, the authors found it to be user-friendly and efficient, and hence, wanted to try it out more extensively in a blended learning environment.

Blended learning refers to a learning scenario where a "portion of traditional face-to-face instruction is replaced by web-based online learning" (Blended Learning Toolkit, n.d.). However, McGee and Reis (2012) point out that there is no absolute agreement within higher education on the exact features of a blended course. The term *blended* generally refers to some combination of on-campus class meeting and online activities. Maxwell (2016) concurs and elaborates that "[w]ithout a universal definition of blended learning there is no shared language by which the education field can describe the phenomena or address its opportunities and challenges" (para.1). Meanwhile, Teach Thought Staff (2018) states that blended learning generally means a mix of online learning and face-to-face learning – meaning that most students would have had some experience in blended learning for years already.

Out of the three primary models of blended learning by Teaching with Technology Online Workshop Series 2012-2013, the chosen model for the *Drama in English* course was the supplemental model. In this model, activities created online were supplements to face-to-face interactions. "The big idea here is *supplementing* – critical learning objectives are met entirely in one space while the 'opposite' space provides the student with specific supplementing experiences that the other did not or could not provide." (Teach Thought Staff, 2018, para. 11).

Different ways are used to describe and identify key features of blended learning. Carman (2002) identifies five key ingredients on how to organize different elements of online and traditional instructions for blended learning. They are *Life Events*, *Self-Paced Learning*, *Collaboration*, *Assessment* and *Support Materials* (p. 2). *Live events* encompass video conferences and synchronous chat sessions, while *self-paced learning* includes internet-based tutorial tasks to be completed at the students' own time. As for *collaboration* to take place, e-mails and threaded discussions are two common platforms. Meanwhile, *assessment* is the measure of learners' knowledge for pre-assessment and post-assessment to measure learning transfer. Finally, the last ingredient *support materials*, both virtual and physical, aid learning retention and transfer. All these five ingredients are rather similar to examples of effective blended teaching strategies as suggested by Pape et al. (2012) which are, online classroom teaching resources for students to review after class, online assessments for self-check, online collaborations after class and homework help through synchronous chat.

A 2010 meta-analysis published by the U.S. Department of Education suggests that "students exposed to both face-to-face and online education were more successful than students who were entirely in one camp or the other" (as cited in Teach Thought Staff, Does Blended Learning Work? 2015, para. 2). Hence, the *Drama in English* course was planned based on the supplemental blended learning model and the five blended learning ingredients by Carman (2002) to maximize learning.

LITERATURE REVIEW

Edmodo was created for K-12 use (Edmodo, 2018) but its use has been extended to many at tertiary level. Given the popularity of Edmodo "the world's largest learning network for K-12 students, teachers, administrators, and parents" (Edmodo, 2018), it is important to establish the advantages on using Edmodo as a platform for blended learning at the tertiary level. The review is divided into three sections. The first section examines the past researches on perception and preference towards the use of Edmodo, followed by the effects of Edmodo on student achievement and the potential of Edmodo to help students enhance their learning. Next, the review examines the characteristics of the students which are relevant for the construction of online activities. The review closes with an explanation on the model and strategy used in the course.

Past Research on Edmodo

In a case study conducted on 24 tertiary students' experience using Edmodo in language learning by Zuraina (2015) in a technical university in Malaysia, the findings showed a mixed review. An interview, and an observation on students' behavior when Edmodo was used in class, were conducted to gauge the students' perception on the use of the platform for language learning to supplement face-to-face discussions in English language classes. The positive reviews were the ease of a resource sharing platform and the user-friendly interface. In addition, they were of the opinion that Edmodo was able to assist in the teaching of the writing skills. To them, collaboration was easy and the use of colors aided them in correcting their mistakes. However, they found that the use of Edmodo did not contribute much to their learning of English, particularly in their presentation skills. Some rejected the idea of using Edmodo for discussions as they were worried that there would be misinterpretations along the way. Hence, they suggested having face-to-face discussions after doing the virtual discussions. Nevertheless, two themes were identified. Firstly, for the benefit of the students, educators need to be constantly updating themselves with the use of the platform before implementing it. Secondly, this research contributes to the knowledge of educators on how to integrate Web. 2.0 technologies into language classes.

In a study on students' preference, Balasubramanian, Jaykumar. and Fukey (2014) conducted a quantitative study on the use of information and communication technology via Edmodo. Survey questions were developed using the RASE (Resource, Activity, Support and Evaluation) model. Two hundred eighty-five undergraduates participated in the study, out of which two hundred forty-nine were taken for the analysis as they were Edmodo users. Both, student engagement and responsible learning were evident when particular Edmodo features were employed. Findings showed that students' preference of using Edmodo was mainly towards the resources, support and communication such as forum, discussions and online activities. They found the platform user-friendly and they enjoyed working on an online class. However, the researchers discovered that the content resources provided by the lecturers were not sufficient and that the resources needed to be used in an engaging manner. Support for students was another issue to be addressed in order for them to solve any difficulties independently. Finally, evaluation to inform students' progress and as a learning tool for the educators to improve teaching was also important to create a responsible learning environment.

There are many interactive platforms for learning but the problem lies in whether the platforms can sufficiently accommodate the dynamics of educational instruction, accountability and the emergence of new technology in K-12. Hankins (2015) investigated the effects of Edmodo on student achievement in Middle Schools in Miami. The observational study failed to identify a significant effect of Edmodo on students' reading achievement except for English Language Learners. However, Edmodo was useful as a platform for interaction, to upload and share links, form groups and work collaboratively (Hankins, 2015; Shashko & Koteska, 2017).

Social media might be used for educational purposes but the use of social platforms presented many problems as their inventions were geared towards social networking purposes (Mokhtar, 2018). For example, Facebook does not help in setting boundaries between professional and personal matters while Twitter has its weaknesses in its character count limitations and buttons that are designed only for social media purposes. Therefore, Mokhtar (2018) focused her study on utilizing Edmodo's potential in helping students to enhance their learning curve and to find ways to assist students in their learning experiences. This study was relevant as the 4 undergraduates were to be future teachers and they would have some hands-on experience on how students feel and experience as a user of the online platform. At the end of the six-month study, the undergraduates were interviewed and requested to draw their own concept maps on their views on Edmodo as a tool for students' education. The findings showed some positive views on "the pedagogical implications pertaining to how Edmodo breaks barriers by enabling students to access learning resources beyond the classroom and empowering student-centered learning through novelty of tasks, suitable learning environment, and mobility" (p. 61).

Characteristics of the Students

It is pertinent to consider the characteristics of the students in the planning of this course. Analyses by Fell (2016) and her research team shared five main characteristics that could be applied to the students of this course. They were:

- Social – with the emergence of flipped classrooms, the learning of content is done outside of class but the essential practice and engagement is done in the social context of the classroom.
- Mobile – today's students are mobile in their use of technology for communication, future jobs and homes.
- Global – likely to work in multiple countries and they are the most globally connected generation.
- Digital – they are also called the *Generation Glass* – where iPad and screen dominate communication through visuals more than the written form.
- Visual – learning styles are heading towards visual and hands-on learning, and away from the traditional auditory delivery form.

These students interact in the post literate community where texts and twits are brief. For them, visual and videos are more effective and preferred forms of communication. Based on the characteristics, activities that appealed to the students were created as supplemental blended learning in the course conducted. These activities enhanced and engaged students for a more effective learning experience.

Model and Strategy

Supplemental Blended Learning

"As digital and social media become more and more prevalent in the life of learners, it was only a matter of time before learning became 'blended' by necessity" (Teach Thought Staff, 2018, para. 11). This is especially so in learning at a tertiary level. Hence, looking for the right platform to facilitate blended learning would be the main concern of educators of this century. Out of the three primary blended models by Teaching with Technology (2012-2013), supplemental blended learning best fits a traditional classroom. Online activities are added as supplements to improve and enhance learning. These activities are planned to complement the lessons taught during the face-to-face interaction with different instructional goals in mind. Some of the goals identified are:

- **Deepen Interaction:** Enable online collaboration to deepen classroom interactions.
- **Increase Time Efficiency:** Demonstrate key concepts or foundational material online to make better use of class time.
- **Offer Flexibility:** On-demand review learning when they enter the classroom.
- **Enable Review and Repetition:** Use online modules to cover complex topics for students to access on demand. (Teaching with Technology, 2012-2013, para. 1)

Users of supplemental blended learning, however, need to take note of the "course and a half" dilemma of not overloading and encroaching into students' learning time. This is because seat time is not reduced even though online components have been added (Teaching with Technology, 2012-2013).

Blended Learning Ingredients

A brief description of the five ingredients on blended learning by Carman (2002) and blended learning strategies by Pape et al., which were the basis of how the online activities were planned, are shown in Table 1. The online activities were planned accordingly except for the first key ingredient. In the course conducted, Live Events was considered synonymous to the face-to-face classroom interactions that were conducted.

Most of the researches conducted on Edmodo are on the perceptions and preferences of students towards Edmodo. Indeed, Edmodo is a popular online platform for communication and the storing of materials. With blended learning becoming a norm in the educational world, especially so in the tertiary level, more researches on how to use the features in Edmodo from the pedagogical view are needed to fill the gap. With the basis of an appropriate blended learning model and blended learning strategies, Edmodo would become the preferred online learning platform to accommodate 21st century learning.

Table 1. Comparison of the key ingredients on blended learning and blended teaching strategies

No.	Blended Learning Key Ingredients (Carman, 2002)	Blended Teaching Strategies (Pape et al., 2012)
1.	Live Events: A live online classroom	
2.	Self-paced Learning: Individual internet-based learning experiences at his own speed and on his own time.	Online collaborations after class Online assessments for self-check Homework help through synchronous chat.
3.	Collaboration: Learners communicate with others using online platforms	Online collaborations after class
4.	Assessment: A measure of learners' knowledge for pre-assessments before live or self-paced events, to determine prior knowledge, and post-assessments following live or self-paced learning events, to measure learning transfer.	Online assessments for self-check
5.	Performance Support Materials: Online materials that enhance learning retention and transfer	Online classroom teaching resources for students to review after class

MAIN FOCUS OF THE CHAPTER

Implementation of Strategies

A virtual class by the name 2016 *TSL1064* was created for the *Drama in English* course in Edmodo in addition to the usual face-to-face class activities. The focus of this chapter is on the online activities administered via Edmodo as the platform for blended learning. Four features of Edmodo (post, poll, small group and assignment) were used to manage the online activities in addition to three other digital applications (Kahoot, Word Search and Crossword Makers) found in the internet. The implementation of online activities based on the blended learning ingredients postulated by Carman (2002) and the rationales for each step taken is indicated in Table 2.

Edmodo was used from the start till the end of the course to facilitate online activities such as sharing of resources, self-check assessments, collaborations after class as well as synchronous and asynchronous homework chat in the course. Online communication and interaction between lecturers and students were easier and more organized with the use of Edmodo. The device agnostic quality of Edmodo enabled access and utilization through various means; hence, there was no problem of access for the students or lecturers. Another plus to the use of Edmodo is its ability to support Google Apps for Education and Microsoft One Note and Microsoft Office (Edmodo, 2017). The four main features of Edmodo utilized in this course – *posts, small groups, assignment tool, and polling tool* are discussed in the following paragraphs.

Feature 1: Posts

Purpose: Lecturers posted announcements and instructions to students so that students could look them up at any time, and as many times as needed. This placed more responsibility on the students to take charge of their own learning. There were no more *missed* announcements or excuses of not getting any announcements from the lecturers.

Table 2. Online activities based on the blended learning ingredients by Carman (2002)

No.	Online Activity	Carman's Blended Learning Ingredient	Edmodo Feature/ Digital Application	Rationale
1.	Let's Watch Shakespeare	Support materials	Small group	To provide some background knowledge on Shakespeare
2.	Let's Twit Shakespeare	Self-paced learning Collaboration	Small group	To alleviate students' fear of learning Shakespeare and motivate the shy ones to be more confident in asking questions and responding to others
3.	Let's Pick the Video and Comment	Self-paced learning	Polling tool	To encourage students to share their point of view
4.	Chain Response to *Prince Lear*	Self-paced learning Collaboration	Post	To provide an avenue for students to be more critical
5.	Quizzes and Online Games	Assessment	Kahoot, Word Search, Crossword Puzzle	To spark some interest in the students through competition
6.	Discussions with group members	Collaboration	Post	To encourage students to share their points of view
7.	Respond to Drama Workshop attended at Performing Arts Centre in Penang	Assessment	Post	To provide an avenue for students to be more critical
8.	Creating online posters to promote the drama presentation	Collaboration	Assignment tool Google Slides & Docs	To instill creativity in the students
9.	Upload tutorial tasks for sharing	Support Materials	Post Padlet	To share tutorial tasks with everyone in the class for easy access
10.	Upload Tutorial tasks for record	Support Materials	Assignment tool	To be able to retrieve students' work easily (for lecturers)

Advantage: This feature was a good centralized storage place for all messages. Prior to the use of this feature in Edmodo, messages that were transmitted through the use of Short Messaging System and WhatsApp had to be deleted after some time due to the limited space in the mobile phones (Jones, 2017). With Edmodo, there was no constrain of space, and hence there was no need for any post to be deleted to accommodate further uploads. In fact, all posts can be retained and referred to at any time and any place. This was a good way for keeping track of the events that happened during the course. Figure 1. shows an example of a post with an announcement which could still be retrieved till today.

Feature 2: Small Groups

Purpose: Ten small groups were created for different purposes in the virtual classroom. As there were 6 tutorial groups in the course TSL1064, *Groups 1-6* were created for each of the tutorial group which consists of 4-5 students. Four other small groups with interesting tasks which are familiar and current to 21st century learning were created. They were named *Let's Tweet Shakespeare, Let's Watch Shakespeare, Online Tasks and Photo Gallery.*

Advantage: These small groups paved the way for an organized way of implementing online activities for the students. The small groups formed are displayed in Figure 2.

The descriptions of the small groups (Groups 11-6, Let's Tweet Shakespeare, Let's Watch Shakespeare, Online Tasks and Photo Gallery) are explained below.

Figure 1. Posts in Edmodo. This figure illustrates an announcement in the post.

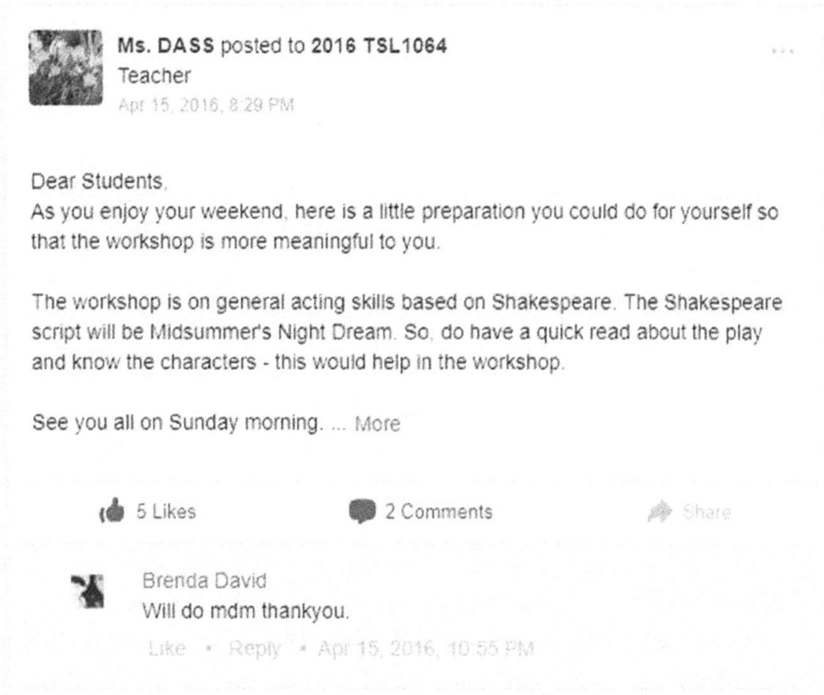

Figure 2. Small groups in Edmodo. This figure illustrates the ten small groups created.

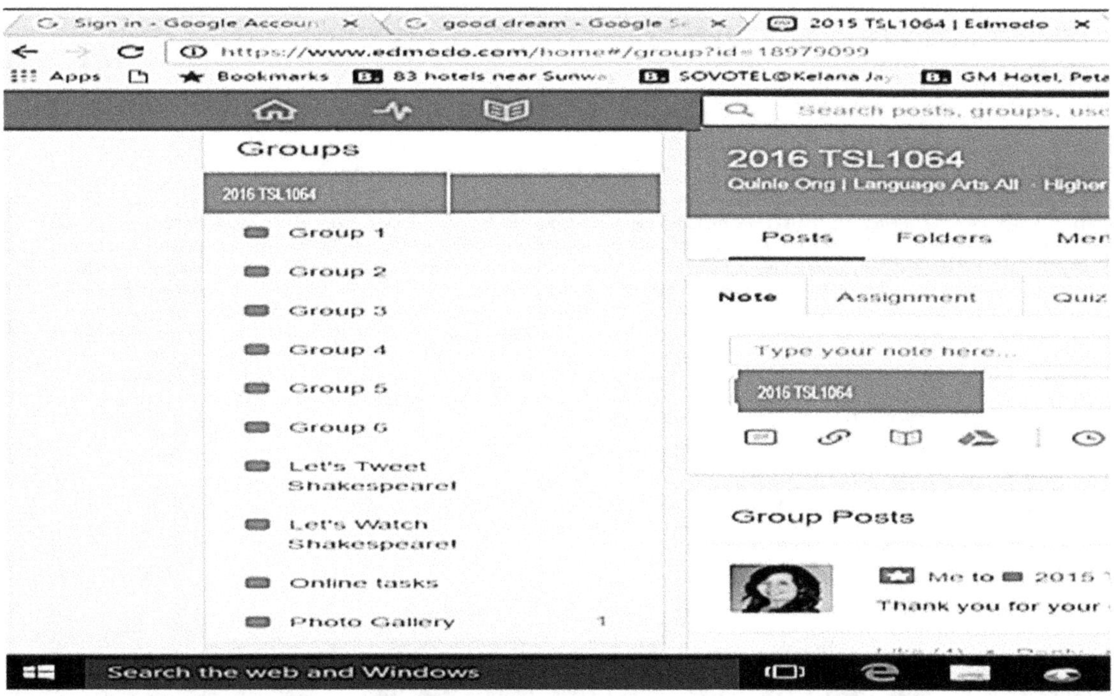

Groups 1 to 6

Purpose: In the course conducted, there were six tutorial groups so six small groups were created for each tutorial group formed. In these small groups, students posted discussions in their respective groups. It was exclusive as only those who were in the group could access the information and discuss with each other. Students were free to voice their opinions among themselves and not feel threatened or embarrassed when they said something wrong in this small group. It was also in this platform that they sent their weekly tutorial tasks for the lecturers to check. An example is shown in Figure 3.

Advantage: This gave the shy students more confidence to participate in the group discussions. They could also get feedback from their lecturers online.

Let's Tweet Shakespeare

Purpose: Let's Tweet Shakespeare (Figure 4) was created with the aim of alleviating the fear of learning Shakespeare. These students have never done any play by Shakespeare before. The little they knew about Shakespeare was that the language in the text looked different and difficult. Hence, to address their initial fear and to lower their anxiety, this small group provided the avenue for the students to channel their initial thoughts and feelings on learning Shakespeare. To make it more interactive, the lecturers played the role of Shakespeare by responding to the "twits". The lecturers acknowledged their feelings and thoughts, and provided early encouragement and support as they began their study of Shakespeare.

Figure 3. The first small group in Edmodo. This figure illustrates one of the tutorial tasks created and the interaction between a lecturer and a student.

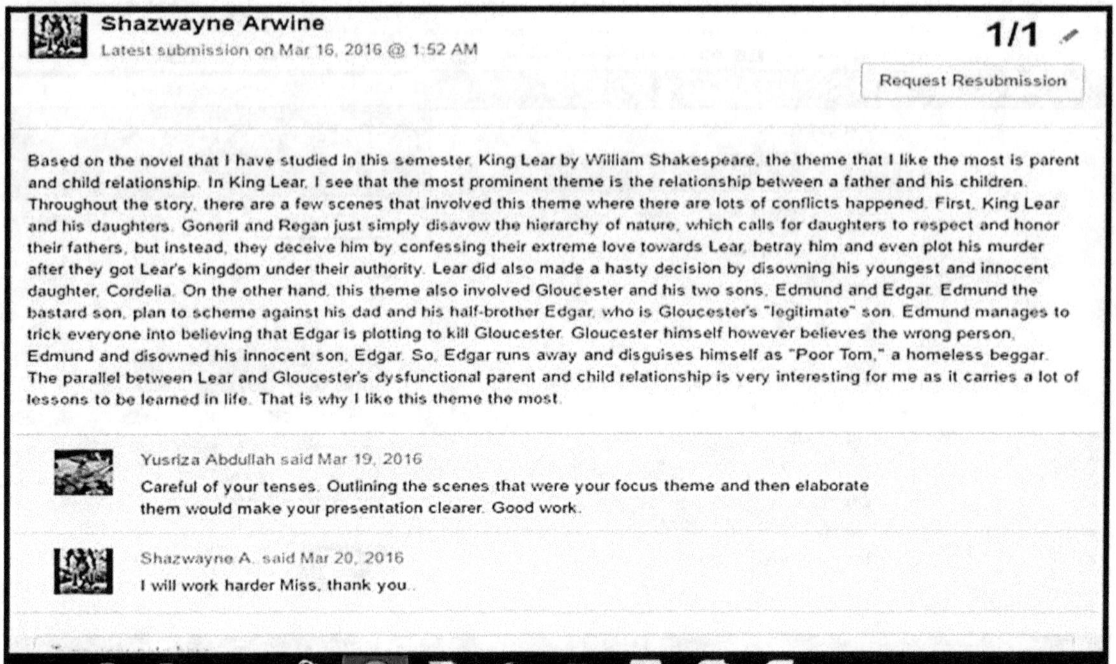

Advantage: This was an excellent opportunity especially for the more reserved students to express and share their reservations. It was also less stressful for students to pose questions and responses about Shakespeare. Students who were reluctant to respond in class were responsive here as they felt less fearful tweeting. "Tweeting" was something they liked doing.

Let's Watch Shakespeare

Purpose: In *Let's Watch Shakespeare* (Figure 5), six videos were frontloaded for the students to watch at the beginning of the course. The purpose was to provide some prior knowledge to the drama that they were going to study. There were a variety of videos which ranged from contemporary dramas to cartoons and original movies. After watching the videos, online tasks were given. One of the tasks was to pick the video they liked best by completing a poll which was created using the *polling tool,* in Edmodo. After voting, they had to justify their answers in the posts online.

Advantage: This Generation Glass welcomed these viewings as they were in line with their characteristics as defined by Fell (2016), i.e., their preference for videos as compared to texts.

Figure 4. The second small group in Edmodo. This figure illustrates the online task, 'Let's Tweet Shakespeare'.

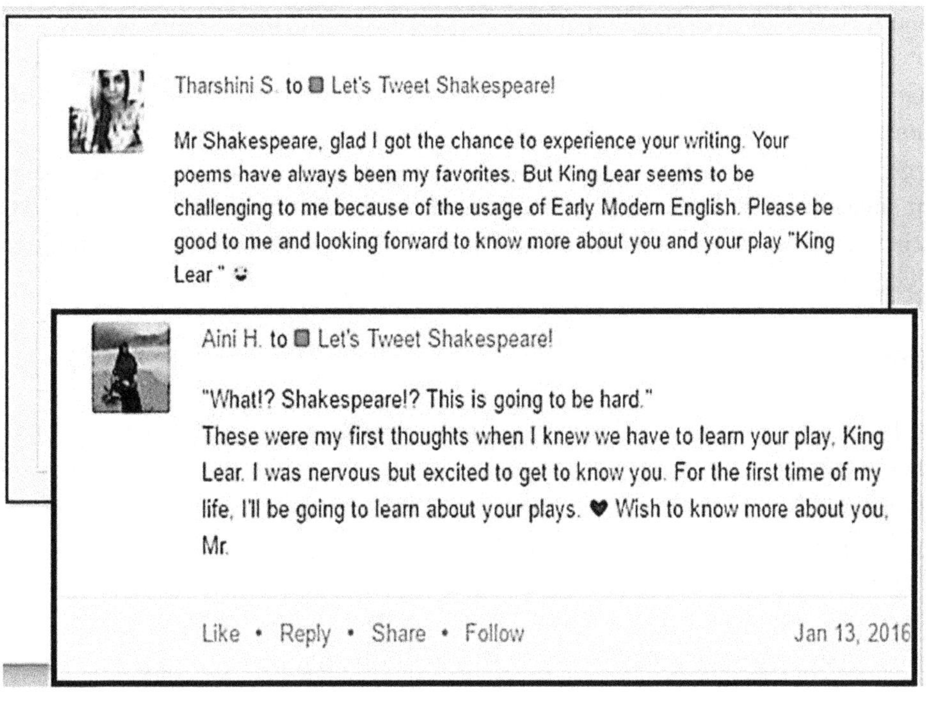

Figure 5. The third small group in Edmodo. The figure illustrates the online task, "Let's Watch Shakespeare!"

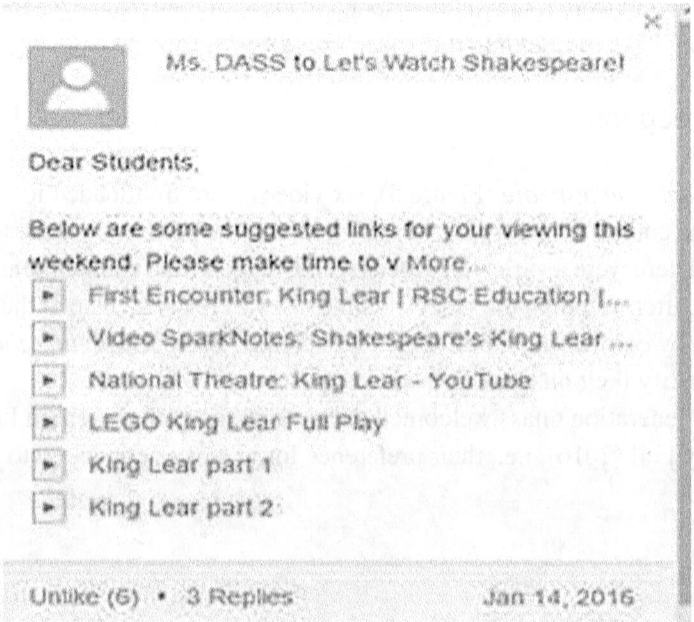

Online Tasks

Purpose: In a special *Online Tasks* group, tasks were created using other online applications like word search maker, crossword maker and Kahoot. In this group too, students uploaded their tutorial tasks onto Padlet. Padlet is an application that is used to create online bulletin boards that images, links, videos, and documents can be added.

Advantage: The quizzes and word games were very appealing to the students and this clearly enhanced their engagement with the text they had to read. As for the use of Padlet, it enabled the students to share their tasks with everyone in the class. Figure 6 shows some of the other online applications used in the course.

Photo Gallery

Purpose: In *Photo Gallery* (Figure 7), students uploaded photos taken during the staging of the drama and also photos taken during class, workshops and theatre visits. As this small group was linked to Google Photos, videos and photos were stored in cloud storage which made the retrieval of videos and photos easy. The purpose of this group is to create a digital storage for their viewing pleasure and easy retrieval of photos and videos.

Advantage: Since visual and videos were preferred forms of communication for these students, they were eager to post and share their photos and videos in the group.

Figure 6. The fourth small group in Edmodo. The figure illustrates word games from other online applications.

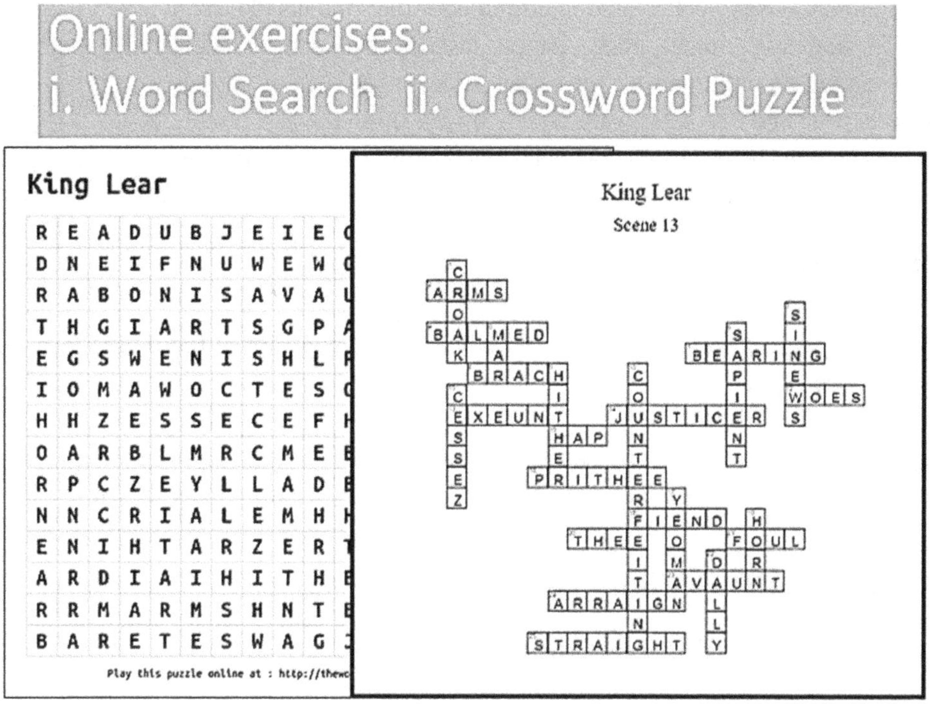

Figure 7. The fifth small group in Edmodo. The figure illustrates photos posted in Edmodo which were linked to Google Photo.

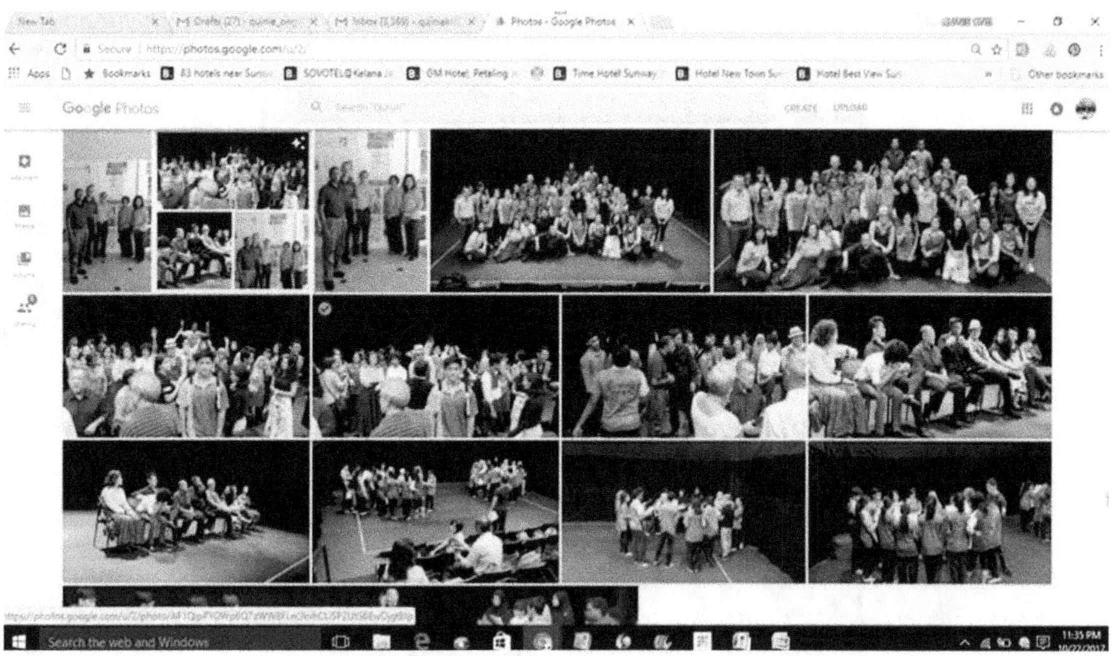

Feature 3: Assignment Tool

Purpose: Assignment boxes were created in Edmodo for the students to send in their assignments online individually. The four assignment tasks were on describing characters, identifying themes, clarifying symbols and creating an online poster for a play to be staged. The questions for the assignment tasks were easily accessible and due dates were clearly displayed in the assignment boxes (Figure 8).

Advantage: Personalized learning was evident as the lecturers were able to communicate with the students personally. Students could then resubmit their work after discussions with the lecturer (Figure 9).

Feature 4: Polling Tool

Purpose: Students get to pick their favorite video via the polling tool in the online task, *Let's Watch Shakespeare.* Besides that, data collection on students' views on the use of Edmodo was also done using the polling tool.

Advantage: Using the polling tool, students got to vote without any fear or threat as anonymity was assured. In addition, tabulation of data was automatically done in Edmodo as shown in Figure 10.

Figure 8. Assignment tool in Edmodo. The figure illustrates a platform for submitting assignments.

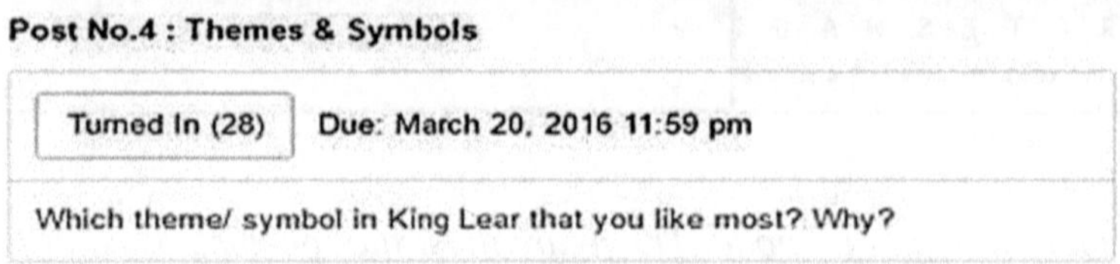

Figure 9. Assignment tool in Edmodo. The figure illustrates homework help through asynchronous chat.

Figure 10. Polling tool in Edmodo. The figure illustrates two of the questions posted on their views of Edmodo.

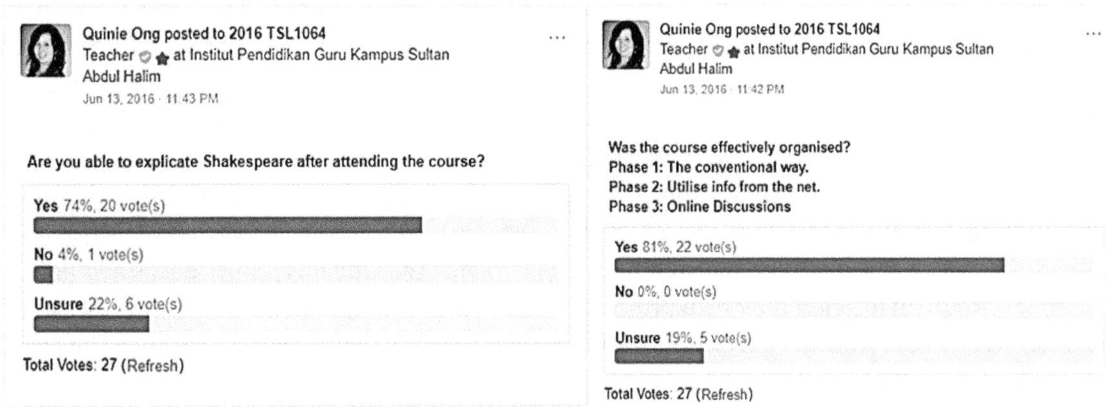

FINDINGS AND DISCUSSIONS

Six questions were posted in the Edmodo poll and the students made their choices by clicking on one response - *Yes, No* or *Unsure*. A summary of the results is tabulated in Table 3.

Based on the online poll done in Edmodo, 81% of the students found that the course was effectively organized and 74% would recommend the use of Edmodo in future courses. The LMS Edmodo was able to accommodate and facilitate the dynamic nature of this course from the beginning till the end. It served the changing needs, purposes and activities of the course very well.

In response to the question on whether they were able to explicate Shakespeare after attending the course, 74% responded "Yes." From the evaluation given by the students, they have gained the skills to help them understand a Shakespeare's play without fear or boredom. Their motivation and engagement were sustained, resulting in a more effective learning experience. Contrasting this to the traditional

Table 3. Students' responses at the completion of the drama in English course

Question	Yes (%)	Maybe/ Unsure(%)	No(%)
1. Was the course effectively organized?	81	19	-
2. Are you able to explicate Shakespeare after attending the course?	74	22	4
3. Would you recommend the use of Edmodo in another course?	74	15	11
4. Would you recommend the use of Padlet in another course?	59	26	15
5. Is the use of Google Apps (Docs, Slides) beneficial?	63	19	19
6. Can the digital applications used in the course be relevant to your own teaching practice in the future?	93	4	4

classroom instruction of chalk and talk, we can say that blended learning is an innovation to promote the learning of Shakespeare.

The students have also given positive responses on the digital applications that were used in the course whereby 63% agreed that the Google Apps (Docs and Slides) were beneficial. Meanwhile, 59% would recommend the use of Padlet in another course. Finally, 93% of the students felt that the applications used in the course would be relevant to their own teaching practice in the future. This is particularly encouraging because these students are going to be practitioners in their own classrooms. Their evaluation on their newly acquired skills and knowledge of the digital applications contributes to the building of best practices in their classrooms.

Blended learning together with the web tools has enriched the students' experiences and provided good models for them to begin their own professional learning. Through blended learning, the students gained the best of both worlds - the meeting point and the balancing of the human touch in the face-to-face interaction and the virtual world - an amalgamation of both for an enhanced learning experience.

OBSERVATIONS AND REFLECTIONS

Using Edmodo in this course was significant for both – the authors and their students. It was the first time that a course was conducted in a more comprehensive blended learning mode using Edmodo utilizing four of its features as opposed to the previously limited use which was used to store course materials only. This time, the duration of the course, the number of features used, and the frequency and quality of interactions were more organized. Students found Edmodo to be an effective and efficient LMS. Since Edmodo share similar functions with Facebook, the students embraced it without a hitch. The features of Edmodo were clearly indicated and helpful for all users. There were no overlaps of functions and this made it easy for students to navigate and locate what they needed. Hence, Edmodo was useful in facilitating blended learning.

This was a new learning experience for the authors themselves because they never had any online or digital learning experience – neither during their schooling nor undergraduate studies. It was also the first time that the authors used Edmodo beyond the basic upload of course material as previously done. This extensive use of Edmodo has compelled the authors to continue to keep themselves abreast with current technology.

Similarly, for the students, it was their first time that a course was conducted in a more comprehensive blended learning mode using Edmodo. It went beyond the usual storing and downloading of course materials. Although these students were apt at digital tools, they had never experienced blended learning as students themselves. This course opened up a new dimension of learning experience to them. Now, they have first-hand experience of what a blended learning mode is like and what it entails. This hands-on experience was another opportunity to raise their awareness and equip them with the knowledge and skills in carrying out blended learning in their own classes in the future. This blended learning using Edmodo is especially beneficial to the students as they are going to be primary school teachers; and having a personal experience and a wider repertoire of LMSs that are more suited to primary school level will help them make informed choices on the most suitable LMS for their own pupils.

CONCLUSION

Finally, it is crucial that the activities for blended learning are planned well. The activities should complement the lessons taught during the face-to-face interaction with specific instructional goals in mind. Otherwise, it would just be using the LMS as a platform for storing of resources and communication. As asserted by Sloan Consortium, blended education uses online technology to not just supplement, but transform and improve the learning process (as cited in Teach Thought Staff, 2015). Therefore, the use of Edmodo and its four features, posts, small groups, assignment tool, and polling tool can be utilised to maximise learning in a blended learning environment by adopting the blended learning ingredients with instructional goals in mind.

FUTURE RESEARCH DIRECTIONS

Although students gave positive feedback about Edmodo, some of them still needed constant reminders to look into the posts. The sense of enthusiasm was less as compared to their interest and enthusiasm to the use of social media. Hence future studies should look into why and how LMS use can be more attractive as social media is to students.

In line with the 4th Industrial Revolution, blended learning and online learning platforms are becoming more of a necessity than a choice (Teach Thought Staff, 2018). The number of online learning platforms is on the rise but which is more relevant and of use, and in which context? Further researches should be done to identify a suitable platform for every level from primary to tertiary level. In order to have optimum use of the platforms for blended learning, researchers should not only be focused on the students' views but also on the perspectives of the educators who are among the major stakeholders. "Teachers must be committed to and well trained in blended education and its technologies, and students must have a clear understanding of what is expected of them in this new environment." (Teach Thought Staff, Is there a Catch section, 2015, para. 1).

This research received no specific grant from any funding agency in the public, commercial, or not-for-profit sectors.

REFERENCES

Ali, Z. (2015). A case study of tertiary students' experiences using Edmodo in language learning. *International Journal of Language Education and Applied Linguistics.*, *2*, 39–48.

Arroyo, G. (2011). *On-Line Social Network: Innovative Ways towards the Boost of Collaborative Language Learning*. Paper presented at the International Conference "ICT for Language Learning", Florence, Italy. Retrieved from https://conference.pixel-online.net/conferences/ICT4LL2011/common/download/Paper_pdf/CLL16-428-FP-Gonzalez-ICT4LL2011.pdf

Balasubramanian, K., Jaykumar, V., & Fukey, L. N. (2014). A study on "Student preference towards the use of Edmodo as a learning platform to create responsible learning environment". *Procedia: Social and Behavioral Sciences*, *144*, 416–422. doi:10.1016/j.sbspro.2014.07.311

Banados, E. (2006). A blended-learning pedagogical model for teaching and learning EFL successfully through an online interactive multimedia environment. *CALICO Journal*, *23*(3), 534.

Blended Learning Toolkit. (n.d.). *What is blended learning?* Retrieved from https://blended.online.ucf.edu/about/what-is-blended-learning/

BusinessWire. (2014). *Edmodo releases app for Windows 8.1 at ISTE 2014*. Retrieved from https://www.businesswire.com/news/home/20140628005006/en/Edmodo-Releases-App-Windows-8.1-ISTE-2014

CAE Computers Aided USA Corp. (2018). *What is an LMS and why do I need one?* Retrieved from https://www.cae.net/what-is-an-lms

Capterra. (2018a). *Best LMS software/ 2018 Review of the most popular systems*. Retrieved from https://www.capterra.com/learning-management-system-software/#infographic

Capterra. (2018b). *Edmodo reviews and pricing*. Retrieved from https://www.capterra.com/p/129886/Edmodo/

Carman, J. M. (2002). *Blended learning design: five key elements*. Retrieved from http://blended2010.pbworks.com/f/Carman.pdf

Cauley, P. (n.d.). *A guide to explain it all*. Retrieved from http://www.csub.edu/~tfernandez_ulloa/edmodo%20user%20guide.pdf

Delacruz, S. (2013). Online reading response using Edmodo. *The Florida Reading Journal*, *49*(2), 9–12.

Edmodo. (2013, March 5). Edmodo acquires education technology start-up root-1 [Web log post]. Retrieved from https://blog.edmodo.com

Edmodo. (2014). *20 ways to use Edmodo (teacher)*. Retrieved from https://support.edmodo.com/home#entries/21720784-20-ways-to-use-edmodo teacher

Edmodo. (2017). Platform overview. *Edmodo Inc.* Retrieved from https://developers.edmodo.com/resources/overview/

Edmodo. (2018). *Get more from your classroom*. Retrieved from https://go.edmodo.com/

Ekici, D. I. (2017). The use of Edmodo in creating an online learning community of practice for learning to teach science. *Malaysian Online Journal of Educational Sciences 2017, 5*(2). Retrieved from https://eric.ed.gov/?id=EJ1142512

Fell, A. (2016). *Generation Z defined; The five characteristics of today's students*. Retrieved on https://www.youtube.com/watch?v=QvCwcQmnyfE

Garrison, D. R., & Vaughan, N. D. (2008). *Blended learning in higher education: Framework, principles, and guidelines*. San Francisco, CA: Jossey-Bass.

Jones, B. (2017). *Is WhatsApp using too much space on your phone?* Retrieved from https://www.psafe.com/en/blog/whatsapp-using-much-space-phone/

Kongchan, C. (2013). How Edmodo and Google Docs can change traditional classrooms. *Proceedings of The European Conference on Language Learning 2013*, 629-637.

Maxwell, C. (2016). *What blended learning is - and isn't*. Retrieved from https://www.blendedlearning.org/what-blended-learning-is-and-isnt/

McGee, P., & Reis, A. (2012). Blended course design: A synthesis of best practices. *Journal of Asynchronous Learning Networks*, *16*(4), 7–22.

Mills, K., & Chandra, V. (2011). Microblogging as a literacy practice for educational communities. *Journal of Adolescent & Adult Literacy*, *55*(1), 35–45. doi:10.1598/JAAL.54.1.4

Mokhtar, F. A. (2018). Breaking barriers through Edmodo: A qualitative approach on the perceptions of University of Malaya undergraduates. *Online Learning*, *22*(1), 61–80. doi:10.24059/olj.v22i1.1026

Pape, L., Sheehan, T., & Worrell, C. (2012). How to do more lessons from online learning? *Learning and Leading with Technology*, *39*(6), 18–22.

Pappas, C. (2017). What is a learning management system? LMS basic functions and features you must know. *eLearning Industry*. Retrieved from https://elearningindustry.com/what-is-an-lms-learning-management-system-basic-functions-features

Purnawarman, P., Susilawati, S., & Sundayana, W. (2016). The use of edmodo in teaching writing in a blended learning setting. *Indonesian Journal of Applied Linguistics*, *5*(2), 242–252. doi:10.17509/ijal.v5i2.1348

Quigley, A. (2018). *What is an LMS?* Retrieved from https://www.linkedin.com/learning/learning-management-systems-lms-quick-start/what-is-an-lms?autoplay=true&trk=course_tocItem&upsellOrderOrigin=trk_default_learning

Renard, L. (2016, Dec 23). Top-8 online learning platforms [Web log post]. Retrieved from https://www.bookwidgets.com/blog/2016/12/top-8-online-learning-platforms

Shashko, A., & Koteska, G. (2017). The virtual classroom Edmodo as a teaching tool as part of the flipped classroom method in the English language classroom. *TEACHER International Journal of Education*. 54-58.

Teaching with Technology Online Workshop Series. (2012-2013). *Blended Learning Models & Examples*. Retrieved from https://teachingwithtech.lss.wisc.edu/m3w1.htm

TeachThought Staff. (2015). *The definition of blended learning*. Retrieved from https://www.teachthought.com/learning/the-definition-of-blended-learning/

TeachThought Staff. (2018). *Find the model that works for you: 12 types of blended learning*. Retrieved from https://www.teachthought.com/learning/12-types-of-blended-learning/

Tomlinson, B., & Whittaker, C. (Eds.). (2013). *Introduction. Blended in English language teaching: Course design and implementation*. London, UK: British Council.

Witherspoon, A. (2011). *Edmodo a learning management system*. Retrieved from http://www.poweredwithtechnology.com/2011/01/edmodoa-learning-management-system.html

Yagci, T. (2015) Mobile Social Media in Higher Education & Implementation of "EDMODO" in reading classes. *Proceedings of 6th International Visible Conference on Educational Studies and Applied Linguistics*, 436-442.

ADDITIONAL READING

Ajjan, H., & Hartshorne, R. (2008). Investigating faculty decisions to adopt Web 2.0 technologies: Theory and empirical tests. *The Internet and Higher Education, 11*(2), 71–80. doi:10.1016/j.iheduc.2008.05.002

Balcikanli, C. (2010). *The effects of social networking on pre-service English teachers' metacognitive awareness and teaching practice* (Unpublished doctoral dissertation). Gazi University Institute of Educational Sciences, Ankara.

Banyen, W., Viriyavejakul, C. and Ratanaolarn, T. (2016). *A Blended Learning Model for Learning Achievement Enhancement of Thai Undergraduate Students*. Retrieved from doi:10.3991/ijet.v11i04.5325

Brady, K. P., Holcomb, L. B., & Smith, B. V. (2010). The use of alternative social networking sites in higher educational settings: A case study of the e-learning benefits of Ning in education. *Journal of Interactive Online Learning*, *9*(2), 151–170.

Dudeney, G., & Hockly, N. (2007). *How to teach English with technology*. Harlow: Pearson Education Limited.

Durak, G. (2017). Using Social Learning Networks (SLNs) in Higher Education: Edmodo Through the Lenses of Academics. *International Review of Research in Open and Distributed Learning*, *18*(1). doi:10.19173/irrodl.v18i1.2623

Garner, B., & Oke, L. (n.d.). *Blended learning: Theoretical foundations*. Retrieved from https://www.indwes.edu/academics/centers/cli/the-learning-academy/blended-learning--theoretical-foundations/blended-theoretical.pdf

Gruba, P., & Hinkelman, D. (2012). *Blended technologies in second language classrooms*. Basingstoke: Palgrave Macmillan. doi:10.1057/9780230356825

Huang, R., Ma, D., & Zhang, H. (2008). Towards a design theory of blended learning. In *Proceedings of Hybrid Learning and Education: First International Conference* (pp.66-78). doi: 10.1007/978-3-540-85170-7_6

KEY TERMS AND DEFINITIONS

Blended Learning: A combination of face-to-face learning and web-mediated learning.

Digital Applications: Applications in the internet which have specific functions and can be accessed from any computer with an Internet connection.

Digital Communication Tool: An online tool used by educators to communicate with their students.

Digital Platform: An avenue to create a virtual classroom for educators and their students.

Edmodo: An online learning platform which has similar features to Facebook and educators use it to manage a virtual classroom.

Explicate: To give a detailed and clear explanation.

Facilitate: To make the process of blended learning easier.

K-12: A term used in the United States and Canada for public schools from kindergarten to the 12th grade.

Learning Management System: An LMS offers classroom management for instructor-led training for blended learning.

Meta-Analysis: A statistical approach to combine the results from multiple studies.

Online Learning: Learning through online activities created using the features in Edmodo and other digital applications in the internet.

Supplement: Activities added to the traditional classroom interaction in order to complete or enhance learning.

Chapter 9
Gamification Meets Mobile Learning:
Soft-Skills Enhancement

Affah Mohd Apandi
Universiti Utara Malaysia, Malaysia

ABSTRACT

Higher education in the era of Industrial Revolution 4.0 (IR 4.0) has given a brand new drive on Malaysian educational transformation. The impact of a numerous technology innovations in ICT towards Malaysian education is noticeable. Some existing jobs will be eliminated in this era, but more new jobs that require digital literacy are going to be created. Therefore, Education 4.0 must be able to produce graduates with excellent communication and collaborative skills. However, Malaysian Digest states that fresh graduates lack soft skills. Several approaches have been proposed by the Ministry of Higher Education for the purpose of enhancing students' soft skills. Therefore, by having technologies to help overcome this problem is sought after. This chapter describes the utilization of mobile learning and gamification in enhancing learner's soft skills. It elaborates the effectiveness of using gamification in mobile learning technique to develop learners' higher order thinking skills to help them succeed in a global economy.

INTRODUCTION

The term "Industry 4.0" or known as the fourth Industrial Revolution (4IR) was initially introduced in Germany and became publicly known in 2011. It is comprised of many components that refers to the shift towards digitalisation, as advancements in technology such as the Internet of Things (IoT), cloud computing, big data, mobile, data communication technologies, programmable logic controllers, sensors and actuators all play a key role. 4IR is a broad vision with clear framework mainly characterized by bridging of physical industrial assets and digital technologies in so-called cyber-physical systems (CPS) that produce smart industry and ecosystems of industrial innovation and collaboration. This is a new era where the CPS alongside IoT interconnect in a way where the combination of Artificial Intelligence (AI), and automation are poised to dramatically change the way we live our lives. 4IR is set to

DOI: 10.4018/978-1-5225-7832-1.ch009

Copyright © 2019, IGI Global. Copying or distributing in print or electronic forms without written permission of IGI Global is prohibited.

send ripples of change throughout the world, and will require both big and small businesses to innovate, to stay relevant and competitive in the global market or risk becoming defunct. These emerging technologies will no doubt have a negative impact on employees and there will be a focus on companies to look at how workers and machines can work together symbiotically. Looking to the changes that will be sparked by the rapid advancements in technology, the education sector is likely to undergo a difficult period of adjustment. Therefore, it is important for us to look at what needs to change in education to ensure it is as effective and accessible as possible in order to develop the skills to work in the future.

In Malaysia, the theme "Higher Education 4.0: Knowledge, Industry and Humanity" created by the former Higher Education Minister, Datuk Seri Idris Jusoh as a recognition that the fourth industrial revolution will have an impact on our Higher Education system. There will be a requirement for a skill change where higher education must be prepared to train graduates to embrace the fourth industrial revolution and handle new technologies (Zulita Mustafa, 2018). Hence, higher education in the fourth industrial revolution can potentially transform society for the better. Datuk Seri Idris Jusoh said that the ministry will implement the Professional Development Programme 4.0 for lecturers. This programme aims to have 30 percent of teaching and learning at public higher education be aligned to the fourth industrial revolution by 2020. The ministry also targets to transform public universities into smart campuses (Rozana Sani, 2018)

It is estimated that 65% of kids enrolling in primary school today will end up working in jobs that haven't been created yet (Study Malaysia, 2018). This means that the number of job opportunities in the new industries will continue to increase relative to the increasing number of young people in the future even if the traditional jobs begin to disappear. Despite having all technological changes in terms of production and delivery in these new future industries, business will still be run by people. The business will still be owned, managed and directed by people. Customers, the organisations, staff who produce products and services will always be people. No matter how many robots or computers there may be in a factory, it must be handled by people. A a result, to only possess hard skills such as e-literacy and e-skills that can come from experience and formal education seems not sufficient enough to land someone a job. Therefore, soft and social-skills are also a demand. These are crucial skills for students to succeed in their life after the classroom. Students must be able to create and share knowledge, communicate and socialize in their workplace, maintain motivated with their work.

However, Malaysian Higher Learning Institutions have been held accountable for the perceived unsatisfactory development of soft skills leading to an increase in graduate unemployment by not meeting the globalized 21st Graduate skills requirement since 2006 (Chelvi Murugiah, 2015). Despite of the growing demand for graduates with high hard (technical) skills, one of the main complaints from the industry is the lack of soft (personal) skills. The requirement skill-set for the "jobs-of-the-future" would transform so fast. Technical skills alone are no longer enough for employees especially in this highly competitive marketplace of the 21st century. Hence, the need for individual capabilities specifically soft skills is greater than before. Personal attributes such as soft-skills embraces many other significant skills. The skills that has taken on heightened importance are including problem-solving skills, 'taking charge' attitude, leadership, oral and written communication, teamwork, sense of urgency, time management, and attention to detail (Emmer & Brunhoeffer, 2017). These abilities really give great influence towards one's motivation, achievement, and is probable to work harder, participate and pursue challenging objectives, and able to complete task successfully (Multon, Brown, & Lent, 1991; Bandura, 1997). Some experts referred these abilities as 21st century skills which falls into higher-order thinking skills category by means deeper learning outcomes.

4IR has produced a new model of education for the future, Education 4.0 has transformed the landscape of educational technology where education in 21st century is all about embracing technology. According to Blair (2012), "a dramatic shift is sweeping through our schools". We can see that even elementary students using cell phones, navigating iPad better than adults. Nowadays, middle school students have a strong audience following their blogs, tweeter, and even YouTube. Therefore, educators need to mind that these learners are not the same from the first part of the 21st century. They evolved in order to gain access to new knowledge, demanding more engaging learning. Education 4.0 is preparing graduates for future life and employment where smart robots will replace humans in certain job sectors. Therefore, Education 4.0 must harness on appropriate current skills and knowledge to learners so that they could not be substituted by robots. Technology disruption that produces Education 4.0 that focuses on educational innovation and agility has made future learning more personalized, hyper, interactive, mobile, global and virtual. This reflects that the nature of education will need to change. The function of our tertiary level should somehow be different particularly in teaching and learning activities. Otherwise, educators will face a lot of challenges where they need to attend the millennial kids with digital intelligence and cyber-talent. These digital natives are extremely different from those who started the millennium. They communicate in a language that older generation may not fully understand. They have their own ways of interpretation and expression. This generation also prefer an interactive approach in learning. Learning that blends with games, simulation-animation, integration information with complex montage, images, sound, video, icons, and artificial intelligence which connects them to a different learning experience are highly recommended to engaged and motivated them. Beyond the 21st century skills, digital agility and innovation such as artificial intelligence, cloud computing and mobile solution, social media, the internet of things, Virtual Reality (VR) and Augmented Reality (AR) with digital games are fundamentally changing the dynamics of education and instructional technologies landscape into new form of digital pedagogy.

Hence, for plenty of educators this has become a challenging task when these learners are not the same from the first part of the 21st century. They evolved, demanding more engaging learning strategies in order to gain access to new knowledge. Blair (2012) mentions that it is no longer enough for these students to just watch videos during class, play an internet game, or even take turns at an interactive whiteboard. In order to move in that tide, teachers need to continuously seeking novel instructional approaches, need to explore new and innovative ways in using educational technology to redesign future learning. Lee and Hammer (2011) agreed that today's educational institutions have problems with student motivation and engagement. The approach of gamification as one of the learning strategies is much needed to engage and motivate this 21st century learners. This kind of approach is desirable since it is not only reinforcing knowledge but also emphasizing important soft skills such as collaboration, problem-solving, creative and critical thinking, and communication. According to Kapp (2012), games have incredible motivational power that encourage people to engage with them even without any rewards, the enthusiasm normally comes just for the pleasure of playing and the desire to win. Nevertheless, it requires certain technical infrastructure and appropriate pedagogical integration to make sure the usage of gamification in classroom is effective. In this regard, this chapter postulates the need for educators to rethink their old conceptions of teaching and learning and redesign their students' learning experiences so that they meet the requirements of Education 4.0. In addition, this chapter will also explore on how to use right educational technology to not only develop students' technological skills but also helps them to develop their soft skills. In developing these skills and emphasizing them in students learning, it will help graduates to rise above the job market competition in the future.

MOBILE LEARNING

We live in an era of technology revolution where the pace of technological development requires the fastest and cheapest technique of generating and transmitting knowledge. One of the alternative strategies in education in this era is the use of mobile devices in teaching and learning since the usage of mobile devices among the younger generation are increasing hence makes it a potential for teaching-learning purposes. In Malaysia, it was reported that the penetration rate of mobile in Malaysia for the year 2010 is 106%, which implies multiple subscriptions among users (MCMC). In Malaysia, more than 30% mobile subscribers are from youth community (Ng, Goi, & Voges, 2010). Before that, it is necessary to understand the concept of using mobile learning in education. Numerous literatures in the mobile learning, also called m-learning, have discussed on various approaches and ways in which the idea of mobility should be implemented. The idea of m-learning has been around for some time now with several studies have reported their findings on respondents' readiness for mobile learning in terms of their ownership of mobile devices (Quinn, 2000; Sharples, 2002; Alexander, 2004; Wang, 2004; Trifonova, 2006; Corbeil, & Valdes-Corbeil, 2007). M-learning is a distinctive way of learning as learners frequently move from place to place. They can simply access learning material, get instructions and other content course related applications anytime and anywhere. Students want to access resources that are open where they can learn and study anytime and anywhere and without any barriers to access or interact. They look for every information that is copiable, remixable, and the most important it is free (Johnson et al., 2013).

Study done by Issham, Rozhan and Thenmolli (2010) proved that the advantages of using phone are what students really enjoy with. They are using mobile phone as a communication tools for learning purposes to check timetable easily, to get information on class if it is delay or postpone and getting current class or university updates without having to go out. This reflects that in order to keep up with the demand of students' everyday life they take mobile phones as a very important tool where it is not only for the sake of on-the-go communication, they also need mobile phone for information sharing, training and learning. There are many digital learning materials or applications available for the delivery and management options. Hence, it is important to ensure appropriate selection of teaching and learning methods for achieving particular objectives. It is also necessary to select methods that can provide the maximum benefits to students. It can be concluded that m-learning could be a possible platform to create not only joyful learning atmosphere, it could also serve as an additional learning medium that fosters the way learners access information, create innovative learners and good problem-solvers.

GAMIFICATION

Gamification is an application that contains all the fun and addicting game elements to real-world or productive non-game activities (Nah, Telaprolu, Rallapali, & Venkata, 2013; Werbach & Hunter, 2012; Zichermann & Cunningham, 2011; Deterding, Dixon, Khaled, & Nacke, 2011;). Gamification is a form of active learning that has been very useful to 21st century education. It is usually included the use of digital badges to encourage learning and recognize learner's abilities. According to Icard (2014), there are humorous elements that distributed in games that interspersed to three distinctive concepts particularly in dynamics, mechanics, and components. Structured as a pyramid, components are the base element, with mechanics in the middle, and dynamics at the top. Figure 1 shows the elements of each category.

Figure 1. Gamification

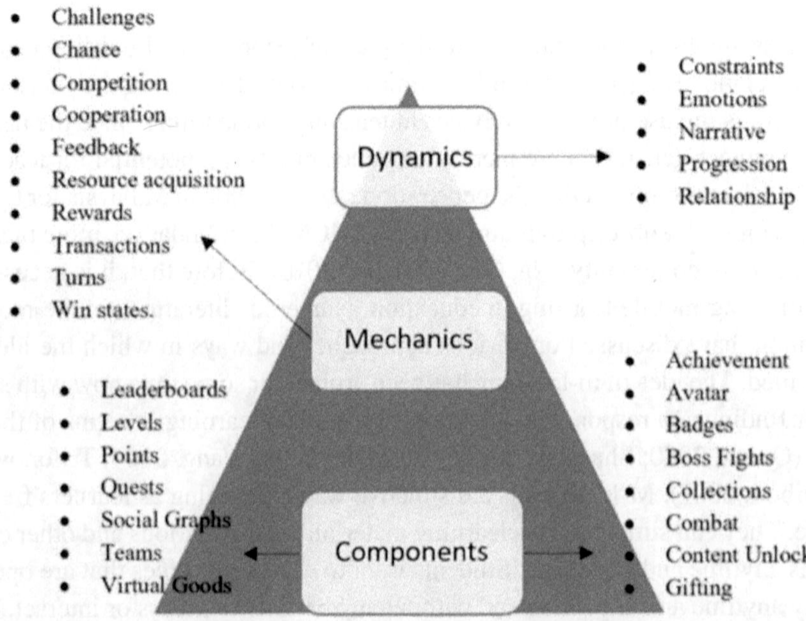

Dynamics is at the top of the pyramid that comprehends the big picture parts of a gamified system. They are the highest level of conceptual elements in a gamified system. Even if this part doesn't really enter directly into the game itself, it is absolutely important. When analysing games, the game dynamics is the highest level of abstraction that must be considered when developing a game or gamifying a system. Though, the components should not be noticeable. When these five elements and design process are integrated, gamification occurs organically and forming the abstract notion of a game.

Mechanics is the second group of the elements that drive users to engage with the content and continue to drive the action forward. Game mechanics include rules which control what a play can and cannot do. There are ten mechanics elements that ties to one or more dynamics. It is important to distinguish the elements whichever is chosen it must match the objectives and goals of the situation.

The largest group of game elements is components. The components are more specific form of either mechanics or dynamics. These elements are less abstract than the first two categories and lead to actual tools that can be employed to begin to incorporate gamification in the environment of interest. Game components also include levels, badges, points and many other elements that work according to specified rules.

These elements addressed personally, designed elaborately, and include increasingly difficult questions that requires cognitive effort and enhanced skills. Hence, increased in the participation of users. Therefore, gamification is one of teaching and learning method which is required in 21st century classroom to create students that can think creative and critically, to create a competitive environment and to create alternative atmospheres that enhanced learning (Huseyin Bicen, & Senay Kocakoyun, 2018). The main objective of incorporating gamification with education is to release effective and attractive learning experience for the student (Figueroa, 2015). According to Kingsley and Grabner-Hagen (2015), gamification in education that integrates game elements and rewards mechanisms can motivate and engage students at the same time promote healthy competition in learning. Having this kind of activities in the classroom, not only help the students learn a concept but also help them practice their skills as if they were playing games. Hence, makes teaching and learning experience more challenging and fun and at the same time motivates them to move forward (Vassileva, 2008).

GAMIFICATION IN MOBILE-LEARNING ENHANCE LEARNER'S SOFT SKILLS

Dicheva and Dichev (2017) highlighted the gamification categories for teaching and learning as shown in Figure 2. For this chapter, authors will touch on gamification on mobile learning category only.

Smartphone is a priority gadget for people nowadays, even for students. Mobile applications can be good learning way for students to make learning more attractive. Playing games using mobile applications can make students engage in their learning process. This gamification method can make them feel like they were playing games and win the challenges, complete the levels to get badges and trophies, as well as make the students eager to try and earn the rewards. (Chadyiwa & Mgutshini, 2015; Khaddage, Lattemann, & Acosta-Díaz, 2014). Based on Hamari, Koivisto, and Sarsa (2014), mobile applications have been a trending object and subject for user engagement and have helped to improve positive patterns in their learning.

Education system previously focus on giving knowledge and facts. Therefore, it is very challenging for students to learn soft-skills if this type of teaching and learning still be used by teachers especially in university. Students are not sufficiently prepared for a working life after they are graduated since they don't learn soft-skills in university if teachers still promote a boring traditional teacher-centred style of teaching and learning. University students' level are in between adulthood and childhood where the best

Figure 2. Gamification categories for teaching and learning

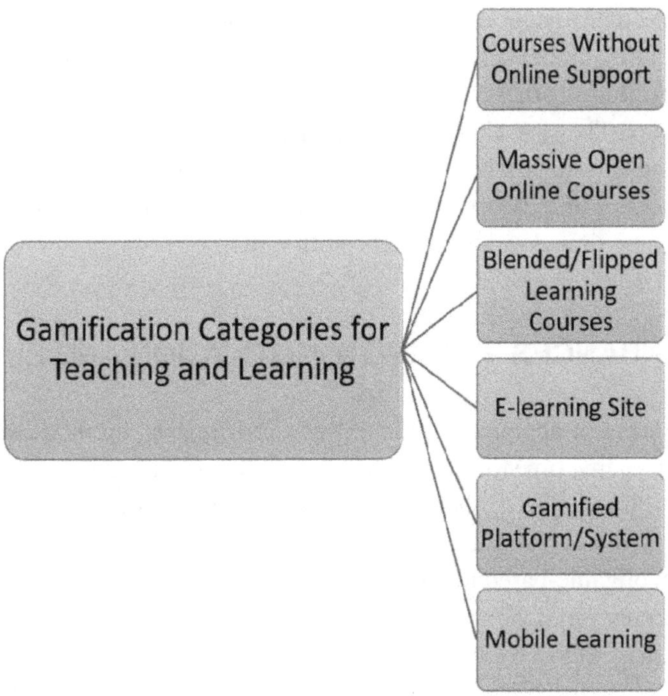

way to integrate soft-skills is by using gamification in a very edgy way such as by using mobile phone. It is proven that gamification can actually develop learners' metacognitive abilities, promote empathy, and builds teamwork skills (Wang, Zhu & Satre, 2016). So, by combining mobile-learning that create special environment which is more to student-centred, with the right content of learning, with game elements in an easy and fun way, teachers can create an optimal stage of learning. This combination of gamification and mobile-learning will develop great learning experiences that enhances learners' soft-skills.

Gamification can be regarded as an effective educational tool that can enhance learners' soft-skills over the use of game. It develops collaborative learning when learners involve with a multi-player game designed. This type of approach enhances and develop learner's interpersonal competencies such as negotiation, creative problem resolution and collaborative decision-making. This will allow a realistic collaborative "learning by doing" approach that avoids real-life risks (Leemkuil, de Jong, de Hoog, & Christoph, 2003). Herz (2001) stated that by playing game that expose with collaborative task appears to be a powerful incentive for students where they can actually transform knowledge into social capital because of peer acknowledgement.

McDonald (2017) indicated that gamification can improve learners' soft skills where in this case, critical thinking skills. This also helps in the development of higher order cognitive skills (Stahl, 2007). Through an interactive teaching method using gamification together with problem-solving activity, it helps to enhance students' critical thinking. The study helps to understand the importance of problem-solving in the context of an entrepreneurial setting. He found that solving a problem in a time limited setting will emphasis students' critical thinking skills where the result is an improvement of this skill.

A study conducted by Bodnar and Clark (2017) that focuses on investigating whether gamification could be a great tool for enhancing students' written and oral communication skills. The findings show that the use of gamification towards engineering students could enhance their oral and written communication skills as long as it is carefully implemented. Therefore, gamification should be designed according to both the pedagogical and technical points of view. This is to ensure the monitoring of actions for each player and the group in order to help reach their learning goals and at the same time enhance learning quality and their soft skills. Students like to learn in a fun and exciting environment. So, when game is incorporated into learning, learners are less likely to be anxious because the learning environment will be informal and relax. They are likely to be responsive and active to participate in class. Moreover, gamification is engaging and mobile-learning is convenient, both are effective and popular among 21st century students. As a result of these two meeting, educators are able to produce an awesome learning experience and at the same time develop students' soft-skills.

GAMIFICATION ENHANCES STUDENTS' MOTIVATION AND ENGAGEMENT

Gamification is the concept of applying game mechanics to engage and motivate students in learning. It has been used for quite some time to enhance learners' motivation and improve learning engagement. According to Zarzycka-Piskorz (2016), there is strong evidence shows that motivation does increase when game-playing is implemented in learning. It has been found that learners' motivation and engagement increased in fun factors of game-based learning lesson. A gamification study by Huckabee and Bissette (2014) found that, not only physical awards could internally motivate learners but the objectives and capability of learners also could give important effects on how they engage in work. It helps learners to have natural desires for socializing, self-expression, be competitive and at the same time learning. It is more than just playing games and have fun that does not actually involve playing games at all. It is a concept of applying game-design thinking to non-game applications and the use of game elements and techniques in non-game contexts.

Gibson, Ostashewski, Flintoff, Grant, and Knight (2013) stated that by creating competitions that have goal attainment, achievement, and status can be powerful when badges are used with points and leader boards. Using badges in a game can actually motivate learners to expand their performance through greater skill acquisition, higher engagement, and time spent on learning. Santos, Almeida, Pedro, Aresta, and Koch-Grunberg (2013) explored students' experiences while earning badges in a learning process and they found that more than 90% of the students agreed that the badge system made them more motivated, focused, and engaged. They found that by using badges in education, it offers students a sense of achievement during classroom activities and are considered as "symbols of recognition".

Barata, Gama, Jorge, and Goncalves (2013) gamified a course in Information System and Computer Engineering by presenting numerous game design elements into the course design. This is including badges, challenges, leader boards, experience points, and levels. They found that students are more engage based on the number of downloads and participation in online. de Freitas and de Freitas (2013) applied the concept of gamification in a computer science class for undergraduate students into a system called "Classroom Live". The authors considered various game design elements including levels, experience points, and in-game rewards. The findings show that students' engagement was improved and the students was more enjoyable.

Figueroa (2016) stated that some of the benefits that a 21st century learner encounter in a gamified class include: fun and engaging classroom, a constant motivation for the student to complete the learning activities, promotes that the student becomes more focus and attentive to their learning, and promote healthy competition between peers. Others researchers (Chee-wei et al., 2017; Nicholson, 2015) stated that gamification can improve motivation and engage student learning.

GAMIFICATION METHODS AND STRATEGIES

In order to have a successful gamification teaching and learning experience, where 21st century learner becomes engaged and motivated, the educator needs to plan accordingly. This requires that every instructor follow the five-step model educational gamification, which is presented in the work of Huang and Soman (2013).

Each step of the model presented by Huang and Soman (2013) includes the following:

Figure 3. Educational gamification five step model

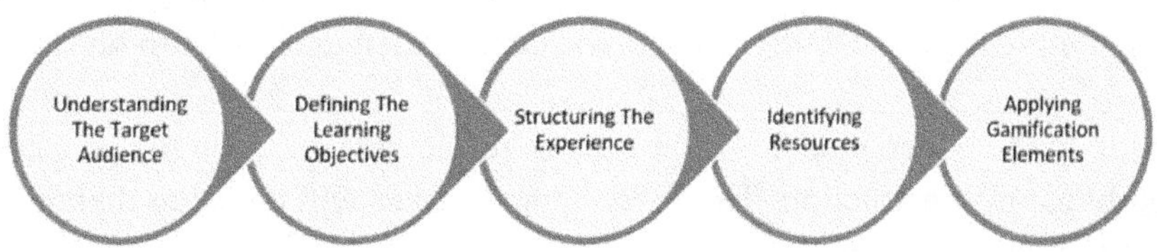

Step 1 - Understanding the Target Audience and the Context: The instructor needs to know who are the students including other several factors such as group size, environment skills sequence, and length. The pain points (real problem) are exposed in this step such as motivation, focus, pride, skills, learning environment and nature of the course, and also mental, physical, and emotional factors. The educator will be ready to determine the gamification elements that need to implemented when these points are understood.

Step 2 - Defining the Learning Objectives: Focus on either one or two objectives of learning. Before creating a game, make sure to focus on the learning goals. Try to limit the main goals and objectives of the lesson. This will make the game easier to develop and design. This is to make sure that the game and the entire course is mobile-friendly, if applicable. Creating a game takes creativity and hard work, therefore we need to be extra careful that the resulting game is focused and purposeful. The quest can be played by one or in team player according to the goals that have been made. To enhance students' confidence, create a one player game. Games that need a team normally focus on enhancing collaboration and team work skills.

Step 3 - Structuring the Experience: The educational program needs to be created from simple to complex. To make sure the students stay engaged and motivated, the educator needs to start the task with the easier milestones. Creating a scenario-based games will allow learners to develop specific skills by making them perform the same actions in real-life situations. Whenever learners manage to accomplish the goals in the game or meet checkpoints that have been placed try to provide rewards. It is important to offer achievable rewards that resonate with the whole team in order to motivate students to participate. This requires thought and attention. Do consider surveying rewards that are particularly motivating, so that when they get a reward they will do it again, and again. It reinforces a habit or behaviour. This will motive them to succeed which will then be a motivation for them to translate it to their on-the-job performance.

Step 4 - Identifying Resources: The instructor will have to complete assurance of which stage can or can't be gamified and reflect in regards to several aspects that need to be considered such as tracking mechanism, currency, levels, rules, and feedback.

Step 5 - Applying Gamification Elements: The educator needs to decide the two elements which are self and social elements should be applied before conducting a game. Mobile devices and gamification provide these self and social opportunities. Self-elements focus on students to compete with themselves and recognize self-achievement by using badges, levels and time restrictions. Meanwhile, social-elements are the interactive competitions along with cooperation. Making learners feel like part of a team or community on their journey will certainly create loyalty and a positive learner experience. Take this advantage by using leader boards, multiple players, team grouping and instant message feature in creating activities. This will incorporate collaboration, communication and competition for learners. The multiple leader boards will show the position of each player, this will allow them to see how their fellow competitors are doing. When they see the position of other players, it will motivate them to do better in the next challenge in order to be succeed.

After presenting the five-step model is important to reflect and envision the educational objectives in order to properly gamify instruction. The 21st century learner will envision them as challenges and will be motivated to move from one level to another, or one stage to another. This could become part of a learning outcome.

GAMIFICATION RESOURCES

The use of gamification and m-learning have been widely spread especially in the form of student response systems such as "iClicker" and "Poll Everywhere". A new generation of gamification meets m-learning is the game-based student response systems (GSRS) (Plump & LaRosa, 2017; Wang, 2015). Nowadays, there are plenty of accessible resources that can be implemented by instructors interested in using gamification as part of their strategies towards the 21st century learner. Most are available in software, online, and mobile application. Some of the gamification resources are Class Dojo, Duolingo, Edmodo, Socrative and Kahoot!. The following are a list of gamification resources to implement in the teaching and learning experience and suited for the 21st century learner.

1. **Class Dojo:** This application is more to student behaviour management platform where it provides the instructor to connect with students' parents so that they can get involved on the students' learning. Instructor and parents can share, track, and evaluate student participation along with immediate feedback. Learning strategies that combine points, avatars, and leader boards help motivating learners. It can be accessed via the Web interface or an Android or iOS app.
2. **Duolingo:** Educators can use this application as part of daily homework. It is a language learning translation platform that covers the areas of speaking, listening, grammar and vocabulary necessary for second language learning. It has six languages including English, Spanish, Portuguese, Italian, German, and French. Learner can easily track progress and receive immediate feedback. It shows users progress through several levels and motivates student-driven work along with communication and collaboration. It works for iPhone, iPod Touch, iPad, and Android.
3. **Edmodo:** An educational social networking platform with gamification elements. It has an interface that is similar to Facebook where users can comment on post, submit assignments, and track their progress. Educator can use this platform as an extension of the classroom by posting polls, design quizzes, open discussion, and even post assignments. It can be used by all educational levels. It promotes collaborative learning, and teamwork so it is great as a motivating tool in learning. Educator also can give feedback to parents who have Edmodo account. It works with any iPhone, iPad, iPod Touch, Android or Web browser,
4. **Socrative:** This is a dynamic student response system that engage students via mobile devices. Educators can assess students either for formative or summative assessment. Students can answer questions without feeling stress or anxious since it is more to a game-based activity. It gives immediate feedback, live results, and effortless data analysis. It also allows the users to import images to the question items and it has gamification elements that motivate and engage students to participate.
5. **Kahoot!:** It is a platform where teachers and students could communicate by transforming the classroom into a game show and the teacher be the host. By using the existing infrastructure, Kahoot! is likely engage students through competitive knowledge tasks. Students will earn points when they answer related questions correctly. They could compete who will get the highest points and be a winner. A winner would be announced at the end of a game session. To create quizzes in Kahoot! teachers do need an account whereby students do not need to create one. They can just join the game through any mobile device that have a web browser, such as Chromebook, iPad or Android device.

A boring classroom can change to a fun and exciting classroom by using GSRS. It creates a positive learning environment by bringing collaboration, emotion and play together. It facilitates learners' social experience by creating a fundamentally more social and engaged learning environment. GSRS is an excellent tool as an ice breaker in the first lesson where teacher can understand students' weaknesses before he or she starts the class. This can help the teacher to plan reinforcement on students' weak areas. It is also an excellent tool to get quick responds of students' feedback either on the lesson given or on the method of instruction. It helps to engage with students' attention or as a break when students seem bored during a long lecture. It is an easy tool to collect students answers and in an organized manner and to evaluate their own understanding and reinforce learning. Lastly, GSRS is a great tool for designing instant polls. This competitive but yet safe, fun and engaging environment created are beneficial either for academic or psychological purposes. Students can grow in confidence when they are able to express their ideas and do well on the quiz. In addition, this will help to boost students' self-esteem when they are being recognized and appreciated by their teachers and peers.

This GSRS that is taken as one of the supplements of pedagogical practices that allows teachers and learners to work together over competitive knowledge games. GSRS focuses on motivation and engagement through gamification which promotes learning (Wang, Zhu, & Satre, 2016). GSRS engages students through game-like activities for instance, impromptu quizzes, discussions and even surveys played by the whole class in real time (Byrne, 2013; Cross, 2014; Thomas, 2014; Licorish, Owen, Daniel, & George, 2018; Huseyin Bicen & Senay Kocakoyun, 2018; Tan, Ganapathy Malini, & Manjet Kaur, 2018). It is also widely use at the end of a class for assessing students' understanding of a lesson. A research conducted by Tan, Ganapathy and Manjet Kaur (2018) exposed students of English for the Media in University Sains Malaysia to the use of GSRS found that gamification develops learners' motivation, engagement, metacognitive, fostering and reinforcing learning, promotes empathy and builds teamwork skills. Thus, making this GSRS relevance with this chapter of combining gamification and m-learning to enhance students' soft skills including motivation and engagement.

Grammar is one of the language components which is challenging to teach as students' attention towards learning grammar is difficult to retain. It is argued that through GSRS this retention can be achieved. In a study by Iaremenko (2017) examined ways in which Kahoot! can be used to learn grammar. Based on the findings, it was found that the students were engaged in their learning as they compete to win the game. With this learning environment, students' motivation to learn increased and they were able to learn grammar effectively. This learning environment is also applicable for increasing students' vocabulary knowledge. Medina and Hurtado (2017) explored the use of GSRS to teach vocabulary to undergraduate students of various CEFR levels. The findings showed that the students' motivation in language learning increased and they became engaged in their language learning. As a result, their interaction and acquisition of vocabulary improved.

GSRS is also applicable for assessment purposes. Ismail and Mohammad (2017), conducted a survey to investigate students' perceptions of Kahoot! as a formative assessment tool. The participants were first year undergraduate medical students from one public university in Malaysia. The findings showed that students perceived Kahoot! as a fun and effective tool as they were focused in their learning through the immediate feedback that they received. Feedbacks have great influences on people (Hattie, 2016). In the case of Kahoot!, the immediate feedbacks the students receive can boost their motivation and enhance their language learning.

Based on the studies discussed, it can be seen that GSRS has many benefits in teaching and learning of 21st century learners. It not only creates fun learning environment, but also retains students' attention in the classroom. This chapter also explores university students' perceptions of using Kahoot! and the impacts of Kahoot! on their motivation and engagement.

FINDINGS AND DISCUSSION

In this study, a total of 35 students who attended an English course in Semester II, 2017/2018 were recruited as the participants. Their fields of study included engineering, business management and applied sciences. After the semester ended, the participants were given a questionnaire which contained a set of open-ended questions.

The figure above shows participants' previous experience in using Kahoot!. Based on Figure 4, 30 students have the experience of using Kahoot! before this. Those who had previous experience and who were familiar with the tool agreed that the tool created fun and enjoyable learning environment. It helped the participants to be focused on what to learn. In addition, the tool also required the students to think quickly but accurately within a limited time.

After using Kahoot! for two terms in the class appears to have facilitated a greater degree of motivation among students during sessions. They really looking forward to play Kahoot! at the end of each lesson. What the authors impressed was on how the students pay attention on the lecture and they seemed to want to understand the content so they could do well in the Kahoot! quiz at the end. The authors believe that this game has helped with student engagement. It also gave students opportunity to communicate with the lecturer and their peers. They were actively participated either in individual or group work. One of the participants stated that:

Figure 4. Experience in using Kahoot!

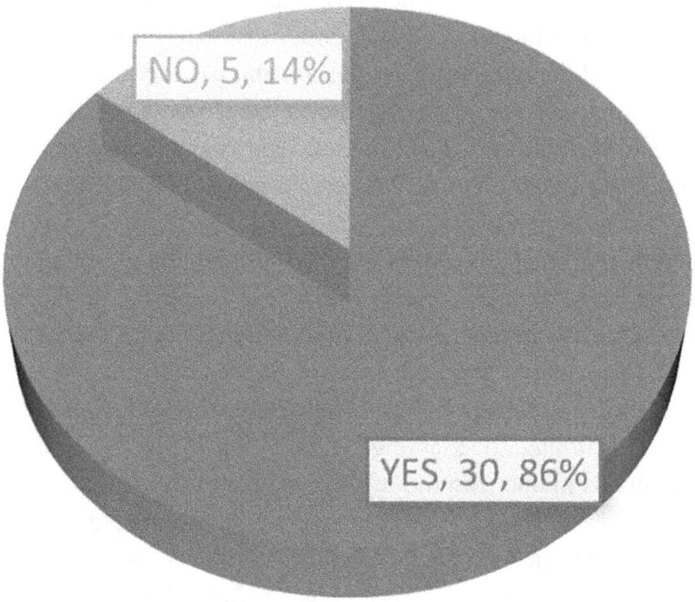

This is the best class ever. I enjoy it very much. I can communicate freely with my friends and give full cooperation to win the game. [P14]

From past experience in a class that no game-based learning has been implemented, normally students would have disengaged in the middle of a lecture. However, when they were introduced to Kahoot! they were really engaged until the very end. They were waiting for the quiz to commence at the end of the lecture. Another interesting finding is, every week there was a different winner. These winners came from different groups of students where normally when lecturer ask questions in a class, it would be the same students to raise their hands and answer the question. By looking into this situation, the authors found that this type of learning helped a wider range of students to engage in class. It really fostered wider and active participation. They interacted more with peers around them than they typically would do in any other lecture. They felt comfortable either with the lecturer or their classmates. One of the participants stated that:

This class is not boring. Need to focus more to win the game make me give full attention to the lecture. I like it very much. [P7]

The positive impacts found in this study are similar to those mentioned in the literature (see Bodnar and Clark, 2017; Ismail & Mohammad, 2017; Merdina & Hurtado, 2017; Iaremenko, 2017; Huckabee & Bissette, 2014).

From Figure 4 also shows that there were 5 participants who do not have previous experience in using Kahoot! before they enter this class. Out of these 5, two of them were positive about the using tool although they had some difficulties in getting use to the tool earlier. They believed that the tool helped to sharpen their thinking as they needed to provide the answers within a short time. The other three participants felt that using Kahoot! was a waste of time as they felt that they were not able to focus on their learning. Thus, teaching and learning should be conducted without the use of any application. This suggests a kind of resistance towards the use of gamification in English language teaching and learning. Gamification is one of the e-learning tools which is part and parcel of teaching in the 21st century and students who do not prefer to use this kind of tool, either inside or outside the classroom, may be left behind in their learning. One of the reasons for the participants to have this perception may be due to the time limit. A number of participants suggested that they needed more time to decide on the answers and therefore, time limit should be increased. The time limit may have increased the participants' anxiety, making them unable to focus on their learning. Further investigation is required to provide insights into the reasons for resisting the use of gamification.

The participants have highlighted three significant issues in using gamification in the class. Firstly, the internet connection needs to be strong. Kahoot! is a real time game application and students need to provide their answers within a short time limit. Weak internet connection may cause some students to lose the game because their answers were not detected by the application. This may cause frustration and lead to rejection of using any e-learning tools. Secondly, although Kahoot! is suitable to be used as a tool for assessment, it is limited to assessing students' knowledge about the language and not the language skills. This finding is in line with the study conducted by Ismail and Mohammad (2017) who found that gamification is not suitable to be used to assess complex courses or topics. Thirdly, students need to be familiar with any gamification applications before a game or a quiz is conducted. One of the participants who did not have previous experience in using Kahoot! stated that:

Maybe next semester when some lecturers use this again, I can know what the benefits of the game are. But now, don't see it yet [P27]

The participant was uncertain as to whether gamification helped him in his learning. In the case of unfamiliarity, students may make technical errors when they attempt to answer the questions within the limited time. As a result, they may feel frustrated and resist using any e-learning tools.

The interesting points appeared from the observation and survey were the importance of discussion and competition to the students where it encourages them to think critically. This is what the authors like about this game, the competitive element. The increasing of students' participation in the class, their energy levels in competing with each other really created a lively classroom dynamic. They are motivated to perform better and better for every quiz so that they can be the top on the scoreboard. Having a desire to be the best in the class encouraged students to always be prepared and engaged in the class.

CONCLUSION

In conclusion, gamification is a learning approach that can become an excellent strategy for the 21st century learner which merely facilitating the social learning experience. It is primarily about generating a positive learning environment by bringing collaboration, emotion and play together which creates a fun and competitive environment in teaching and learning. It gives benefits both to educators and students. This paper aimed to encourage more educators to apply and integrate gamification and m-learning for teaching and learning in assisting their students to prepare them for future jobs. Using gamification in a classroom setting by challenging students to create and design game, facilitate them with the gaming experience and let them play the game with their classmates can really expose them to develop different skills. This encourages them to do thorough research, critical and creative thinking and enhances their presentation skills. It may help educators to make changes from a traditional classroom where it is typically teacher-centred to a 21st Century classroom which is more learner-centred. Students can have a better learning experience, better learning environment, give instant feedback, encourage behavioural change, and enhance soft-skills. This certainly creates a fundamentally more social and engaged learning environment. Hence, an effective gamification must cover these concepts which are capturing learners' attention, engaging and entertaining, it must be challenging and at the end teaching them what they need to know. Having all these concepts will definitely makes lessons a successful one. Nevertheless, more studies can be done on m-learning and gamification usage in classes on how it can produce different outcomes in students' learning ability and on how it helps our students to foster these skills to prepare them for future jobs.

Despite all the benefits of using gamification, there are still some concerns that need to be addressed. It's difficult to create games or gamify instruction without proper understanding and knowledge of the approaches. In addition, extrinsic and intrinsic motivation of the 21st century learner is something that needs to be considered at the moment of designing new curriculum using this approach. There is still plenty of research that needs to be done in regards to the digital native, the approaches, and the teaching and learning experiences. More studies can be performed on gamification usage to determine the different outcomes in students' learning abilities and on how it helps our students learn and adapt in the new learning environments especially in Malaysian context.

REFERENCES

Alexander, B. (2004). Going nomadic: mobile learning in higher education. *EDUCASE Review, 39*(5), 28-35. Retrieved from http://www.educause.edu/ero/article/going-nomadic-mobile-learning-higher-education

Bandura, A. (1997). *Self-efficacy: The exercise of control.* Macmillan.

Barata, G., Gama, S., Jorge, J., & Goncalves, D. (2013). Engaging Engineering Students with Gamification. *Proceedings of the 5th International Conference on Games and Virtual Worlds for Serious Applications*, 1-8.

Bicen, H., & Kocakoyun, S. (2018). Perceptions of Students for Gamification Approach: Kahoot as a Case Study. *International Journal of Emerging Technologies, 13*(2), 72–93. doi:10.3991/ijet.v13i02.7467

Blair, N. (2012, January). Technology Integration for the New 21st Century Learner. *Principal.*

Bodnar, C. A., & Clark, R. M. (2017). Can Game-Based Learning Enhance Engineering Communication Skills? *IEEE Transactions on Professional Communication, 13*(1), 24–41. doi:10.1109/TPC.2016.2632838

Byrne, R. (2013). *Free technology for teachers: Kahoot! - create quizzes and surveys your students can answer on any device.* Retrieved from http://www.freetech4teachers.com/2013/11/kahoot-create-quizzes-and-surveys-your.html#.VLnc78buzuU

Chadyiwa, M., & Mgutshini, T. (2015). Using mobile handheld devices as tools of learning and teaching for student EHPS: A blessing or a curse? *International Journal of Humanities, Arts and Social Sciences, 1*(2), 85–91.

Chee-wei, T., Pei-duo, Y., Ling, L., Chung-kit, F., Chun-kiu, L., & Yanru, C. (2017). *Teaching computational thinking by gamification of k-12 mathematics: Mobile app math games in mathematics and computer science tournament. In Proceedings of Computational Thinking Education 2017* (pp. 55–59). Hong Kong, China: The Education University of Hong Kong.

Corbeil, J. R., & Valdes-Corbeil, M. E. (2007). Are You Ready for Mobile Learning? *EDUCAUSE Quarterly, 30*(2).

Cross, J. (2014). *Introduction to Kahoot for your classroom assessments.* Retrieved from https://www.youtube.com/watch?v=PYfoRRtLXys

de Freitas, A. A., & de Freitas, M. M. (2013). Classroom Live: A Software-assisted Gamification Tool. *Computer Science Education, 23*(2), 186–206. doi:10.1080/08993408.2013.780449

Deterding, S., Dixon, D., Khaled, R., & Nacke, L. (2011). From game design elements to gamefulness: defining "gamification". In *Proceedings of the 15th International Academic MindTrek Conference: Envisioning future media environments* (pp. 9–15). Tampere, Finland: ACM. 10.1145/2181037.2181040

Dichev, C., & Dicheva, D. (2017). Gamifying education: What is known, what is believed and what remains uncertain: A critical review. *International Journal of Educational Technoilogy in Higher Education, 14*(9), 1–36.

Emmer, M. J., & Brunhoeffer, G. C. F. (2015). Knowledge and attributes of forecasting index: Self-assessment for graduating Construction Management students. *Proceedings of the 2015 Associated Schools of Construction Annual International Conference*.

Figueroa, J. F. (2015). Using gamification to enhance second language learning. *Digital Education Review*, *27*, 32–54.

Figueroa, J. F. (2016). Gamification and Game-based Learning: Two Strategies for the 21st Century Learner. *World Journal of Education Reseacrh*, *3*(2), 507–522. doi:10.22158/wjer.v3n2p507

Gibson, D., Ostashewski, N., Flintoff, K., Grant, S., & Knight, E. (2013). *Digital Badges in Education. Education and Information Technology*. New York: Springer.

Hamari, J., Koivisto, J., & Sarsa, H. (2014). Does gamification work?-A literature review of empirical studies on gamification. In *Proceedings of the 47th Hawaii International Conference on System Sciences (HICSS)* (pp. 3025–3034). Waikoloa. 10.1109/HICSS.2014.377

Hattie, J. (2016). *On Formative Assessment: Readings from Educational Leadership*. Alexandria, VA: ASCD.

Herz, C. J. (2001). *Gaming the system: What higher education can learn from multiplayer online worlds*. The Internet and the University, Educause Forum on the Future of Higher Education. Retrieved from http://www.educause.edu/ir/library/pdf/ffpiu019.pdf

Huang, H. Y. W., & Soman, D. (2013). A Practitioner's Guide to Gamification of Education. Research Report Series: Behavioral Economics in Action. University of Toronto –Rotman School of Management.

Huckabee, I., & Bissette, T. (2014). *Learning made fun. Training Industry Magazine*. Retrieved from http://www.nxtbook.com/nxtbooks/trainingindustry/tiq_2014spring/

Iaremenko, N. V. (2017). Enhancing English language learners' motivation through online games. *Information Technologies and Learning Tools*, *59*(3), 126–132.

Icard, S. B. (2014). Educational technology best practices. *International Journal of Instructional Technology and Distance Learning.*, *11*(3), 37–41.

Ismail, M. A.-A., & Mohammad, J. A.-M. (2017). Kahoot: A promising tool for formative assessment in medical education. *Education in Medicine Journal*, *9*(2), 19–26.

Issham, I., Rozhan, M.I., & Thenmolli, G. (2010). *Motivation, psychology and language effect on mobile learning in University Sains Malaysia*. Academic Press.

Johnson, L., Adams, S., Cummins, M., Estrada, V., Freeman, A., & Ludgate, H. (2013). *The NMC Horizon Report: 2013 Higher Education Edition*. Austin, TX: The New Media Consortium. Retrieved from http://www.nmc.org/pdf/2013-horizon-report-HE.pdf

Kahoot! Game-based blended learning & classroom response system. (2014). Retrieved from https://getkahoot.com

Kapp, K. M. (2012). *The gamification of learning and instruction: game-based methods and strategies for training and education*. New York, NY: John Wiley & Sons.

Khaddage, F., Lattemann, C., & Acosta-Díaz, R. (2014). Mobile gamification in education engage, educate and entertain via gamified mobile apps. In *Proceedings of Society for Information Technology & Teacher Education International Conference* (pp. 1654–1660). Jacksonville, FL: Association for the Advancement of Computing in Education (AACE).

Kingsley, T., & Grabner-Hagen, M. (2015). Gamification questing to integrate context, knowledge, literacy, and 21st century learning. *Journal of Adolescent & Adult Literacy*, *59*(1), 51–61. doi:10.1002/jaal.426

Lee, J. J., & Hammer, J. (2011). Gamification in education: What, how, why bother? *Academic Exchange Quarterly*, *15*(2).

Leemkuil, H., de Jong, T., de Hoog, R., & Christoph, N. (2003). KM Quest: A Collaborative Internet Based Simulation Game. *Simulation & Gaming*, *34*(1), 89–111. doi:10.1177/1046878102250605

Licorish, S. A., Owen, H. E., Daniel, B., & George, J. L. (2018). Students' perception of Kahoot! influence on teaching and learning. *Research and Practice in Technology Enhanced Learning*, *13*(9), 1–23. PMID:30595732

Malaysian Digest. (2016, January 19). *Lack Of Self-Initiative, Soft Skills Leads To Unemployed Fresh Graduates*. Retrieved from http://www.malaysiandigest.com/frontpage/29-4-tile/590693-lack-of-self-initiative-soft-skills-leads-to-unemployed-fresh-graduates.html

McDonald, S. D. (2017). Enhanced critical thinking skills through problem-solving games in secondary schools. *Interdisciplinary Journal of e-Skills and Lifelong Learning, 13*, 79-96. Retrieved from http://www.informingscience.org/Publications/3711

MCMC. (2008). *Fact & Figures (Statistics & Record)*. Retrieved from http://www.skmm.gov.my/facts_figures/stats/index.asp

Medina, E. G. L., & Hurtado, C. P. R. (2017). Kahoot! A digital tool for learning vocabulary in a language classroom. *Revista Publicando*, *4*(12), 441–449.

Multon, K. D., Brown, S. D., & Lent, R. W. (1991). Relation of self-efficacy beliefs to academic outcomes: A meta-analytic investigation. *Journal of Counseling Psychology*, *38*(1), 30–38. doi:10.1037/0022-0167.38.1.30

Murugiah, C. (2015). *Article Review Paper on Soft Skills Development in Malaysia Through Shared-value Partnerships*. Retrieved from https://www.researchgate.net/publication/283297939_Articles_Review_on_Development_Soft_Skills_in_Malaysia

Mustafa, Z. (2018, April 25). Way forward for higher education in 4IR era. *News Straits Times*. Retrieved from https://www.nst.com.my/education/2018/04/361452/way-forward-higher-education-4ir-era

Nah, F., Telaprolu, V., Rallapalli, S., & Venkata, P. (2013). Gamification of education using computer games. In S. Yamamoto (Ed.), *HCI 2013, Part III* (Vol. 8018, pp. 99–107). LNCS.

Ng, P.Y., Goi, C.L., & Voges, K. (2010). *The importance of mobile phone applications to young consumers: an exploratory study in Malaysia.* Retrieved from http://anzmac2010.org/proceedings/ pdf/anzmac10Final00103.pdf

Nicholson, S. (2015). *A recipe for meaningful gamification. In Gamification in Education and Business* (pp. 1–20). Berlin, Germany: Springer.

Plump, C. M., & LaRosa, J. (2017). Using Kahoot! in the classroom to create engagement and active learning: A game-based technology solution for e-learning novices. *Management Teaching Review*, *2*(2), 151–158. doi:10.1177/2379298116689783

Quinn, C. (2000, Fall). *mLearning: mobile, wireless, in-your-pocket learning.* Retrieved from http://linezine.com/2.1/features/cqmmwiyp.htm

Sani, R. (2018, January 17). Embracing Industry 4.0. *News Straits Times*. Retrieved from https://www.nst.com.my/education/2018/01/325914/embracing-industry-40

Santos, C., Almeida, S., Pedro, L., Aresta, M., & Koch-Grunberg, T. (2013). Students' Perspectives on Badges in Educational Social Media Platforms: The Case of SAPO Campus Tutorial Badges. *Proceedings of the IEEE 13th International Conference on Advanced Learning Technologies*, 351–353. 10.1109/ICALT.2013.108

Sharpless, M. (2002). The design and implementation of a mobile learning resource. *Journal of Personal and Ubiquitous Computing*, *6*(3), 220–234. doi:10.1007007790200021

Stahl, G. (2007). *Meaning making in CSCL: Conditions and preconditions for cognitive processes by groups*. Paper presented at the international conference on Computer Support for Collaborative Learning (CSCL 2007), New Brunswick, NJ. 10.3115/1599600.1599723

Study Malaysia. (2018, May 2). *The Fourth Industrial Revolution (IR 4.0) and what it means for students like you.* Retrieved from https://www.studymalaysia.com/education/top-stories/the-fourth-industrial-revolution-ir-4.0-and-what-it-means-for-students-like-you

Tan, Malini, & Kaur. (2018). Kahoot! It: Gamification in Higher Education. *Pertanika Journal of Social Science & Humanities*, *26*(1), 565–582.

Thomas, C. (2014). *Kahoot!* Retrieved from https://www.graphite.org/website/kahoot

Trifonova, A., Georgieva, E., & Ronchetti, M. (2006). Determining students' readiness for mobile learning. In *Proceedings of the 5th WSEAS International Conference on E-activities* (pp. 84-89). Venice, Italy: Academic Press.

Vassileva, J. (2008). Toward social learning environment. *IEEE Transactions on Learning Technologies*, *1*(4), 199–213. doi:10.1109/TLT.2009.4

Wang, A. L. (2015). The wear out effect of a game-based student response system. *Computer Education*, *82*, 217–227. doi:10.1016/j.compedu.2014.11.004

Wang, A. L., Zhu, M., & Satre, R. (2016). The effect of digitizing and gamifying in classrooms. In *Proceedings of the 10th European Conference on Game Based Learning*. University of the West of Scotland.

Wang, Y. (2004). Context awareness and adaptation in mobile learning. *Proceedings of the International Workshop on Wireless and Mobile Technologies in Education*, 154-158.

Werbach, K., & Hunter, D. (2012). *For the win: How game thinking can revolutionize your business*. Philadelphia: Wharton Digital Press.

Zarzycka-Piskorz, E. (2016). Kahoot it or not? Can games be motivating in learning grammar? *Teaching English with Technology, 16*(3), 17–36.

Zichermann, G., & Cunningham, C. (2011). *Gamification by design*. Sebastopol, CA: O'Reilly Media.

ADDITIONAL READING

Arbaugh, J. B. (2000). How classroom environment and student engagement affect learning in internet-based MBA courses. *Business Communication Quarterly, 63*(4), 8–26. doi:10.1177/108056990006300402

Azad, A. M. (2013). The impact of mobiles on language learning on the part of English Foreign Language (EFL) university students. *Proceedings of the Social and Behavioral Science*, (pp. 104-108).

Chantorn, C., Vatcharaporn, E., & Jocelyn, W. (2011). *Designing mobile communication tools: A framework to enhance motivation in online learning environment*. Mobile and Contextual Learning.

Chien, Y. T., Chang, Y. H., & Chang, C. Y. (2016). Do we click in the right way? A meta-analytic review of clicker-integrated instruction. *Educational Research Review, 17*, 1–18. doi:10.1016/j.edurev.2015.10.003

Chien-Hung, L., Yu-Chang, L., Bin-Shyan, J., & Yen-Teh, H. (2014). Adding social elements to game-based learning. *International Journal of Emerging Technologies in Learning, 9*(3), 12–15. doi:10.3991/ijet.v9i3.3294

Coller, B., & Scott, M. (2009). Effectiveness of using a video game to teach a course in mechanical engineering. *Computers & Education, 53*(3), 900–912. doi:10.1016/j.compedu.2009.05.012

Hunsu, N. J., Adesope, O., & Bayly, D. J. (2016). A meta-analysis of the effects of audience response systems (clicker-based technologies) on cognition and affect. *Computers & Education, 94*, 102–119. doi:10.1016/j.compedu.2015.11.013

Johnson, L., Adams, S., & Cummins, M. (2012). *The NMC Horizon Report: 2012 Higher Education Edition*. Austin, Texas: The New Media Consortium; http://www.nmc.org/pdf/2012-horizon-report-HE.pdf

Jones, A., Issroff, K., Scanlon, E., Clough, G., & McAndrew, P. (2006). Using mobile devices for learning in informal settings: is it motivating? *Proceedings of IADIS*. Kahoot! Game-based blended learning & classroom response system. (2014). Retrieved from https://getkahoot.com

Spitzer, D. R. (1996). Motivation: The neglected factor in instructional design. *Educational Technology, 36*(3), 45–49.

Section 3
Digital Technologies and Assessment in Higher Education 4.0

Chapter 10
The Future of Product Design Education Industry 4.0

Jennifer Loy
University of Technology Sydney, Australia

James I. Novak
https://orcid.org/0000-0003-4082-4322
University of Technology Sydney, Australia

ABSTRACT

When a society is undergoing transformational change, it is a challenge for all involved to step outside their immediate context sufficiently to evaluate its implications. In the current digital revolution driving Industry 4.0, the pace of change is rapid, and its scale and complexity can inhibit a proactive, rather than reactive, response. Yet if it were possible to return to the first industrial revolution, armed with twenty-first century knowledge and historical perspective, planning for a healthy society and the future of work could have been very different. This chapter aims to support educational leadership in the development of proactive strategies to respond to the challenges and opportunities of Industry 4.0 to inform the future of work, industry, and society. This is framed through the lens of product design, with its unique position at the nexus of engineering and the humanities, and directly tied to changes affecting manufacturing in the fourth industrial revolution.

INTRODUCTION

From disruptive technology to disruptive ideas, the early decades of the twenty-first century can be characterized as a period of non-conformity and new direction. Whilst globalization and amalgamation dominated economic strategy at the turn of the century, the predicted homogenization into a world without borders has been fractured by a backlash of nationalism and separatism, burgeoning entrepreneurship and new business practices based on a sharing economy. In this setting, the expectation could be that product design education would be working through a corresponding period of radical change, yet in many universities, with higher numbers and reduced funding, there are increasing pressures to conform to a modularized education system; design has become a linear process, and limited approaches to creativity

DOI: 10.4018/978-1-5225-7832-1.ch010

Copyright © 2019, IGI Global. Copying or distributing in print or electronic forms without written permission of IGI Global is prohibited.

are prescribed with diluted outcomes defined by business school drivers and design thinking approaches to interdisciplinary practice. This chapter questions the dominant influences in prevailing product design education, challenging existing thinking in the disciplines to call for educational reform in the face of outdated conventions and thinking. If the discipline is to remain relevant in this era of digital disruption and rapid technological advancements, then pedagogy must be subject to critical scrutiny in order to ensure that the pathways aligned with this approach are not restricted by existing practices and closed thinking. This chapter argues for academics to learn from the past and present in designing pedagogy and curriculum, informing the need for change to ensure authentic learning for product designers in the twenty-first century.

BACKGROUND

Product design emerged as a profession during the eighteenth century in response to the drive towards mass production and the design challenges it created. According to design historian Adrian Forty, in his seminal book, *Objects of Desire* (1992), the earliest product designers in the UK employed by innovators, such as Josiah Wedgewood, were educated on European trends in art and architecture, as well as the reduction of a design into repeatable components. Production uniformity and aesthetic conformity became megatrends of the time and were central to design development, allowing manufacturers to produce standardized products within an orderly system of centralized manufacturing. The impact on the organization of labor, and subsequently lifestyle and the urbanization of the population, was relatively gradual. During the late nineteenth and early twentieth century, however, as Ford established the moving assembly line, the pace of urbanization changed with large-scale factories drawing in workers from large distances (Sparke, 2013). For designers, their role became increasingly constrained by mass-manufacturing processes and practices, with these being the driving technologies of the times. Design had to conform to assembly rules, and generally the lower the cost of components the higher the margins for business. For workers, the transition to working within a system where labor was divided to its most basic action became common place. The impact on the organization of society was immense.

The transition can be characterized as a shift from traditional hand production to massive industrial machinery and factory production. To a large extent it was stimulated by the invention of large-scale manufacturing processes fuelled by the discovery of new methodologies for exploiting the energy stored in huge iron and coal deposits. The subsequent access to apparently unlimited energy and human resources engendered by the rise of capitalism and individual and corporate entrepreneurship and innovation, marked a major transition in human affairs. The Industrial Revolution was the socioeconomic equivalent of the Big Bang. (West, 2017, p.211)

Following the Second World War, building the economy was seen as paramount, with marketing and mass production scaled up. This drive meant that designers during the twentieth century were frequently locked into the contradictory practice of trying to design the best possible outcomes yet encourage obsolescence for repeat sales. Dissenters, such as Victor Papanek, argued for a sense of moral responsibility in design. He was shunned during his early career following the publication of his seminal work, *Design for the Real World* in 1971 (revised edition 2005), but as the social and environmental impacts of mass production began to be understood towards the end of the century, attitudes to his work within the design community began to alter. By the nineteen-nineties, Papanek's book was required reading in design schools as environmentally-conscious, socially aware design gained momentum. The straightforward design – or redesign – of commercial objects for mainstream retailers was replaced in design schools by the responsible design of products for specific challenges facing society, with the lifecycle analysis of the product integral to the work. A century after the role was introduced as a taught profession (attributed to the Deutscher Werkbund established in 1907, precursor to the Bauhaus (Droste, 2015)), the environmental and social impact of the work of the profession and the manufacturing and marketing parameters that it had been established under, were, to a large extent, discredited (Hawken, 2010). Sustainable design became central to the teaching of product design in universities with lifecycle assessment, extended producer responsibility and designing for the circular economy core to industrial design education for the twenty-first century (Walker, 2006).

Whilst the terms 'product design' and 'industrial design' are often used interchangeably, the latter is arguably more systems oriented than the first, but both are closely linked to the emergence of mass production and consumer culture in the early 20th century (Fiell & Fiell, 2006; Walker, 2006). Over the last forty years, the perception of the role of the product designer in higher education has changed significantly (Loy, 2012) as the social and environmental impacts of design for production have become better understood. As a discipline synonymous with large-scale manufacturing and distribution, product design has shaped the physical world through the creation of objects constrained by economic manufacturing drivers, such as standardization and repeatability. Based on this, Walker (2006) argues that the very terms product design and industrial design therefore carry with them an associated baggage that may no longer be suitable in the twenty-first century as new technologies disrupt traditional modes of design and production. This is part of broader digital transformation impacting manufacturing today, termed Industry 4.0. Much like the earlier Industrial Revolution's embrace of the latest technologies of the times, Industry 4.0 encompasses today's latest technologies including artificial intelligence (AI), machine learning, the Internet of Things (IoT), and ubiquitous computing, and will radically alter the world of work, education and consumer relationships with products.

Designers no longer focus on purely human-centered design and the functioning of the product and its stylistic appeal, but rather on the product service system it operates within. The agency of the product is recognized as impacting behavior and sustainability, and is as core to the design as its economic viability. As Tatum discussed in his *Design Issues* (2004) article on design responsibility, every design decision, however small, has consequences because of the cumulative effects of incremental change. From an environmental perspective, the next twenty to fifty years will be critical, and design education needs to anticipate the challenges coming. Founder and Executive Chairman of the World Economic Forum, Schwab (2017) explains that the fourth industrial revolution:

began at the turn of this century and builds on the digital revolution. It is characterized by a much more ubiquitous and mobile internet, by smaller and more powerful sensors that have become cheaper, and by artificial intelligence and machine learning... By enabling "smart factories," the fourth industrial revolution creates a world in which virtual and physical systems of manufacturing globally cooperate with each other in a flexible way. This enables the absolute customization of products and the creation of new operating models. (2017, p. 7)

Product design must necessarily evolve from its roots in the first industrial revolution to maintain relevancy in this new era of digital technology. With boundaries between the physical and digital worlds blurring through digital fabrication technologies, such as additive manufacturing, the demand for product designers to standardize manufacturing and develop one-size-fits-all products is being removed. Additive manufacturing (AM), also known as 3D printing (3DP), is comparatively new; it refers to a range of technologies fed directly by three-dimensional computer-aided design (CAD) files. Described by Gibson, Rosen and Stucker (2015, p. 2), the object is "fabricated directly without the need for process planning... the key to how additive manufacturing works is that parts are made by adding material in layers; each layer is a thin cross-section of the part derived from the original CAD data." This contrasts traditional plastics manufacturing technologies, such as injection molding, which requires significant investment in machining molds before a single product can be produced, with manufacturers needing to sell high volumes of parts from each mold in order to recoup their initial investment and make a profit. The significance of additive manufacturing is that the creation of molds and tooling, and therefore the need to design and sell standardized products, is removed. This has led to a realization of mass-customization, also referred to as mass-personalization, (Hu, 2013; Tseng, Jiao, & Wang, 2010) whereby the 3D CAD file describing a product is altered to suit the unique functional or aesthetic needs of an individual. Objects of similar dimension yet different detail cost approximately the same to 3D print. This is resulting in the 3D printing of high-value customized products such as hearing aids, medical implants and orthotics.

In the twenty-first century, a confluence of innovation based on digital technologies is creating a paradigm shift in industrial practice. Whilst AM is just one example of a technology contributing to the fourth industrial revolution, it is the catalyst for a shift in thinking from mass production to print-on-demand and representative of the changes facing designers. In Industry 4.0, factory automation is increased, along with the monitoring of different facets of production. This includes machines, the progress of the part and the quality of the part at each stage of manufacture. The supply of materials and additional components could be integrated into an automated extension of the 'just in time' (JIT) approach. However, when the practice extends beyond the factory and the collection of data, the analysis of data, and automated responses to changes in the system, the approach moves beyond incremental manufacturing efficiencies. Where distribution and product lifecycle are monitored as part of product service systems, this data is fed back to the organization, providing insights on product performance, adaptation and interaction as part of a connected workflow. This approach is enabled by the drop in cost of monitoring systems, increased sophistication of algorithms and communication within systems (termed the Internet of Things), due to developments in 'Cloud' storage (Seldon & Abidoye, 2018, p. 137) and wireless connectivity.

On a practical level, this shift requires serious up-skilling of academics and upgrading of facilities in higher education to prepare students for their new role within a cyberphysical manufacturing system. However, more fundamentally, academia needs to lead an understanding of the implications of transformational change in this sector to develop strategies and policy for the future of work and industry. The changes currently occurring have the potential to impact societies, the environment and economies. Product design programs and research in higher education are central to these changes and therefore academics need to be providing leadership in addressing the long-term implications of decisions made now, informed by history and a reasoned mapping of the changes about to come.

EDUCATIONAL FRAMEWORK

Learning From History

With the benefit of experience gained by looking back at the long-term consequences of earlier industrial revolutions, both positive and negative, the recognition of a coming paradigm shift should theoretically allow societies to be better prepared for the widespread change that accompanies an industrial revolution. The impact of technological change on the organization of work and the subsequent impact on societies should be central to discussion on the implementation of technological interventions to production systems, such as automation and machine learning, characteristic of Industry 4.0. However, in his book *Are Robots Taking Our Jobs* (2017), Cameron suggests that when considering the changes brought about by digital transformation, people - and therefore organizations - have an inability to comprehend the scale, scope and speed of change. This is because its impact and complexity will be on a different scale because of global connectivity, and it is therefore challenging to view in its entirety. It is also difficult to view different aspects through a narrow lens to try to build a larger perspective, because of the lack of conventional organizational borders and hierarchy in thinking it engenders, and its challenge to existing heuristics. New ways of thinking about the world will be required to provide education with accessible strategies to assess the implications of the paradigm shift, and to attempt to mitigate problems it could create, as well as frame and address problems it could solve.

Higher education has a responsibility to prepare future citizens for their role in society and as the future workforce. With Industry 4.0, this will involve training to ensure the workforce are equipped to cope with the new technology, but more fundamentally, it should involve developing values and aspirations for society informed by the opportunities and challenges digital industrialization will bring. It will also involve attempting to anticipate the changes to people's lives on a daily basis, such as the need to change infrastructure and ways of working. In general, as the concept of Industry 4.0 emerged from the German car industry, it is seen as increased automation and maximized throughput. However, the term more broadly refers to an increased ubiquity of computing capability across manufacturing and product service systems. Increased monitoring systems, generating data, analyzed by algorithms as the basis for operations within the factory changes the way manufacturing operates. However, when monitoring is expanded out to the distribution of the product and its interactions during its lifetime, this informs the factory on future demand and working as part of a system to feedback information on its operation and performance, and changes the way people and products interact.

In disciplines based on the design of manufactured products and product service systems, such as product design, a simple analogy would be moving from 2D paper-based drawings to 3D computer models, as has happened over the last thirty years. For designers, and design educators, it has required a different way of understanding and manipulating form. Expertise in traditional 2D working drawings was no guarantee of expertise in computer 3D modeling – it could even be a hindrance as it predetermines the expectations of the designer. In a similar way, Industry 4.0 heralds a shift from 3D to 4D product modeling, where the entire life of a product is considered during design, with the product possibly having the capacity to update or adapt through time as needs and circumstances change (Novak & Loy, 2017). This is because the digital technologies that inform developments in Industry 4.0 allow for connected products to also be responsive, and therefore develop into complex systems. For industry and academia to respond, thinking needs to be disruptive and informed by new knowledge. Furthermore, Schwab describes that:

While the profound uncertainty surrounding the development and adoption of emerging technologies means that we do not yet know how the transformations driven by this industrial revolution will unfold, their complexity and interconnectedness across sectors imply that all stakeholders of global society - governments, business, academia, and civil society - have a responsibility to work together to better understand the emerging trends. (2017, p. 2)

The challenge to educational strategists is to step outside traditional discipline boundaries, as suggested by Schwab, and build objective, new perspectives that respond to its transformative potential. Design education is by definition forward looking and requires strategies for facing the unknown, engaging in initial research, and creating new, objective views of the situation being faced that allow realistic problem-framing and solution development. Product design academics should therefore be ideally placed to envisage new perspectives for Industry 4.0 education to respond not only to the technological developments in isolation, but also their broader societal, environmental and economic implications.

In learning from the past, Walker (2006, p. 9) talks about previous design movements that have "challenged prevailing stereotypes and stimulated and influenced subsequent designers." Examples include the De Stijl group in The Netherlands, Bauhaus and the Italian Memphis designers. However, he also points out that:

the issues and agendas to which they were responding are not our issues and agendas. Today, we are facing new challenges associated with the globalization of industrial capitalism, the environment, national and transnational socio-economic disparities, and rapidly evolving scientific and technological developments. (Walker, 2006, p. 9)

Educators today need to learn from the achievements of the past, particularly in creating social change, but whilst the ideas need to be viewed in context, equally, they cannot be embraced by the current generation without critical review. This is demonstrated by Fuad-Luke (2009), where he considers movements in design history through the lens of sustainability in his book *Design Activism: Beautiful Strangeness for a Sustainable World*. This approach provides a critical examination of the past, in order to inform current thinking. The necessity of a revised approach to higher education responding directly to the potential future of work enabled by Industry 4.0 needs to be recognized. Product design provides the basis for this discussion because of its position at the nexus of engineering and the humanities, but the challenge to build a radically different educational response to the paradigm shift this industrial revolution effects extends across all disciplines – in whatever future form they emerge as traditional academic boundaries merge. There is a need for strategies to address educational change to keep pace with developments in digital technology but, more fundamentally, for higher education to prepare future citizens to recognize and respond to the issue of human development in a digital era.

Learning From the Present

The digital revolution at the turn of the century has transformed communication and interaction across the globe. Consequently, digital technology is changing learning and teaching for all disciplines. For industrial and product design, the digital revolution is exactly that – a revolution in ideas, aspirations, technological opportunity and direction. It challenges conventions in product design, manufacturing, development, the designer-producer-user relationship, and has the potential to allow for major restructuring in manufacturing systems around the world. This is particularly driven by the sustainability imperative, illustrated by the discovery of plastic waste in remote regions, including from arctic ice-core samples.

The highest density of plastic ever to have been documented was on an uninhabited coral atoll called Henderson Island in the South Pacific. Scientists studying the island estimated there were over 38 million pieces of plastic, with items from Germany, Canada and other distant places recorded. (McCallum, 2018, p.27)

Rethinking the future based on an in-depth, researched understanding of complex situations, and research into the opportunities and challenges provided by new technologies and their relationship to existing ones, whilst keeping the human experience, social and environmental responsibilities to the fore is surely fundamental to the product design discipline. But is this actually the case? Taking a critical view, is it the reality that in working to become an established academic field of research and education over the last thirty years in a competitive tertiary education environment, the discipline has been compromised? Is creativity education being reduced to predictable, manageable, linear exercises due to practical reasons or for accountability? Has a drive to employ academics with a decent H-index pushed industrial and product design educators out or under the auspices of an engineer? In a study into the relationship between research and teaching activities in tertiary institutions, Gunn (2018) found that the need to meet the requirements for producing research outputs and bring in research funding was distorting the priorities of the universities. Gunn (2018, p.153) argued that it was "an anomaly" in terms of a business practice, as the dominance of research focus directed resources away from serving the "customers" in the "core business." He explains how the environment in which UK universities have operated over the last twenty-five years has been transformed and there is a need to refocus on teaching to meet the requirements of a new educational era. The extent of the disruption of digital technologies to society and the future of work and industry has only begun to emerge. However, based on examples such as the online communication platforms supporting Uber and Airbnb, disruption is likely to have far reaching consequences. Academics will need to rethink the curriculum and pedagogy in response to the complex nature of the changes and prepare students for the unknown and the unexpected. However, has the very ability to respond positively and imaginatively to the unknown (a critical need for society in the face of disruptive change and surely characteristic of design), been rendered insignificant and even obsolete over that time because it is hard for institutions to quantify - or for those from other disciplines to properly replicate? Industrial and product design disciplines will need to step outside the confines of dominant, established academic practices and attitudes and take an objective look at the role of the designer in the twenty-first century, and an unfettered look at educational development that supports the discipline ideal irrespective of external influences and current ideas of what tertiary education and research should be. This is a time of change, and change is what design should be all about.

Over the last twenty years, elements of a design approach to problem framing and problem solving have been hived off by business, health, engineering and transdisciplinary programs. There have been two major problems with this: Firstly, those teaching within these programs are frequently not fully-fledged designers and therefore not adequately equipped to teach the subject effectively. Imagine if designers started to teach engineering without qualifications or professional experience; there would be considerable objections raised. Yet it appears that it is acceptable within tertiary education for this practice, which speaks to the current understanding of what design is, and the status of design programs in universities based on the current priorities of those tertiary institutions. Secondly, the elements of the design process that have been adopted by other programs are limited, resulting in a narrow interpretation of design and its application. In particular the systemization of a design process, as frequently presented in engineering programs and publications, is reminiscent of presenting a health care plan that does not consider the human element. The complications and subtleties of the situation are lost. Design is reduced to incremental and predictable actions.

Many design philosophers, such as Hugh-Aldersey Williams (2011), have drawn parallels between life after the digital revolution and life before the industrial revolution. The seminal work by economic philosopher Adam Smith, *An Inquiry into the Nature and Causes of the Wealth of Nations,* proposed that efficiency would be improved by the division of labor into the smallest possible increments within manufacturing (Forty, 1992). This approach has dominated mass production over the last hundred years, supported by a stock market driven economy where individual, short term gain has taken precedence over a holistic view of the human experience, contraction and convergence equity issues and long-term environmental impacts. Yet, as West (2017, p.211) points out, this approach is unsustainable: "As presently constituted, our continued success requires the supply of coal, gas, oil, fresh water, iron, copper, molybdenum, titanium, ruthenium, platinum, phosphorous, nitrogen, and much, much more to be available at an exponentially increasing rate." Digital disruptions include algorithms for stock market trading, the displacement of conventional banking with blockchain, and the highlighting of the inadequacies of existing regulatory systems to address changes in business practice, healthcare and manufacturing. These changes are happening quickly. As Seldon and Abidoye (2018, p.169) point out, academia's understanding of the changes in educational access with developments in AI can be outdated and its reaction slow: "Google is the greatest pedagogical success and it is a piece of AI."

In universities, even the introduction of programs can take a number of years, irrespective of public demand, because of the protective systems in place. When a program is added, it often must be justified by evidence of similar, successful programs elsewhere. It also needs to fit within the established program offerings and the organization of faculties, schools and departments. These processes make agile changes difficult. The reduction of risk in the introduction of programs also reduces the ability of a university to change direction quickly or to offer diversity in the programs that operate across traditional organizational boundaries. Given the rapid pace of technological change, this inability to adapt at a similar pace is cause for concern. To date, research universities have been secure in their position as educational providers. Yet complacency has a history of preventing sufficient preparation in the face of unanticipated or under estimated change. This is a dangerous attitude for all disciplines to have, including, as discussed, long-established and regulated disciplines such as engineering, but for design, there is really no excuse. Digital technologies have the potential to cause far more disruption than appears to be currently being predicted within educational institutions. "A growing national and institutional policy focus on teaching, and anxiety over the student experience, presents the chance for new narratives of investment in learning and teaching to develop beyond the hegemonic reach of research excellence and research-intensive universities" (Charles, 2018, p.28). Genuine, informed, engaged design education is not hampered by the weight of formulaic restrictions and discipline legacy. It has not only the freedom, but the imperative, to keep a watching brief for change and respond energetically to researching the possible futures on the horizon. However, in order to maintain integrity in this aim, industrial and product design disciplines need to disentangle themselves from the reflected status of disciplines that are currently the longest established, and most recognized as authoritarian in universities at this time. Design education needs to be founded on working with the unknown, and the constant development of new practices and approaches to explore, understand and address forever new problems and opportunities. Students and academics need to re-embrace that spirit of improvisation, and tertiary education needs to understand its value. Ways of working founded on this approach need to be revisited and formally developed, with an alignment perhaps closer to acting studies than current design education programs that are aligned to business or engineering.

Learning for the Future

According to Seldon and Abidoye (2018, p.169) "AI is impacting everyone, every day, but schools and universities are not consciously using it to exploit teaching and learning or adequately to prepare their learners for the AI-driven workforce they will subsequently encounter." They continue:

Nothing but nothing is more important than education in ensuring that AI works in the interests of all humanity. We need to re-imagine our schools from the ground up to teach our young to be more fully human and not to be content any longer with giving them just 'factory era' skills." (Seldon & Abidoye, 2018, p.312)

Seldon and Abidoye point out that in the UK, the Prime Minister's Industrial Strategy, announced in November 2017, has AI as the leading component, and start-ups in London have grown from 16 in 2008 to 6000 in 2018. There have been fairly radical shifts for product design education over the last forty years – a relatively short time compared to other, more traditional disciplines, such as engineering and architecture. Yet the changes facing all of higher education in the rapidly emerging digital context are likely to disrupt product design education even more significantly because of their intrinsic connection to manufacturing and therefore AI. Seldon and Abidoye argue:

There is no more important issue facing education, or humanity at large, than the fast approaching revolution of Artificial Intelligence, or AI. This book is a call to educators everywhere, in primary, secondary, further and higher education (HE), and in all countries, to open our eyes to what is coming towards us. If we do so, then the future will be shaped by us in the interests of all. If not, others, the large tech companies, governments and even the bad guys will decide, and we will have only ourselves to blame. (2018, p.1)

The suggestion is that educational models will change and that governments must lead changes to modernize education. (Seldon & Abidoye, 2018) Artificial Intelligence in this context does not refer solely – or even predominantly – to humanoid robots, but to the automation of labor and its impact on the future of work and industry. For higher education, the challenges are both to prepare graduates for a digitally connected future, but also to radically evolve education, to create relevant pedagogy and curriculum in a digital age. The context and intent of a curriculum for the coming digital era needs to be informed by the dichotomy that is emerging in response to the rapid changes digital technologies are bringing. This dichotomy is exemplified by the contrasting opinions of Greenfield (2017) and Kelly (2016) on technology and society issues such as the 'Quantified Self' (Kelly, 2016, p.238, Greenfield, 2017, p.33). It is also illustrated for the general public in exhibitions such as 'The Future Starts Here' at the Victoria and Albert museum in London (Hyde & Pestana, 2018). Instead of a science-fiction, all white, technology-worshipping ideology, the exhibition presented a curiously dystopian view of developments in a vaguely post-apocalyptic setting. In a darkened room, with exhibition stands that look like the relics of a Bladerunner 2049 film set, cutting-edge research, such as the MIT 4D pneumatics prototypes, were shown alongside examples of digital-based interference in politics (both Trump and Brexit) and fake news. An example of 3D printing being used by designer Sarah Hendren and anthropologist Caitrin Lynch to support the individual needs of an amputee in the design of low-cost, bespoke assistive technology (Galloway, 2018) was positioned alongside a stand devoted to those who are working to extend life through strategies such as cryogenics and downloading the human brain.

For product design education, Industry 4.0-specific education is critical for future graduates. As products increasingly become connected to the digital world, and are created through entirely digital means, the questions for product designers have become bigger, shifting from how digital technologies such as sensors, the Internet of Things and additive manufacturing will affect products, to how it will affect product systems, industries, their organization and the future workforce. Education with regards to Industry 4.0 will need to comprise of two things: The first is the means to critically research and interrogate the opportunities and implications of technological change. The second is the ability to constantly update skills and understandings fluidly in a continuous state of change. Both are difficult to achieve in the face of the educational heritage that has built up over the last fifty years as education itself became more of a business, with changes to funding and the focus of universities generally leading to larger class sizes and more standardized learning. Traditionally a designer has gone through an education system where the content generally changes only incrementally and has worked in a career where the tools of designing, the processes and manufacturing technologies, change slowly, if at all. Whilst an occasional updating of skills might have been necessary – lifelong learning has not. However, design is currently in flux, linked closely with the burgeoning developments in digital technology: "Technology is humanity's accelerant. Because of technology everything we make is always in the process of becoming. Every kind of thing is becoming something else, while it churns from "might" to "is." All is flux. Nothing is finished. Nothing is done" (Kelly, 2016, p. 6). The education of designers must also now be in flux – subjects taught the same way as they were only a few years ago would arguably be setting students up for failure, as the knowledge they would graduate with would be outdated before they began their working career. Brown and Alder (2008) built on this idea describing that:

In the twentieth century, the dominant approach to education focused on helping students to build stocks of knowledge and cognitive skills that could be deployed later in appropriate situations. This approach to education worked well in a relatively stable, slowly changing world in which careers typically lasted a lifetime. But the twenty-first century is quite different. The world is evolving at an increasing pace. When jobs change, as they are likely to do, we can no longer expect to send someone back to school to be retrained. By the time that happens, the domain of inquiry is likely to have morphed yet again. (Brown & Adler, 2008, p. 30)

Therefore, lifelong learning must be embedded in design – it must be synonymous with the act of designing. Gore (2013) makes a similar point, arguing there is still too much emphasis in education on the memorization of facts: "Yet in a world where all facts are constantly at our fingertips, we can afford to spend more time teaching the skills necessary to not only learn facts but also learn the connections among them" (Gore, 2013, p. 67). The challenge is less about reshaping education for the students, as each new influx of students knows little about what has been taught in previous years, but rather how to transition design staff to embrace change, particularly if they have past industry experience in product design based in traditional mas-manufacturing technologies. Many academics (not just design) have been experts in their field, but as change brought on by digital technologies such as AI accelerates, many established practitioners and academics simply do not want to adapt and remain relevant. At what point does the need to prepare students for the future outweigh this experience of the past? Petko, Egger, Cantieni and Wespi (2015, p.49) observe that "research has shown that successful technology adoption does not so much rely on hardware and software but, more importantly, on teachers' skills and beliefs." Leadership in academia tends to be older because of the traditional hierarchical promotion system, and industry professors are still rare, though not unheard of (for example, journalist Professor Peter Fray at the University of Technology Sydney in 2018). Whilst age is not a barrier to the adoption of new technology, it is arguably more difficult to adapt to new ways of working when expertise is built on traditional working practice. Large scale operations, such as universities, are difficult to redirect. Examples of disruptive change brought about by management decisions can be seen at the University of Melbourne though, for example, where first year studies were integrated, and trimesters introduced. Brown and Alder (2008) introduced "Learning 2.0," linked to the Open Educational Resources (OER) movement:

In a traditional Cartesian educational system, students may spend years learning about a subject; only after amassing sufficient (explicit) knowledge are they expected to start acquiring the (tacit) knowledge or practice of how to be an active practitioner/professional in a field. But viewing learning as the process of joining a community of practice reverses this pattern and allows new students to engage in "learning to be" even as they are mastering the content of a field. (Brown & Adler, 2008, p. 20)

This prepares students for lifelong learning. In many ways this idea is more closely linked with an apprenticeship for design whereby learning is practice-led and conducted in the process of designing – thus learning and designing become intertwined (Novak, 2018). However, the rapid pace of change and the radically different ways of working that are emerging with Industry 4.0 require more proactive strategies than relying on current practitioners to lead the generation. Arguably it will be the other way around, as young people elect entrepreneurial practice over conventional employment, influenced by the success of companies such as Google, which started in a garage in 1998 and has 4.2 billion search requests daily in 2018, and YouTube, which started in a room above a pizzeria in 2005 and today has 8.8 billion videos online (Seldon & Abidoye, 2018). According to Seldon and Abidoye (2018, p. 137) "the IoT facilitates the collection of big data on a scale we are still unable fully to assess or exploit, as the sheer volume can militate against sifting the quality from the unreliable evidence and forming solid conclusions," and China is working to establish itself as the world's primary AI innovation center by 2030 (p. 314). Relying on conventional innovation practices within established industries responding to the challenges and opportunities of Industry 4.0 seems a dangerous strategy for the rest of the world. However, the rise in interest in entrepreneurship (for example 40% of students at the University of Technology Sydney expressed a preference for being an entrepreneur over being conventionally employed (Conroy, 2016)), and the opportunities to foster it that digital connectivity provides, may provide new direction. In order to maximize this trend, the university system needs to become nimble in its offerings, dynamic, connected and practice-led. This can only be achieved with a framework of practice to support academics in moving to this new way of thinking.

Responding to these changes, a proposed model of education that blends elements of the flipped classroom model (Altemueller & Lindquist, 2017; Gavriel, 2015) with a 'Build-Measure-Learn' lean startup approach may offer an appropriate strategy whereby students take responsibility for their own learning, guided by a facilitator. This is diagrammatically shown in Figure 1. The traditional roles of teacher and learner are reversed, with students bringing new knowledge to classes, which is disseminated to fellow students and facilitator alike. This is significant in a time where technology is evolving so rapidly that it is impossible for a single lecturer or facilitator to know everything. Combined with a lean startup approach, students are encouraged through projects to consider their products as part of larger business models, analyzing investment strategies, testing prototypes on the market, and even developing Kickstarter campaigns (for example a course called Innovation and Entrepreneurship run by the authors at Griffith University). This approach is agile and sets students up with lifelong learning skills as they participate in the practice of learning alongside peers and mentors.

The industrial design discipline by definition is about imagining, planning and realizing the future. Even when engaged with rationalizing the details of a specific manufacturing requirement for an individual company, the designer is part of a discipline focused on innovation, understanding and effecting progress. As designers will constantly face new challenges, new technologies and changing aspirations for society that impact design decisions, design education necessarily needs to prepare graduates to be lifelong learners and proactive, critical thinkers. Lynam and Cachia (2018) argue that in an increasingly digital, interconnected world, information is readily accessible and becomes rapidly outdated, and therefore future employers will value graduates who are critical thinkers rather than just knowledgeable. They highlight the ability to source and evaluate credible information as necessary for effective problem-solving and decision-making, and these skills are an important part of the framework described by Figure 1 as "out of class" activities. Sterling (2015) also argues that in preparation for real-world, complex problem solving, students need to be educated not about change, but rather *for* change and this requires a pedagogical shift from transmissive education to transformative education.

In the early decades of the twenty-first century, digital technologies are disrupting traditional business and industrial practices. Online communication platforms are fostering more realistic entrepreneurial ambitions than were previously possible, ubiquitous computing and the Internet of Things are forcing a complete rethink of product service systems, and additive manufacturing and the digitalization of inventory are enabling global distributed manufacturing. Whereas the development of industrial and product design practice over the last century could be characterized as incremental, building a body of

Figure 1. Proposed framework of learning based on the flipped classroom and lean startup 'build-measure-learn' models (Novak 2018)

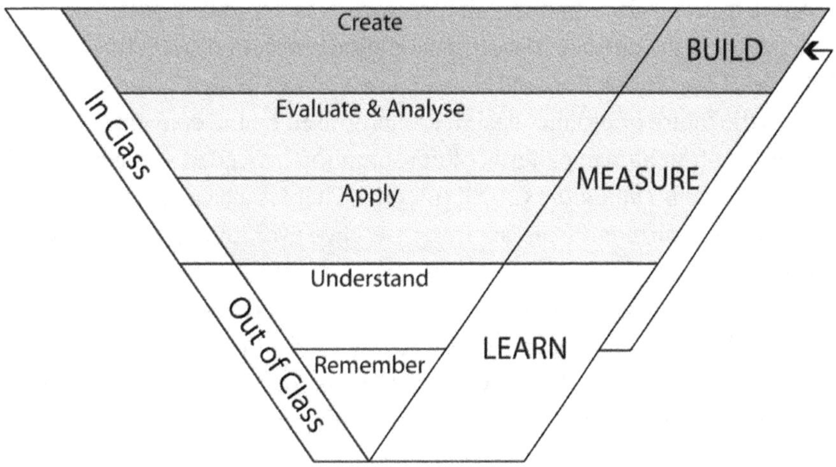

knowledge on production practices and materials that informed design education, recent developments are creating a paradigm shift in the place of design in society and the role of the designer that contrasts with accepted practice. As digital disruption increasingly challenges established hierarchies and conventions, educational conformity may itself be under threat and traditional disciplines may benefit from a rethink on the role of product design education. Only, however, if product design education itself recognizes the emerging opportunities for change and breaks away from the constraints it has accepted will it gain acceptance in the current academic value system.

AUTHENTIC LEARNING IN THE TWENTY-FIRST CENTURY

Schools and universities in 2018 are little changed from 1600, for all the ubiquitous (non-AI) technologies. The teacher or lecturer address rows of students from the front of the class, who advances at the same pace. Learning is not personalized and the teacher is weighed down by the burden of preparing material, setting and marking assignments, and reporting on student performance. (Seldon & Abidoye, 2018 p. 140)

As university funding has changed, so have university student numbers increased. The organization of university education, its monitoring and regulation, have become driven by administrative realities that are created to work across large organizations. There are advantages in this approach, for example the ability for students to work more easily across modular programs to differentiate themselves through specializations, and the ability for students to retake single subjects or study part time in order to support themselves financially or negotiate other external commitments. However, the overall result is the modularization of the learning experience. Academics have to break learning into distinct events. The student is expected to put the experience together in their own minds to become well rounded graduates. It is clear from the rise in importance of graduate attributes from a university perspective that there are concerns already with this approach, and considering the large body of educational research arguing for authentic learning in recent years (e.g. Dee Fink (2013), Weimer (2013)), it is important to question whether organizational constraints are in fact distilling the educational experience. Before modularization, design was predominantly taught as a series of projects. If research shows that this approach is more effective for design teaching in the twenty-first century, will universities support the change? If the modularization of learning for administrative reasons was set to one side, and an objective, fresh view of design education in the current context, post digital revolution, with the growing sustainability imperative and the need for contraction and convergence and extended producer responsibility, what would it look like? Is the future of product design compromised by the conventions of past learning at a time when a paradigm shift in human patterns of consumption is needed and new thinking is required? As Gore discussed in his book *The Future* (2013, p. xv)): "There is a clear consensus that the future now emerging will be extremely different from anything we have ever known in the past. It is a difference not of degree but of kind."

Design academics need to engage more intensively with research into learning to build a new body of knowledge specifically on design teaching for the twenty-first century. It may be that teaching intensively in short bursts over the year best prepares students for the stresses and pressures involved in their role as designers in the future, or it may be that providing no factual knowledge to students during their degree is a strategy, where they have to gain first-hand knowledge of factors, such as materials and processes, at every level.

Students need to prepare for individualized learning programs. Schools and universities have to catch up with their students quickly. Equally, they must prepare for life-long learning, which will spread rapidly. Credit transfer will become the norm rather than the exception. Students should all be taught computer, digital and AI literacy, and how to understand the difference between the human and the machine. The Welsh government is ahead of other UK nations on this. (Seldon & Abidoye, 2018. p. 313)

Research into design studio assessments needs to challenge the underlying assumptions that tertiary educational accountability builds in. It may be that assessment should never be based on the individual, that a more genuine engagement in teaching group work and differentiating roles is essential for effective operation as a designer in the future. This may include the use of online hubs to bring together geographically separated individuals to work collaboratively, rather than relying on the traditional physical classroom, with Massive Open Online Courses (MOOC's) and collaborative websites like GitHub, providing new ways for individuals to learn and engage in real-world projects. Design is a real-world activity, and assessments will have more meaning for the students if they are current and real-world relevant. According to research by Lynam and Cachia:

Student-focused assessments were preferred by students and encouraged engagement. Learners appreciated assessments that built on their skill set; involved an element of choice and creativity; and were associated with a balanced workload. Students particularly valued the benefit of having assessments that were relevant to their career ambitions and developed their skill set. (Lynam & Cachia, 2018, p.231)

The disruptions that technological developments are creating in the world need to be acknowledged in university industrial and product design education. The disciplines need to throw off the academic cultural cringe that has resulted from coming late to an established game, with rules that did not take them into account, and competing on a playing field that is far from level. This is not for the sake of the disciplines, but for future generations who need the benefit of design education that is not watered down but has genuine integrity that responds to the context it is constructed within. This century is crucial for the future of mankind and the health of the environment. Designers are needed, and design academics must prepare.

There is a dominant link in education between the idea of interdisciplinarity in education and industry 4.0. Seldon and Abidoye (2018) argue for problem solving, open ended problems, collaboration and the reduction or delaying of specialization. In design, project-based learning is common-practice in higher education, and the concern Seldon and Abidoye express over the dominance of 'right' answers in education does not realistically apply in design as there are rarely 'right' answers. Projects are already commonly marked based on process as much as, or instead of, end products. Interestingly, interdisciplinarity is more contentious. Whilst it is frequently argued that projects break disciplinary boundaries and require an interdisciplinary team, it is more difficult to teach in this way across disciplines beyond merely bringing students from different domains together. Transdisciplinary programs, such as the ones run at the University of Technology Sydney, appear to work well for service design problems. However, if the focus is on Industry 4.0 and preparing for the future role of product designers to work within the subsequent systems, there are some differences. The reality is that, just as earlier manufacturing processes had constraints and opportunities, so too do the processes within the systems of Industry 4.0. There are difficult technical challenges in designing for these production processes, for example for additive manufacturing; this is a challenging technology to design for and if it is not taught properly it will become solely the domain of engineers using topological optimization software, or growth software using algorithms capable of 'designing' products automatically without the need for designers at all. Further changes are affecting the communication of concepts, with renderings taken from CAD programs replacing concept drawings, and virtual reality experiences emerging to replace these renderings.

Product designers need to change their thinking to looking at the industrial system, the product service system but also now the full life of the product which itself can change over time through embedded electronics or 4D materials. The idea of a static product has been debunked because of the agency attached to it. Culturally products have impact, and the future of work and industry is impacted by design. As Rosling, Rosling and Rosling Ronnlund (2018) point out, the current worldview is outdated and many people are working on incorrect facts on important issues, such as crime statistics, healthcare costs and employment statistics: "Would you be happy if your doctor was using cutting-edge research from 1965 to suggest your diagnosis and treatment?" (Rosling, Rosling, & Ronnlund, 2018, p. 26). According to Cameron (2017) and Bregman (2017), the worldview has to change in response to the digital revolution. Industry 4.0, encompassing AI and machine learning, will change not only the organization of labor but also its nature: "It's time for a new labor movement. One that fights not only for more jobs and higher wages, but more importantly for work that has higher intrinsic value" (Bregman, 2017, p.261). Academics need to recognize this possibility and address it within higher education in all disciplines, and product design is no exception.

RECOMMENDATIONS

Raising the next generation of designers requires academics to question current educational models, practice and content. The impact of Industry 4.0 on their educational experience should be radical, not incremental, yet this is difficult for academics immersed in current systems and ontology. In preparing students for a future in design, academics need to focus not solely on skills, but on preparation for lifelong learning, and, more essentially, for understanding their role in human development in a digital era. An educational framework for the emerging discipline is suggested, based on key points summarized here:

Table 1. Proposed framework of learning based on the flipped classroom and lean startup 'build-measure-learn' models

Learning From History	Learning From the Present	Learning for the Future
Reduction in poverty through industrialization	*Widening educational access and changing educational models*	*Lifelong learning and agility in educational offerings*
Social planning problems due to centralized manufacturing	*Fragility of policy and political decisions*	*Distributed manufacturing*
Impacts of pollution and the failure of current systems	*Extended producer responsibility*	*Circular economy*
Waste of natural resources	*Product service systems*	*Entrepreneurship*

This chapter addresses issues relating to disruptive change for higher education signified by Industry 4.0, using the product design discipline as an example, and outlines the paradigm shift needed for education to remain relevant. It highlights the importance of education that integrates the humanities with engineering for the future of society and draws on a historical perspective of the impact of societal and environmental change brought about by earlier industrial initiatives. Through this approach, the chapter calls for a greater emphasis on human development in a digital era and environmental, social and economic responsibilities embedded in the implementation of Industry

CONCLUSION

Over the last twenty years, the proportion of the global population living in extreme poverty has halved. This is absolutely revolutionary. I consider it to be the most important change that has happened in the world in my lifetime. It is also a pretty basic fact to know about life on earth. But people do not know it. On average only 7 percent – less than one in ten! - get it right. (Rosling et al., 2018 p.6)

Product designers need to be non-conformists. They need to be able to consider the world and its systems with an informed objectivity, and not be held back by outdated ideas or knowledge that has been superseded. Based on the impact of mass production on the environment, and, in particular, the throw-away society of the twentieth century, it is interesting to consider how the benefit of hindsight could have altered production and disposal practices that characterized the era. Education on the cultural changes involved could have contributed to the development of different values and systems whilst change was in its infancy, and it was possible to influence its direction. Product designers need to be intrinsically motivated to keep updating their understanding of the world and their impact on it. It is the role of product design educators to foster that motivation by constantly questioning educational practice and engaging in pedagogical research. The future of work and industry will be very different in the way it operates to how it did last century. "Time is money. Economic growth can yield either more leisure or more consumption. From 1850 until 1980, we got both, but since then, it is mostly consumption that has increased" (Bregman, 2017 p. 139). The digital revolution is changing interaction, identity, community and economics. Universities as an institution are under threat with competition from open access education and alternative educational providers after the 2010 restrictions on for-profit private universities entering the sector were relaxed.

Coming late to the academic community, industrial and product design educators have strived for acceptance and credibility within established hierarchies and systems. However, from disruptive technology to disruptive ideas, the early decades of the twenty-first century are creating a period of paradigm change and design education should not align itself with entrenched, complacent ill-prepared academia. An educational rebellion is needed for design learning to remain relevant, and to shoulder its responsibilities not only in preparing the designers of the future, but also in leading a rethink of education that is not based on the assumption that accepted practice is unassailable and should not be challenged. As West observes, "regardless of the diversity of our actions, material desires, and well-bring, we all want to have meaningful and fulfilling lives" (West, 2017, p. 211). Pedagogy must be subject to constant critical, unbiased scrutiny and product design academics must never forget they are designers at heart.

REFERENCES

Aldersey-Williams, H. (2011). *The new tin ear: Manufacturing, materials and the rise of the user-maker*. London, UK: RSA Projects.

Altemueller, L., & Lindquist, C. (2017). Flipped classroom instruction for inclusive learning. *British Journal of Special Education*, *44*(3), 341–358. doi:10.1111/1467-8578.12177

Bregman, R. (2017). *Utopia for realists and how we can get there*. London, UK: Bloomsbury.

Brown, J., & Adler, R. (2008). Minds on fire: Open education, the long tail, and learning 2.0. *EDUCAUSE Review*, *43*(1), 16–32.

Cameron, N. (2017). *Will robots take your job?* Cambridge, UK: Polity.

Charles, M. (2018). Teaching, in spite of excellence: Recovering a practice of teaching-led research. *Studies in Philosophy and Education*, *37*(1), 15–29. doi:10.100711217-017-9568-1

Conroy, G. (2016). *Kick-starting student start-ups*. Retrieved from http://sciencemeetsbusiness.com.au/kick-starting-student-startups/

Dee Fink, L. (2013). *Creating significant learning experiences: An integrated approach to designing college courses, revised and updated*. San Francisco, CA: Jossey-Bass.

Droste, M. (2015). *Bauhaus*. Cologne, Germany: Taschen.

Fiell, C., & Fiell, P. (2006). *Industrial design A-Z*. Cologne, Germany: Taschen.

Forty, A. (1992). *Objects of desire: Design and society since 1750*. London, UK: Thames & Hudson.

Fuad-Luke, A. (2009). *Design activism: Beautiful strangeness for a sustainable world*. London, UK: Routledge.

Galloway, A. (2018). Engineering at home. In The future starts here. London, UK: V&A Publishing.

Gavriel, J. (2015). The flipped classroom. *Education for Primary Care*, *26*(6), 424–425. doi:10.1080/14739879.2015.1109809 PMID:26808941

Gibson, I., Rosen, D., & Stucker, B. (2014). *Additive manufacturing technologies: Rapid prototyping to direct digital manufacturing* (2nd ed.). New York: Springer.

Gore, A. (2013). *The future*. New York: Random House.

Greenfield, A. (2017). *Radical technologies: The design of everyday life*. London, UK: Verso.

Gunn, A. (2018). Metrics and methodologies for measuring teaching quality in higher education: Developing the Teaching Excellence Framework (TEF). *Educational Review*, *70*(2), 129–148. doi:10.1080/00131911.2017.1410106

Hawken, P. (2010). The ecology of commerce revised edition: A declaration of sustainability. New York: Harper Business.

Hu, S. J. (2013). Evolving Paradigms of Manufacturing: From Mass Production to Mass Customization and Personalization. *Procedia CIRP*, *7*, 3–8. doi:10.1016/j.procir.2013.05.002

Hyde, R., & Pestana, M. (Eds.). (2018). The Future Starts Here. London, UK: V&A Publishing.

Kelly, K. (2016). *The inevitable: Understanding the 12 technological forces that will shape our future*. New York: Penguin Books.

Loy, J. (2012). Creating confidence in an alienating educational environment. In *Proceedings International conference on Engineering and Design education*. Artesis University College.

Lynam, S., & Cachia, M. (2018). Students' perceptions of the role of assessments at higher education. *Assessment & Evaluation in Higher Education*, *43*(2), 223–234. doi:10.1080/02602938.2017.1329928

McCallum, W. (2018). *How to give up plastic: A guide to changing the world, one plastic bottle at a time*. London, UK: Penguin Life.

Novak, J. (2018). Self-directed learning in the age of open source, open hardware and 3D printing. In E. Ossiannilsson (Ed.), *Ubiquitous inclusive learning in a digital era* (pp. 154–178). Hershey, PA: IGI Global.

Novak, J., & Loy, J. (2017). Digital technologies and 4D customized design: Challenging conventions with responsive design. In V. C. Bryan, A. T. Musgrove, & J. R. Powers (Eds.), *Handbook of research on human development in the digital age* (pp. 403–426). Hershey, PA: IGI Global.

Papanek, V. (2005). *Design for the real world: Human ecology and social change* (revised ed.). Chicago: Chicago Review Press.

Petko, D., Egger, N., Cantieni, A., & Wesoi, B. (2015). Digital media adoption in schools: Bottom-up, top-down, complementary or optional? *Computers & Education*, *84*(May), 46–61.

Rosling, H., Rosling, O., & Rosling Ronnlend, A. (2018). *Factfulness: Ten reasons why we're wrong about the world and why things are better than you think*. London, UK: Sceptre.

Schwab, K. (2017). *The fourth industrial Revolution*. New York: Crown Business.

Seldon, A., & Abidoye, O. (2018). *The Fourth education revolution: Will artificial intelligence liberate or infantilise humanity*. Buckingham, UK: University of Buckingham Press.

Sparke, P. (2013). *An introduction to design and culture: 1900 to the present* (3rd ed.). London, UK: Routledge. doi:10.4324/9780203129999

Sterling, S. (2015). *Sustainable Education: Revisioning learning and change (Schumacher Briefings)*. Cambridge, UK: Green Books UIT.

Tatum, J. (2004). The challenge of responsible design. Design Issues, 20(3), 66-80.

Tseng, M. M., Jiao, R. J., & Wang, C. (2010). Design for mass personalization. *CIRP Annals - Manufacturing Technology, 59*(1), 175-178. doi:10.1016/j.cirp.2010.03.097

Walker, S. (2006). *Sustainable by design: Explorations in theory and practice*. London, UK: Routledge.

Weimer, M. (2013). *Learner-centered teaching: Five key changes to practice* (2nd ed.). San Francisco, CA: Jossey Bass.

West, G. (2017). *Scale: The universal laws of life and death in organisms, cities and companies*. London, UK: Weidenfeld & Nicholson.

ADDITIONAL READING

Beetham, H., & Sharpe, R. (2013). *Rethinking pedagogy for a digital age: Designing for 21st century learning* (2nd ed.). New York, New York: Taylor and Francis.

Fleischmann, K. (2015). The democratisation of design and design learning: How do we educate the next-generation designer. *The International Journal of the Arts in Society*, *8*(6), 101–108.

Global Trends 2030: Alternative Worlds. (2012). Retrieved from http://www.dni.gov/index.php/about/organization/national-intelligence-council-global-trends

Greenfield, A. (2006). *Everyware: The dawning age of ubiquitous computing*. Berkeley, California: New Riders.

Novak, J. (2018). Re-educating the educators: Collaborative 3D printing education. In I. M. Santos, N. Ali, & S. Areepattamannil (Eds.), *Interdisciplinary and international perspectives on 3D printing in education*. Hershey, PA, USA: IGI Global.

Petrova, M. (2014, 4-5/9/2014). *Educating Designers from Generation Y – Challenges and Alternatives*. Paper presented at the DS 78: Proceedings of the 16th International conference on Engineering and Product Design Education (E&PDE14), Design Education and Human Technology Relations, University of Twente, The Netherlands.

KEY TERMS AND DEFINITIONS

3D Printing (Additive Manufacturing): A digital fabrication technology that allows the production of an object by adding material layer-by-layer in three dimensions.

Computer-Aided Design (CAD): The use of computer systems to assist in the creation, modification, analysis or optimization of a design in 2D or 3D.

Flipped Classroom: This is a teaching methodology that encourages students to access lecture material outside of class, devoting class time to hands-on problem solving and the application of knowledge. The teacher's role shifts to that of a facilitator, and collaborative learning and problem-based learning are important features of the flipped classroom.

Industry 4.0: Also known as the "fourth industrial revolution," this describes the current trend for increased automation in manufacturing, communication and machine-to-machine and human-to-machine relationships more broadly.

Product and Industrial Design: Disciplines tightly linked to mass production and the design of goods to be manufactured for consumption.

Chapter 11
Active Use of Digital Technologies in Mathematical Problem Solving

Duncan Symons
The University of Melbourne, Australia

Robyn Pierce
The University of Melbourne, Australia

ABSTRACT

The need for students to engage in critical thinking, collaboration, communication, and creativity when utilizing digital technologies is well established given the current push to prepare our students for future work expectations, often referred to as Industry 4.0. In this chapter, 54 Grade 5 students participated in online collaborative mathematical problem solving over a period of 9 weeks. Text-based discussion and software derived artefacts were investigated in an effort to understand how this approach aligns with these expectations. An examination of the frequency/density of technical mathematical vocabulary use and identified examples, Talk Types is used as a means to understand how often students within the online environment are likely to be engaged in work that might be considered productive. The findings of this study suggest that utilizing online synchronous and asynchronous collaborative learning platforms for the purposes of mathematical problem solving is a small but important approach to preparing students for the era of Industry 4.0.

INTRODUCTION

Developments in technology and in particular "intelligent" technology are transforming the way we will live. This is why the current Australian mathematics curriculum mandates that technology be utilised to support students to "investigate, create and communicate mathematical ideas and concepts" (ACARA, 2014). However, research to date has found that use of digital technologies in Australian primary mathematics classrooms is often superficial and focusses on the lower-order drilling of algorithms and basic facts (Day, 2013). These are skills that are already commonly assigned to technology for speed

DOI: 10.4018/978-1-5225-7832-1.ch011

Copyright © 2019, IGI Global. Copying or distributing in print or electronic forms without written permission of IGI Global is prohibited.

and accuracy. To prepare our students for the automated, networked and data driven nature of future work expectations, often referred to as "Industry 4.0", students need to move beyond algorithmic skills and engage in critical thinking, collaboration, communication and creativity (Griffin, McGaw, & Care, 2012) when utilising digital technologies. Whilst there are some recent efforts towards describing the conditions needed in education to prepare students for this future (see for example, Loveder (2017); Matthews, Garratt, and Macdonald (2018); Murphy and Irwin (2018) and Wanyama, Singh, and Centea (2018)) the lack of alignment between research-based evidence and the written and enacted curriculum suggests uncertainty about the pedagogical use of technology within primary mathematics now and moving into the future.

The aim of the study discussed in this chapter was to develop and trial an approach to the integration of digital technologies within the primary mathematics classroom that actively promotes the Australian Curriculum cross curricular priorities of Critical and Creative Thinking. For this purpose, a nine-week teaching and learning intervention was planned. This approach involved students in a face-to-face class followed by a week when their "homework" was to collaboratively, in small online groups, solve open, context based, mathematical problems. The research also aimed to investigate the opportunities this alternative use of technology in primary mathematics might present for researchers, teachers and students. If we can guide young students in richer and deeper engagement with technologies, we may offer greater preparation for the new and in part unanticipated consequences of Industry 4.0. It is expected that the outcomes of this project may provide evidence-based advice and examples for teachers who do not feel confident to use technology within the mathematics classroom for uses other than drill and practice. With these aims in mind the following research question was investigated:

- How does student engagement with online mathematical problem solving align with the Australian Curriculum priority of Critical and Creative Thinking?

This chapter draws on work initially disseminated in Jazby and Symons (2015); Symons (2017) and Symons and Pierce (2015). We will first provide some discussion of the importance that the Australian Curriculum places on critical thinking. Following this, a framework developed by Perkins and Murphy (2006) will be used to explore whether different aspects of critical thinking were evident in student online discussion. Following use of this tool, student discussion is reanalysed through the lens of 'Talk Types' (Mercer & Wegerif, 1999) with the aim of showing that where *Exploratory talk* was evident, student critical thinking also occurred. It is hypothesised that critical thinking may be evident through the presence of a greater density of specific technical mathematical vocabulary use in identified examples of *Exploratory talk.*

LITERATURE REVIEW

Technology integration in primary level mathematics, commonly, at best emphasizes fluency rather than understanding, reasoning and problem solving (BECTA, 2004; Day, 2013; Wachira & Keengwe, 2011). At worst, when not carefully monitored, it is simply an exercise in keeping students busy with little emphasis on mathematical concept development. This approach to technology integration is at odds with the development of the 21st century skills students will require to prosper in the era of Industry 4.0.

The learning outcomes set for the intervention discussed in this study focused on students being able to demonstrate technology supported collaborative problem solving, reasoning and critical thinking. Various components of such higher order thinking are discussed in the literature reviewed below.

Industry 4.0

The call for changes in education is not recent. Makrakis (2005) calls for a transformation in approaches to technology integration within education as we move further into the 21st century and prepare students for the impact of Industry 4.0. He highlights the need for a focus on critical thinking, learning by discovery and authentic assessment. This is a theme that was initially foreshadowed in Prensky's (2001) seminal paper. Whilst, the dichotomy Prensky (2001) suggests between digital natives and digital immigrants has been widely criticised as oversimplification, (see for example, Brown and Czerniewicz (2010) and Selwyn (2009)) his view that a 'singularity' has taken place whereby "the arrival and dissemination of digital technology in the last decades of the 20th century" (Prensky, 2001, p. 1) has fundamentally changed the way that we must approach teaching and learning to prepare our students for as yet unknown and uncertain futures remains prescient.

Critical Thinking

Critical Thinking is highlighted as a key component of higher order thinking. The need for development of critical thinking is seen as fundamental to education for the era of Industry. Therefore, throughout this review connections and parallels will be drawn between the two terms. Facione (2013, p. 26) defines critical thinking as "purposeful, self-regulatory judgment which results in interpretation, analysis, evaluation, and inference, as well as explanation of the evidential, conceptual, methodological, criteriological, or contextual considerations upon which that judgment is based". Lesh and Zawojewski (2007) take a summary perspective and propose 'Higher-Order Thinking' as an umbrella term incorporating 3 specific components; 'Metacognition', 'Beliefs and Dispositions' and 'Habits of Mind'. We will consider these three themes in our discussion of higher order thinking.

Metacognition

Flavell (1976) is first associated with defining 'metacognition'. A definition he proposed is as follows:

Metacognition refers to one's knowledge concerning one's own cognitive processes or anything related to them, e.g. the learning-relevant properties of information or data. For example, we are engaging in metacognition... if we notice that we are having more trouble learning A than B; if it strikes me that we should double-check C before accepting it as a fact; if it occurs to me that we should scrutinize each and every alternative in a multiple-choice task before deciding which is the best one.... Metacognition refers, among other things, to the active monitoring and consequent regulation and orchestration of those processes in relation to the cognitive objects or data on which they bear, usually in the service of some concrete [problem solving] goal or objective (p. 232).

Facione's (2013) description of critical thinking involving purposeful self-regulatory judgments overlaps with the definition of metacognition provided above.

Schoenfeld (1992) describes the relationship between self-regulation and metacognition; "the issue is one of resource allocation during cognitive activity and problem solving" (p. 57). Lesh and Zawojewski (2007) describe self-regulation as "the mechanism that enables problem solvers to break up larger and more complex problems into subtasks, prioritize the subtasks, efficiently sequence each subtask, and finally do each subtask" (p. 767). Within this study analysis of online discourse allows some insight into students' metacognitive processes.

Beliefs and Dispositions

In addition to seeing self-regulation within the area of metacognition as being essential to higher order thinking, Lesh and Zawojewski (2007) also identify beliefs and dispositions as key components of this area. Schoenfeld (1992) further delineates 'beliefs' as 'student beliefs', 'teacher beliefs' and 'societal beliefs' but in this review, we will focus only on 'student beliefs' as these will be particularly relevant to this study.

Student beliefs (in the context of mathematical problem-solving) are extrinsically mediated through the process of adoption of various classroom-based cultures within their educational experience. Lampert (1990) describes the well-recognised process of admission into this culture and how;

"cultural assumptions are shaped by school experience, in which doing mathematics means following the rules laid down by the teacher; knowing mathematics means remembering and applying the correct rule when the teacher asks a question; and mathematical truth is determined when the answer is ratified by the teacher. Beliefs about how to do mathematics and what it means to know it in school are acquired through years of watching, listening, and practicing" (p. 32)

There are some clear parallels between the implicit rules of the mathematics classroom described by Lampert (1990) and the 'didactic contract' as interpreted by Pierce, Stacey, and Wander (2010). Too many long-held beliefs about school mathematics hold true today and form barriers to students applying mathematical thinking when engaging in the reflective and integrated thinking required to solve real world problems. Still relevant is Schoenfeld's (1992) list of the following beliefs that potentially have a negative impact on mathematical thinking:

- Mathematics problems have one and only one right answer.
- There is only one correct way to solve any mathematics problem -- usually the rule the teacher has most recently demonstrated to the class.
- Ordinary students cannot expect to understand mathematics; they expect simply to memorize it, and apply what they have learned mechanically and without understanding.
- Mathematics is a solitary activity, done by individuals in isolation.
- Students who have understood the mathematics they have studied will be able to solve any assigned problem in five minutes or less.
- The mathematics learned in school has little or nothing to do with the real world.
- Formal proof is irrelevant to processes of discovery or invention (p. 69)

Habits of Mind

Cuoco, Goldenberg, and Mark (1996) lament that "it has been shown possible for students to learn the facts and techniques that mathematicians (historians, auto diagnosticians, etc.) have developed without ever understanding how mathematicians (or these others) think" (p. 377). This statement makes reference to the issue already recounted, of the development of procedurally driven skills in mathematics at the expense of a broad conceptual understanding of mathematical processes. Cuoco et al. (1996) suggest that high school students must develop a range of general habits of mind in order to counter this disconnected understanding of mathematics. They describe 'habits of mind'as "mental habits that allow students to develop a repertoire of general heuristics and approaches that can be applied in many different situations" (p. 378).

In contrast to the beliefs which Schoenfeld saw as barriers Cuoco et al (1996) outlined the following list of enabling 'mental habits' that they saw as leading to the development of an approach to solving mathematical problems in a manner similar to that of a mathematician:

- Students should be pattern sniffers
- Students should be experimenters
- Students should be describers
- Students should be tinkerers
- Students should be inventors
- Students should be visualisers
- Students should be conjecturers
- Students should be guessers (p. 378)

More recently, these authors have published work suggesting that the same principles should be applied to an upper elementary (primary) school setting. They state, "developing mathematical habits of mind in the middle grades is essential for students who are making the critical transition from arithmetic to algebra" (Mark, Cuoco, Goldenberg, & Sword, 2010, p. 505).

Clarification Assessment Inference Strategies

Identifying elements of critical thinking for research or assessment purposes requires a framework. The Clarification Assessment Inference Strategies (CAIS) model developed by Perkins and Murphy (2006) (see Figure 1 below) is a practical model for identifying engagement in critical thinking. This model for the assessment and analysis of critical thinking was chosen because of the rigorous approach to its development. The researchers (Perkins & Murphy, 2006) evaluated a range of existing Critical Thinking models (see for example Brace-Govan and Clulow (2001); Bullen (1997); Garrison, Anderson, and Archer (2001); Henri (1992); Newman, Webb, and Cochrane (1995); Norris and Ennis (1989)) and synthesised the approaches taken within the research in order to develop their model. Additionally, the CAIS model was developed to assess and analyse teaching and learning within online, asynchronous learning, which is the approach taken within this study. Therefore, the CAIS model is very well suited to the interrogation of student meaning making within this study.

Perkins and Murphy (2006) identified the increasing use of Computer Supported Collaborative Learning (CSCL), in the form of asynchronous message boards, across a range of educational contexts. They described a belief that having students engage in discussion in these online environments may lead to development of higher order thinking, specifically in the area of critical thinking and developed the CAIS model as an approach to help test this conjecture. In this study their approach is used to analyse the online work of primary school students.

As will be seen in Table 1 and Figure 1 below, there are similarities between the indicators of critical thinking identified by Perkins and Murphy (2006) and the delineated developmental continuum for Critical and Creative Thinking in the Australian Curriculum (ACARA, 2017). Both emphasise student clarification and organisation of concepts and ideas, the generation of different strategies ideas and approaches, reflecting on and analysing and evaluating decisions taken and strategies utilised.

The Study Background

The intervention and study were conducted in a middle-outer suburban state primary school in south eastern Australia. The school has an average level of socio-educational advantage when compared against the rest of the state of Victoria. Few students have a high level of socio-economic advantage andfew students are considerably disadvantaged.

Across the school, 43% of students are from a language background other than English (LBOTE), while in the sample of students participating in the study 40% were LBOTE.

Table 1. Perkins & Murphy's clarification, assessment, inference, strategies model (2006, p. 301)

CLARIFICATION				
All aspects of stating, clarifying, describing (but not explaining) or defining the issue being discussed.				
Proposes an issue for debate.	Analyses, negotiates or discusses the meaning of the issue.	Identifies one or more underlying assumptions in a statement in the discussion.	Identifies relationships among the statements or assumptions.	Defines or criticizes the definition of relevant terms.
ASSESSMENT				
Evaluating some aspect of the debate; making judgments on a situation, proposing evidence for an argument or for links with other issues.				
Provides or asks for reasons that proffered evidence is valid.	Provides or asks for reasons that proffered evidence is relevant.	Specifies assessment criteria, such as the credibility of the source.	Makes a value judgment on the assessment criteria or a situation or topic.	Gives evidence for choice of assessment criteria.
INFERENCE				
Showing connections among ideas; drawing appropriate conclusions by deduction or induction, generalizing, explaining (but not describing), and hypothesizing.				
Makes appropriate deductions.	Makes appropriate inferences.	Arrives at a conclusion.	Makes generalisations.	Deduces relationships among ideas.
STRATEGIES				
Proposing, discussing, or evaluating possible actions.				
Takes action.	Describes possible actions.	Evaluates possible actions.	Predicts outcomes of proposed actions.	

The Participants

Participants in the intervention were 54 Year 5 students (ranging in age between 10 and 12 years old). There were 26 boys and 28 girls across two classes. For this project the 54 students were allocated to 10 mixed ability groups within the online space. These groups were created on the advice of their teachers, since students were already classified as performing either below level, at level or above level in mathematics. The teachers had previously made this classification based on a series of tests (not associated with this project) which assessed the students' level of procedural and algorithmic fluency and general understanding across key areas of mathematics.

The Researcher as Facilitator

While, the first author of this study was not one of the students' regular teachers, he facilitated the intervention. In addition to setting up and monitoring, but not entering, the online environment, he conducted one-hour face-to-face sessions with the students each week. These face-to-face sessions were conducted with all 54 students present.

Problem solving that involves students communicating mathematical ideas was not included in the current topics in their regular mathematics classes. These face-to-face sessions had several goals, one of which was to pre-empt potential student difficulties by introducing some new mathematical language as needed for each new weekly topic. It was expected that the language and literacy based demands of expressing mathematical thinking via asynchronous text-based communication could be challenging for all students and that it may prove to be especially challenging for LBOTE students.

Throughout the course of the nine-week intervention, the classroom based sessions took the format of a short review and discussion of the online interaction that occurred in the previous week followed by a discussion of the new problem to be solved throughout the following week, and then a short discussion of any additional technical information that may have been helpful, for example how to use the software such as Edmodo, MS Excel, MS Word. In the first weeks, considerable time was spent outlining appropriate online behaviours and utilising the work of Alexander (2006) to facilitate productive dialogic interactions as opposed to discussion that might be of a more superficial nature. The level of support offered to students throughout these discussions diminished as students became more confident working in the online environment and engaging in mathematical problem solving in this way.

The final two weeks of the intervention (weeks 8 and 9) were approached slightly differently. We were interested to know how students would engage in the online space with no classroom-based support provided. We hypothesised that after working collaboratively online for some weeks, students may have developed the range of necessary skills to allow them to work with less facilitator or teacher based support. Therefore, in weeks 8 and 9 of the intervention, instead of the usual class-based discussion preceding their online mathematical problem solving, the hour of class time was used by the students immediately to start their work in the online space. As a result, any data collected from weeks 8 and 9 were largely free of influence from the researcher (facilitator).

The Weekly Activities

Each week during the intervention the students were set a problem to solve. We aimed to choose problems that were authentic for 10-12-year-old students. These were either real world problems or ones which would appeal to their imagination. Problems were adapted from Holton and Symons (2015). No expectations were placed on students in terms of the amount of time they should spend working on the activities. We utilised a very narrow range of technology in this study. The four pieces of software familiar to the students were utilised: Edmodo, MS Paint, MS Excel and MS Word. Of these four pieces of software there were only two for which students required considerable extra instruction (MS Excel, and Edmodo).

Table 1 provides details about the problem titles and the mathematical content area in focus each week throughout the period of data collection. Below the table an example of the instructions provided for one of the weekly problems (*see* Figure 2) is provided with some discussion of how it was approached by students. As can be seen the mathematical content focus for each week varied. All focus areas link explicitly with the Australian Curriculum: Mathematics (ACARA, 2014). As the aim of the study was on the development of problem solving, reasoning and critical thinking it was not important to maintain a single, consistent mathematical content area in the study. We chose to give students experience in problem solving across the curriculum content.

Example Problem: Animal Ages

See Figure 1.

Problem Purpose

The purpose of this problem was to help students think critically about the concept of age and to develop an understanding of proportional reasoning. It also afforded the opportunity to use technology to enhance communication through the use of Microsoft Excel to produce interactive tables and visual representations, for example, creating proportional pictograms. (E.g. 1 dog = 7 dog years).

Table 2. Mathematical content area in weeks of data collection

Week	Mathematical content area	Problem Name
Week 1	Number & Algebra	Toilet Paper Rolls
Week 2	Measurement & Geometry	Working Days
Week 3	Measurement & Geometry	Wallpaper Symmetry
Week 4	Statistics & Probability	How Big is a Dog?
Week 5	Number & Algebra	Animal Ages
Week 6	Measurement & Geometry	Shapes
Week 7	Statistics & Probability	Pet Names
Week 8	Number & Algebra	Mr Mac's Iphone
Week 9	Number & Algebra	Geese

Figure 1. Week 5 Problem - Animal Ages

It is sometimes said that each year a human lives is equivalent to 5 cat years. What does this mean? How old is your cat in human years (If you don't have one pretend!)?

- *Create a 'cat age' table in Excel. This should include columns for a cat's age and a column for the equivalent human age (up to 100 years).*
- *Create a line graph in Excel representing these data up to 100 human years.*
- *Like last week you may copy both the table and graphs that you have created into word. In word summarise what the graph shows. You may also include images.*
- *Every year a human lives, is said to be the same as 2 years for a Camel. Repeat the above process for Camels (add to your chart). How old are you in Camel years?*
- *Some lorikeets live to be 25 years old. Add a column to your table that makes it easy to convert between lorikeet and human ages.*
- *Use the table to tell how old the members of your family would be if they were lorikeets.*
- *Create a 'pictograph' using Excel displaying the various animal data up to 10 'human' years.*
- *Why in your opinion would Mr Symons have asked you to make this chart only up to 10 human years?*

Research Method

Figure 2 (below) shows the cyclical approach taken to investigating student critical thinking in this chapter. The lens through which critical thinking is initially examined occurs at a macro level first using the framework of the Australian Curriculum then applying the CAIS framework to explore how one group of students within the study engaged in different aspects of critical thinking. This framework allows inferences to be made about how various members of the group position themselves, according to which of the categories of critical thinking they display more or less strongly. The analysis of student online discussion then allows a micro level analysis of the "types of talk" (Mercer & Wegerif, 1999) that were more likely to be productive and associated with critical thinking. The density of technical mathematical vocabulary in Exploratory talk was observed and noted. A final step was to align the identification of Exploratory Talk and a higher density of mathematical vocabulary use back to the prevalence of the Australian Curriculum prescribed general capability of Critical and Creative thinking (ACARA, 2017).

Figure 2. Progression of analysis of critical thinking in online space

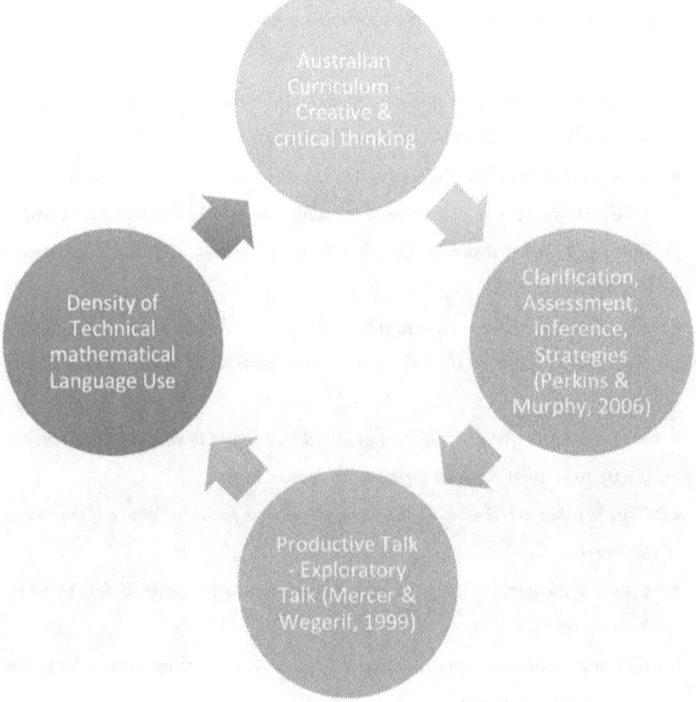

Critical Thinking in the Australian Curriculum

The Australian Curriculum lists seven 'general capabilities'. When describing the capabilities ACARA (2017) states:

General capabilities are identified where they are developed or applied in the content descriptions. They are also identified where they offer opportunities to add depth and richness to student learning via the content elaborations, which are provided to give teachers ideas about how they might teach the content. Teachers are expected to teach and assess general capabilities to the extent that they are incorporated within learning area content (ACARA, 2017).

Teachers are expected to embed and assess development of the general capabilities throughout their teaching of all content areas, including mathematics. 'Critical and Creative Thinking' is one of the listed general capabilities. The Australian Curriculum provides the following broad discussion of Critical and Creative Thinking within the content area of Mathematics:

In the Australian Curriculum: Mathematics, students develop critical and creative thinking as they learn to generate and evaluate knowledge, ideas and possibilities, and use them when seeking solutions. Engaging students in reasoning and thinking about solutions to problems and the strategies needed to find these solutions are core parts of the Australian Curriculum: Mathematics.

Table 3. Australian curriculum critical and creative thinking content descriptions (ACARA, 2017)

Critical and Creative Thinking Content Descriptions			
Inquiring – identifying, exploring and organising information and ideas	**Generating ideas, possibilities and actions**	**Reflecting on thinking and processes**	**Analysing, synthesising and evaluating reasoning and procedures**
Pose questions Pose questions to clarify and interpret information and probe for causes and consequences	**Imagine possibilities and connect ideas** Combine ideas in a variety of ways and from a range of sources to create new possibilities	**Think about thinking (metacognition)** Reflect on assumptions made, consider reasonable criticism and adjust their thinking if necessary	**Apply logic and reasoning** Assess whether there is adequate reasoning and evidence to justify a claim, conclusion or outcome
Identify and clarify information and ideas Identify and clarify relevant information and prioritise ideas	**Consider alternatives** Identify situations where current approaches do not work, challenge existing ideas and generate alternative solutions	**Reflect on processes** Identify and justify the thinking behind choices they have made	**Draw conclusions and design a course of action** Scrutinise ideas or concepts, test conclusions and modify actions when designing a course of action
Organise and process information Analyse, condense and combine relevant information from multiple sources	**Seek solutions and put ideas into action** Assess and test options to identify the most effective solution and to put ideas into action	**Transfer knowledge into new contexts** Apply knowledge gained from one context to another unrelated context and identify new meaning	**Evaluate procedures and outcomes** Evaluate the effectiveness of ideas, products, performances, methods and courses of action against given criteria

Students are encouraged to be critical thinkers when justifying their choice of a calculation strategy or identifying relevant questions during a statistical investigation. They are encouraged to look for alternative ways to approach mathematical problems; for example, identifying when a problem is similar to a previous one, drawing diagrams or simplifying a problem to control some variables (ACARA, 2017).

In addition to the broad statement of Critical and Creative thinking within mathematics, the Australian Curriculum provides the following delineated expectations (see Figure 4) of the level to which students should have developed in this area by the conclusion of primary school.

In the following sub-section, a discussion of Perkins and Murphy's (2006) CAIS framework will be provided. When closely analysed, it is evident that the indicators of critical thinking embedded within the Australian Curriculum (ACARA, 2017) content descriptions (in Figure 4) are represented in the CAIS framework. Therefore, the CAIS framework is used as a method of aligning and assessing whether or not, and also how, students engaged with the Australian Curriculum expectation of critical thinking.

Data Collection, Coding, and Analysis

The analysis of critical thinking using CAIS discussed in this chapter is illustrated by data from one representative group of four students who participated in the CSCL environment. The group included a mix of genders; students from each of the three teacher assigned ability groups and students from both LBOTE and English-speaking backgrounds. Perkins and Murphy's (2006) CAIS framework was used to code students' online posts, where each post is treated as an utterance. Coded utterances were classified as relating to critical thinking and then broken down into the four categories used by Perkins and

Murphy. Uncoded utterances, for example "Is anybody online?" could not be related to critical thinking. The assessment framework was applied to the online records of weeks two to eight. In week 1 the focus was on familiarising students with the use of the Edmodo platform for posting comments and files.

RESULTS AND DISCUSSION

Animal Ages: Summary of Responses

Students had various levels of success with the animal ages problem (Figure 1). At the simplest level students were able to identify and record growing patterns (or recursive equations). For example, they knew that each successive dog year would be 7 human years more than the previous dog year. Some students laboriously entered each multiple of 7 into each of the subsequent cells, whilst others realized the opportunity to use Excel's 'fill down' feature, which would recognize the 'recursive expression' and pre-fill subsequent cells with these data.

Students working at a higher level of sophistication were able to create a series of 'explicit equations' allowing subsequent terms to be obtained. For example, they understood that multiplying the given number of human years by 5, 2 and 4 would provide the number of cat, camel and lorikeet years. They were then able to transfer this understanding into basic algebraic formulae (e.g. cat years = B7 x 5, Camel Years = B7 x 2 and Lorikeet Years = B7 x 4). This understanding, represents a significant level of achievement for students of this age/ year level. For example, The Australian Curriculum (ACARA, 2014) anticipates that within 'Patterns and Algebra' at level 7 students should *Create algebraic expressions and evaluate them by substituting a given value for each variable*

Technology also supported differentiation for students' understanding of graphs. Students working at a lower level represented the linear growth of the different animal ages using basic line graphs or histograms. Students with more sophisticated understanding (in addition to creating the more 'basic' graph types) were also able to create pictograms in Excel where images of animals were proportionally representative.

Student Focus of Critical Thinking

Table 2 summarises the percentage of utterances coded as *Clarification, Assessment, Inference and Strategies* posted by students within this group. The total number of coded and uncoded utterances for each student is also provided in the last two columns.

Two students, Chaz and Igor, made the smallest number of posts and contributed less than 10% to the group's posts exhibiting critical thinking. As a consequence, it is difficult to identify which of the four critical thinking categories they engaged most or least with. The majority of the group's critical thinking posts were made by Zaid and Olive. Zaid and Olive contributed an almost equal proportion of *Assessment* related posts (42% and 46% respectively). Olive contributed 73% of the group's *Clarification* posts while Zaid posted 64% of the *Inference* and 63% *Strategies* posts. This indicates the prominent role that two students took within their group. Olive's high level of *Clarification* based posts indicates that she was more likely to question the assumptions made by others within the group, seek clarity around shared definitions and to draw together important relationships (mathematical or otherwise). Interest-

Table 4. Assessment of students' critical thinking evident in students' utterances

Student	Aspect of critical thinking				Coded utterances N=128	Uncoded utterances N=104
	Clarification (%)	Assessment (%)	Inference (%)	Strategies (%)		
Chaz	8	4	0	2	5	10
Igor	3	8	7	6	7	13
Olive	73	46	29	29	58	32
Zaid	16	42	64	63	58	49
	100	100	100	100		

ingly, Olive had been 'streamed' by her teacher into the lowest achieving mathematics group for her daily mathematics instruction. Tests assessing her ability to perform routine mathematical skills and procedures were used to inform this placement. Yet, in the CSCL environment, Olive's contribution to online learning is significant. Zaid more commonly made utterances aligning with the *Inference* and *Strategies* classifications. Zaid was more likely to pose questions and help the group co-construct definitions. His collaborative role within the group was to take the evidence and data provided by himself and others within the group and draw conclusions, make deductions and generalise ideas and concepts. Whilst showing some leadership traits he would also be more likely to either suggest or take action without allowing all of the group to contribute. The patterns of interaction that were apparent in the online discussion helped the teacher identify not only students' proficiency in problem solving but also their skills for working collaboratively.

The use of the framework not only provided summative data about the group's general critical thinking based participation within the group, but also provided valuable information about those aspects of critical thinking that each student exhibited to a greater or lesser extent. This analysis provides us with the ability to consider the Australian Curriculum (2017) general capability of Critical and Creative thinking and ensure, firstly that critical thinking, as prescribed by the curriculum, is evident in this online approach to mathematical problem solving and secondly to understand how the individual members of a small group focus their attention on different aspects of critical thinking thereby ensuring an effective collaborate approach.

Talk Types

Whilst the CAIS framework made visible the fact that students were engaging in critical thinking, it did not allow us to understand the way they used 'talk' (online discussion in this case) to construct their learning. The following sub-sections will examine how the types of talk that students engaged in within the space, allow us to understand which Talk Types could be seen to be more productive and thus more likely to lead to critical thinking.

The three types of talk of special interest are Mercer and Wegerif's (1999) *Disputational Talk*, *Cumulative Talk*, and *Exploratory Talk*. These authors analysed many hours of videotaped discussion amongst British primary age students and defined *Disputational Talk*, *Cumulative Talk* and *Exploratory Talk as follows:*

Disputational talk, *which is characterised by disagreement and individualised decision making. There are few attempts to pool resources, or to offer constructive criticism of suggestions. Disputational talk also has some characteristic discourse features – short exchanges consisting of assertions and challenges or counter-assertions.*

Cumulative talk, *in which speakers build positively but uncritically on what the other has said. Partners use talk to construct a 'common knowledge' by accumulation. Cumulative discourse is characterised by repetitions, confirmations and elaborations.*

Exploratory talk, *in which partners engage critically but constructively with each other's ideas. Statements and suggestions are offered for joint consideration. These may be challenged and counter-challenged, but challenges are justified and alternative hypotheses are offered. Compared with the other two types, in exploratory talk knowledge is made more publicly accountable and reasoning is more visible in the talk (p. 85).*

Unlike the study by Mercer and Wegerif (1999), the discourse analysed in this chapter occurred in an online environment, and occurred asynchronously in a discussion board. However, we continue to feel justified to use the term 'utterances' when referring to students' statements. Like Mercer and Wegerif (1999), we are interested in gaining an understanding of how students might jointly construct knowledge as they work to solve a series of mathematical problems. We hypothesised that, given that *Exploratory Talk* is represented by talk with public accountability, critical thinking and reasoning will be visible. We also expected to see a greater density of technical mathematical vocabulary, words with a domain-specific usage. They are generally the most difficult words for students to acquire because of the very limited opportunities students have to experiment with them. In the context of mathematics, technical vocabulary would include such terms as: 'formula', 'equation', 'symmetry', 'median'.

Data Collection, Coding, and Analysis

Talk Types were coded using two approaches. We first adopted a course-grained approach to the coding of data, whereby a general code was allocated for each small group discussion, for each problem. Given that students completed 9 problems and there were 10 small groups, this meant that there were approximately 90 individual passages coded using this approach. Despite many of the passages containing discourse to exemplify two or even three of the different Talk Types, we selected the talk type that was **most** evident in the passage. Coding in this way provided a general sense of the more and less prevalent Talk Types occurring within the space.

The second approach to coding was fine grained. Each individual 'turn' taken or 'utterance' made by participants was coded using this approach. This method provided a great deal of information that was not available using the first approach. For instance, *Disputational Talk* was not detected when taking the course-grained approach, however a substantial number of examples of this talk type were identified when coding using a more 'fine grained' method. This showed that whilst *Disputational Talk* was unlikely to represent the dominant talk type within a passage, consistently short bursts of this talk type would occur within a passage that was more broadly characterised as either *Cumulative* or *Exploratory Talk*.

In addition to adhering to the definitions provided by Mercer and Wegerif (1999), the following complementary criteria were used for the (fine grained coding) of Talk Types in this context:

Disputational talk:

- Students becoming frustrated and/or annoyed.
- Students deliberately engaging in antagonism
- Some degree of conflict, generally not of a 'constructive' nature

Cumulative talk:

- Conversation is of a more general nature. Questions are equally non-specific. E.g. "Can you help me? we don't get it."
- Does not involve the operations, digits, numbers or mathematical representations.
- Students provide an answer only. No explanation of thinking is provided.
- Seemingly, polite, non-specific feedback. E.g. "That's great work", "Fantastic!"
- Talk is vaguely (but not specifically) related to the task.

Exploratory talk:

- Critiques are constructive and contain logical arguments
- Feedback may be polite, however includes specific observation. E.g. "we like how you included 3rd order rotational symmetry in your wallpaper"
- Questions are not of a general nature. Questioner asks questions indicating a specific need that requires addressing. E.g. when trying to find out the labels for mean median and mode a participant might ask, "we can't remember the three m words. What were they again?"
- Answer to problems includes some description of thinking behind contribution.

We also used the following guidelines when coding for technical mathematical vocabulary:

- Specific mathematical vocabulary
- Misspelt words were included
- Seemingly 'simple' words with specific mathematical meaning were included. E.g. 'shape', 'pattern', 'height', 'centimetre'.
- Homonyms, where the meaning conveys mathematical meaning were included. E.g. "we have used the animal names as headings for my **table**."

RESULTS AND DISCUSSION

Table 5 shows an example illustrative of discussion from the online message board coded as *exploratory talk* with technical mathematical vocabulary bolded. The discussion in Figure 5 (below) is provided verbatim from the online space. In this example students worked on a problem where they were required to make a conjecture, gather evidence and support a claim about whether cats' names or dogs' names are generally longer. The students researched a number of the most common cats' names and dogs' names, calculated the mean, median and mode of these data, graphed their results using Microsoft Excel then discussed their results.

Table 5. Example of discussion coded as exploratory talk

Sally	*we think dogs and cats are the same number of letters because in my* ***graph*** *it came up with 8 fives and eight fives each.* *So that my Information* *Please reply* *Thanks guys*
Sally	*Hi Sunny, where is your* ***graph****?*
Sally	*Hi everyone, we have done the exel spread sheet and the names that we have got are feamale and male. we am neally compleated.*
Holly	*Hey guys what do you do after you have writen down all the names and numbers?*
Sally	*hi everyone. What are the three words that we have to do. They are the M words. What are they?*
Sam	*Same we forgot about those m words we think one was maintain.* *we am not sure about if it is right. Please reply under* *Thanks guys*
Sam	*Guys we know the m words they are* ***mean, mode*** *and* ***median****.* *we just remembered today. we hope this helps you in you* ***bar graph, column graph*** *and lastly* ***line graph*** *etc. Please reply if you are on edmodo.* *Thanks guys* *And see you tommorow*

In this example of *Exploratory Talk*, we see the students attempting to decide on appropriate mathematical vocabulary to describe the three common measures of central tendency. One student offers the word 'maintain' as a possibility to describe one of the measures. Eventually though, they are able to arrive at the conclusion that the words 'mean', 'median' and 'mode' are the words that they have been seeking to find. This suggests that students' vocabulary may benefit from the co-negotiation of definitions, trialling and experimentation with new terms that the context of this setting allows. The two transcripts that follow are provided verbatim and in an unedited format.

Table 6. Example of discussion coded as cumulative talk

Annie	*hey guys do we need to do male and female cats names if you do please post*
Kevin	*do you what me to rshoq [means 'research']*
Annie	*what is rshoq?*
Kevin	*that mans resuch*
Annie	*no we was thinking we have already done the female and the male cat names and they are french names is that alright with you guys*
Annie	*and you spell research like this.*
Sheldon	*hi*
Annie	*hi just tell me if you guys want to know the names and we will tell you*
Annie	*we will tell you anyway the female names are:* *Sassy Misty Princess Samantha Kitty Puss Fluffy Molly Daisy Ginger* *Midnight Precious Maggie Lucy Cleo Whiskers Chloe Sophie Lily Coco*
Annie	*And my male names are:* *Max Sam Tigger Tiger Sooty Smokey Lucky Patch Simba Smudge Oreo Milo Oscar Oliver Buddy Boots Harley Gizmo Charlie Toby*

Table 7. Frequency of technical mathematical vocabulary use in cumulative and exploratory talk

Talk Type	No. Technical mathematical words	Identified examples of Talk Type	Average No. technical mathematical words per Example
Cumulative Talk	361	49	7.4
Exploratory Talk	284	27	10.5

Table 6 (below) shows an example from the online discussion where *Cumulative Talk* is apparent. This discussion is taken from a different small group of students attempting to solve collaboratively the 'Pet Names' problem. In this example, Annie positions herself as leader within the group. She repeatedly rephrases her desire for suggestions or agreements related to whether she should provide information about the various pet names. No constructive criticism is present; however eventually we see some common knowledge emerging. This excerpt of discussion shows an absence of technical mathematical vocabulary. The group did not present any analysis of the length of their pet names.

Table 7 shows the number of examples of Talk Types identified, from the coarse- grained analysis and the number of examples of technical mathematical vocabulary. The dominant talk type throughout all discussion during the nine weeks of data collection was *Cumulative Talk* as shown in Table 3. Forty-nine examples of *Cumulative Talk* were identified, whilst only 27 examples of *Exploratory Talk* were identified. Across the data we see an average of between seven and eight technical mathematical words used during examples of *Cumulative Talk* discussions, whilst there are between ten and eleven examples of this type of vocabulary used in examples of *Exploratory Talk*. This shows that students engaged in *Exploratory Talk* were more likely to use technical mathematical vocabulary than when they were engaged in *Cumulative Talk*.

Aligning Critical Thinking, CAIS, Talk Types and Vocabulary

Table Table 8 (below) provides evidence for the drawing together of the frameworks (see Table 6) underpinning this chapter. Perkins and Murphy's (2006) CAIS framework was initially utilised as a means to provide evidence of student critical thinking within the CSCL environment. We have shown that it can also provide insight into how students focussed the critical thinking they communicated within their small groups. Mercer and Wegerif's (1999) Talk Types were then used as a means to examine the discourse that students engaged in. The density of technical mathematical vocabulary use within each of the three Talk Types was examined. A greater density of technical mathematical vocabulary was found in examples of *Exploratory Talk*, reinforcing Mercer and Wegerif's (1999) suggestion that this Talk Type is generally more productive. It is here that we see evidence of critical thinking and reasoning.

Table 4 (below) represents coding of Talk Types using the fine grained 'utterance' level analysis. It shows that of the three Talk Types, there is generally a greater incidence of all four CAIS categories of critical thinking where *Exploratory Talk* is present. Thus, our data indicate that where *Exploratory Talk* is evident students will engage in greater levels of critical thinking. Whilst each of the four categories were more evident where *Exploratory Talk* was evident, the CAIS (Perkins & Murphy, 2006) category of *Inference* was the most obvious example of this. Table 8 shows that of the identified examples of *Inference,* 84% occurred when students were engaged in *Exploratory Talk*. It is likely that the more collaborative, dialogic nature of *Exploratory Talk* provided students the opportunity to engage in the important work

Table 8. CAIS occurrence in three talk types

CAIS Category	*Cumulative Talk*	*Disputational Talk*	*Exploratory Talk*	*Exploratory Talk* (% of Total)
Assessment	28	8	65	64
Clarification	59	7	131	66
Inference	8	0	42	84
Strategies	111	1	161	59

of deducing appropriately, inferring appropriately, arriving at conclusions, making generalizations and finding relationships amongst concepts and ideas.

The data suggest that students will use technical mathematical vocabulary more regularly when engaged in *Exploratory Talk* than when engaged in *Cumulative Talk*. It has also been shown that *Cumulative Talk* is likely to be the dominant talk type, given the conditions described. We suggest that it may be beneficial to specifically encourage the engagement of students in *Exploratory Talk* for two reasons. Firstly, in order to prompt students to more regularly experiment with newly acquired vocabulary. Secondly, as a means to encourage critical thinking within their collaborative mathematical problem solving. Explicitly teaching students about the three Talk Types and discussing their various attributes and characteristics, including why *Exploratory Talk* might be the most productive talk type, may promote this. Such teaching would include an explanation of the importance of building a repertoire of technical mathematical vocabulary. Approaches like this may result in groups 'self-regulating' their discussion and being aware of when talk had become less productive.

Exploratory Talk as discussed by Mercer and Wegerif (1999) appears to be closely linked to critical thinking. The facets of critical thinking, as indicated by both the CAIS framework (Perkins & Murphy, 2006) and the indicators of Critical and Creative Thinking as provided by the Australian Curriculum (ACARA, 2017) are represented when students engage in discussions involving critiquing, justifying, clarifying, analysing that *Exploratory Talk* describes. Having established that *Exploratory Talk* is likely to exhibit a greater density of technical mathematical language use, we can also make the claim that it is likely that when students utilise a greater density of this type of vocabulary that they are also more likely to engage in critical thinking. This is supported by the greater prevalence of the CAIS categories identified in *Exploratory Talk*.

Active Use of Technology in Mathematical Problem Solving and Industry 4.0

In this study we see evidence of students' critical thinking and reasoning in mathematics supported by technology. Evidence of individual students developing critical thinking and reasoning and collaboration skills can also be monitored by their teacher at a level not possible in the hurly burly of classroom collaborative group work. With practice a teacher can quickly look over the online discourse, with the students' artefacts, and identify exploratory talk and technical mathematical language.

Students' online collaborative mathematical problem solving contributes to their experience of decentralised decision making. They learn to work together with other students in an asynchronous manner while also working with their computers as contributors to the discussion. Technology is being used as an available expert to provide requested information, as a tool for calculation and to both amplify and clarify communication using diagrams, graphs and colour. Again, the teacher can quickly see which skills need specific attention in the face-to-face lessons.

Taking advantage of "cognitive computing" and "cloud computing" requires experience with, and understanding of, decision making processes, development of reasoning and communication skills, in particular dialogue that is mediated through the synchronous and asynchronous use of technology. Working on mathematical problem solving in a computer supported collaborative learning environment is a small step but gives students real experiences not available when computers are used only to encourage skills practice.

REFERENCES

ACARA. (2014). *The Australian Curriculum V5.2*. Australian Curriculum and Reporting Authority. Retrieved from http://www.australiancurriculum.edu.au/

ACARA. (2017). *The Australian Curriculum: General Capabilities: Critical and Creative Thinking*. Retrieved 30/5/2017, from ACARA: http://www.australiancurriculum.edu.au/generalcapabilities/critical-and-creative-thinking/introduction/in-the-learning-areas

Alexander, R. (2006). *Towards dialogic teaching: Rethinking classroom talk*. Dialogos York.

BECTA. (2004). *A review of the research literature on barriers to the uptake of ICT by teachers*. BECTA ICT Research. Retrieved from https://ezp.lib.unimelb.edu.au/login?url=https://search.ebscohost.com/login.aspx?direct=true&db=edsoai&AN=edsoai.771640667&site=eds-live&scope=site

Brace-Govan, J., & Clulow, V. (2001). Comparing face-to-face with online: Learners' perspective. *Academic Exchange Quarterly*, *5*(4), 112–118.

Brown, C., & Czerniewicz, L. (2010). Debunking the 'digital native': Beyond digital apartheid, towards digital democracy. *Journal of Computer Assisted Learning*, *26*(5), 357–369. doi:10.1111/j.1365-2729.2010.00369.x

Bullen, M. (1997). *A case study of participation and critical thinking in a university-level course delivered by computer conferencing*. University of British Columbia.

Cuoco, A., Goldenberg, P., & Mark, J. (1996). Habits of mind: An organizing principle for mathematics curricula. *The Journal of Mathematical Behavior*, *15*(4), 375–402. doi:10.1016/S0732-3123(96)90023-1

Day, L. (2013). A snapshot of the use of ICT in primary mathematics classrooms in Western Australia. *Australian Primary Mathematics Classroom*, *18*(1), 16–24.

Facione, P. A. (2013). *Critical thinking: What it is and why it counts*. Retrieved from http://www.insightassessment.com

Flavell, J. H. (1976). Metacognitive aspects of problem solving. *The Nature of Intelligence, 12*, 231-235.

Garrison, D. R., Anderson, T., & Archer, W. (2001). Critical thinking, cognitive presence, and computer conferencing in distance education. *American Journal of Distance Education*, *15*(1), 7–23. doi:10.1080/08923640109527071

Griffin, P., McGaw, B., & Care, E. (2012). *Assessment and teaching of 21st century skills*. Springer. doi:10.1007/978-94-007-2324-5

Henri, F. (1992). Computer conferencing and content analysis. In *Collaborative learning through computer conferencing* (pp. 117–136). Springer. doi:10.1007/978-3-642-77684-7_8

Holton, D., & Symons, D. (2015). *Creative Activities in Mathematics: Book 2. Problem-Based Maths Investigations for Upper Primary*. ACER.

Jazby, D., & Symons, D. (2015). Mathematical Problem Solving Online: Opportunities for Participation and Assesment. In *The 39th Conference for the Psychology of Mathematics Education* (Vol. 3, pp. 105-112). PME.

Lampert, M. (1990). When the problem is not the question and the solution is not the answer: Mathematical knowing and teaching. *American Educational Research Journal*, *27*(1), 29–63. doi:10.3102/00028312027001029

Lesh, R., & Zawojewski, J. (2007). Problem solving and modeling. Second handbook of research on mathematics teaching and learning, 2, 763-804.

Loveder, P. (2017). Australian Apprenticeships: Trends, Challenges and Future Opportunities for Dealing with Industry 4.0. Conference Paper. *National Centre for Vocational Education Research (NCVER)*.

Makrakis, V. (2005). Training teachers for new roles in the new era: Experiences from the United Arab Emirates ICT program. *Proceedings of the 3rd Pan-Hellenic Conference on Didactics of Informatics*.

Mark, J., Cuoco, A., Goldenberg, E. P., & Sword, S. (2010). Developing Mathematical Habits of Mind. *Mathematics Teaching in the Middle School*, *15*(9), 505–509.

Matthews, K. E., Garratt, C., & Macdonald, D. (2018). *The higher education landscape: trends and implications*. Academic Press.

Mercer, N., & Wegerif, R. (1999). Is exploratory talk productive talk? In K. Littleton & P. Light (Eds.), *Learning with computers: Analysing productive interaction* (pp. 79–101). London: Routledge.

Murphy, J., & Irwin, F. (2018). Design thinking: Can it prepare students for industry 4.0? Independence, 43(1), 46.

Newman, D. R., Webb, B., & Cochrane, C. (1995). A content analysis method to measure critical thinking in face-to-face and computer supported group learning. *Interpersonal Computing and Technology*, *3*(2), 56–77.

Norris, S., & Ennis, R. (1989). What is critical thinking. *The practitioner's guide to teaching thinking series: Evaluating critical thinking*, 1-26.

Perkins, C., & Murphy, E. (2006). Identifying and measuring individual engagement in critical thinking in online discussions: An exploratory case study. *Journal of Educational Technology & Society*, *9*(1).

Pierce, R., Stacey, K., & Wander, R. (2010). Examining the didactic contract when handheld technology is permitted in the mathematics classroom. *ZDM, 42*(7), 683–695. doi:10.100711858-010-0271-8

Prensky, M. (2001). Digital natives, digital immigrants part 1. *On the Horizon, 9*(5), 1–6. doi:10.1108/10748120110424816

Schoenfeld, A. H. (1992). *Learning to think mathematically: Problem solving, metacognition, and sense making in mathematics*. Information Age Publishing Inc.

Selwyn, N. (2009). The digital native–myth and reality. *Aslib Proceedings*.

Symons, D., & Pierce, R. (2015). Mathematical Language Development and Talk Types in Computer Supported Collaborative Learning Environments. In *The 38th Annual Conference of the Mathematics Education Research Group of Australasia* (pp. 579-586). MERGA.

Wachira, P., & Keengwe, J. (2011). Technology Integration Barriers: Urban School Mathematics Teachers Perspectives. *Journal of Science Education and Technology, 20*(1), 17–25. doi:10.100710956-010-9230-y

Wanyama, T., Singh, I., & Centea, D. (2018). A Practical Approach to Teaching Industry 4.0 Technologies. In *Online Engineering & Internet of Things* (pp. 794–808). Springer. doi:10.1007/978-3-319-64352-6_74

Chapter 12
Operationalizing Computerized Testing in Mathematics Competition

S. Kanageswari Suppiah Shanmugam
Universiti Utara Malaysia, Malaysia

Liew-Kee Kor
University Teknologi MARA, Malaysia

Mohan Chinnappan
University of South Australia, Australia

ABSTRACT

This chapter seeks to take a modest and yet formative approach in proposing computer-based test (CBT) as the inevitable next wave in digital assessment. The major section reports on the practical design of developing and administering CBT in a mathematics competition by HOTS educational innovation development (HEID) in a mathematics competition. This chapter describes the CBT developmental process undertaken by HEID, which focuses on constructing multiple-choice test items, administering CBT, and scoring the test. While CBT offers the possibility of improving assessment, the process of preparing items and conducting online competition have nevertheless encountered several challenges such as getting quality test items, school and teacher readiness, seat time in front of computer, computer compatibility, and internet speed. As with many other educational reforms at infancy, more research in the area of implementing CBT to facilitate international competitions is needed in landscaping the assessment platform of Education 4.0.

DOI: 10.4018/978-1-5225-7832-1.ch012

Copyright © 2019, IGI Global. Copying or distributing in print or electronic forms without written permission of IGI Global is prohibited.

INTRODUCTION

Internet of things, artificial intelligence, cloud computing, and robots as learning partners are a few of the drivers that characterise Education 4.0 as a result of Industry 4.0 that has revolutionized global workforce and talent requirements in the digital economy. It is estimated that 65 percent of primary students schooling today will be employed in jobs, which are currently not in existence. This in turn spurs the need to treat students, their learning environment, and their learning tools as one cognitive system and in the present context of Education 4.0, the direction taken cannot afford to stray from technology. Accordingly, assessment in the 21st century can no longer just be designed to testing students' knowledge per se. The need to answer the clarion call of extending its capacity to effectively and efficiently evaluate students' knowledge application using technology support, particularly computer-assisted assessments can no longer be ignored. This urgency to design CBT, however, need to be meticulously planned and conducted, without compromising test reliability, validity and fairness, especially in view of battling against the challenges of implementing CBT in the global context of only four billion having access to internet connection among a population of seven billion. Ultimately, re-inventing assessment practices and aligning it with Education 4.0 is shaping the assessment landscape of this century as assessment shapes what, when and how students study. Assessment determines how much effort and work students are prepared to invest and the approach they will take in achieving the desired outcome. Therefore, the assessment design is one of the most influential aspects to consider in the pursuit of enhancing the quality of learning achieved at the end of any learning programme. This is needless to say that improving student learning begins by firstly improving the assessment design (Higher Education Academy, 2012).

Understanding Assessment in the Contexts of 21st Century and Education 4.0

Advocates of High Impact Educational Practices emphasise that where students study is not as important as what they do while they are there (Kuh & Schneider, 2008). Thus, within the context of associating assessment to the essence of Education 4.0, Schneider (2008) argues that the definition of 'student success' needs to encompass evidence of student learning as a measure of the quality of education that has been received. This is because of the nature of learning "always entails participation in relationship and community transformation both of the person and of the social world" (Packer & Goicoechea, 2000, p227). Therefore, the quality of the education can only be assured if the education is valued by the society and empowers the individual (Schneider, 2008).

Retention and graduation are best described as partial indicators of student success—necessary, but scarcely sufficient. The college degree is meaningful, after all, only when it represents forms of learning that are both valued by society and empowering to the individual. Twenty-first Century metrics for student success need to capture that reality. They need to address evidence about the quality of learning as well as the evidence about persistence and completion (Schneider, 2008, p. 2).

In an attempt to determine what is deemed valued, National Association of Colleges and Employers [NACE] in 2016 commissioned a survey, *Job Outlook 2016* to ascertain what graduate attributes were sought for employability. Findings revealed that the academic skills such as written (70.2%) and oral communication (68.9%), problem solving skills (70.2%), analytical or quantitative skills (62.7%), technical skills (59.6%) and computer skills (55.3%) were deemed important. The employers were also asked to indicate factors that influenced their decisions to hire graduates. Interestingly, a non-academic factor, which is *active participation in co-curricular activities* raked the third top position among eight other attributes (National Association of Colleges and Employers, 2016).

It is clearly apparent that measuring diversity in learning (problem solving, analytical and technical skills and, computer literacy) have emerged relevant in the present context of Industrial Revolution 4.0 (IR 4.0) and Education 4.0 and, seem to garner superiority in securing jobs. Even more surprising is that student participation in co-curricular activities such as competition is slowly gaining recognition among employers. Accordingly, in aligning learning and assessment for sustainability, assessments that measure application of student learning is needless to say is a requisite as revealed by a survey conducted by the Association of American Colleges and Universities (AAC & U) in 2008. Specifically, the results concluded:

When it comes to the assessment practices that employers trust to indicate a graduate's level of knowledge and potential to succeed in the job world employers dismiss tests of general content knowledge in favor of assessments of real-world and applied-learning approaches. (Maki, 2009, pp. 16- 17).

What is evident from the perspective of assessment is that paper-pencil tests fail to capture the essence of student success' in the wake of Education 4.0, resulting in a pressing need to search for assessments (Maki, 2009) that can capture the robustness and realistic evidence of students' diverse learning. With the emerging trend of international assessments implementing computer-based assessment [Programme for International Student Assessment (PISA) 2006 for Science and, Trends in International Mathematics and Science Study (TIMSS) 2019), operationalizing CBT in any magnitude is a move in the right direction. Accordingly, infusing technology into assessment is inevitable in addressing the importance of capturing the multiple complex representations of learning in Education 4.0 and therefore, suggesting technology-based assessment as the way forward in remaining at par to the call made in Education 4.0 and relevant to marketing graduates of the 21st century.

TECHNOLOGY-BASED ASSESSMENT

There is an emerging consensus the technology-driven assessment confers considerable advantages over traditional modes of assessment from a logistic perspective such as user-friendliness, validity, reliability, rapid access to results and ease of interpretation by both learners and students. What is less examined in this equation, is how and how well does CBT-based models of assessment address pedagogical issues that are central to aims of assessment, more generally. 21 century skills demand students to acquire, draw on and utilize prior mathematical knowledge flexibly in the solution of complex problems. Chinnappan and colleagues (Tajudin & Chinnappan, 2016; Lawson & Chinnappan, 2015) owed that mathematical knowledge that drives high levels of performance is complex and multifaceted. Equally, the flexibility in the use that knowledge during search for solution is a function of the degree of quality and quantity of connections that leaners are able construct among the different strands of their mathematical knowledge. These latter knowledge structures were captured in the framework of schemas. For examples, students' use of geometric knowledge is guided by the sophistications and robustness of their geometric schemas (Chinnappan, 1998). If one extends the schema perspective to the diversity of skills and complexity of mathematical content knowledge that are required to tackle problems of the 21st Century, learners have to acquire interdisciplinary knowledge schemas. Thus, 21st Century skills include students having to access and apply knowledge of concepts from within and across other domains such as science and history.

What are the implications and challenges for the design and implementation of CBT-based assessment tools in light of the above schema-based analysis of mathematical content and general skills that underpin competencies of 21st Century skills where the emphasis is less on what knowledge students have acquired to more on what they can do with that knowledge. Another feature of 21st Century learning the shift from individual to collective learning. In the 21st century students will be expected to tackle real-life problems that are more complex and open-ended with multiple solutions which have to be evaluated. The first challenge for technology-based assessment would be to provide accurate and immediate data on basic knowledge of the content of mathematics that students have acquired. However, as this knowledge is connected and structured, the items in the test must be sensitive and pliable enough to identify concepts and *links* among concepts. Equally, the links must me measured in both quantitative and qualitative terms. The second challenge will be measuring skills that students could deploy in the solution of problems. As these skills are not specific and could vary depending on the problem at hand, CBT assessment tasks could assume problem-based scenarios. Logistically, the presentation of the problem scenario and evaluation of skills in developing solutions could present a challenge on-line or other IT-based devices. With the evaluation of problem-based skills, we are moving into more of a subjective mode of assessment. The question arises as to how and who will decide on the range, quality and appropriateness of skills. A potential solution could be to develop a set a priori skills for the problems of interest on the basis of advice from experienced teachers and built that data into the assessment system. Such a model of IT-based assessment tool can be refined and expanded for a bigger and more complex pool of problems.

The third and final challenge as we anticipate in this analysis involves assessment of *collective* knowledge and skills for 21st Century students. We are now moving into a relatively uncharted territory. Again, in order to generate meaningful data about knowledge and skills, we could make progress by drawing boundaries around problems, and the solution of problems by a cohort of students of age that is similar to the ones who are assessed. For example, let's analyse a problem that asks a group of five Grade 9 students to investigate recycling paper waste at a site such as school. The solutions to this problem would depend how the individual in the group understand the problem and their collective prior knowledge and experiences that is brought to bear on the problem. For the purpose of assessment, we could and need to generate a set of knowledge and skills that individual students working as a member of a group could bring to and share with other members of that group. That information could be integrated into and used subsequently to assess against 21st Century group skills.

HOTS Educational Innovation Development (HEID)

HOTS Educational Innovation Development (HEID) is a private organizing body focusing on the creative and innovative learning of science and mathematics. It is set up to promote critical and analytical learning of science and mathematics among the higher primary and lower secondary school students. This is in tandem with the national educational policy of incorporating Higher Order Thinking Skills in all curricula. To achieve this end, series of competitions, teaching programmes, learning camps, seminars and workshops are regularly organized for students, teachers and parents who are interested to develop further in the areas of science, mathematics, robotic, IT and AI. HEID envisions nurturing a generation of creative and innovative individuals who are adept at critical thinking and problem solving through our diverse range of programmes and activities. It organizes world-class competitions to provide a healthy and competitive environment for the students to challenge their limits as well as to pitch their abilities against other students for improvement and growth. All programmes, activities and competitions are professionally developed and customized to achieve its primary goal – to enhance and maximize active and innovative learning of students in the fields of science, mathematics, and technology.

Computer-Based Educational Assessments

Computer-based educational assessments are developed with reference to the characteristics and needs of students and computerized based test (CBT) is one of such assessments (Martin & Lazendic, 2017). This paper presents CBT in a slightly different manner. It discusses engaging CBT in a competition outside the context of traditional school assessment. The mode of using computer based testing discussed in this paper mirrors criterion-referenced test. The criterion of reference here is the domain of subject matter according to the grade level in school that the competition is designed to assess. This paper outlines a CBT test development process adopted by HEID from preparing to publishing test items in a mathematics competition. It focuses on (1) constructing multiple-choice test items; (2) administering CBT, and (3) scoring the test. Issues and challenges of operationalizing CBT in the completion are also discussed.

Relationship Between Evaluation, Assessment, Measurement, and HEID Competition

Designing academic competition has a close relationship between evaluation, assessment, and measurement. Adapting from Zimmaro's (2016, p.4) summary of how evaluation, assessment, measurement and testing terms are related, the following flow chart shows the course of testing in a competition:

Based on Figure 1, we conclude that preparing test items for academic competition undergoes the similar process as preparing test items for standardized school assessment.

HEID's CBT Framework

HEID's CBT framework in the implementation of CAT in the Malaysia Mathematics Competition (MMC)is composed of three main foci.

Focus 1: Constructing Multiple-Choice Test Items for Competition

In most conventional paper-and-pencil or computer-based tests, items are typically developed to cater for the "average" ability respondent. It is very common to find items that are often too easy for respondents who are above average and too difficult for respondents who are below average on the aspect being assessed (Martin & Lazendic, 2017; Way et al., 2010). Multiple-choice questions (MCQ) are usually more difficult to build than essay items. A well-constructed multiple choice test is capable of assessing the extent of a student's working knowledge that is sufficient for the tested content. Examination experts estimate a quality MCQ requires 20 to 60 minutes to construct (Case & Swanson, 2002; Palmer & Devitt, 2007). According to Ory (2009) a professional item writer produces only 9 - 10 good MCQ in a day's time.

Figure 1. Relationship between evaluation, assessment, measurement and testing

Evaluation
Examining information about various components of the mathematics competition and comparing or judging its quality, worth or effectiveness in order to make decisions

based on ↓

Assessment
The process of gathering, describing, or quantifying information about performance

Includes ↓

Measurement
Process of assigning number to the quality of the answers to the test item according to category and analyzing that data based on statistical theory

specific way to measure performance is ↓

Testing (in a Competition)
A method used to measure the level of achievement in a particular competition according to students' cognitive ability in the subject

HEID administers multiple-choice test items in the MMC. One of the reasons for such choice is MCQ have the advantage of ease of scoring, especially if scoring is automated (Towns, 2014). HEID is also stringent on the quality of the test items and is aware that item-writing flaws may cause competition questions easier or more difficult than intended. Hence, in constructing the test items, steps were taken to minimize items flaws that allow unprepared students to guess the correct answer or prevent prepared students from demonstrating their knowledge. One of the steps is engaging experts in item construction to vet the items.

MMC is designed to accommodate four categories of contestants according to school grade: Lower Primary (Standard 1 and Standard 2); Middle Primary (Standard 3 and Standard 4); Upper Primary (Standard 5 and Standard 6); and Junior Secondary (Form 1, Form 2 and Form 3). In the process of building item banks, HEID requires item developers and experts to observe the following:

Validity and Reliability

Validity is crucial in developing reliable and valid testing. As mentioned above, HEID has engaged several item writers to contribute to the same category in the competition. Each independent item writer was invited to build items for lower primary, middle primary, upper primary, and the junior category respectively. To check the validity of the items and item flaws, a group of mathematics experts (consisting of university professors, lecturers, and senior school teachers with subject teaching experience of more than 10 years) was appointed to go through each item thoroughly. The entire task was conducted in three phases.

Figure 2. An example of classification and calibration of test items

	Remember	Understand	Apply	Analyze	Evaluate	Create
Topic A	*Easy*		*Moderate*	*Hard*		
Item 1	√					
Item 2			√			
⋮						
Topic B						
Item 1		√				
Item 2	√					
⋮				√		
No. of items						

Test Blueprint

In phase 1, item writers prepared a set of 50 questions for each assigned category. Using a test blueprint, also known as test specifications table in a matrix form, the experts helped to verify the alignment of the test items with the objectives and skills specified according to topics in the Malaysia School Curriculum. The experts also went through the test specifications to classify items into easy, moderate, and hard category. They referred to Anderson's et al. (2001) cognitive revised domain (Remember, Understand, Apply, Analyze, Evaluate, and Create. See Figure 2) to classify items based upon the type of knowledge required and the level at which the question is asked (see Figure 3).

Item Quality

In phase 2, the group of experts inspected the content of the item in respect to the context and language (e.g. grammar, punctuation, spelling, mathematical symbols, diagram, figures, table, etc.) of the stem of each MCQ. Finally, in phase 3, the experts verified the answers for each item after all newly written items were piloted tested. They also helped to check the list of possible response options and the suitability of the distractors.

Focus 2: Administering CBT in Mathematics Competition

The competition procedure is divided into two main parts: registration and implementation. Similar to Chuesathuchon's (2008) procedure, the HEID procedure (see Figure 3) includes functions such as the exam registration, programme provider's data, contestant's data, and test instruction. The contestant's data required a student's information, i.e., student email/username and password (security system).

HEID adopts multi-stage strategies to administer the mathematics competition which contains many items of different levels of difficulty. Under this strategy the competition starts with initial 10 easy items, follows by 10 moderate items, and lastly five difficult items. The given time to answer all questions is 60 minutes. And the items are written in two languages: English and Bahasa Malaysia. Once the specifications for all the necessary components have been determined, as well as any additional algorithms is included, the final CBT was published.

HEID's Computer Procedure

1. Log into the Exam Page (see Figure 4)

Go to https://hotseducational.runexam.com and sign in with your assigned username and password given by administrator.

OR

Go to http://hotseducational.com and click on Online Test to be redirected to the exam page (see Figure 5).

2. Look for Upcoming Exam (see Figures 6 and 7)
3. Read Test Instruction (see Figure 8)
4. The Interface (see Figure 9)

An example of a test item interface showing the essential buttons on the default screen:

1. **Next Button**: To go next question.
2. **Previous Button:** To go previous question.
3. **Question List:** Browsing the questions.
4. **Current Question (Navigation):** Showing current question out of total questions.
5. **Candidate & Exam Name:** Showing candidate and exam name.

Figure 3. Procedure of HEID's CBT

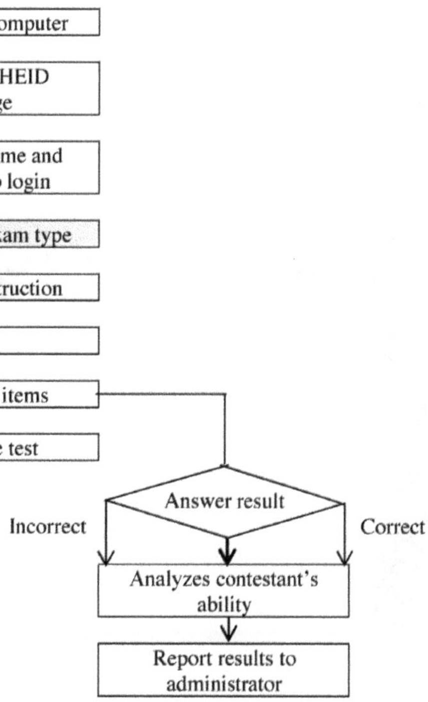

Figure 4.

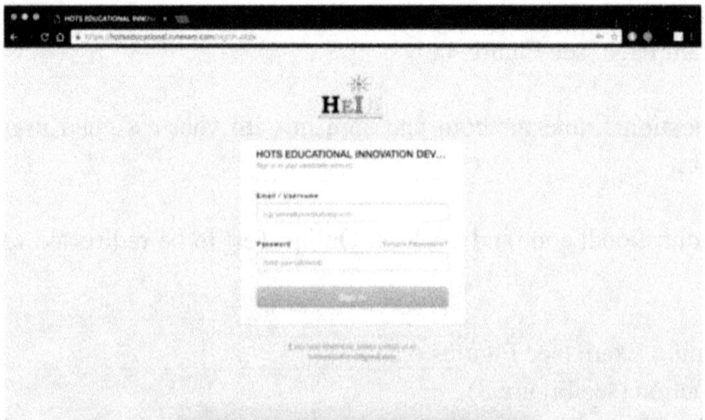

Figure 5.

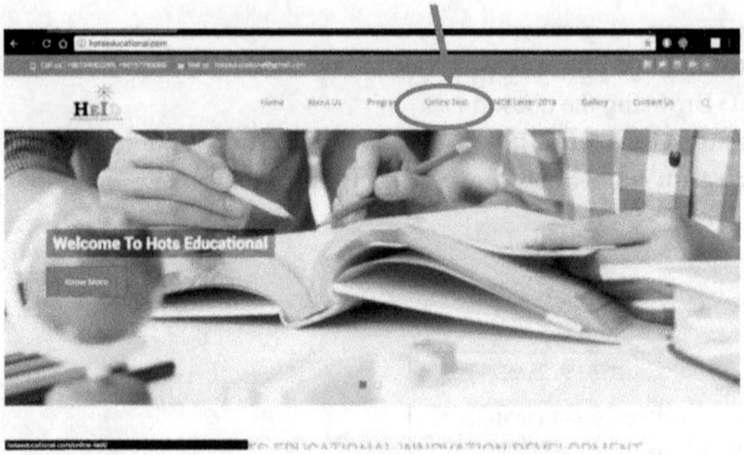

Figure 6.

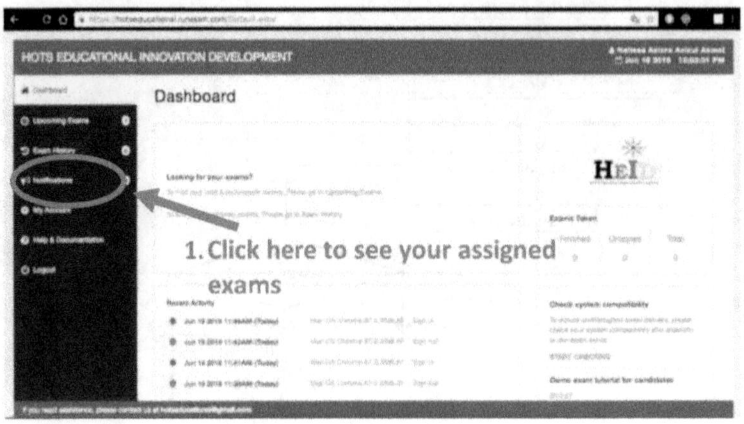

Figure 7.

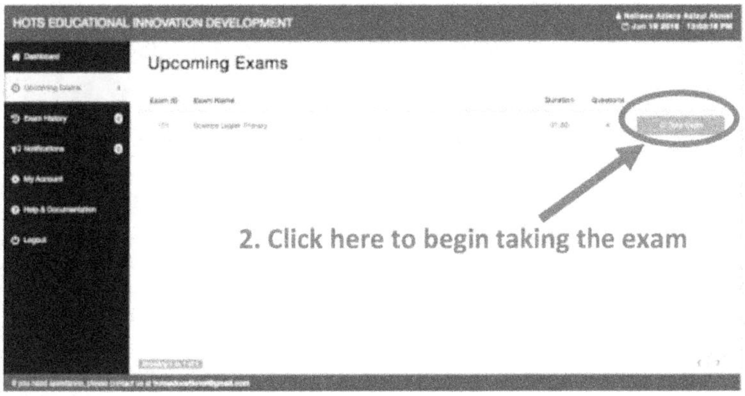

Figure 8.

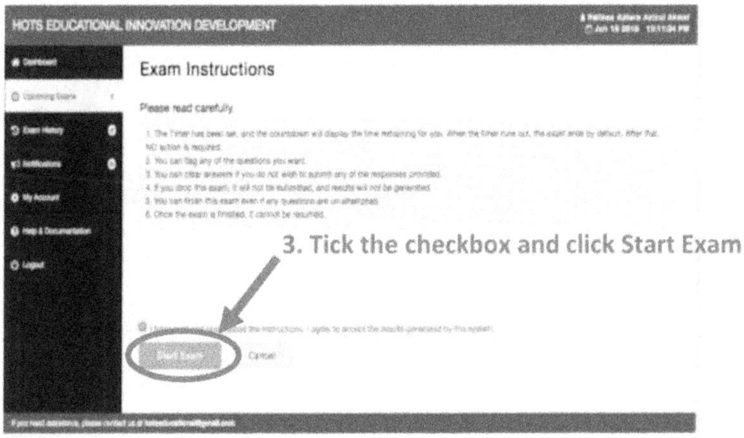

6. **Section Name:** Showing section name.
7. **Clear Button:** Clear the response from here.
8. **Flag Button:** Flag (Review / Mark) question from here.
9. **Drop Button:** To drop (cancel) an exam, use this button. Once the contestant drops the exam, he/she cannot continue with the exam once the duration time has ended.
10. **Feedback Button:** To submit content related feedback to the administrator (e.g. "This question has grammatical errors.")

Figure 10 shows how to check the time remaining for a test.

During the test, when the duration time is about to end with 60 seconds still remaining, the timer colour will change from blue to red at the last question. Click "Finish" button will end the exam. The colour pallets on the question list (on the left) indicate the status of the particular question:

- Gray – indicates the question is not read yet.
- No pallet – indicates the question has been read and attempted.

Figure 9.

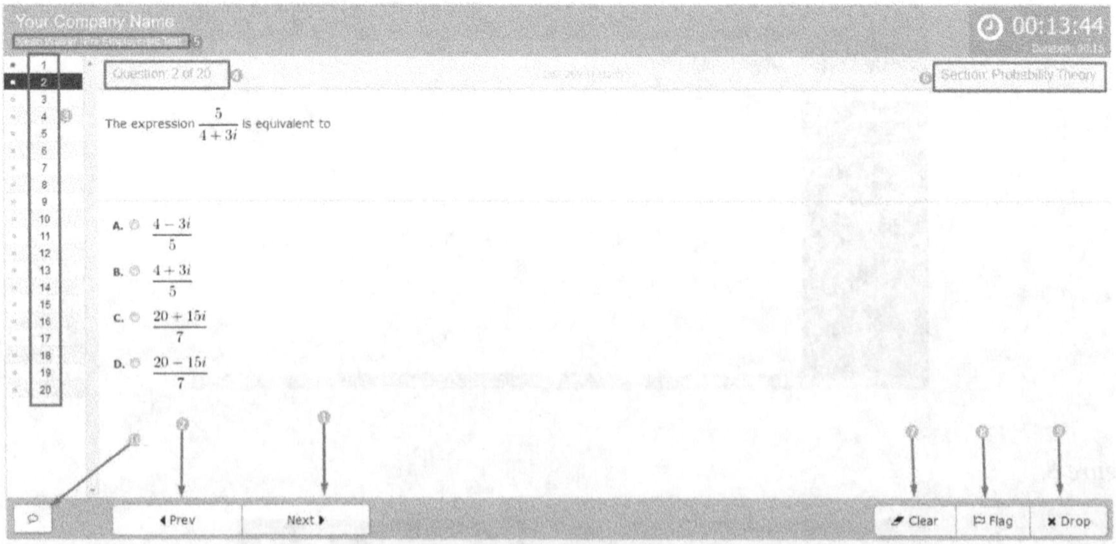

Figure 10.

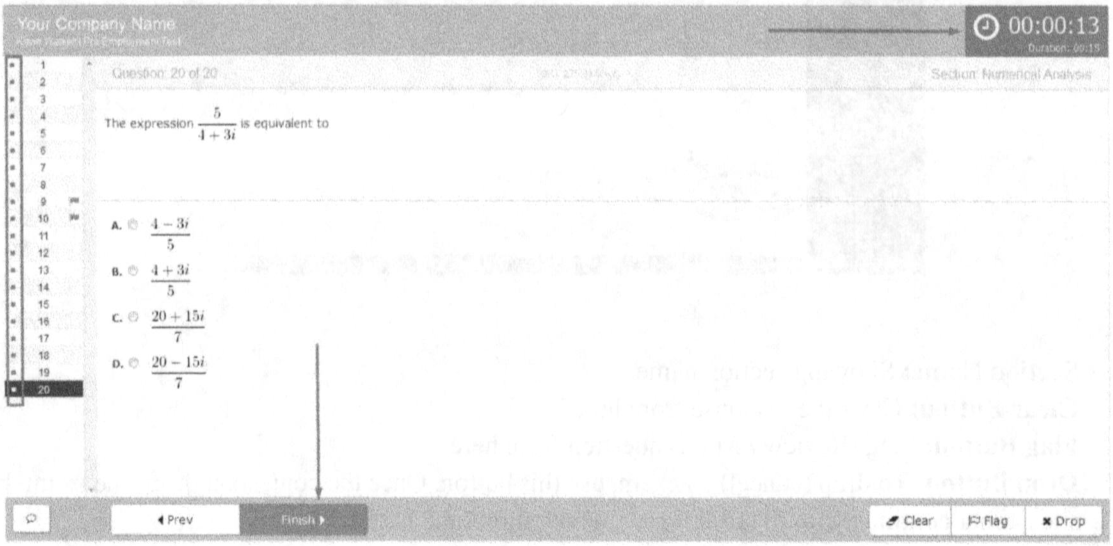

- Dark Gray – indicates the question has been read but NOT attempted.

Focus 3: Scoring the test

HEID uses conventional scoring method to score the multiple choice tests. The test items were scored using a conventional number right (NR) scoring whereby correct answers were scored with a positive value and incorrect answers and absent or omitted answers with a value of zero (Lesage, Valcke, & Sabbe, 2013). The sum of the scores for correct responses is the test score.

Marks for Scoring

The MMC scoring guide adopted by HEID is as follow:

There are four categories (Lower Primary, Middle Primary, Upper Primary, and Junior) of contestants. Testlet for each category contains 25 questions with different scores.

- Questions 1 to 10 (Easy level) are worth 3 marks each.
- Questions 11 to 20 (Moderate level) are worth 5 marks each.
- Questions 21 to 25 (Hard level) are worth 8 marks each.

Each category gives a total of 120 marks. Final results are based on the total score.

Benchmark for Achievement Awards

In this type of competition, contestants with the same number of correct answers may obtain different scores due to different levels of difficulty. The cut-off marks for each award will also differ therein. The comparison of achievement is done within the same category and within the same country in the same year of competition as follow:

- Scorers of top 5% – Gold Award
- Scorers of top 5% to 15% – Silver Award
- Scorers of top 15% to 30% – Bronze Award
- Scorers of 30% to 55% – Distinction
- Scorers of 55% to 85% – Merit

ISSUES AND CHALLENGES

Undoubtedly, CBT is becoming conceptually appealing that educators might become enamored of the idea and inspire to proceed (Thompson & Weiss, 2011). HEID concurs with Thompson and Weiss (2011) that the first stage in CBT development begins not with the decision to implement CBT, but rather when the question is raised as to whether CBT approach is even feasible for a testing programme. This is a question on feasibility, applicability, and planning studies that interested stakeholders and test organizers will have to contemplate.

In this competition, a major concern about MCQ scoring method is that contestants without the ability to solve a particular item may answer correctly through guessing. Lesage, et al. (2013) concur that guessing introduces a random factor into test scores that lowers reliability and validity. To overcome this problem, HEID test developers are constantly advised to be meticulous in introducing distractors in the response set. This is to minimize students getting correct answers based on a guess and not on knowledge mastery.

In the near future, HEID may have to endorse item-writing guidelines to create assessments that are reliable and valid, and able to discriminate between high and low achieving students. These guidelines hold several advices given by Town (2014) which are noteworthy and beneficial to MCQ test developers. Among few, to develop a numerical response set, use common computational errors that students make to generate distractors. These numerical responses should be arranged in ascending or descending order. Town believes that using common errors to design the distractors in the response set increases the likelihood that the options are plausible to students. And for verbal responses, the responses should be kept to nearly the same length and arranged logically. This is to remove cues that test-wise students may use in order to determine the correct response or to eliminate answers from the response set.

Another major concern for HEID to contemplate is getting access to seat time in front of computers for massive numbers of students. This concern was also highlighted by Gewertz (2011) who warns of the challenges posed by another commonly used computer adaptive testing. HEID anticipated that in a school with 200 students participating in a HEID competition at a given time and day, a one-hour test will require 10 hours of seat time in front of a computer in a computer lab with a capacity for 20 students. Obviously such setting poses a difficult design issue. However, if HEID allows students to use their own computers or mobile phone to attempt the test, then this may raise other issue such as test security. At this juncture, HEID will have to seriously decide between computer-based or paper-based assessment, or conduct both modes or move forward to computer adaptive testing in the future. Wang and Shin (2010) warn that both the administration mode and paradigm effect on examinees' performance should be examined to ensure the comparability of the CBT and its paper-based counterpart.

There are other issues about compatibility of school computers, speed of internet, and the readiness of teacher-in-charge. HEID has to take necessary actions to prepare for an uninterrupted test environment for the upcoming competitions. There are several steps that HEID has undertaken:

1. Instruction for teacher-in-charge of the competition – HEID has developed SpeedExam User Manual to guide the teachers on the rules and regulation of competitions. For example, before starting the exam teachers are advised to check system compatibility to ensure an uninterrupted exam delivery. The specified system compatibility is shown in Figure 11.
2. Teachers in charge are to make sure that each candidate has his/her own log in ID and password to access to the examination page.
3. HEID to offer a free trial run for all participating schools before the actual day of competition to get the teachers in charge familiarize with the administration and delivery mode.

CONCLUSION

This paper describes a novel attempt of CBT taken for a MMT by HEID., which mirrors a criterion-referenced test. The criterion of reference here is the domain of subject matter according to the grade level that the competition is designed to assess. It focuses on (1) constructing multiple-choice test items; (2) administering CBT, and (3) scoring the test. The CBT testing paradigm presented in this paper is different from CAT administration, even though CAT could be a promising avenue in the near future for HEID. Instead of allowing test items that are tailored to each test taker's ability to be selected and administered, testlets in the competition focuses on the knowledge domain according to grade level (Lower Primary, Middle Primary, Upper Primary, and Junior). Even though this paper reports on CBT conducted among primary and secondary students, the argument is that the implementation of CBT can be successfully replicated to any institutions of higher education, especially in light of its major benefits of instilling HOTs. Particularly, by incorporating into the Learning Management System, the back-end data such as

Figure 11. System compatibility

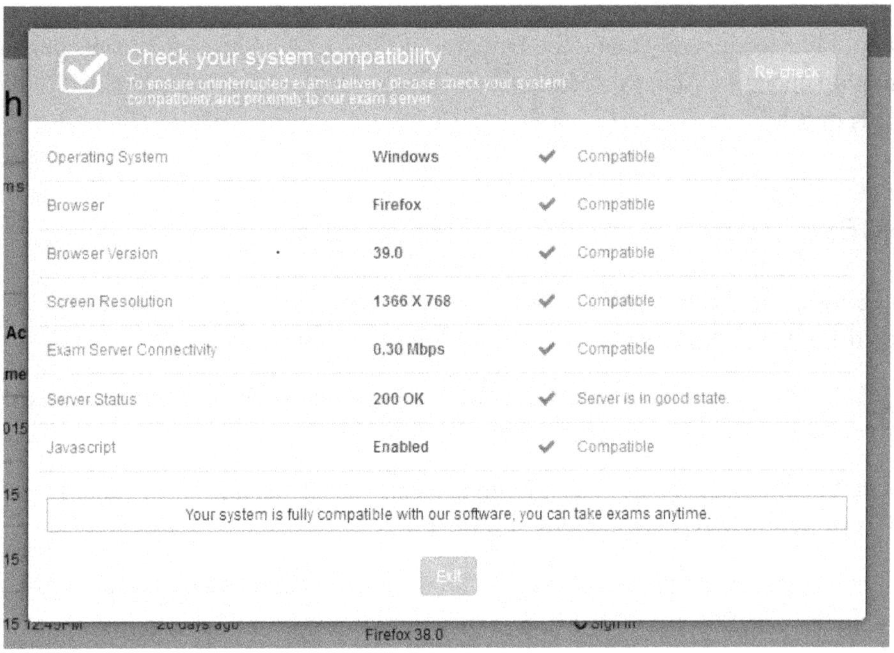

frequency of log-ins and discussions posts can be analysed in relation to the students assessment scores for a more meaningful interpretation of their performance and learning progress.

While computer-based testing offers the possibility of improving assessment, getting the appropriate testing paradigm presents significant challenges. The process of preparing items and conducting the competition online has nevertheless encountered several problems such as getting quality test items, school and teacher readiness, seat time in front of computer, computer compatibility, internet speed, and etc. Meanwhile, running a competition in a non-profit organization and without financial aids is a costly affair. HEID has to deliberate over the practical and business considerations. In order to safeguard the organization from running into the danger monetary loss, HEID will constantly go over Thompson and Weiss's (2011, p.3) four checklist questions: (1) Does the organization have the psychometric expertise, or is it able to afford it if an external consultant is used? (2) Does the organization have the capacity to develop extensive item banks? (3) Is an affordable CBT delivery engine available for use, or does the organization having the resources to develop its own? (4) If CBT costs more and does not substantially decrease seat time, is that fact sufficiently offset by the increase in precision and security to make it worthwhile for the organization? HEID acknowledges that further studies are needed to know exactly what type and amount of instruction is optimal to the stakeholders and the test-takers before taking a computerized test.

In conclusion, technology enhances the testing experience but it cannot compensate a poorly designed test. The authors are not advocating that CBT is a panacea as ultimately, it depends on the test objectives and design so that the obtained test scores are reliable and valid in measuring what they were intended to measure. However in the era of 21st Century testing, the potential of CBT needs to be recognized as the way forward to realize the aspirations of Education 4.0 in higher education in producing globally marketable graduates.

ACKNOWLEDGMENT

The author would like to sincerely thank the Director of HEID, Mr. Alex Tan, for his generosity in providing information about the Malaysia Mathematics Competition (MMC) that has led to the possibility to complete this paper.

REFERENCES

Case, S. M., & Swanson, D. B. (2002). *Constructing written test questions for the basic and clinical sciences* (3rd ed.). Philadelphia: National Board of Medical Examiners.

Chinnappan, M. (1998). Schemas and mental models in geometry problem solving. *Educational Studies in Mathematics*, *36*(3), 201–217. doi:10.1023/A:1003134323371

Chuesathuchon, C. (2008). *Computerized adaptive testing in mathematics for primary schools in Thailand*. Retrieved from http://ro.ecu.edu.au/theses/1591

Chunyu, L. I. U., Meng, P., Zhang, Z., & Pan, Y. (2010). Research on Computerized Adaptive Testing System Based on IRT. *2010 Seventh International Conference on Fuzzy Systems and Knowledge Discovery (FSKD 2010)*. Retrieved from https://ieeexplore.ieee.org/stamp/stamp.jsp?tp=&arnumber=5569425

Davey, T. (2011). *A guide to computer adaptive testing systems*. Educational Testing Service. Retrieved from https://files.eric.ed.gov/fulltext/ED543317.pdf

Gewertz, C. (2011). *Computer-adaptive testing poses challenges, expert warns*. Retrieved from http://blogs.edweek.org/edweek/curriculum/2011/05/computer-adaptive_testing_pose.html?qs=Reckase

Higher Education Academy. (2012). *A Marked improvement: Transforming assessment in higher education*. Retrieved from https://www.cardiff.ac.uk/__data/assets/pdf_file/0007/583990/A-Marked-Improvement-transforming-assessment-in-higher-education-HEA.pdf

Kimura, T. (2017). The impacts of computer adaptive testing from a variety of perspectives. *Journal of Educational Evaluation for Health Professions*, *14*(12). doi:10.3352/jeehp.2017.14.12 PMID:28552899

Kuh, G. D., & Schneider, C. G. (2008). *High impact educational practices: What they are, who has access to them, and why they matter*. Washington, DC: Association of American Colleges and Universities.

Lawson, M. J., & Chinnappan, M. (2015). Knowledge connectedness and the quality of student and teacher mathematical knowledge. In E. A. Silver & P. A. Kenney (Eds.), *More lessons learned from research* (pp. 272–277). Reston, VA: National Council of Teachers of Mathematics.

Lesage, E., Valcke, M., & Sabbe, E. (2013). Scoring Methods for Multiple Choice Assessment in Higher Education – Is it still a Matter of Number Right Scoring or Negative Marking? *Studies in Educational Evaluation*, *39*(3), 188–193. doi:10.1016/j.stueduc.2013.07.001

Maki, P. L. (2009). Moving beyond a national habit in the call for accountability. *Peer Review: Emerging Trends and Key Debates in Undergraduate Education, 11*(1), 13–14.

Martin, A. J., & Lazendic, G. (2017). Computer-adaptive testing: Implications for students' achievement, motivation, engagement, and subjective test experience. *Journal of Educational Psychology*. doi:10.1037/edu0000205

National Association of Colleges and Employers. (2016*). Job Outlook 2016: Attributes Employers Want to See on New College Graduates' Resumes.* Retrieved from http://www.naceweb.org/s11182015/employers-look-for-in-new-hires.aspx

Ory, C. J. (2009). *Improving Your Test Questions, Evaluation and Examination Service*. University of Iowa.

Packer, M. J., & Goicoechea, J. (2000). Sociocultural and constructivist theories of learning: Ontology not just epistemology. *Educational Psychologist, 35*(4), 227–242. doi:10.1207/S15326985EP3504_02

Palmer, E. J., & Devitt, P. G. (2007). Assessment of higher order cognitive skills in undergraduate education: Modified essay or multiple choice questions? Research paper. *BMC Medical Education, 7*(49), 1–7. doi:10.1186/1472-6920-7-49 PMID:18045500

Plajner & Vomlel. (2016). Student Skill Models in Adaptive Testing. *JMLR: Workshop and Conference Proceedings, 52*, 403-414. Retrieved from http://proceedings.mlr.press/v52/plajner16.pdf

Schneider, C. G. (2008). *Liberal education and high-impact practices: Making excellence—for once and for all—inclusive. In Forward to High impact educational practices: What they are, who has access to them, and why they matter*. Washington, DC: Association of American Colleges and Universities.

Tajudin, N. M., & Chinnappan, M. (2016). The link between higher-order thinking skills, representation and concepts in enhancing TIMSS tasks. *International Journal of Instruction, 9*(2), 199–214. doi:10.12973/iji.2016.9214a

Thompson, N. A., & Prometric, T. A. (2007). Practitioner's guide for variable-length computerized classification testing. *Practical Assessment, Research & Evaluation, 12*, 1–13.

Thompson, N. A., & Weiss, D. J. (2011). A framework for the development of computerized adaptive tests. *Practical Assessment, Research & Evaluation, 16*(1), 1–9. Retrieved from http://pareonline.net/getvn.asp?v=16&n=1

Torres, C., Lopes, A. P., Babo, L., & Azevedo, A. M. (2009). Developing multiple-choice questions in mathematics. *Proceedings of ICERI 2009 - International Conference of Education, Research and Innovation*, 6218-6229.

Towns, M. H. (2014). Guide to developing high-quality, reliable, and valid multiple-choice assessments. *Journal of Chemical Education, 91*(9), 1426–1431. doi:10.1021/ed500076x

Wang, H., & Shin, C. D. (2010). Comparability of Computerized Adaptive and Paper-Pencil Tests. *Test, Measurement & Research Services Bulletin*, 13. Retrieved from https://pdfs.semanticscholar.org/6f1e/bcee90a66fcc969c4caba1362bedb39d505a.pdf

Way, W. D., Twing, J. S., Camara, W., Sweeney, K., Lazer, S., & Maeo, J. (2010). *Some considerations related to the use of adaptive testing for the common core assessments*. Educational Testing Service.

Weiss, D. J., & Kingsbury, G. (1984). Application of computerized adaptive testing to educational problems. *Journal of Educational Measurement*, *21*(4), 361–375. doi:10.1111/j.1745-3984.1984.tb01040.x

Zimmaro, D. M. (2016). *Writing good multiple-choice exams*. Retrieved from https://facultyinnovate.utexas.edu/sites/default/files/writing-good-multiple-choice-exams-fic-120116.pdf

Chapter 13
Alternatives to the Traditional Doctoral Dissertation:
A Research Literature and Policy Review

Gary Berg
University of Phoenix, USA

ABSTRACT

In response to increased criticism of the utility of the traditional doctoral dissertation, some institutions have incorporated additional options for students such as articles, portfolios, and industry-specific projects. This trend towards allowing alternatives for doctoral students coincides with the rise internationally of what are variously called professional, applied, or practitioner doctorates in various disciplines. The goal of this review of the relevant research literature and policy documents is to understand the evolution of professional doctoral degrees, and how rigor and quality are evaluated specifically with non-dissertation capstone projects. Findings suggest that universities, accrediting agencies, and professional associations recognize a need for new standards for alternatives to the traditional dissertation in order to make such options rigorous and relevant for the students.

INTRODUCTION

Some scholars and practitioners have criticized the dissertation-only format in doctoral degree programs as not meeting the needs of neither students nor employers. Alternatives used by some institutions include a structured or negotiated equivalent of linked research papers, teaching portfolios, joint research projects, service, public lectures, translation of academic texts, and some sort of original digital product. A number of institutions have also incorporated practitioner-oriented industry-specific capstones, such as the execution of a business plan, as alternatives to the traditional dissertation. For institutions intending to offer such substitutions there is a strong imperative to understand how to best structure and evaluate doctoral capstone alternatives. The goal of this review of research literature and policy documents is to understand the evolution of professional research doctoral degrees, and how rigor and quality are assured with non-dissertation capstone projects. There are two main groups of primary documentation

DOI: 10.4018/978-1-5225-7832-1.ch013

Copyright © 2019, IGI Global. Copying or distributing in print or electronic forms without written permission of IGI Global is prohibited.

and research literature relevant to this project: research on the rise of professional doctoral degrees and dissertation alternatives internationally, and quality assurance and accreditation considerations.

This chapter examines changes in doctoral instruction and is especially germane to the overall book topic on redesigning higher education. The need for academia to better understand the impact of rapidly changing technologies and industries is clear, especially at the graduate level. This chapter considers the readiness and challenges that trends represent for higher education, and provides specific examples of alternatives emerging in countries around the world.

HISTORY AND EVOLUTION OF DOCTORAL EDUCATION

Unlike in Europe, early American higher education focused on undergraduate education (Archbald, 2011), and did not focus on graduate degrees with seriousness until the late 19th century. The first doctorate was awarded in the United States in 1861 at Yale, which conferred three doctorates in that year (Storr, 1953). In 1876, influenced heavily by the German research universities, Johns Hopkins University was established and for the next 50 years changed higher education by influencing the formation of American graduate schools. Academic authority was placed with faculty and an Academic Senate, and administrative control distributed in academic departments (Horton, 1940). Johns Hopkins' first graduate program by design required a thesis: to "show the candidate's mastery of his subject, his powers of independent thought as well as careful research, and his ability to express, in clear and systematic order, and in the appropriate language, the results of his study." (Johns Hopkins University, 1878, p. 27.)

By 1900, doctoral programs in America had developed in both private and public institutions including Hopkins, Chicago, Harvard, Columbia, Yale, Cornell, California, Michigan, and Wisconsin. The main structural features of the doctoral program were established requiring full-time matriculation, two to three years of coursework, and several years writing a dissertation. Typically, doctoral funding subsidized students in apprentice-like research or teaching assistantships to cover costs of living. The traditional doctorate's function was thought of as preparation for a research career (Archbald, 2011). This traditional structure remains little changed since its inception. The growth in the number of doctoral degrees granted in the United States alone reached 40,000 annually at beginning of the 21st century (Golde & Walker, 2006).

Walker et al. (2008) identified five major stages in the development of American doctoral education. The first stage beginning in the late 19th century involved scholars traveling to Germany, and bringing the traditions back to American when founding American doctoral programs with a particular emphasis on combining research and teaching (Kehm & Teichler, 2016; Osakwe et al., 2015; Walker et al., 2008). In stage two after World War II, graduate education expanded greatly in America with reportedly new doctoral schools opening at the rate of one per week, and the number of doctoral students exploded funded largely by government support for research within universities. The third stage was marked by federal financial support receding, and the doctoral student population declined pressuring changes including expanding doctoral education to non-traditional populations including older students, lower income students, and many programs became more applied, focused on solving local and national problems. Stage four was characterized by increased diversification of the student population starting in the 1980s. While the numbers of degree programs once again started increasing, the expected retirement of older faculty failed to materialize, which resulted in a tight job market for doctoral program graduates. As a result, rather than entering into academia, doctoral grads in increasing numbers entered into industry and the labor force. The fifth stage of development started in the 1990s with increasing global competition and scholars studying the changes in direction (Gokhberg, Shmatko, & Auriol, 2016; Lee & Danby, 2012; Nerad & Evans, 2014).

As more doctorates enter into industry instead of academia, nation states recognize the impact of doctorates on Gross National Product and economic stability (Nerad & Evans, 2014). Additionally, doctorates have become increasingly mobile, moving within countries and between countries, with more international policies (such as the European Bologna Process) impacting a global understanding of higher education (Kehm, 2010; Kehm, & Teichler, 2016; Lee & Danby, 2012; Nerad & Evans, 2014). Finally, doctoral education has become increasingly tied to industry, especially considering most doctorates are how hired into industry instead of academia post-graduation (Gokhberg, Shmatko, & Auriol, 2016).

Some scholars have focused on challenging Boyer's notion of a domain of discovery (Boyer, 1996) in traditional dissertations within the sciences that have focused exclusively on the search for new knowledge and the dissemination through publication and presentation (Osakwe, Keavey, Uzoka, Fedoruk, & Osuji, 2015). These scholars challenge the solitary focus of research on discovery and encourage recognition of research that is applied as well.

Kehm and Teichler (2016) note nine general categories of doctorates: the research doctorate, the taught doctorate, Ph.D. by published work, the professional doctorate, the practice-based doctorate, the 'new' doctorate, two models of joint doctorate, the cooperative doctorate, and the industrial doctorate. The challenge arises when the doctorates are held in the same research-focus as the Ph.D. which uniquely emphasizes research and research methodology, which don't align well with the expected program outcomes within the practitioner doctoral programs.

According to Rominger (2018), practice-based output includes dissertation and capstone projects which go beyond traditional research including action research projects within specific organizations, creation and execution of business plans, and professional practice projects in medical professions. Rominger, consistent the Council of Graduate Schools report (2016), argues that rather than creating a book-length written document, students might be encouraged to work on a more profession-appropriate output such as publication of multiple peer-reviewed articles, engagement in collaborative research, creation of pertinent business documents (organizational analysis, business plans, etc.), or creation of community-based programs. Specific disciplines such as architecture have struggled with these alternatives to the dissertation (Rice & Matthews, 2012). Finally, Rominger (2018) notes that graduate students must be able to communicate professionally within their professional domain and it is important to incorporate a capstone project which further develops these skills.

Pressure to Change the Doctoral Degree

Many scholars and organizations commented with increasing fervor throughout the 20th century on problems in the doctoral program, and the dissertation requirement (Cude, 2001). *The Responsive Ph.D.: Innovations in U.S. Doctoral Education from The Woodrow Wilson National Fellowship Foundation* (2005) was especially influential. Another of the key studies published at the turn of the 21st century by the The Pew Charitable Trusts (Golde & Dore, 2001) found: "The training doctoral students receive is not what they want, nor does it prepare them for the jobs they take," and "Many students do not clearly understand what doctoral study entails, how the process works and how to navigate it effectively." More broadly, in a large international survey of students, Noble (1994) found that the most frequent issue identified was a problem in the definition of what constitutes appropriate doctoral research.

A study by the Carnegie Foundation (Golde & Walker, 2006) asked leading scholars in specific disciplines what constitutes knowledge and stewardship in the discipline, and how Ph.D.'s should be educated and prepared. Responding scholars noted that doctoral training often does not align with career practice, nor provide a public benefit. The general problem identified was that doctoral education doesn't link to the post-PhD reality for students as well as it should. In addition, the overall context and structure of higher education prevents change. The report recommended that the key mentoring relationship between faculty and doctoral students be examined in terms of professional relationships and funding. In conclusion the authors of this report argued that the purpose of doctoral education is first research and stewardship of the discipline. Voicing a traditionalist point of view, the respondents surveyed believed that doctoral graduates must demonstrate the ability to conduct research and make a unique contribution to the field.

A meta-analysis of the research literature on evolving practices and pressures on the doctoral degree identified three main categories including "Connection with Non-Doctoral Education Groups;" "Revisioning the Doctoral Education Process;" and "Aligning Dissertation/Capstone Projects with Type of Doctoral Program" (Rominger, 2018). The analysis found the doctoral programs increasingly influenced by specific industries as the main employer of graduations.

Across disciplines there have been active discussions about needed changes. For example, Smith (2016) in *Manifesto for the Humanities: Transforming Doctoral Education in Good Enough Times* asserted a five-point argument for making changes in scholarship practices, teaching practice to inform research, resisting overspecialization, and recognizing that not all students follow traditional tenure-track faculty pursuits. One of the most significantly relevant documents urging changes in doctoral programs came from the Modern Language Association (MLA) (Force, 2014). The MLA Task Force on Doctoral Study in Modern Language and Literature report focused on concerns about the future of humanistic study. In the light of persistent criticism from within the academy and from a larger public, the authors argued that the legitimacy of doctoral study needs reformulation to meet key challenges including a median time to degree of around nine years for language and literature doctoral recipients, and a long-term academic job market that provides tenure-track employment for only around sixty percent of doctorate recipients. Specific recommendations from the MLA Task Force included reconsidering: "the character, purpose, and timing of the prospectus for the dissertation project". Examples of an expanded repertoire mentioned are a linked group of essays on a common theme, web-based projects, translations, public projects that include collaboration with people in other cultural institutions, and the treatment of texts in terms of their pedagogical value.

Shulman, Golde, Bueschel, and Garabedian (2006) pointed out that the problems of the Education doctorates specifically are chronic and crippling. The purposes of preparing scholars and practitioners are confused, and as a result, neither is done well. They proposed two distinct models: the research doctorate (the Ph.D.), with its strong links to practice, and development of a robust and distinct practice doctorate (the P.P.D.) with a distinctive scholarly base. Unlike most current education Ph.D.s and Ed.D.s, the two degrees would serve different purposes, and like their medical analogs—the biomedical Ph.D. and the M.D.—would have distinct curricula and assessments.

Some scholars have focused on challenges in the traditional doctoral program especially for traditional underserved groups. For instance, Dinkins and Sorrell (2014) conveyed stories of women's dissertation journeys and note specific barriers in the process such as the common practice of faculty advisers giving students copies of approved dissertations, which have the effect of both intimidating students and limiting their original thinking.

The Rise of the Professional Doctoral Degree and Policy Efforts

Scholars (Storey, 2016) have observed the international trend to create what are variously called professional doctorates, applied doctorates, practitioner doctorates, or clinical doctorates in various disciplines. Typical degrees of this kind include the doctor of education (EdD), doctor of psychology (PsyD), doctor of engineering (EngD), doctor of music art (DMA), doctor of dental surgery (DDS), doctor of juridical science (SJD), and doctor of public health (DPH). Some researchers have sought to catalog and categorize professional doctoral degrees. For instance, Bourner, Bowqden and Laing (1999) developed a profile of research degree awards using data from individual universities, exploring standardization, coherence and clarity. Alternately, Pappas and Jerman (2011) defined the nontraditional doctorate as characterized by non-traditional student populations, often part-time attendance, delivered in compressed or alternative delivery formats, an overall practitioner curricular emphasis, and a lessened focus on subsequent career paths as researchers. Common issues that arise internationally in this discussion include the specific degree purpose, admission requirements, program outcomes and learning objectives, career directions, quality assurance, capstones, and examinations. Professional doctorates were developed by some institutions in the face of rising criticism about the relevance of doctoral research to practice and the changing context and content of knowledge.

Professional doctorates, particularly in Australia and the United Kingdom, grew significantly in number over the last 30 years. Park (2005) noted that key drivers for change in the United Kingdom included a new emphasis on skills and training, quality of supervision, changes in the examination of the thesis, and the introduction of national benchmarking. Banerjee and Morley (2013) discussed the emergence of professional doctorates in management education and their contribution to a more practice-based approach to doctoral education in the doctor of business administration degree (DBA). Neumann (2005) analyzed professional doctorates in Australia in education, management, law and the creative arts, remarking on similarities and differences found between PhD and professional doctorate programs, which typically provide additional insight into practice. Three specific areas were analyzed by Neumann. The first was the recruitment and selection of students, student choice of professional doctorates and perceived career benefits. The second area was the structure and organization of PhD and professional doctorate programs, including the identification of the research topic. The third area was the perceived status of professional doctorates vis-à-vis the PhD.

One of the central documents in the research literature on professional doctoral programs is a Task Force Report created over a decade ago by the Council of Graduate Schools (2007). This comprehensive report recognized the positive trend to make doctoral programs relevant to societal needs and prepare a workforce, yet also noted fears that financial pressures were leading to degree inflation, and a lack of sufficient rigor. The report identified three core characteristics of the professional doctoral degree: a focus on area of professional practice not met by other degrees, an emphasis on applied research, and including leaders of the profession who drive development of practices and standards. The authors claimed that primary distinctions from the Ph.D. are the length of study, involvement in practical/clinical experiences, and the provision of alternatives to the dissertation for students. The report concluded that graduate colleges and deans should play leading roles in establishing standards, and that the institutional review processes should emphasize the role of professional doctorates in relationship to the mission of the specific institution. Finally, the report noted the need for explicit criteria for accreditation of professional doctoral degrees and national standards.

Some leading organizations have developed a specific focus on addressing needed changes in doctoral programs, beginning in specific disciplines such as Education and the Health Sciences. One such organization is the Carnegie Project on the Education Doctorate (CPED) (https://cpedinitiative.site-ym.com/), a consortium committed to undertaking the effort to redesign the doctorate in Education. Key understandings of the group included that the development of a professional practice preparation program in Education needs to be both purposeful as well as fluid, and that no one-size-fits-all model of preparation will meet the diverse needs throughout the country. Furthermore, the measure of quality and success in professional practice preparation should be the subsequent graduate's impact on educational practice. To achieve these goals, the Consortium produced a definition of the EdD as a set of program principles and design concepts that will serve as the scaffolding for any EdD program design. The dissertation in practice is a scholarly endeavor that impacts a complex problem of practice.

Scholars have considered discipline-specific issues in connection with the professional doctoral degrees. For instance, Margolin (2010) examined education in Design at the doctoral level, finding that at issue are questions related to doctoral education and its usefulness in the context of design. One conclusion to draw from this analysis is that doctorates in Design need to have focus. Other researchers have looked at issues linked to serving specific student populations, as in Ravitch and Lytle (2016) who presented a case study arguing for a re-envisioned dissertation that offers opportunity for mid-career leaders to engage in research applied to real settings.

Alternatives to the Traditional Dissertation

A central part of the argument for changes in the doctoral degree often has to do with criticisms of the traditional dissertation as a capstone requirement. Watson and Nehls (2016) observed that very few doctoral graduates publish whole or even part of their dissertations. The authors argue that the dissertation project rarely leads to dissemination of work, and the project fails to prepare candidates for the type of writing and research that they will be engaged in subsequent careers. To help confront this reality, options recommended by these authors include the multiple paper format, with variations including two or three manuscripts supporting linked theme, and a portfolio option for professional degrees. Similarly, Duke and Beck (1999) criticize the traditional dissertation for its limited audience and dissemination, and lack of generalizability, and instead argue for a multiple paper, "ready for publication" alternative. However, they warn against requiring publication because of difficulty of such a goal, even for worthy papers.

Sanders (1997) conducted a large study to determine if there were viable alternative research strategies, other than the traditional doctoral dissertation, which would be more beneficial to future doctoral graduates in order to compete in their future professional roles in an information-intensive society. A synthesis of data resulted in a list of viable alternatives to the traditional doctoral dissertation, a list of future research needs, and recommendations from those experts whose institutions have successfully implemented alternatives to the traditional doctoral dissertation.

Drake and Heath (2011) described professional doctoral program studies as "insider research" characterized as inquiry in which the scholar has some special experience into the worlds being investigated, and familiarity with the community studied. The new knowledge created in this kind of doctoral program comes from combining understandings from professional practice with the individual study. Professional doctoral degrees undertake research with the idea of an outcome that will positively impact professional practice. Maxwell and Kupczyk-Romanczuk (2009) likewise advocated for applying research specifically to professional practice.

Much of the research and practice regarding alternatives to the traditional dissertation focuses on discipline specific solutions. For instance, Smith (2016) made a five-point argument for expanding models for the humanities dissertation recognizing that not all students pursue traditional tenure track-faculty positions. The author suggested alternatives such as a suite of three or four essays, or a portfolio involving joint research, teaching, public lectures, a translation, or a digital product.

One large study (Auerback, 2011) in the California State University system concentrated on EdD programs found that some programs stressed scholarly rigor similar to a Ph.D., while others advocated a more relevant, practitioner-oriented approach. Overall, there was a sense of faculty still feeling its way regarding dissertation expectations. A combination of legislation, accreditation requirements, and faculty's own doctoral experience tend to inhibit the development of applied approaches to research requirements. Similarly, Archbald (2010) described a case study of an alternative model in an educational leadership program and found a central challenge was to develop an option that involved authentic problem solving. The author noted that faculty in a program had firmly-held assumptions and expectations which were a challenge to providing this new alternative.

Discussion: Quality Assurance and Accreditation

In terms of quality assurance internationally, Byrne, Jørgensen and Loukkola (2013) studied the ARDE project considering how quality assurance for doctoral education has been implemented in European universities. As the Bologna Process has developed, universities have put great effort into professionalizing their quality assurance and found that doctoral education was being managed more professionally with institutions giving more attention to accountability and quality enhancement.

Internationally, professional doctorates tend to have very different learning outcomes than research doctorates, where there are more part-time doctoral students, and a greater supply of graduates from such programs making the landscape more competitive (Blessinger & Stockley, 2016). Scholars noted challenges include time to degree, program rigor and relevance, cost, and employment prospects. Dawson and Kumar (2016) argued for three quality principles in dissertation alternatives: that they address critical problems of practice, demonstrate rigorous research skills, and show proof of the impact of the research. Although there is limited research literature on professional doctorates in Asia, Sarros, Willis and Hardie (2004) examined the Doctor of Business Administration offered in Hong Kong, Singapore, China, and Kuala Lumpur with a focus on sustainability.

American regional accrediting agencies generally have stayed away from direct guidance on professional doctoral degrees. Although there are differences among the six regional agencies who evaluate doctoral degree granting institutions, the typical focus during the review process is on the institutional level, avoiding specific proscriptions in terms of pedagogical issues. However, The Higher Learning Commission North Central Association of Colleges and Schools in 2006 formed a task force on professional doctoral degrees and published an important report (The Higher Learning Commission North Central Association of Colleges and Schools, 2006). In the report, the Commission noted that it tends to rely on its familiarity with institutional contexts in which research doctorates are delivered, and that this approach does not work very well for professional doctoral degree programs. The recommendations of the Task Force were organized into three overarching issues: Institutional context, program content, and shared quality assurance. Furthermore, the authors of the report pointed out that in 2006 professional associations were thus far not providing consistent standards for quality assurance of professional degrees that accrediting agencies could turn to for reference.

In terms of specific policy at the American accrediting agencies, the New England Association of Schools and Colleges, Commission on Institutions of Higher Education in Standard 4.26 addresses professional doctoral degrees as a category noting that "Professional, performance, or practice-oriented programs at the doctoral or master's degree levels are designed to prepare students for professional careers involving the application or transmission of existing knowledge or the development of new applications of knowledge within their field" (https://cihe.neasc.org/standards-policies/standards-accreditation/standards-effective-july-1-2016#standard_four). The other five regional accrediting agencies do not specifically reference professional doctoral degrees at this time.

CONCLUSION

This review of relevant research literature and policy documents reveals pressures on higher education institutions to change and differentiate doctoral degree programs. While there is a long history of criticism about traditional doctoral degree requirements as not meeting the needs of students and the field, there remain concerns about maintaining academic rigor. Accrediting agencies for the most part appear cautious in changing on the policy level how alternatives to the traditional doctoral degree and alternatives to the dissertation might be evaluated in the institutional review process. Yet there is a need for explicit criteria for standards for professional doctoral degrees and dissertation alternatives.

It is clear that there is wisdom in the Council of Graduate School's recommendation that leadership on developing standards for alternatives to the dissertation need to come from specific institutional academic colleges and deans. Furthermore, as the Higher Learning Commission suggests, the role of professional associations could be important by providing consistent standards for quality assurance of professional degrees, and alternatives to the dissertation.

Further research on this emerging practice is especially needed in terms of outcomes and effectiveness. Are alternatives to the dissertation in fact more effective in meeting desired degree outcomes, and the subsequent career objectives of graduates? Some options offered, such as a series of published papers, may inadvertently add unnecessary complexity and time to degree for the students. While there is certainly a history of success in the performance of professional doctoral degree programs, the benefits of alternatives to the dissertation in the traditional research doctorate have not yet been well documented.

REFERENCES

Archbald, D. (2010). Breaking the mold in the dissertation. *Journal of Continuing Higher Education*, *58*(2), 99-107.

Archbald, D. (2011). The emergence of the nontraditional doctorate: A historical overview. *New Directions for Adult and Continuing Education*, *2011*(129), 7–19. doi:10.1002/ace.396

Auerback, S. (2011). It's not just going to collect dust on a shelf: Faculty perceptions of the applied dissertation in the new CSU. *Journal of Research on Leadership Education*, *6*(3), 59-82.

Banerjee, S., & Morley, C. (2013). Professional Doctorates in Management: Toward a Practice-Based Approach to Doctoral Education. *Academy of Management Learning & Education, 12*(2), 173–193. doi:10.5465/amle.2012.0159

Blessinger, P., & Stockley, D. (Eds.). (2016). *Emerging directions in doctoral education.* Bingley, UK: Emerald. doi:10.1108/S2055-364120166

Bourner, T., Bowqden, R., & Laing, S. (1999). A National Profile of Research Degree Awards: Innovation, Clarity and Coherence. *Higher Education Quarterly, 53*(3), 264–280. doi:10.1111/1468-2273.00130

Boyer, E. L. (1996). From scholarship reconsidered to scholarship assessed. *Quest, 48*(2), 129–139. doi:10.1080/00336297.1996.10484184

Byrne, J., Jørgensen, T., & Loukkola, T. (2013). *Quality assurance in doctoral education: Results of the ARDE Project.* EUA Publications. Retrieved from https://brjk.amu.edu.pl/__data/assets/pdf_file/0018/224262/EUA_ARDE_Publication.pdf

Council of Graduate Schools. (2007). *CGS Task Force Report on the Professional Doctorate.* Washington, DC: Council of Graduate Schools.

Cude, W. (2001). *The Ph.D. trap revisited.* Toronto, Canada: The Dundurn Group.

Dawson, K., & Kumar, S. (2016). Guiding principles for quality professional practice dissertations. In *Contemporary approaches to dissertation development and research methods.* Hershey, PA: IGI Global.

Dinkins, C. S., & Sorrell, J. M. (2014). *Our dissertations, ourselves: Shared stories of women's dissertation journeys.* New York, NY: Palgrave Macmillan.

Drake, P., & Heath, L. (2011). *Practitioner research at doctoral level.* London, UK: Routledge.

Duke, N. K., & Beck, S. W. (1999). Education should consider alternative formats for the dissertation. *Educational Researcher, 28*(3), 31–36.

Force, M. T. (2014). *Report of the MLA Task Force on Doctoral Study in Modern Language and Literature.* Modern Language Association.

Gokhberg, L., Shmatko, N., & Auriol, L. (Eds.). (2016). *The science and technology labor force: The value of doctorate holders and development of professional careers.* Springer International. doi:10.1007/978-3-319-27210-8

Golde, C. M., & Dore, T. M. (2001). *At Cross Purposes: What the experiences of doctoral students reveal about doctoral education.* Philadelphia, PA: A Report Prepared for The Pew Charitable Trusts.

Golde, C. M., & Walker, G. E. (2006). *Envision the future of doctoral education.* San Francisco, CA: Jossey-Bass.

Horton, B. J. (1940). *The graduate school.* New York, NY: New York University.

Johns Hopkins University. (1878). *Third Annual Report.* Author.

Kehm, B. M. (2010). Quality in higher European education: The influence of the Bologna Process. *Change*, *42*(May/June), 40–46. doi:10.1080/00091381003704677

Kehm, B. M., & Teichler, U. (2016). Doctoral education and labor market: Policy questions and data needs. In L. Gokhberg, N. Shmatko, & K. Auriol (Eds.), *The Science and Technology Labor Force* (pp. 11–29). Springer International; doi:10.1007/978-3-319-27210-8_2

Lee & Danby. (Eds.). (2012). Reshaping doctoral education: International approaches and pedagogies. New York, NY: Routledge.

Margolin, V. (2010). Doctoral Education in Design: Problems and Prospects. *Design Issues*, *26*(3), 70–78. doi:10.1162/DESI_a_00031

Maxwell, T. W., & Kupczyk-Romanczuk, G. (2009). Producing the professional doctorate: the portfolio as a legitimate alternative to the dissertation. *Innovations in Education and Teaching International*, *46*(2), 135-145.

Nerad, M., & Evans, B. (Eds.). (2014). *Globalization and its impacts on the quality of PhD education: Forces and forms in doctoral education worldwide*. Boston, MA: Sense Publishers/Center for Innovation and Research in Graduate Education. doi:10.1007/978-94-6209-569-4

Neumann, R. (2005). Doctoral Differences: Professional doctorates and PhDs compared. *Journal of Higher Education Policy and Management*, *27*(2), 173–188. doi:10.1080/13600800500120027

New England Association of Schools and Colleges, Commission on Institutions of Higher Education. (2018). *Standard Four: The Academic Program*. Retrieved September 28, 2018, from: https://cihe.neasc.org/standards-policies/standards-accreditation/standards-effective-july-1-2016#standard_four

Noble, K. A. (1994). *Changing doctoral degrees: An international perspective*. Buckingham, UK: Open University Press.

Osakwe, C., Keavey, K., Uzoka, F. M., Fedoruk, A., & Osuji, J. (2015). The relative importance of academic activities: Autonomous values from the Canadian professoriate. *Canadian Journal of Higher Education*, *45*(2), 1–22.

Pappas, J. P., & Jerman, J. (Eds.). (2011). *Meeting adult learner needs through the nontraditional doctoral degree*. San Francisco, CA: Jossey-Bass.

Park, C. (2005). New Variant PhD: The changing nature of the doctorate in the UK. *Journal of Higher Education Policy and Management*, *27*(2), 189–207. doi:10.1080/13600800500120068

Ravitch, S. M., & Lytle, S. (2016). Becoming practitioner-scholars: The role of practice-based inquiry dissertations in the development of educational leaders. In *Contemporary approaches to dissertation development and research methods*. Hershey, PA: IGI Global.

Rice, C., & Matthews, L. (2012). Designing (in) the PhD in architecture. In A. Lee & S. Danby (Eds.), *Reshaping doctoral education: International approaches and pedagogies* (pp. 99–112). New York, NY: Routledge.

Rominger, R. (2018). Building bridges and growing wings through doctoral education: A literary meta-analysis. *International Journal of Higher Education*, *7*(3), 171–182. doi:10.5430/ijhe.v7n3p171

Sanders, K. A. (1997). *Identifying research strategies for the future: Alternatives to the traditional doctoral dissertation* (Order No. 3006632). Available from ProQuest Dissertations & Theses Global. (304390109). Retrieved from https://search-proquest-com.contentproxy.phoenix.edu/docview/304390109?accountid=35812

Sarros, J. C., Willis, R. J., & Hardie, T. (2004). The DBA in Australia and the Asia Pacific: Opportunities and challenges. *International Journal of Organisational Behaviour*, *7*(8), 440–455.

Shulman, L. S., Golde, C. M., Conklin, A., Bueschel, A. C., & Garabedian, K. J. (2006). Reclaiming education's doctorates: A critique and a proposal. *Educational Researcher*, *35*(3), 25–32. doi:10.3102/0013189X035003025

Smith, S. (2016). *Manifesto for the humanities: Transforming doctoral education in good enough times*. Ann Arbor, MI: University of Michigan Press.

Storey, V. A. (Ed.). (2016). *International perspectives on designing professional practice doctorates*. New York, NY: Palgrave Macmillan. doi:10.1057/9781137527066

Storr, R. J. (1953). *The beginnings of graduate education in America*. Chicago, IL: The University of Chicago Press.

The Carnegie Project on the Education Doctorate (CPED). (n.d.). Retrieved from https://cpedinitiative.site-ym.com/

The Higher Learning Commission North Central Association of Colleges and Schools. (2017). *A Report to the Board of Trustees from the Task Force on the Professional Doctorate. Received by the Board of Trustees of The Higher Learning Commission, June 6, 2006*. The Higher Learning Commission North Central Association of Colleges and Schools. Retrieved December 18, 2017, from: http://www.hlcommission.org/Publications/publications-list.html

The Woodrow Wilson National Fellowship Foundation. (2005). *The responsive Ph.D.: Innovations in U.S. doctoral education*. Princeton, NJ: Author.

Walker, G. E., Golde, C. M., Jones, L., Conklin Bueschel, A., & Hutchings, P. (2008). *The formation of scholars: Rethinking doctoral education for the twenty-first century*. San Francisco, CA: The Carnegie Foundation for the Advancement of Teaching.

Watson, D. L., & Nehls, K. (2016). Alternative dissertation formats: Preparing scholars for the academy and beyond. In *Contemporary approaches to dissertation development and research methods*. Hershey, PA: IGI Global.

Chapter 14
Redesigning Prediction Algorithms for At-Risk Students in Higher Education:
The Opportunities and Challenges of Using Classification Techniques in a University Academic Writing

Dennis Foung
The Hong Kong Polytechnic University, Hong Kong

ABSTRACT

Use of algorithms and data mining approaches are not new to Industry 4.0. However, these may not be common for students and educators in higher education. This chapter compares various classification techniques: classification tree, logistic regression, and artificial neural networks (ANN). The comparison focuses on each method's accuracy, algorithm, and practicality in higher education. This study made use of a dataset from two academic writing courses in a university in Hong Kong with more than 5,000 records. Results suggest that classification trees and logistic regression can be easily used in the higher education context, but ANN may not be applicable in higher educational settings. The research team suggests that higher education administrators take this research forward and design platforms to realize these classification algorithms to predict at-risk students.

INTRODUCTION

Industry 4.0 is a term that "designates the increasing interlinking of production and information as well as communications technology" (Graube & Mammes, 2017, p. 844). With the emergence of Industry 4.0, there is a pressing need for higher education faculty members to develop algorithms that fit into changing need of intelligent systems.

DOI: 10.4018/978-1-5225-7832-1.ch014

Copyright © 2019, IGI Global. Copying or distributing in print or electronic forms without written permission of IGI Global is prohibited.

At the university level, faculty members in higher education have started using new technologies that have emerged in the context of Industry 4.0. Some have started with a simple motion-detected lighting system in their classrooms, or a timed air conditioning system in their office. Others have started with more adaptive learning management systems that can cater to the needs of students at different knowledge levels (Ciolacu & Beer, 2016). These systems are only starting points, and most faculty members are aware of the greater scope of emerging Industry 4.0 technologies for use in higher education.

There are course-level opportunities for faculty members to make use of Industry 4.0 technologies; establishing algorithms to identify at-risk students is one of them. In the previous generation, the use of mass lecture was prevalent in higher education and this happens in all classrooms, including a language classroom (Abdullah, Ramli, & Rafek, 2017). Due to the high number of students in mass lectures, it is hard for lecturers to give sufficient feedback and attention to individual students (Chingos & Whitehurst, 2011). It is possible that students enjoy having a good chat with their lecturers and that they may be facing problems in their studies. Without the support of computers or algorithms, these students may end up failing the course, with lecturers only realizing problems at the end of the term. This gives rise to an emerging line of research, "actionable intelligence" which includes identifying students who are struggling in a course. It also includes predicting students' academic success (Park, Yu, & Jo, 2016).

A great range of studies have outlined the benefits of using learning analytics to identify at-risk students. For example, Klüsener and Fortenbacher (2015) and Bainbridge et al. (2018) highlighted the need for learning analytics to predict at-risk students in an online environment. Lawson, Beer, Rossi, Moore, and Fleming (2016) reported on a university-wide early prediction system. Essa and Ayad (2012) built a similar system and concluded that the early warning system is a bridge between algorithms and support interventions. Several researchers believe that using algorithms to predict students' performance will be common and useful in the future, when there are more online (or blended learning) courses and when data are available for exploration. It is therefore time for faculty members in Industry 4.0 to take this one step forward and identify and support students who are at-risk.

With emergence of the trend for predicting which students may be at risk, more applications are available (Clow, 2013; Wolff, Zdrahal, Nikolov, & Pantucek, 2013), but not all are useful. One of the leading learning management providers, Blackboard, offers a tool called Retention Center that derives algorithms and rules to help identify at-risk students. A student is marked at-risk if he or she misses a deadline or has scores that are lower than average (Blackboard, 2018). The algorithms, which Blackboard calls rules, seem to overlook the differences among students across cohorts and place unnecessary emphasis on certain incidental factors (e.g. submitting the assignment late by a few hours). These flaws may affect accuracy in predicting at-risk students.

In the era of Industry 4.0, huge datasets are readily available from learning systems used in higher education institutions. According to Pinkham (2017), Industry 4.0 technologies can facilitate the development of more advanced algorithms for identifying at-risk students. Therefore, it is an important initiative for faculty members to develop a smarter system to identify at-risk students instead of using readily-made warning systems.

Developing algorithms using huge datasets is not new. Other industries have already started developing algorithms in a similar manner. In the field of physiology, for example, Preece et al. (2009) conducted a study on the use of algorithms to predict the daily activities of human beings based on data collected from body sensors. Similarly, Tsai and Lu (2009) published a study on predicting customer attrition using various data mining techniques. In a third example, Marcano-Cedeno, Quintanilla-Domínguez, and Andina (2011) described the use of prediction algorithms to correctly predict that a person will get breast cancer. The success of these studies should motivate the higher education sector to make good use of data and to develop appropriate algorithms for identifying at-risk students.

Despite the success of prediction in other disciplines, there has not been consensus on which technique has advantages over others. Dreiseitl and Ohno-Machado (2002) compared the use of two types of algorithms: logistic regression and an artificial neural network. Their studies, conducted with a biomedical dataset, led to the conclusion that different methods have their individual merits. Paliwal and Kumar (2009) conducted a review on the use of artificial neural networks in articles across various disciplines and praised the use of the methodology due to its ability to approximate non-linear functions. Sarle (1994) asserted that methodologies should not be completing with each other and that there can be overlaps among methodologies. In fact, Dreiseitl and Ohno-Machado (2002) believe that the quality of prediction results depend on the quality of the dataset employed, the parameters chosen for modeling, and the evaluation criteria used. This shows that performance of prediction methodologies can vary across disciplines. Within an educational setting, a systematic review was conducted by Shahiri, Husain, and Rashid (2015), i.e. not empirical comparisons, and their reviews placed emphasis on how different algorithms are being used in an educational setting. The authors call for further research in this area. Therefore, it is the inconclusive results in previous studies, and review-based studies in the educational sector that drives the current study, which is designed to explore which type of algorithms can be better adopted in an educational setting.

This study describes how various methods of developing advanced algorithms can predict at-risk students in higher education more effectively than previous models have been able to do. It also describes the challenges and opportunities present in these endeavors.

LITERATURE REVIEW

Overview of Data Mining

The term *algorithm* is actually a layman term that represents the outcome of data mining. Data mining, therefore, is the process of developing algorithms (Wu et al., 2008). In fact, data mining and algorithm development are not new concepts and have been popular for years (Sahoo, 2013). To understand these two notions, it is important to first understand traditional statistical techniques that are deeply rooted in modern research. Traditional statistics focuses on hypothesis testing and is hypothesis-driven (Baepler & Murdoch, 2010). Researchers first establish the research questions based on theoretical foundation and then collect relevant data. Next, researchers run certain types of preplanned statistical tests with the dataset to determine if the hypothesis is rejected or not. For example, to understand the reading skills of students after an educational intervention, researchers may conduct a reading test (i.e., pretest) with students before the intervention. Then, intervention can take place and a post-test is conducted. Finally, a hypothesis test, a paired-sample t-test is administered so as to find out if there are statistically significant differences in mean between the pretest and post-test (Woodrow, 2014). In contrast, data mining is a "knowledge discovery process" (Finlay, Pears, & Connor, 2014; Al-Maqaleh, Abdullah, 2017). Datasets are obtained from given sources due to the readily available data sources in recent decades. Then, data mining techniques are used to explore interesting patterns in the datasets without any presuppositions. This is why the metaphor of mining is used; data mining is the process of looking for interesting patterns. These interesting patterns eventually become algorithms that describe the notable relationships between variables (i.e., patterns) and can be used for specific purposes.

Llorente and Morant (2011) introduced a four-step approach to data mining: data collection, data preparation, modelling, and evaluation. In their approach, data collection, which is common in traditional research, is followed by data preparation. This step refers to the process of merging multiple datasets, dealing with missing data, and detecting and removing outliners. The complexity of this process varies according to the needs of different data mining techniques. After the preparation stage, the dataset will be divided into two or three groups: training dataset, validation dataset, and testing dataset. The modelling stage makes use of the training dataset and validation dataset to build and optimize the model. In other words, the modelling process is the process of machine learning. The computer will build the model and algorithms using the training dataset, and it will fine-tune the model with the validation dataset. After creating a polished algorithm, the computer uses the testing dataset (i.e., a dataset that has not been processed by the computer) to test the efficacy of the model.

Within the spectrum of data mining, a range of purposes can be achieved and is summarized in Table 1 below:

Overview of Classification Techniques

The current study is interested in predicting at-risk students through classification techniques. To summarize the definition by Goyal and Vohra (2012), classification technique is the prediction of class membership using some given variables. Tan, Steinbach and Kumar (2014) define classification as a task of learning a target function *f* that maps each attribute set *x* to one of the predefined class labels *y*. If the target variable is a numeric one (e.g., age, rating scale), other techniques (e.g., multiple linear regression) should be used. Classification techniques are used to predict the class membership (i.e., target label) of a new dataset.

Table 1. Purposes of data mining

Data Mining Techniques and Purposes	Definitions by Goyal and Vohra (2012)	Possible Application in Higher Education
Anomaly detection (outlier/ change/deviation detection)	Identification of unusual data records that might be interesting, or of data errors which require further investigation	Look for students who experience a sudden change in emotion (e.g., low course grades over the entire semester).
Association rule learning (dependency modelling)	Method for finding relationships between variables	Look for rules: mature students who do badly in Course A are more likely to struggle in Course B
Clustering	Discovering groups and structures in the data that are similar in some way “”without using known structures in the data	Group students into different profiles based on demographic information
Classification	Generalization of known structures to apply to new data	Identify at-risk students
Regression	Technique that attempts to find a function that models the data with the least amount of error	Predict student test scores
Summarization	A compact representation of the data set, including visualization and report generation	Create a visualization of when students complete a certain online activity

It is essential to understand the nature of the different elements that comprise classification. Here, the target class labels should be nominal (e.g., Wealthy / Not wealthy; or occupation: Student, Teacher, and Driver), instead of a numeric variable (e.g., age). In practice, of course, a numeric value is assigned to each label (e.g., Student = 0, Teacher = 1, and Driver = 2), but a nominal label means that a mathematical calculation cannot be performed in a meaningful manner. Also, with classification techniques, it is important to note that the target label can range from two to many. If there are only two labels, the target label is a binary variable (e.g., Wealthy / Not Wealthy). However, it is possible to have more labels (see the occupation example above). Due to the binary nature of identifying at-risk students (i.e., At-risk or Not at-risk), all analysis performed in this study are based on a binary target variable.

After understanding the nature of the classification techniques, it is now important to survey various commonly used data mining techniques. Dreiseitla and Ohno-Machado (2002) listed several important classification techniques (Table 2):

Even though some attempts have been made to compare the effectiveness of these classification techniques, not much analysis has been done in the context of higher education. For example, Dreiseitla and Ohno-Machado (2002) compared the use of Logistic regression and artificial network techniques in their biomedical study. Paliwal, and Kumar (2009) have conducted a review to compare various classification techniques. Even though researchers have conducted general studies on educational data mining as well as and case studies on the classification techniques to predict at-risk students, few studies on comparing techniques have been published until recently. Due to the strengths and weaknesses of different classification techniques, the higher education sector needs to look for the best algorithms amidst the wave of prediction technology in the era of Industry 4.0; this study is conducted to achieve this goal. Summarizing the review results from Shahiri, Husain and Rashida (2015) and Paliwal and Kumar (2009), this paper adopted the Decision tree analysis, Logistic regression, and Artificial neural network for the basis of comparison. Specifically, this study aims to answer the following questions:

1. How accurate is the prediction rate of different classification techniques?
2. What will the challenges and opportunities be in using certain classification technique in the higher education sphere?

Table 2. Different classification techniques

Classification Techniques	Introduction by Dreiseitla and Ohno-Machado (2002)
Support vector machine	Model that builds optimal separating boundaries between data sets by solving a constrained quadratic optimization problem
k-Nearest neighbour	Algorithm that uses the data directly for classification without building a model first
Decision tree	Algorithm that repeatedly splits the data set according to a criterion that maximizes the separation of the data, resulting in a tree-like structure
Logistic regression	The drawing of a regression line to predict the probability of success of a binary variable
Artificial neural network	System that attempts to model the capabilities of the human brain

METHODOLOGY

Participants

The dataset is drawn from the learning management system of the university in question, which encompasses a total of 5,993 records (i.e., students) taking two academic English courses at a university in Hong Kong. The first course is a basic English course for academic purposes, and the second is an advanced English course. Most students are enrolled in an undergraduate programme, whereas others are completing a sub-degree programme. Researchers believe that the differences in programmes and proficiency levels do not affect the results of the data mining process.

Variables: Course Assessments for Data Mining

Students completed various assessments after taking two courses, thus the assessment results become important elements for the analysis. In the first course, English for University Studies (EUS), students must complete three assignments: a 500-word problem-solution essay, an 800-word discursive essay, and an academic oral presentation. In the second course, Advanced English for University Studies (AEUS), students must complete three assessments: a 600-word draft argumentative essay, a final 1200-word argumentative essay, and an oral defence. The writing assignments across both courses are graded based on four criteria: content, organization, language, and referencing. Meanwhile, the speaking assignments across both courses are graded based on these criteria: content, delivery, language, and pronunciation/fluency. Both English courses include an online component (i.e., online exercises) and students must complete them. To achieve the goal of early identification, this study aims to predict the overall course performance of the second course using the assessment results of (1) all assessments in EUS, (2) first assessments in AEUS, and (3) scores of the online component for the first course.

The two courses adopt a common assessment scheme. Each assessment component is assessed using a rating scale from 0 to 4.5. Teachers assign a scale to each assessment component. Table 3 presents the details of these rating scales:

Table 4 summarizes the variables to be mined in this study:

Data Preparation and Cleaning

To apply appropriate data mining techniques, several steps of data preparation and cleaning must be implemented. First, only students who have completed all assessments are included; for this reason, 10 records were deleted. Next, the overall grade of AEUS was recoded to be a binary variable. If students receive a C+ grade (2.5) or below, they are considered to be at-risk students and marked as "1". Those who attained a higher grade were marked as "0". Then, the dataset was divided into two groups for

Table 3. Grading scheme of the research site

Fail	Inadequate		Satisfactory		Good		Outstanding	
0	1	1.5	2	2.5	3	3.5	4	4.5

Table 4. Assessment components/research variables

Course Assessments and Assessment Components	Variable Type (Range)	Acronym
First English Course (EUS)		
500-word problem-solution essay - Content - Organization - Language - Referencing	Rating Scale (0-4.5)	ealc ealo eall ealr
800-word discursive essay - Content - Organization - Language - Referencing	Rating Scale (0-4.5)	ea2c ea2o ea2l ea2e
Academic oral presentation - Content - Delivery - Language - Pronunciation/Fluency	Rating Scale (0-4.5)	ea3c ea3d ea3l ea3p
Online Component	0-1	elndi
Overall Course grade	Rating Scale (0-4.5)	eover
Second English Course (EUS)		
600-word draft positive-argumentative essay - Content - Organization - Language - Referencing	Rating Scale (0-4.5)	aalc aalo aall aalr
Final Course Grade	Rating Scale (0-4.5)	Target variable

outliner detection. The multivariable outliners in each group were removed based on Mahalanobis distance (MD), i.e., the MD for each group was computed separately. If the MD of a record is greater than the 99th percentile of a chi-square distribution with 18 degrees of freedom, it is considered an outliner; thus, 390 records across two groups were removed. The dataset was then combined for further analysis.

As a standard procedure for model building and validation, the dataset was divided into two parts: training dataset and testing dataset. In the current study, two-thirds of the cleaned dataset was grouped as the training set (3,729 records) and the remaining records are grouped as the testing dataset (1,864 records). The significance of dividing the dataset into two groups is discussed in the results section.

Data Analysis: To Identify At-Risk Students

The main goal of the classification technique is to predict positive/negative outcomes in a course of students (e.g. the final course grade). Since the aim of the chapter is to describe and introduce different classification techniques, the details of the classification and the parameters used are presented in the results section. Four rounds of analysis will be conducted and analysed in the current study: (1) classification tree; (2) classification tree (with random forest); (3) logistic regression, and (4) artificial neural network. After presenting the results of these four rounds of analysis, the researcher will discuss the results and how they can be applied in a higher education context.

A free software, R (version i386 3.4.3) was used in the entire study. Several libraries of R were downloaded and installed for specific classification techniques, such as "rpart" and "rpart.plot" for decision tree analysis and "nnet" for artificial neural network.

As data mining is a computation intensive process, it is important to highlight that all analysis in the current study was conducted with the computer having an Intel® Core ™ i5-4460 CPU @3.20 GHz and 8.00 GB ram and operating under the Window 10 system.

Data Analysis: To Compare Different Methodologies

To evaluate the effectiveness of different methods, some simple comparisons were made. First, the accuracy of the algorithms derived is compared. A confusion matrix is used to present the error rate of each technique. Also, the amount of coding needed to generate of results was presented. The researcher only used necessary codes to ensure the comparison is meaningful. Third, the information provided in each algorithm will be presented in discussed. The goal of the discussion is to understand if the information provided in each algorithm is bring challenges/drawbacks in a higher education setting.

Research Ethics

This project received approval from the University's Human Subjects Ethics Application Review System.

RESULTS AND DISCUSSION

Classification Tree: Prediction Algorithms

The classification tree presents the results of the prediction algorithm in a simple way. Figure 1 below presents the tree plot generated by the software (R). Based on the figure, at-risk students can be identified in no more than three steps. In other words, a teacher can know if the student is at-risk by look for answers for no more than three questions. First, if students attain content scores of 2.8 or above (out of 4.5) in the AEUS Assessment 1 (A1), they are not likely to be at-risk students. If they get a lower score than that, a second question is needed. If, after the second question, the student *does not* attain a language score of 2.2 or more, they are quite likely to be at-risk. If a student does not get a content score in AEUS A1 of more than 2.8 but achieves a language score of more than 2.2, the third and final question is needed. After this question, does the student score 2.2 or more in referencing for AEUS A1? If so, the student is not likely to be at-risk. If not, the student is likely to be at risk. This is represented in Figure 1.

Based on the tree built by the software, it seems that a student's assessment score from the AEUS is a better predictor than scores from their first English course (EUS), even though assessment scores for both courses were entered in the model for selection by the computer.

Classification Tree: Prediction Accuracy

Table 5 and Table 6 below present the error rates of the tree during the training and testing stages. Generally, the error rate was around 24%. This means that among 100 students, the algorithm may misclassify 24 of them incorrectly, as at-risk when they are not at-risk or vice versa.

Figure 1. Classification tree

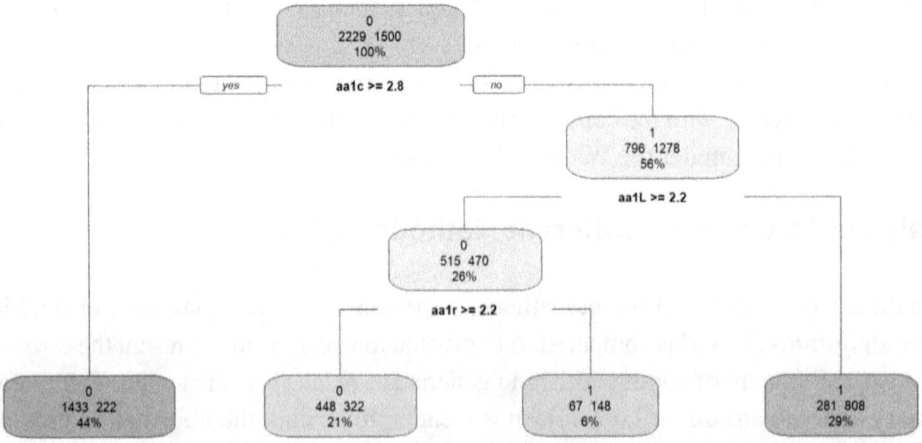

When predicting at-risk students, it is important to take note of false-negative errors. False-negative means that the students were actually at-risk but were predicted as not at-risk. This is problematic in the higher education context because false-negative results mean that students in need may not get the support they require. The cost of this is much higher than false-positives (i.e. students who do not need help but receive some extra help). In the current analysis, the false-negative rate was around 9%.

Error Rate (Training): (348+544) / 3729 = 23.92%

Another important element to note is the difference in error rates between the two datasets. If the error rate is quite low in the training stage but much higher in the testing stage, it means that the algorithm is

Table 5. Classification tree: error rates (training)

PREDICTED	Actual		TOTAL
	At-risk	Not At-Risk	
At-risk	1881	544	2425
Not At-risk	348	956	1304
TOTAL	2229	1500	3729

Table 6. Classification tree: error rates (testing)

PREDICTED	Actual		TOTAL
	At-risk	Not At-Risk	
At-risk	920	279	1199
Not At-risk	171	494	665
TOTAL	1091	773	1864

Error Rate (Testing): (171+279) / 1864 = 24.14%

over-fitted; this affects the accuracy if the algorithm is already in use, facing other datasets that have not been processed by the computer before. A general rule-of-thumb is that the error rates for both datasets should be close, and this is the case in the current analysis.

Classification Tree: Practical Concerns

In a higher education context, the classification tree seems to be appropriate. First, it is easy for faculty members with no statistical background to interpret it and use it to act. Also, it makes it possible for teaching staff to tell students what action can be taken to avoid being at-risk. In this case, for example, teachers can ask students to pay attention to their AEUS A1 content score first, followed by their language and referencing scores.

Logistics Regression: Prediction Algorithms

Other than classification trees, logistics regression is the next method to be evaluated. In fact, multiple linear regression is an even more popular technique in modern research. The goal of regression is to draw a line that can help researchers predict the target variable based on independent variables. Multiple regression can help predict a continuous variable with a straight line, while logistic regression, in simple terms, is used to predict a binary variable using a non-straight line.

In the current study, as a data mining approach, a manual backward approach was adopted to derive the prediction algorithm. First, all 18 independent variables were entered into R for analysis and only the statistically significant variable was retained. In the current analysis, Table 7 presents variables retained.

It is not easy to interpret the results of the logistic regression. Based on the results from the software, each retained variable has a value that shows its impact on the prediction algorithm. Unlike multiple linear regression, this beta value cannot be directly interpreted because different beta values may have different levels of impact to the exponential function of the logistic regression. It is more appropriate for the researcher to assume some basic scenarios to understand the algorithm, as presented in Table 8 below. After doing some simple calculations with the help of Microsoft Excel, if a student receives "Satisfactory" scores in all retained assessment components and a score of 0.5 for the online component (i.e. the minimum score required), the probability of them being at-risk is 90.5%, i.e. very likely. Similarly, if a student gets a "Good" grade in all retained assessment components and 0.5 in the online component, the probability is 10.8%, i.e. not very likely.

Table 7. Logistic regression: retained variable

Assessment Components	Acronyms	Beta
EUS Academic Presentation - Delivery	ea3d	-0.5573
EUS Online Component Score	eIndi	-1.1679
AEUS Draft Essay - Content	aa1c	-1.2801
AEUS Draft Essay – Organization	aa1o	-0.6643
AEUS Draft Essay – Language	aa1L	-1.1287
AEUS Draft Essay – Referencing	aa1r	-0.7420

Table 8. Impact of retained variables

Item	Beta	All "Satisfactory" /"2"	All "2.5"	All "Good" /"3"
ea3d	-0.5573	2	2.5	3
*eIndi	-1.1679	0.5	0.5	0.5
aa1c	-1.2801	2	2.5	3
aa1o	-0.6643	2	2.5	3
aa1L	-1.1287	2	2.5	3
aa1r	-0.7420	2	2.5	3
Intercept	11.587			
Probability of being at-risk)		90.5%	51.8%	10.8%

*eIndi: The Online component score is a score from 0-1. Students MUST get 0.5 or above in order not to receive overall grade penalty.

Other than the above simple analysis, it is also possible to conduct a What-If Analysis (i.e. manipulate one score to see how the grade is affected; See Table 9) and see the effect. This can be done with a simple Excel spreadsheet. For example, right before the completion of EUS, a student scores 2.5 in the delivery of a presentation (ea3d) and foresees himself scoring 2.5 in all the component grades in AEUS (i.e. aa1c, aa1o, aa1L and aa1r). At the moment, he has obtained 0.5 in the online component. In this case, the student can be told that he must get a higher score in the online component, say from 0.5 to 0.9. Then the probability of being at-risk will be reduced by 14.3%. See Table 8 for details. Such an obvious change in probability can motivate students to do well in a certain component.

Logistic Regression: Prediction Accuracy

Table 10 and Table 11 illustrate the error rates of the training and testing data. The error rates are quite stable, at around 22%. The error rate is similar to that of the classification tree. The false-negative rate is also similar, at around 9% to 10%. In other words, the prediction accuracy of the use of logistic regression is quite similar to that of the classification tree.

Table 9. What-if analysis

Item	Beta	My Current Result	What-if...
ea3d	-0.5573	2.5	2.5
*eIndi	-1.1679	0.5	0.9
aa1c	-1.2801	2.5	2.5
aa1o	-0.6643	2.5	2.5
aa1L	-1.1287	2.5	2.5
aa1r	-0.7420	2.5	2.5
Intercept	11.587	2.5	2.5
Probability of being at-risk)		51.8%	37.5%
Change in Probability		-14.3%	

Table 10. Logistic regression – error rates (training)

PREDICTED	Actual		TOTAL
	At-risk	Not At-Risk	
At-risk	1867	465	2332
Not At-risk	362	1035	1397
TOTAL	2229	1500	3729

Error Rate (Training): (362+465) / 3729 = 22.18%

Table 11. Logistic regression – error rates (training)

PREDICTED	Actual		TOTAL
	At-risk	Not At-Risk	
At-risk	905	238	1143
Not At-risk	186	535	721
TOTAL	1091	773	1864

Error Rate (Testing): (186+238) / 1864 = 22.75%

Logistic Regression: Practical Concerns

The interpretation of logistic regression is not easy, but it is not impossible. There is a need for a technician with a statistical background to help interpret the results or prepare the respective algorithms for What-If Analysis. These may not be simple tasks for busy faculty members. However, after setting up all the algorithms in an Excel spreadsheet or even an online platform, students can understand how likely they are to be classified as at-risk students. Students may also conduct different What-If Analyses which could, in turn, motivate them to do well in the course. Therefore, logistic regression is not impossible to apply in the higher education context.

Artificial Neural Networks (ANN): Prediction Algorithms

An artificial neural network (ANN) is a more complicated algorithm than the other two techniques for predicting at-risk students. It composes of an input layer, a hidden layer and an output layer. The current study involved 18 independent variables, so there were 18 nodes in the input layer. Also, there *was* a binary output variable, i.e. the target variable. ANN users can choose the size of the hidden layer. It is generally thought that, to avoid overfitting, the hidden layer should not be larger than the input node. When 18 nodes were implemented in the hidden layer, the ANN was found to be overfitted. After testing multiple hidden layer sizes, the researcher decided to place four nodes within it. Figure 2 presents the ANN of the current study.

In the process of building the ANN, no variable was to be removed or added (unlike logistic regression). All nodes in the input layer were to have different degrees of impact on the hidden and output nodes. The model building process was used to derive the beta value (i.e. impact level) of each node. With the beta values and the ANN, one can derive the prediction algorithm shown in Figure 3. X_1-X_{18} refers to input variables; readers can refer to Table 4 for details.

Figure 2. Artificial neural networks

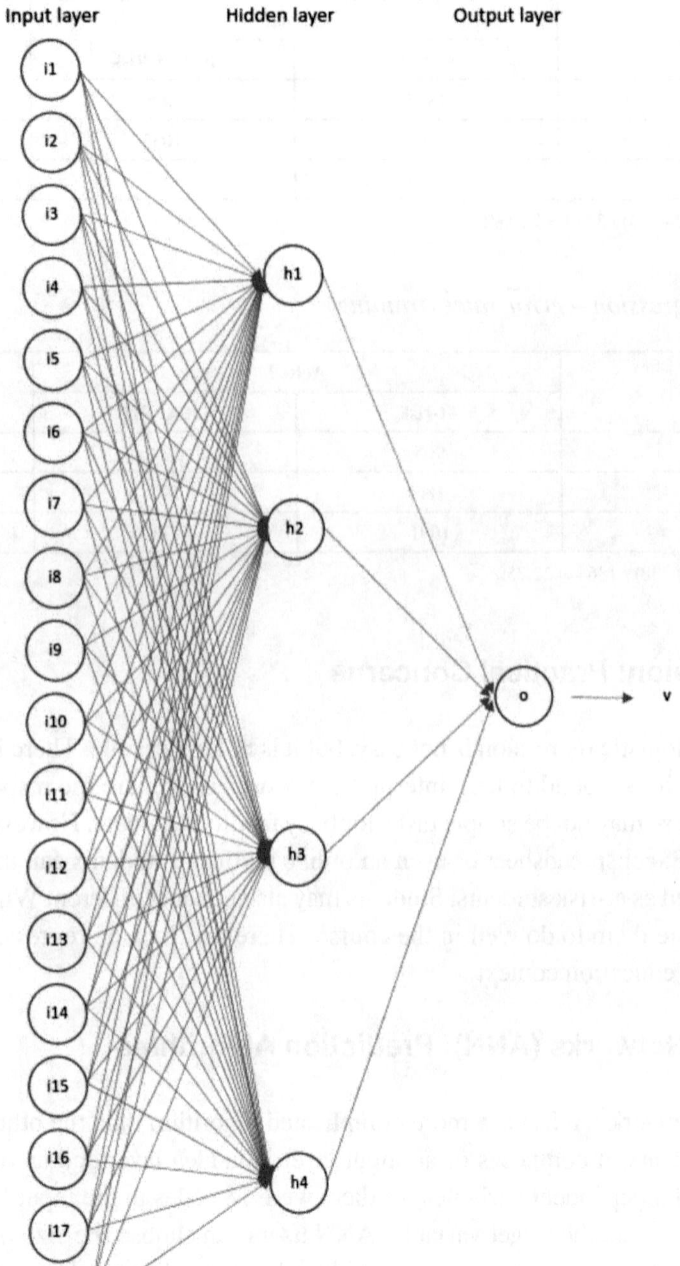

Artificial Neural Networks (ANN): Prediction Accuracy

The ANN has an unstable error rate as well as a large difference in error rates (approximately 19%–25%) between its Training and Testing datasets (Table 12 and Table 13). Researchers believe that the difference is insufficient to declare the ANN over fitted. The false-negative rate is higher in this method than in other two, but the actual differences were small.

Artificial Neural Networks (ANN): Practical Concerns

The problem with ANNs is that while their results are useful for teachers and students to see, it is hard to interpret their meanings. Another important issue is that there may not be any actionable items for students or teachers. As there is no easy way to interpret these results, it is hard for students and teachers

Figure 3. Algorithms for artificial neural networks

$$h1 = 1432.09 + 347.72x_1 - 266.89x_2 + 27.14x_3 - 29.88x_4 + 99.37x_5 + 36.08x_6 - 26.99x_7 - 0.29x_8 + 36.79x_9 - 124.73x_{10} + 192.78x_{11} - 33.89x_{12} - 16.08x_{13} + 39.31x_{14} - 40.68x_{15} - 182.37x_{16} - 548.27x_{17} - 37.54x_{18}$$

$$h2 = -234.74 + 14.95x_1 + 73.84x_2 - 141.58x_3 - 1.43x_4 - 204.00x_5 + 285.85x_6 - 177.23x_7 + 110.13x_8 + 53.95x_9 - 39.04x_{10} + 12.62x_{11} - 86.23x_{12} + 339.59x_{13} - 125.30x_{14} + 61.03x_{15} + 24.74x_{16} - 172.22x_{17} - 209.43x_{18}$$

$$h3 = -4.46 + 0.11x_1 - 0.14x_2 - 0.02x_3 + 0.05x_4 + 0.01x_5 + 0.08x_6 - 0.01x_7 + 0.01x_8 - 0.02x_9 + 0.03x_{10} + 0.09x_{11} - 0.04x_{12} + 0.14x_{13} + 0.17x_{14} + 0.17x_{15} + 0.10x_{16} + 0.05x_{17} + 0.07x_{18}$$

$$h4 = -447.97 + 2.77x_1 + 316.35x_2 + 5.89x_3 - 153.57x_4 - 6.27x_5 - 93.41x_6 + 27.37x_7 + 58.91x_8 + 54.81x_9 - 40.22x_{10} + 29.79x_{11} + 118.21x_{12} - 88.43x_{13} - 72.52x_{14} + 274.87x_{15} - 109.66x_{16} - 202.09x_{17} + 105.85x_{18}$$

$$v = 7.03 + 1.54h_1 + 2.74h_2 - 98.82h_3 - 2.06h_4$$

Table 12. Artificial neural networks: error rates (training)

PREDICTED	Actual		TOTAL
	At-risk	Not At-Risk	
At-risk	1869	378	2247
Not At-risk	360	1122	1482
TOTAL	2229	1500	3729

Error Rate (Training): (360+378) / 3729 = 19.79%

Table 13. Artificial neural networks: error rates (testing)

PREDICTED	Actual		TOTAL
	At-risk	Not At-Risk	
At-risk	873	234	1107
Not At-risk	218	539	757
TOTAL	1091	773	1864

Error Rate (Testing): (218+234) / 1864 = 24.25%

to outline specific improvement measures. This seems to undermine the algorithm's intention to predict at-risk students in higher education. One final point to note with ANNs regards time spent generating results. ANNs are highly computation-intensive and require a significant amount of time (20–30 seconds) to generate a model using 18 variables (i.e. from entering the final line of code to seeing the results). This problem was not observed in other techniques discussed in this paper (which may take less than 1 second to generate results).

General Discussion

There is no one method or algorithm that can meet all needs in higher education, but the following discussion explores how each algorithm performs well in specific contexts. In terms of accuracy, there are no major differences across the three methods (Table 14). The error rates (across three methods) are around 25%. This is not an ideal error rate, but it represents a good starting error rate for developing algorithms. Another important aspect is the variance in false-negative rates; once again, no major difference is observed across the three methods. Therefore, from an accuracy perspective regarding the current dataset, all three methods are acceptable.

Another point of comparison is the coding process. Table 15 presents the number of characters and lines required to code with R. The table suggest that the most time-consuming process is that of data preparation, which involves the largest number of characters and lines. It is most efficient to code and build a classification tree with the smallest number of characters and lines of the three methods, and the second most efficient method is logistic regression. The ANN method seems to be more time-consuming

Table 14. Comparison of prediction accuracy

	Classification Tree	ANN	Logistic Regression
Training	23.92%	19.79%	22.18%
Testing	24.14%	24.25%	22.75%
False negative	9.17%	11.70%	9.98%

than the others, involving significantly more characters and lines. However, these measurements of coding evaluate the coding process without considering inevitable mistakes, errors or debugging, which demand more time. Additionally, the current study assumes that the faculty members are equally competent in coding all necessary processes and methods. According to these results, it is reasonable to state that classification tree and logistic regression methods have advantages over ANNs in coding time.

The final factor under comparison is the practicality of each method in the context pf higher education. Table 16 summarises the main points discussed in the previous sections and assigns a rating of their practicalities. The visual advantage offered by classification trees makes them easy to read without much manipulation. Logistic regression can also be easily interpreted once a knowledgeable person manipulates the results and presents them on a platform. Because of this additional required effort, logistic regression appears less simple than classification. However, one cannot neglect its potential, such as its ability to conduct what-if analyses. Finally, ANNs do seem to be difficult to implement in higher education contexts. The algorithms themselves are hard to interpret and provide few actionable items for both faculty teachers and students to adopt. Therefore, ANNs seem to be least applicable in the higher education sector.

CONCLUSION

Using a real dataset from higher education contexts, this study compares the use of classification trees, logistic regression and ANNs. The comparison focusses on the accuracy and simplicity of algorithms as well as their practicality in higher education. It appears that classification trees can be easily applied to the higher education sector and that logistic regression also has potential. However, ANNs do not seem to be as useful.

This study has several limitations, and readers are recommended to interpret the results with caution. First, its accuracy is highly dependent on the dataset. The current dataset is shown to be noisy, and accuracy may be improved with other datasets. This potential variability affects the reliability of the current study. Second, the coding process and the parameters for different classification techniques have been simplified; the study's accuracy and required coding may change if the parameters do.

Table 15. Comparison of coding process

	Data Preparation	Classification Tree	ANN	Logistic Regression
No. of Characters	497	228	502	348
No. of Lines	30	9	22	12

Table 16. Comparison of practicality

	Classification Tree	Logistic Regression	ANN
Faculty teachers with no Statistics background	Easy	Medium	Difficult
Students	Easy	Medium	Difficult

It is a major task to redesign a prediction algorithm to meet higher educators' wishes to identify at-risk students. To cope with Industry 4.0, educators could attempt to keep themselves up-to-date with the development of the latest algorithms. Additionally, basic coding knowledge is necessary to understand the concepts behind the intelligent systems of Industry 4.0. Most importantly, faculty members can use the the results of these analyses to develop platforms to use these algorithms on a platform. These recommendations will transform higher education.

ACKNOWLEDGMENT

The author thanks colleagues from the English Language Centre and Educational Development Centre (Institutional Assessment and Evaluation Section) at The Hong Kong Polytechnic University, who provided insights and expertise that greatly assisted the research. The author would also like to thank Prof. William Cheung from the Department of Statistics at the Chinese University of Hong Kong as he inspired the author to carry out this data mining project in a language teaching context.

REFERENCES

Abdullah, N. H., Ramli, N. H. L., & Rafek, M. (2017). Mass lecture in language learning: What do the boys and girls think? *Journal of Advances in Humanities and Social Sciences*, *3*(2), 115–123.

Al-Maqaleh, B. M., & Abdullah, A. M. G. (2017). Intelligent predictive system using classification techniques for heart disease diagnosis. *International Journal on Computer Science and Engineering*, *6*(6), 145–151.

Baepler, P., & Murdoch, C. J. (2010). Academic analytics and data mining in higher education. *International Journal for the Scholarship of Teaching and Learning*, *4*(2), 17. doi:10.20429/ijsotl.2010.040217

Bainbridge, J., Melitski, J., Zahradnik, A., Lauría, E. J., Jayaprakash, S., & Baron, J. (2015). Using learning analytics to predict at-risk students in online graduate public affairs and administration education. *Journal of Public Affairs Education*, *21*(2), 247–262. doi:10.1080/15236803.2015.12001831

Blackboard. (2018). *Retention centre.* Available from: https://help.blackboard.com/Learn/Instructor/Performance/Retention_Center

Chingos, M. M., & Whitehurst, G. J. (2011). Choosing blindly: Instructional materials, teacher effectiveness, and the common core. The Brookings Institution.

Ciolacu, M., & Beer, R. (2016). Adaptive user interface for higher education based on web technology. *2016 IEEE 22nd International Symposium for Design and Technology in Electronic Packaging (SIITME)*. 10.1109/SIITME.2016.7777299

Clow, D. (2013). An overview of learning analytics. *Teaching in Higher Education*, *18*(6), 683–695. doi:10.1080/13562517.2013.827653

Dreiseitl, S., & Ohno-Machado, L. (2002). Logistic regression and artificial neural network classification models: A methodology review. *Journal of Biomedical Informatics*, *35*(5-6), 352–359. doi:10.1016/S1532-0464(03)00034-0 PMID:12968784

Essa, A., & Ayad, H. (2012, April). *Student success system: Risk analytics and data visualization using ensembles of predictive models.* Paper presented at the 2nd International Conference on Learning Analytics and Knowledge, Vancouver, Canada. 10.1145/2330601.2330641

Finlay, J., Pears, R., & Connor, A. M. (2014). Data stream mining for predicting software build outcomes using source code metrics. *Information and Software Technology*, *56*(2), 183–198. doi:10.1016/j.infsof.2013.09.001

Goyal, M., & Vohra, R. (2012). Applications of data mining in higher education. *International Journal of Computer Science Issues*, *9*(2), 113–120.

Graube, G., & Mammes, I. (2017). Industry Involvement in Technology Education. In M. J. Vries (Ed.), *Handbook of Technology Education* (pp. 843–856). New York, NY: Springer. doi:10.1007/978-3-319-38889-2_60-1

Klüsener, M., & Fortenbacher, A. (2015, September). *Predicting students' success based on forum activities in MOOCs.* Paper presented at IEEE 8th International Conference, Warsaw, Poland. 10.1109/IDAACS.2015.7341439

Lawson, C., Beer, C., Rossi, D., Moore, T., & Fleming, J. (2016). Identification of 'at risk'students using learning analytics: The ethical dilemmas of intervention strategies in a higher education institution. *Educational Technology Research and Development*, *64*(5), 957–968. doi:10.100711423-016-9459-0

Llorente, R., & Morant, M. (2011). Data mining in higher education. In *New fundamental technologies in data mining* (pp. 201-220). Available from: http://www.intechopen.com/books/new-fundamental-technologies-in-data-mining/data-mining-in-higher-education

Marcano-Cedeño, A., Quintanilla-Domínguez, J., & Andina, D. (2011). WBCD breast cancer database classification applying artificial metaplasticity neural network. *Expert Systems with Applications*, *38*(8), 9573–9579. doi:10.1016/j.eswa.2011.01.167

Paliwal, M., & Kumar, U. A. (2009). Neural networks and statistical techniques: A review of applications. *Expert Systems with Applications*, *36*(1), 2–17. doi:10.1016/j.eswa.2007.10.005

Park, Y., Yu, J. H., & Jo, I. H. (2016). Clustering blended learning courses by online behavior data: A case study in a Korean higher education institute. *The Internet and Higher Education*, *29*, 1–11. doi:10.1016/j.iheduc.2015.11.001

Preece, S. J., Goulermas, J. Y., Kenney, L. P., Howard, D., Meijer, K., & Crompton, R. (2009). Activity identification using body-mounted sensors—A review of classification techniques. *Physiological Measurement*, *30*(4), R1–R33. doi:10.1088/0967-3334/30/4/R01 PMID:19342767

Sahoo, G. (2013). Study of parametric performance evaluation of machine learning and statistical classifiers. *Int. J. Inf. Technol. Comput. Sci.*, *5*, 57–64.

Sarle, W. S. (1994). Neural networks and statistical models. *Proceedings of the 19th annual SAS users group international conference*.

Shahiri, A. M., Husain, W., & Rashid, N. A. (2015). A review on predicting student's performance using data mining techniques. *Procedia Computer Science*, *72*, 414–422. doi:10.1016/j.procs.2015.12.157

Tan, P., Strinbach, M., & Kumar, V. (2013). *Introduction to data mining* (International Edition). Pearson.

Tsai, C. F., & Lu, Y. H. (2009). Customer churn prediction by hybrid neural networks. *Expert Systems with Applications*, *36*(10), 12547–12553. doi:10.1016/j.eswa.2009.05.032

Wolff, A., Zdrahal, Z., Nikolov, A., & Pantucek, M. (2013, April). Improving retention: predicting at-risk students by analysing clicking behaviour in a virtual learning environment. In *Proceedings of the Third International Conference on Learning Analytics and Knowledge* (pp. 145-149). ACM. 10.1145/2460296.2460324

Woodrow, L. (2014). *Writing about quantitative research in applied linguistics*. New York, NY: Springer. doi:10.1057/9780230369955

Wu, X., Kumar, V., Quinlan, J. R., Ghosh, J., Yang, Q., Motoda, H., ... Zhou, Z. H. (2008). Top 10 algorithms in data mining. *Knowledge and Information Systems*, *14*(1), 1–37. doi:10.100710115-007-0114-2

Section 4

Awareness and Challenges in Redesigning Higher Education 4.0 Initiatives

Chapter 15
Industry Revolution 4.0 and Education System in Malaysia:
Are We Ready for the Challenges?

Subramaniam Kolandan
Universiti Utara Malaysia, Malaysia

ABSTRACT

Industry Revolution 4.0 (IR 4) is highly rated and greatly establishing itself. The future of our world looks so promising. The revolution is happening, and the impact is already seen in many sectors. Education is one of the sectors that needs to be in line with the growth of the technology to produce a world class work force. In Malaysia, the government had earlier made many changes in education through its National Higher Education Action Plan 2007-2010 in transforming teaching and learning process focusing on technology. Innovative and entrepreneurial skills must be possessed by new graduates and they must deal with complexity with cognitive flexibility. This is because they need to accept robots as their working colleagues apart from humans. The need for better communication and collaborative skills will be far more important than ever. Graduates must acquire self-learning skills to remain relevant in the era of rapid changes.

INTRODUCTION

Industry Revolution 4.0 (IR 4) is highly rated and greatly establishing itself. The future of our world looks so promising. The revolution is happening and the impact is already being seen in many sectors. What is IR 4? It would be a common question for all of us. IR 4 is defined as a stage in the development of knowledge in which the lines between the physical, digital and biological spheres are being blurred (Schwab, 2016). Ministry of Higher Education draws the four main backbone of IR 4, which are artificial intelligence (AI), cloud computing, big data analytics and internet of things. If we could see the history of industry revolution started from with the introduction of first mechanical loom. 18th century has seen the first industrial revolution which had seen the shift from human or animal power to machines run by water or steam. Introduction of electricity, and the rise of the steel and oil industries occurred between

DOI: 10.4018/978-1-5225-7832-1.ch015

Copyright © 2019, IGI Global. Copying or distributing in print or electronic forms without written permission of IGI Global is prohibited.

1870 to 1914 had triggered the era of mass production of goods and vehicles. In the 1960s, first programmable logistic controllers and early versions of computers been introduced hereby boosting the automation and control of production lines. This marked the extensive use of computer networks, and the eventual birth of the Internet changed the world in ways that no one could have imagined. Industry 4.0 is a German strategic initiative mooted in 2011 under its High-Tech Strategy 2020 and adopted two years later. It is aimed at revolutionising the manufacturing industry, by switching from centralised to decentralised networks under which connected equipment and devices communicate with each other to analyse and respond to information received.

Schwab stated the impact of IR 4 can be seen in four major sectors. Government is one of them whereby the IR 4 would be a helpful impact for the government to manage the people. New technologies enable the government to increase control over population based on good surveillance system. They can also control digital based infrastructure and manage them accordingly. Economy and business would be another one having impact from IR 4. Most be the current jobs are expected to be obsolete and new jobs are in line for an introduction in this era. More technology related jobs been created to accommodate the current young generation to be believed as digital natives. It is believed the IR 4 can also improve the current economic system. Society would be next. The new technologies can connect more people through the emergence of many new applications such as social medias. People can be connected no matter how far they are and how busy they are. The virtual communication system is a betterment of current system which is a sign of better future. Last impact can be seen in education system. The technology plays an important part in education. Education is one of the sector which needs to be in line with the growth of the technology to produce world class work force. To face the IR 4, the education system need be in greater situation to produce the talent needed to drive the industry in a greater way. In Malaysia, government had earlier made many changes in education through its National Higher Education Action Plan 2007-2010 in transforming teaching and learning process focusing on technology. Entrepreneurship education would be a perfect fit for IR 4.

In the Malaysian Education Blueprint for Higher Education 2013-2025, the government is giving priority to entrepreneurship by emphasizing its priorities in Malaysia's higher education system. The first spike of the 10 surges outlined is to produce holistic graduates with entrepreneurial and balanced characteristics. According to Datuk Seri Idris Jusoh, the former Higher Education Minister, a more conducive and holistic entrepreneurial ecosystem has begun to be provided by the government for the development of values and the personality of the students so that they are more confident and willing to act as a job generator or "job creator". IR 4 requires imminent changes in the curriculum and the whole education system. Across disciplines, many new contents are ready to make the entry, so emphasis need to be given on certain skills to accommodate the contents. New educational programmes will have to be developed to meet changing demands. In this era jobs that require creativity are likely to stay whereas many other jobs been predicted disappeared or will be abolished. New education system must be able to produce highly creative graduates with the ability to think critically.

Former Deputy International Trade and Industry Minister Datuk Ahmad Maslan stated that entrepreneurs need to think about the future of their business in line with the IR 4 as this would ensure that they would not lag behind on the global stage in the midst of technological evolution. Citing Alibaba.com, one of the world's largest e-commerce platforms he said it is one of the current and future model of businesses that would keep growing. Innovative and entrepreneurial skills must be possessed by new graduates and they must deal with complexity with cognitive flexibility. This is because they need to accept robots as their working colleagues apart from humans. The need for better communication and collaborative skills will be far more important than ever. Graduates must acquire self-learning skills to remain relevant in the era of rapid changes.

BUSINESS EDUCATION AND ITS IMPACT

Career choice is the most important aspect of one's life because it will determine the pattern or role of the individual in society in the future (Abdo & Anwar, 2016). Career selection is important to ensure that individual financial resources are secure and stable. Basically, the field of entrepreneurship is about the situation in which the value is created to stakeholders (McGraw III, 2007). The stakeholders here mean professionals in the field of entrepreneurship responsible for the community and society. Entrepreneurship is one of the career fields that can be graduated by graduates after graduation either at the Public and Private Institutions of Higher Learning (IPTA) or Private (IPTS). This entrepreneurial activity is a catalyst for the country's economic development and acts to enhance the level of innovation, creativity and competitiveness of a country such as Malaysia. The development of entrepreneurship over the last decade shows the importance of entrepreneurship-based education (Garcia, Douglas, Andreia & Regiane, 2017).

To become a developed nation, education plays a very important role. In line with the government's policy to improve the country's economy, education is an aspect that needs to be emphasized. Students in Malaysia are exposed to entrepreneurial concept at the primary school level in the present time where soft skills involving entrepreneurship are applied in teaching and learning. Furthermore, entrepreneurship is applied in the subject of Trade and Basic Economics in high school at form 4 and 5 levels. While in form six, students are exposed to entrepreneurship in the subject of Business Studies. However, the choice of entrepreneurship career is not motivated by the curriculum itself but by other factors such as economic conditions and market needs (Maziah, 2005). Education about entrepreneurship should take into account the aspect of learning skills and not just emphasizing theoretical learning alone. The entrepreneurial teaching process should suppress the "how" problem of the "what" issue (Gibb, 1993). Timmons et al., (1977), Brenneke (1990) and Klandt (1993) describe aspects of managing and creating a business cannot be learned through conventional teaching methods such as reading, lectures and watching movies. Conventional approaches are an effective way of supplying knowledge about entrepreneurship and some mechanical aspects of setting up businesses to students. But for the knowledge of certain types of consciousness, motivation and certain criteria can only be sown when the students have experience and engage in the process of creating a new business through the "negotiating" method and "working with entrepreneurs" method. Both of these methods are among the best methods of stimulating entrepreneurial characteristics among students and are followed by simulation teaching methods, case studies, role play, paperwork or thesis and writing business plans.

In the field of entrepreneurship, change is a definite and eternal and static (Seow, 2010). Changes in the form of technology change one's needs. To stay ahead in business, businessmen, academics and governments need to be aware of these changes and address challenges and opportunities not to miss (Seow, 2010). The intended changes are in all aspects especially in the field of education. The field of entrepreneurship introduced at the higher education level is to encourage the younger generation to transform themselves as an entrepreneur rather than as an employee or a more precise "wage" term.

For that reason, the curriculum is the basis for student enrolment where what they learn is a guide to what they want to explore. In addition, the curriculum itself determines the direction of a student. The curriculum serves to accommodate the needs of all student levels (Paul, 2014). If a student is unable to pursue tertiary studies, his or her ability to become an entrepreneur should be encouraged by the curriculum he has learned so far. That's what researchers want to study in this study. The curriculum review that is consistent with industry and employer needs by academia is necessary to create quality students (Jackson & Chapman, 2012).

To produce skilled students, the curriculum should play a very high role. Excellent academic results are not sufficient to become competent entrepreneurs, so should be well-mastered of all the elements in soft skills (Mohd Fariiqul Amin, 2009). Skilled here means efficiency, credibility and trained. So skilled students can be regarded as trained, competent and competent entrepreneurs in the field of entrepreneurship. Skills in entrepreneurship are an important aspect of entrepreneurial education (Lans, Hulsink, Baert & Mulder, 2008). Curriculum of the Business subjects must be in line with the current needs highlighted in IR 4. As predicted before, most of the current jobs would become obsolete and new jobs expected to fill in the void. To prepare for the changes, current curriculum unable to sustain and change needed for it. As more innovation been predicted to take place, the curriculum itself should be empowering and not lagging behind the technology "tsunami". Students need knowledge that can match with the evolution that either can make us swim together or sank to the core if they are not prepared.

Theory of Change and Education System in Malaysia

As many theories can relate to education, I would like to focus on Theory of Change (TOC). This particular theory focuses on change needed in any situation or scenario for better prospect or further development. In higher education context in Malaysia, this theory does make sense. According to the TOC mapping, there are six stages to relate to the required changes in any situation.

Long-Term Goals

Ministry of Education has clear vision on their goals about the future of education. It can be seen in many ways and Malaysian Education Blueprint for higher education is one of the evidence. However, goals should be based on current trends which can help for development and should not be seen as burden by the academicians. Current requirement highlighted the necessities to produce paper based research works and dissertation. Priority should be given to IR 4 as it is the current trend which mark the evolution of education system to the next level.

Necessary Requirements

Requirement for this goal is very clear as we need the society to be more open to technology based education rather than the traditional way. Books should be replaced by gadgets and should not be burden with outdated skills such as memorising. Gadgets proved to be worth in bringing out many hidden talents in human and we need to utilise them for better future. Gadgets will help in many ways for students in learning process and will be useful for them in their daily routine.

Assumption About Context

IR 4 context is new to Malaysia and I agree the transition process would not be smooth as expected. As many Malaysian students still lack of proficiency in technology would be a great barrier for this initiatives. Lots of research needed to prove the worth of IR 4 in all sectors including education. As we have seen many outstanding result in industries but in education it is kind a new concept for our country. IR 4 will help in the development of education system in Malaysia as many new technologies will ease the teacher's burden in teaching and learning process.

Interventions

Technology should play major part in education as it will bridge the gap between information and skills. Most of the future jobs involved technology heavily and we need our students to be a well versed task forced and the ability to finish the job perfectly. Students can work smartly and solve many major issues without needed to be in that place. Virtual screen will help them to solve many issues and problems while there are doing their household chores and personal things.

Indicators for Measurement

Achievement should be seen as indicators for measurement. Ability to operate and use technology based gadgets indicates the students' achievement in techno world. IR 4 certainly needed the students to think out of the box and stand out performer from the crowd as most of the current jobs may become irrelevant in future.

Explanation

Basically, IR 4 plays an important part in redesigning the standard of education in Malaysian context. We need to place ourselves in line with the technology for us to be successful. Change is inevitable. We need to change our mentality in order to equip with the demand which seems impossible for now but it is actually possible.

BACKGROUND OF TECHNOLOGY BASED EDUCATION

Technology plays an important role especially in the IR 4 learning era. Technology has achieved a high pinnacle in development and has gained the entity of most important tool in our lives. This has an effect on education and its impact can be divided into three phases: namely replacement phase, the transition phase and the reform phase (Willems, Stakenborg & Veugelers, 2000). In replacement phase, the lecturer utilize this as a tool for teaching and learning without alternative teaching methods and approaches, while in the transition phase, the lecturer can stimulate the use of information and communication technology (ICT) to introduce new teaching methods and approaches. Lastly the reform phase also involves a fundamental change in the role of students and lecturers in teaching and learning. Education is a never ending process. In order to develop, we need to educate ourselves with knowledge and continue to feed our needs.

The rapid development of ICT affects almost all aspects of life including education. Developments in technology have also changed the way people learn (Naismith et al., 2004). In addition, the learning process has now surpassed physical space of classrooms (Kukulska - Hulme & Traxler, 2005), the character of globalization and lifelong learning process (Sharples, 2000). In addition, the technology is a linkage between the students' experience and existing knowledge with that of new knowledge which has been taught or learned which is the essence of the theory of constructivism (Jonassen, 2000; Grabe & Grabe, 2004). Hence, the education system in Malaysia is also affected by this problem, particularly in relation to the application of ICT in teaching and learning process.

Teaching and learning process nowadays is no longer focused solely in the classroom, but can occur anywhere and at any time. Earlier, the education system has undergone another evolution in technology education when mobile learning or *Mobile Learning (M-Learning)* was introduced to promote teaching and learning. *M-learning* is a concept implemented in the learning process. It emphasizes the ability to facilitate the transfer of learning process irrespective of the physical location of the learning process (Kukulska-Hulme & Traxler, 2005). In addition, Quinn (2002) defines M-*Learning* as learning through mobile devices such as Palms, PDAs and even mobile phones.

Mobile communication technology has provided a wide array of alternatives and communicative facilities, for example mobile phone is no longer limited for the purpose of receiving and making calls alone. The handset is also equipped with facilities to send and receive *e-mail,* short message service (SMS) and multimedia messaging service (MMS), which is present in the form of a synchronization of music, images, animated sequences and short video recording (Donald, 2003). In addition, by using a smart phone (Smartphone), users can visit the website and browse the various computer applications. Wireless communication facilities have also been provided to enable users to use multiple applications that require internet access.

Learning by *M-Learning* has been proven effective in other countries. Sharing notes among students has been facilitated by the availability of mobile devices such as smart phones among coaches. They no longer need to sit in front of a personal computer to download the notes. Sharing notes has been simplified by using only *wifi* or broadband service provided by the mobile phone service provider subscribed and the process of downloading the notes can be done anywhere and at any time. The ability to use QR code can also help students to download notes during the lecture sessions. This simple process requires the teachers to upload QR-code, and enables students to download the notes using the QR-code. In addition, according to Triantafillou, Georgiadou & Economides (2006), through their study of the test through mobile devices in education, found that the assessment done by testing *M-Learning* was more effective and efficient because it saves time when compared with tests using paper and pencil.

Moreover, according to Wood (2003), the use of *wireless* technology in education can contribute towards solving the digital gap among developing countries, as in general IT equipment such as cell phones and PDAs are cheaper than desktop computers. The use of mobile devices in *M-Learning* helps in increasing students' motivation, improve the skills to organize, promote accountability, promote collaborative learning, and help review students' progress more quickly and effectively (Savill-Smith & Kent, 2003).

The emergence of these technologies are basically because of the weakness in the traditional teaching and learning method. By using these types of technologies, the students have gained various skills in the use of various online applications. In fact, their daily lives are inseparable from the use of various types of gadgets. But, in the real learning situation it has a contrast consequence. They seem to be in the stone age in absence of any sophisticated technology. The learning situation alienated them from being linked*(connected)* to the learning process that is being followed. In the Industry 4.0, this aspect was emphasized for a successful learning process.

Traditional teaching methods or conventional practices in teaching and learning today will not be able to meet the demands of Industry 4.0 education. Learning is no longer focused in the classroom alone, but involved teaching methods which evolved over the pace of ICT in Malaysia. For example, textbooks cannot accommodate the upgrading and renewal of information in line with the times (Chan, 2001). In addition, traditional method to source for materials turned out to take a longer time compared with the use of information and communication technology.

Given that ICT technology can ensure resource information to be obtained easily at anytime and anywhere, thus the use of ICT in teaching and learning as well as management and administration of education is necessary in Malaysia (Chan, 2001). The use of ICT can have a positive impact in the students learning process.

The use of technology in education should not be subjected to limitations or obstacles that may hinder the learning process. The existing constraints in infrastructure and information technology communication in schools or institutions of higher education must be addressed immediately (Ismail et al., 2004). This problem should not be underestimated because only quality education can contribute excellent output of teachers. The lack of infrastructure in education and other contributing factors such as socio-economic background of students, individual development of the students, and other physical and social environment may hinder an individual's success in education and life (Wan Din, 2001).

The evolution saw the beginning of teaching and learning methods using chalk boards; switching to the use of computers, electronic materials, *Liquid Crystal Display* (LCD), video cameras and other technology in the classroom. From time to time, education underwent changes in line with progress in mind, will and instructional technology. This new concept of education or E-Learning is a new learning method which facilitates students to learning when notes and assignments as required by the student are sent by the instructors via electronic mail (e-mail) and *Compact Disc Read Only Memory* (CD-ROM). Through E-Learning, the learning process, critical and creative thinking can occur effectively. In addition, collaborative learning methods, discussions with friends and with other people through the Internet like e-mail, *Face book Messenger, Whatsapp, Wechat, Line* and video conferencing plays an important role. In this way, learning will be more active and effective.

Online learning such as electronic learning *(E-Learning)* has been expanded through *M-Learning.* This is caused by the E-Learning to be unsuccessfully applied fully in the education sector due to several factors. In a study entitled E-Learning governance in institutions of higher learning (IPT) Malaysia by Wahab and Mustafa (2006), they found that the lack of staff, lack of incentive by the IPT to those responsible for implementing E-Learning over IPT, the absence of a clear policy regarding E-Learning and the absence of clear guidelines on the implementation of E-Learning are contributing factors which led to the failure of E-Learning to be implemented fully.

M-learning is a new concept in the learning process and emphasizes the ability to facilitate the transfer of learning process irrespective of the physical location of the learning process (Kukulska-Hulme & Traxler, 2005). In simple words, the learning process can take place anywhere, not in the classroom alone. The advancement of technology indirectly provided opportunity to change the shape of the learning method to create learning communities, to connect people in the virtual world and the real-time, to provide the necessary expertise and support lifelong learning (Sharples, 2000).

THE IMPORTANCE OF RESEARCH

Previously the industrial revolution was first depended on the use of vapour-powered machines. The second industrial revolution involves electric power. Followed by a third industrial revolution based on information technology and computer as its pillar. However, Industry 4.0 covers the discovery of new technologies such as automation, Internet of Things (IoT), analysis and big data, simulations, system integration, robotic and cloud utilization that will accelerate the development of the modern world landscape. This revolution marks the emergence of cyber physical systems involving a whole new capability for human beings, machines and new technology technologies. In other words, the automation technology is seen as a technological capability that does not necessarily involve human beings. Industry 4.0 can also address dependency on energy sources that will significantly change the future of the world of work. So, if the complex new technological challenge is not given due weight, it can cause Malaysia to lag behind in global competition.

Secretary-General of Treasury, Tan Sri Dr Mohd Irwan Serigar Abdullah, said Industry 4.0 that involved all sectors not only increased productivity, but would also lower costs and surged revenue. He also added that Industry 4.0 will modernize all sectors in Malaysia that lead to digitization in line with current needs. With the future looks promising, Malaysia need to step up to catch up with the fast moving trend. Education still a stumbling block to many of us since our education system lacks depth and consistency as it is prone to political moves. Since independence, our education systems been revamped many times without taking into consideration it's impact towards the educators and the students. If a new minister takes over the education ministry, he changed the system according to his style and it continues till today. Inconsistency caused our education system the quality needed to match the demand of worldwide technology.

Through industry 4.0, the education system need some overall overhaul for us to see the desired outcome. Former Higher Education Minister, Datuk Seri Idris Jusoh said in order to achieve the ultimate goal of success in Industry 4.0 era, process of teaching and learning has to be changed. Four aspects need to be emphasized in teaching and learning process. First, the learning spaces should be redesigned. It means that classrooms and lecture halls should be equipped with multi-tiered collaborative tables and the use of smart board should be prioritized. Second, different kind of pedagogies are needed, which are heutagogy (self-determined learning), paragogy (peer-oriented learning) and cybergogy (virtual-based learning). Pedagogies should be interesting and engaging. It shouldn't be dramatic or boring. Innovations such as mobile computing, cloud, social networking such as Facebook, Twitter and Skype and big data have created an opportunity to build a learning ecosystem that allows personalised learning which is independent of time and place. Students will be able to design their own educational pathways based on their personal goals. With a rising level of complexity, it will be highly important to impart deeper learning. This can be achieved with the increased use of blended, project- and scenario-based, and practice-oriented learning. Innovation being key to success, experts suggest that maker space, which is characterised by open source innovation and learning-by-doing, should be utilised as a tool to train graduates (Haseeb, 2018).

Third, the overall curriculum need to be fluid and organic. Idris announced that as of 2018, up to 30 per cent of all university programmes will adopt this concept, enabling them to respond to innovations and new areas of knowledge without being bound by traditional rigid curriculum practices. The fourth aspect is, all of the aforementioned should incorporate the latest learning and teaching technologies.

The virtue concept should be developed as the lesson without lectures might infused lots of discipline to the students. Assessment would be another aspect which need to be look into in this era. The proposed new curriculum should be examination free as students need to be assessed practically and not based solely on examination. The students should be able to transform their knowledge in many ways and should not be judged based on examination results. Many potential students been outclassed by this so called examination based education system and this is the right time for the new move to take place. The development of ideas is based on the improvement done at the fundamental level which would be a good platform for IR 4 to be a hit.

Education 4.0 is suggested to affect all the domains (Cognitive, Affective and Psychomotor) in the Bloom's model (Haseeb, 2018). As per the model, in the cognitive domain, Application, Analysis, Evaluating and Creating will become way more important relative to the lower level cognitive skills. Those skills will open up more possibilities and opportunity for the students in the market. IR 4.0 will require human resources with adequate digital and data literacy. Students across disciplines will, therefore, need to gain digital and data literacy during their studies. Education is playing an important part in helping students to acquire those needed literacies to catch up with the current trend.

THE NEED FOR CHANGES

IR 4 is rapidly growing throughout all over the world. The need for changes is essential as future jobs scope is based on digital literacies. Higher education sector is pressured to accommodate the needs of digital natives. According to the report reported by Ministry of Higher Education, there are six main facts identified to classify digital natives. First one is 20% of the students are addicted to the internet. Basically addicted means using internet for not only education purposes but as a whole. The report also highlighted 90% of children and adults play video games. The smartphone world enables people to use phone to play games. Nowadays playing games can't be considered only for entertainment purpose but many people are playing games while earning online through it. Third fact is 28% of people aged 10-34 engage in literary reading. Fourth one is many students bored by traditional lecturers in the classroom. Fifth fact is there are differences in gender in the use of internet. Females use internet for social reasons while males use it for virtual games. Last fact highlighted by the report is, human manifested 8 seconds of attention span compared to a goldfish with 12 seconds. However, humans are getting smarter and smarter by multitasking.

"Ultimately, technology is a tool. People come first. Its purpose should be centered on humanity and public interest. Where tolerance, respect, care and compassion come into play. Eventually, it boils down to culture and value with Industry Revolution 4.0 ushering into a new cultural renaissance. It needs a collective wisdom". - Former Higher Education Minister, Dato' Seri Idris Jusoh

DIGITAL IMMIGRANTS VS. DIGITAL NATIVES

IR 4 as we all know will take technology as main source in a working environment. To accommodate the needs, workers or people associate with this type technology needed to technology savvy. Currently, people were divided into two main category based on few criteria. Digital immigrants and digital natives.

Figure 1. (a) Digital immigrants and (b) digital natives
(Adapted and redesign from http://reembloogs.blogspot.com/2013/02/digital-natives-digital-immigrants-by.html)

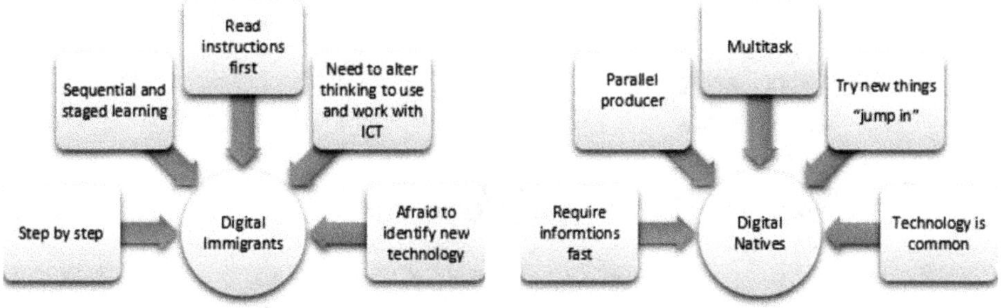

The classification comes with some interesting points although many would argue them. Digital immigrants are those born before 1980 while digital natives are those born after 1980. Digital immigrants needed information step by step whereas digital natives information as fast they could and at this point they will skip some irrelevant steps that might will make them slow. Usually digital immigrants needed an assist or steps sequence to work especially working with technology while younger generations will skip the steps to accommodate their needs. Digital immigrants also will read instructions first before they operate whereas with the knowledge they possessed, digital natives will skip and jump in to try new things. It will make their job much easier rather than follow the instructions. Digital immigrants will also do one work at a time as they wont be drawn into complexity of not finishing their jobs.

Doing one job at a time will ensure they complete the given job perfectly. Digital natives in other hand is good at multitasking. They will do more than one job at a time and will finish perfectly in accordance to the job demand. This will enable them to save lots of time and pressure is off from drawn into "due date". This so called digital immigrants and digital natives needed education for them to functioning, Digital immigrants as we all known are from academic background and they learn through traditional teaching and learning process. Digital natives learn through technological background and been helped by social medias, peer reviews as well as virtual learning. Nowadays people learn lots of simple things through social medias especially YouTube. Just search for any help in fixing and there are thousands of videos to guide us. We no more needed manuals and texts. Graphics make our life easier.

MALAYSIAN HIGHER EDUCATION 4.0 FRAMEWORK

Malaysian Higher Education Ministry has come up with its Higher Education 4.0 Framework which would be a guideline for future education to cope with the needed changes in line with the IR 4. Four aspects given preferences in the framework namely future ready curriculum, agile governance, talent planning and research and innovations. Ministry had come up with this framework as they felt current education system unable to feed the demanding industry requirements as well as to make sure Malaysians fully aware about the importance of IR 4 and its impact towards the country's economic growth. The highlighted areas show the current concerns that we need to look into in order for us to sustain. Future ready curriculum stressed on tailor made curriculum that would fit perfectly to the demanding nature of IR 4. Ministry is committed in redesigning the higher education initiatives as well as fluid and organic

Figure 2.
(Source: www.jpt.mohe.gov.my)

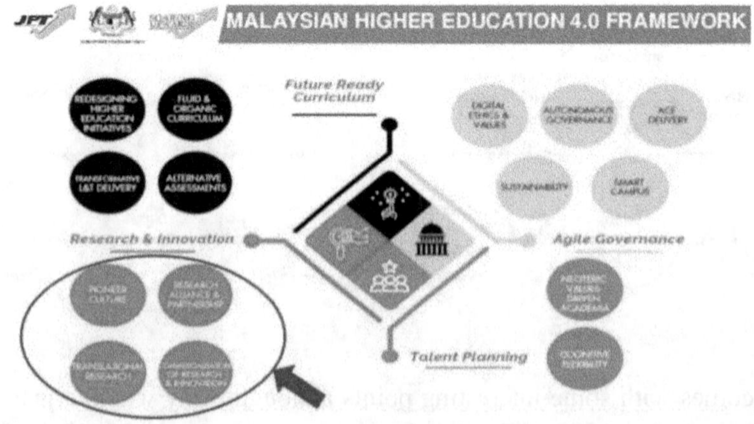

curriculum. These are necessary to equip the students with the competitiveness among students which will enables them to be knowledgeable and skilful worker. Apart from that transforming the teaching and learning delivery from traditional method to more virtue way and alternative assessment which is not examination oriented are other goals. Under agile governance, the ministry focused on digital ethics and values, autonomous governance, ace delivery, sustainability and smart campus agendas. Ministry's talent planning includes neoteric values – driven academia and cognitive flexibility. Under research and innovation field, pioneer culture, research alliance and partnership, transitional research, commercialization of research and innovation. These are the criteria drawn by the ministry under this framework as a guideline for all higher learning institutions to follow in order to produce world class talents.

CONCLUSION

As we all know, the current education system is not competent enough and need changes in order for us to see better future. With the emergence of technology "tsunami", education need to be in line with it in order for us to produce world class talent. Current curriculum unable to stand with the sophisticated standard which is required. Many changes need to be done because they future looks so bright and these changes are in accordance to the trend. We might be living in hi tech world in the future and it education system need to be flexible and open.

REFERENCES

Abdo, A. A. M. (2016). *Factors affecting career choice among undergraduate students in Universitas Indonesia* (Doctoral dissertation). Universiti Utara Malaysia.

Brenneke, J. S. (1990). Case studies and other student based programs in entrepreneurship programs. In Entrepreneurship education: intergrating entrepreneurship in Secondary Curriculum (pp. 231-242). New York: Quorum Books.

Chan, Y., F. (2001). Cabaran Pengurusan Pendidikan di Era Baru. *Jurnal Pengurusan dan Kepimpinan Pendidikan, 11*, 68-79.

Donald, J. L. (2003). *Wireless Messaging Demystified: SMS, EMS, MMS, IM, and Others*. McGraw-Hill Professional.

Garcia, D. H., Leles, A. D., & Romano, R. R. (2017). Program Entrepreneurship and Innovation: Education as the Core of Innovation. In *Advances in The Human Side of Service Engineering* (pp. 235–244). Springer International Publishing. doi:10.1007/978-3-319-41947-3_22

Gibb, A. A. (1993). Enterprise culture and education understanding enterprise education and its links with small business, entrepreneurship and wider educational goals. *International Small Business Journal, 11*(3), 11–34. doi:10.1177/026624269301100301

Grabe, M., & Grabe, C. (2004). *Integrating Technology for Meaningful Learning* (4th ed.). Boston: Houghton Mifflin Company.

Haseeb, A. M. (2018, Jan 10). Higher education in the era of IR 4.0. *New Straits Times*. Retrieved from https://www.nst.com.my

Ismail, B., Ayob, J., Alias, M., & Kaprawi, N. (2004). Penjuruteraan Semula Pendidikan Teknikal dan Vokasional: Satu Cetusan Pemikiran dan Tinjauan. *Jurnal Pendidikan Teknikal., 3*, 24–35.

Jackson, D., & Chapman, E. (2012). Non-technical skill gaps in Australian Business graduates. *Education + Training, 54*(2/3), 95–113. http://dio.org/10.1108/00400911211210224. doi:10.1108/00400911211210224

Jonassen, D. H. (2000). *Computers As Mindtools For Schools: Engaging Critical Thinking* (2nd ed.). Merrill Prentice Hall.

Klandt, H. 1993. Methods of teaching: what is useful for entrepreneurship education? *Internationalizing Entrepreneurship Education and training. Proceedings of the Int. Ent. 93 conference Vienna*, 31-41.

Klaus, S. (2016). The fourth industrial revolution. In *World Economic Forum* (p. 11). Academic Press.

Kukulska-Hulme, A., & Traxler, J. (2005). *Mobile Learning: A Handbook For Educators And Trainers*. London, UK: Routledge.

Lans, T., Hulsink, W. I. M., Baert, H., & Mulder, M. (2008). Entrepreneurship education and training in a small business context: Insights from the competence-based approach. *Journal of Enterprising Culture, 16*(04), 363–383. doi:10.1142/S0218495808000193

Lomax (2014). Current Problems in Business Education. *The Journal of Business Education, 13.*

Maziah, Y. (2005). *Kajian Kecenderungan Pemilihan Kerjaya di Kalangan Siswazah. Jurnal Pengurusan Awam. Jilid 4 bilangan 1.* Jabatan Perkhidmatan Awam Malaysia.

McGraw-III, H. (2007). *Shaping tomorrow's business leader: Principles and practices for a model business ethics programme.* Business Roundtable. Institute of Corporate Ethics.

Mohd, F. A. M. R. (2009). *Persepsi Pelajar SPI Terhadap Amalan Kemahiran Insaniah Dalam Pembelajaran Projek Sarjana Muda UTM.* Tidak diterbitkan.

Naismith, L., Lonsdale, P., Vavoula, G., & Sharples, M. (2004). *Literature Review in Mobile Technologies and Learning.* Academic Press.

Pandiyan, V. (2017, Sept 6). Industry 4.0: The future is here. *The Star.* Retrieved from https://thestar.com.my

Quinn, C. (2002). *MLearning: Mobile, Wireless, in Your Pocket Learning.* Academic Press.

Savill-Smith, C., & Kent, P. (2003). *The Use of Palmtop Computers for Learning: A Review of The Literature.* London, UK: Learning and Skill Development Agency.

Seow, J. (2010, March). Marketing excellence via education and training. *Marketing Magazine,* 10-17.

Sharples, M. (2000). The Design of Personel Mobile Technologies for Lifelong Learning. *Computers & Education, 34*(3-4), 177–193. doi:10.1016/S0360-1315(99)00044-5

Timmons, J. A. (1997). *The entrepreneurship mind.* Prentice-Hall, Inc.

Triantafillou, E., Georgiadou, E., & Economides, A. A. (2006). *He Design and Evaluation of a Computeriaed Adaptive Test on Mobile Devices. In Science Direct* (pp. 1319–1330). Elsevier.

Wan Din, W. C. R. (2001). Wanita dalam pendidikan. *Jurnal Pengurusan dan Kepimpinan Pendidikan, 11,* 1-14.

Willems, G. M., Stakenborg, J. H. J., & Veugelers, W. (2000). *Trends in Dutch Teacher Education.* Garant.

Wood, K. (2003). *Introduction to Mobile Learning.* M Learning.

Chapter 16

Exploring Collaborative Learning for Programming Course to Enhance Students' Interest and Achievement:
A Case Study of Universiti Utara Malaysia

Shafinah Farvin Packeer Mohamed
Universiti Utara Malaysia, Malaysia

ABSTRACT

As programming is one of the core subjects for BSc IT students, it is vital for them to have programming skill. However, students are often terrified of programming since it is a challenging subject. Traditionally, the programming teaching and learning process are limited to lectures and lab exercises. This approach is no longer appropriate in today's learning environment and required skills in the 21st century. Collaborative learning is acknowledged as an approach that fits the programming students. Thus, this chapter has adapted the collaborative approach in Programming 1 class. Three techniques were applied namely problem-based learning, buzz groups, and pair programming. At the end of the semester, an online survey to obtain students' opinions and a focus group discussion were conducted. In addition, the examination results of the repeaters were also compared. Results from the survey and focus group indicate that the collaborative approach is able to increase students' interest and the grade of the repeaters in the Programming 1 subject.

INTRODUCTION

Human capital is an important element in a country's economic growth and it is highly needed in this globalization era. It provides extremely competitive workers towards the nation which in turn promotes the economic growth. The lack of qualified human capital resource will cause a country to become weak. Thus, Malaysia is now working towards producing comprehensive and world class level human capital

DOI: 10.4018/978-1-5225-7832-1.ch016

Copyright © 2019, IGI Global. Copying or distributing in print or electronic forms without written permission of IGI Global is prohibited.

resources. Nowadays, excellent academic achievement is not a guarantee for graduates' employment in Malaysia and also does not prove them to be of highly skilled workforces (Zaliza, Mohd Safari, & Arasinah, 2015). The industry expects them to have the necessary soft skills which are essential in their working life prior to starting their jobs. Unfortunately, there are Malaysian graduates who do not possess those important skills thus it contributes to the unemployment problem in Malaysia. Unemployment is also a global issue since the fresh graduates are not equipped with comprehensive capabilities. The unemployment rate for the year 2017 was 10.8% and it was at its highest, according to a report issued by the MIDF Research (2018). One of the reasons for the unemployment is the inappropriate skills matching for the adolescents aged between 15 to 24 years old.

Zaliza and Mohd Safari (2014) stated that graduates' attributes, lecturers' competency and education quality are the employability factors among graduates. Additionally, Rahmah, Ishak and Lai (2011) highlighted that the cause of unemployment is that there is no balance between supply and demand for graduates. This has caused 30% of the graduates to get salaries that did not match their qualifications. Moreover, the quality of the graduates also plays an important role whereby employers may not satisfy with their skills and work quality that do not meet the requirements. On the other hand, Ranjit (2010) in Zaliza et. al (2015) mentioned that Malaysian graduates lack of ten main weaknesses such as management flaw, problem solving, communication, leadership, creativity, critical thinking, proactive, self-confidence and interaction skills. In short, employers are not only seeking for high grade achievers, but also those who are excellent in soft skills. Therefore, students need to obtain the appropriate soft skills during their studies in the university and not only depend on the academic qualification to survive in the future career world. Eventually, those skills will permit the graduates to compete at the global level (Sarker, Davis, & Tiropanis, 2010).

Similarly, students of the School of Computing (SOC), Universiti Utara Malaysia (UUM) need to have excellent academic qualification as well as the necessary soft skills by the time they graduate. In order to produce high quality students, the teaching and learning process is influenced by many factors, namely the course itself, the level of professionalism besides the lecturers' teaching style and the classroom conditions (Zamali, Fatihah, Munirah, & Tamsin, 2016). Interestingly, lecturers play an important role in increasing students' interest in class (Wasserman, 2013) as well as influencing students' achievement (Zaliza & Mohd Zafari, 2014). Therefore, a suitable approach needs to be adapted in the Bachelor of Science in Information Technology (BSc. IT) students of SOC. A comprehensive and systematic planning need to be done to ensure that the implemented teaching and learning process helps to produce students who can compete globally.

Programming is one of the core subject for students who choose to enroll in the IT undergraduate programs. This has always been the most challenging and difficult subject for students (Isong, 2014). Only a small number of students find that learning programming is fun and enjoyable. Majority are usually not interested in this subject as they have a bad impression that this subject is complex and hard to be understood. As a result, the rate of failure is usually high for this subject as pointed out in the literature (Malik, 2018; Costa, Fonseca, Santana, de Araújo, & Rego, 2017). Nevertheless, students need to acquire and excel in the subject since programming is one of the essential skills for an IT student to survive in the working environment (Law, Lee, & Yu, 2010). Programming skill cannot easily be attained just by reading books without hands-on practice by doing programs. Students should put more effort by doing more programming exercises on their own initiatives particularly during their free time outside from the classroom since in-class practice is not sufficient. Hence, they need to more persistent, motivated and daring in order to make it work.

According to Jenkins (2002), students need to acquire both deep and surface learning style to succeed in programming. Surface learning in programming is needed to memorize the syntax while deep learning is necessary to understand the programming logic and analyze the problem. Unfortunately, most students believe that they can succeed in programming subject by only reading textbooks and understanding the syntax (Gomes & Mendes, 2007). They will just memorize the syntax and do not know exactly when to use them, specifically in solving a problem. For example, when they need to form a complete program, they do not know how to combine the statements for solving a given problem. Eventually, they will get tired and tend to lose their interest to write program especially when they start experiencing repetitive errors in their codes. As a result, the students will produce poor quality codes and submit their assignments with incorrect designs or even incomplete programs which contain errors. Previous studies have revealed that these problems are common among students who are learning programming for the first time (Iqbal & Harsh, 2013; Shuhidan, 2012; Yadin, 2011; Law et al., 2010). This was pointed out by Malik and Coldwell-Neilson (2016) and Fidge, Hogan and Lister (2013) stating that learning programming is a challenging task to students since they have to grasp different sets of skills at the same time such as the syntax and semantics as well as the problem solving skill. However, this problem can be overcome by applying appropriate learning strategy. If students find it challenging to learn programming, lecturers is challenged by the high rate of students failing the subject. Thus, lecturers have to be creative in getting students' attention and interest towards the subject so that they will put more effort in doing programming exercises on their own initiatives without being instructed, during and outside the class hours. Consequently, their programming skills and achievement in the subject can be improved.

Typically, the teaching and learning process of the programming subject will only be conducted through class lectures and lab exercises. In completing the labs exercises, lecturers believe that each student should work individually (Maguire, Maguire, Hyland, & Marshall, 2014). Unfortunately, this approach is no longer appropriate for the 21st century learning environment. Instead, collaborative learning is considered as a better approach that can increase students' interest and engage them with class activities. Collaborative learning approach refers to a learning activity where two or more people learn something together, for example knowledge, skills and competencies (Serrano-Camara, Paredes-Velasco, Alcover, & Velazquez-Iturbide, 2014). In addition, students can acquire the essential soft skills required for today's graduates through this approach. On top of that, it also promotes deep learning, which is desired in learning programming.

Consequently, collaborative learning was promoted to the Programming 1 students in the School of Computing (SOC), Universiti Utara Malaysia (UUM). In the class, students were introduced to Java programming language and the subject was intended for the first semester BSc. IT students. Three collaborative learning techniques were introduced throughout the semester. The three techniques used were buzz group, pair programming and problem based learning (PBL). This paper discusses the outcome from the implementation of those techniques in the class.

BACKGROUND

Academicians in computing are steadily recognizing the collaborative learning as having various benefits (Maguire et al., 2014; Furberg, Kluge & Ludvigsen, 2013; Tsai, Li, Elston & Chen, 2011). For example, Tsai et al. (2011) found that collaborative learning can enhanced student experience in producing Wiki websites. Besides, Maguire et al. (2014) concluded that student engagement and performance can be increased with the use of collaborative learning in the computer science lectures. This learning approach is also widely applied in other fields (Selly, 2013; Young, 2010). There are many important benefits being reported such as stimulating critical thinking, deep learning, and positive attitudes towards learning and team members. It also offers a flexible and open environment to work collaboratively in a group (Wang & Lin, 2007). Among the collaborative learning techniques are the pair programming, buzz group and problem based learning (PBL) activities.

Pair programming is one of the collaborative learning techniques which is increasing its popularity in the field of Computer Science (Eierman, & Iversen, 2018; Umapathy & Ritzhaupt, 2017; Salleh, Mendes & Grundy, 2011; Braught et al., 2008). Through pair programming, two programmers will collaborate to solve the same programming problem by using the same workstation. One programmer becomes the 'driver' and another one becomes the 'navigator'. The driver will have the control of the input devices and type the programming codes. On the other hand, the navigator reviews the codes typed by the driver to check for any flaw, such as syntax or logic errors (McDowell, Werner, Bullock & Fernald, 2006). The navigator will also look into alternative solutions and continuously asks questions (Williams & Kessler, 2000). Both the driver and navigator change their roles frequently. Research has shown that programmers working in pairs produce shorter programs with better design and fewer bugs than those working alone (Cockburn & Williams, 2001). Additionally, previous studies have also found promising evidence on the effectiveness of pair programming for the programming courses (e.g., Eierman, & Iversen, 2018; Umapathy & Ritzhaupt, 2017; Salleh et al, 2011; McDowell, Hanks & Werner, 2003; Williams & Upchurch, 2001), as well as in a primary school setting (Denner, Werner, Campe & Ortiz, 2014; Gallardo-Virgen & DeVillar, 2011).

With pair programming, students feel that they are supported by their friend, since they work collaboratively. Furthermore, when they analyze and critique the code of their pairs, they can learn and reflect with their own style of writing code (Williams & Upchurch, 2001). In addition, they feel more satisfied when they are able to produce a better quality working code and this will motivate them to do programs in contrast to when they produce low quality non-working code (McDowell et al., 2003). This technique is commonly mentioned in texts as a technique that can offer benefits to the first year programming students. Mostly cited benefits are the increment in interest and improvement of the grades (Maguire et al., 2014; Teague & Roe, 2009; Teague & Roe, 2008; Braught et al., 2008; Williams, Mcdowell, Nagappan, Fernald, & Werner, 2003; Chase & Okie, 2000). Other important benefits reported are the quality improvement, shorter completion time, better understanding of the programming process, enriched communication skills, bugs reduction and enhanced learning (Eierman, & Iversen, 2018; Preston, 2005).

Williams and Upchurch (2001) initially validated the positive effects of this technique as compared to individual programming. They found that students are more enjoyable and can learn from each other when working in pairs. Also, there is an improvement in team efficacy and communication, as well as it boosts their confidence. Williams et al. (2003) reported in another study that the lab sessions are more productive, less frustrating and more conducive to advanced, active learning than traditional labs. McDowell et al. (2006) pointed out that paired students are significantly more likely to remain in the course through the final exam (90.8% versus 80.4%). Paired students are also 18% more likely to have interest towards the subsequent courses in programming and also more likely to register for a computer science major. Building on this work, a systematic review by Salleh et al. (2011) concluded that pair programming consistently leads to improved grades and increased student satisfaction.

Buzz group on the other hand is defined as dividing a large group of students into smaller groups which consists of five to ten people to discuss a chosen topic in a limited time (Scannell & Donaldson,

2000). Students can share their own expertise and experiences among their friends by using this technique. Moreover, they have the chance to test their ideas within a smaller group. Additionally, shy or introverted students may feel more comfortable contributing to the discussion in these small groups, especially if they know that a more extroverted member of the group will report the comments or ideas of the group members to the class. Also, students can engage with the activity and become active in class through buzz groups (Young, 2010). Apart from that, Selly (2013) suggested that the learning outcomes can be achieved more efficiently through buzz group.

Likewise, Problem based learning (PBL) is another widely used technique in collaborative learning. PBL is a student centered approach that is used to promote active learning (Hong & Seyal, 2014). In PBL, students are provided with problems that need to be solved in groups collaboratively (Hung, Jonassen, & Liu, 2008). This technique is empirically proven on its benefits for the programming subjects (Lykke, Coto, Jantzen, Mora, & Vandel, 2015; Soares, Fonseca, & Martin, 2015; Hong & Seyal, 2014). Referring to the context of programming subject, the problems given to students are focused on the programming problems. Various benefits are reported on the effectiveness of PBL; it promotes critical thinking and analysis, enables students to manage work as a team, explain the concepts, solve the problems and encourages self-directed learning, as well as increases overall students' achievements (Soares et al., 2015; Hong & Seyal, 2014; Nilson, 2010). Besides, Nuutila, Torma and Malmi (2005) reported that PBL enables students to get motivation and support from their team mates. The three collaborative learning techniques are found to have benefits for the students, thus this study applied them in the Programing 1 class to enhance students' interest and achievement.

METHODOLOGY

The study was conducted through three main activities which are the implementation of collaborative learning, online survey and focus group discussion, as elaborated further in the next sub sections.

Implementation of Collaborative Learning

There were three collaborative techniques employed in the Programming 1 class throughout the semester which are the pair programming, buzz group and PBL. They are explained subsequently.

Pair Programming

The pair programming was performed during the lab sessions. The students were introduced to the pair programming activity and its implementation. Importantly, they were briefed on the role of 'driver' and 'navigator' in the beginning of the lab session. The students were given programming questions to be solved in pairs. They chose their pairs on their own. This is to ensure that they are comfortable to work with each other. Since the number of students in the class was odd, there was one group with three students. The lecturer assisted the students throughout the activity and reminded them to swap their roles by time. At the end, they were asked to present their answers to the class. Students can compare their answer and learn from their friends through presentations since they might present different solutions. Figure 1 illustrates the students who are working in pairs and presenting.

Figure 1. Pair programming activity

Buzz Group

Different activities were implemented to conduct the buzz group technique. Overall, the technique was used at the end of a topic where students had to recall what they had learned about the topic. In one session, the students were given crossword puzzles that needed to be solved in groups. The questions in the crossword puzzles were based on the previous lectures and videos shown in class. The students tried to solve the crossword puzzle in small groups and then shared their answers in the class. They had to understand the important points learnt in order to solve the puzzle. Everyone in the group was responsible to give their idea to promote collaborative learning within small groups. More importantly, they enjoyed the activity and were able to get engaged with the class activity. In another session, the students were asked to form teams to play the Kahoot game. The game consisted of questions regarding topics that they had learnt. Thus, they could recall those topics and work in groups to answer the questions. Alternatively, in another session, the students were asked to create collaborative mind maps. The students created mind maps collaboratively by summarizing those that they have learnt. They were asked to complete the mind maps on the whiteboard which then were used as their notes. Each group had different points to be added into the mind maps and had to contribute in order to complete the mind map. Figure 2 portrays the buzz group activities.

Figure 2. Buzz group activities

Problem Based Learning (PBL)

To implement PBL, the students were given programming problems and asked to design the solutions by using flow charts. They were provided with mahjong paper and marker pen to draw the design. The students had to understand the given problem by themselves through discussion in the group. Then they collaboratively came out with the solution. This had encouraged teamwork, helped to understand concepts better, possessed the ability to solve problems as a team, and encouraged self-directed learning. Then, selected groups were asked to present on their designs. Through presentations, the students can compare their designs with other groups and improve theirs. Figure 3 shows the PBL activity.

Online Survey

An online survey was administered towards the end of the semester to gather the students' feedbacks regarding the collaborative learning techniques implemented in the class. The students were instructed to complete the survey within five days. The instrument was developed by referring to the existing studies, for instance Carvalho (2016) and Hong and Seyal (2014). There were five sections in the instrument, including the demography (personal information and experience in Programming 1 subject), general opinion regarding Programming 1 class (7 questions), pair programming (11 questions), buzz group (5 questions) and problem based learning (9 questions). 7-point semantic differential scale ranging from Strongly Disagree to Strongly Agree was used in most of the questions (Zikmund, Babin, Carr & Griffin, 2010). This scale was used since it is reliable and valid for many research purposes. It is also flexible, easy to be adapted, as well as quick and economical to administer and score, as mentioned by Kerlinger (1973) in Thompson and Stapleton (1979). The respondents were among the students of Programming 1 subject from SOC, UUM. All students had answered the survey (35 of them), which indicated a 100% response rate. The data was analyzed using SPSS software. The analysis mainly involved descriptive statistical analysis.

Figure 3. PBL activities

Focus Group Discussion

Next, a focus group discussion was conducted by gathering seven Programming 1 students, which comprised of students with different level programing abilities in order to get feedbacks from good, average and low ability students. The repeaters of this subject were also chosen. The focus group was used since it can provide valuable feedbacks quickly, as well as it can be conducted easily (Martakis & Daneva, 2013). The discussion which took place at a meeting room in SOC, UUM remained for about 45 minutes. The session was started with a briefing on the objective of the focus group. It was continued with an ice breaking session before the discussion started. The discussion was mainly about the students' opinion on the collaborative learning techniques used in the class. Ten questions were raised towards the students. Figure 4 shows the focus group discussion.

Exam Results Comparison

Finally, the examination results of the repeaters of the subjects were compared to observe their improvements. The comparison was done based on their current grades with their previous grades of the Programming 1 subject.

FINDINGS AND DISCUSSIONS

The findings are elaborated next based on the outcomes from the online survey, focus group discussion and comparison of examination results.

Online Survey Results

The online survey was conducted to gather students' perception on the collaborative learning technique applied in the Programming 1 class. The results are explicated in five sections, as explained further subsequently.

Figure 4. Focus group discussion

Demography

The total of the respondents in the Programming 1 class are 35 students. 51.4% of them are females while 48.6% are males. Majority of them are in semester two to four, who are the repeaters for this subject (88.6%). Among the repeaters, most of them obtained C- (12), 9 of them obtained D+ and the rest obtained D for Programming 1 subject in the previous semester. When asked about the teaching approach used by their previous lecturer, majority of them (69%) replied that their lecturer used only traditional approach in teaching programming subject. Their lecturers used to give lectures and provided them with individual lab exercises, without including collaborative learning activities. The remaining 31% responded that their lecturers used Kahoot application. The demography of the respondents is summarized in Table 1.

General Opinion Regarding Programming 1 Class

To get their insights about the programming subject, the students were asked about their general opinion on the Programming 1 class. The 7-point numerical scale (Zikmund et al., 2010) was used for this question and the remaining questions, which ranges from Strongly Disagree to Strongly Agree. The scale is then mapped to equal intervals, by using the formula found in Ismail, Abedlazeez, and Hussin (2011). The interval values are as represented in Table 2.

Table 3 portrays the students' feedback on this semester's Programming 1 class and regarding the programming subject itself. It shows that they have enjoyed the class as well as the programming subject. They also feel more confident on their programming knowledge by the end of the semester. However, they still feel that programming is a tough subject and hard to be managed since they are confused when looking at the coding. This is a common impression among the programming students, as widely cited in the previous studies, for example Bashir and Hoque (2016) and Rogerson and Scott (2010). This might be the reason why they are less confidence in solving real world problems. As a consequence, they are not much interested to consider a job in the programming field.

Table 1. Demography of respondents

Items	Categories	Frequency	Percentages (%)
Gender	Male Female	17 18	48.6 51.4
Semester	Semester 1 Semester 2 Semester 3 Semester 4	4 29 1 1	11.4 82.8 2.9 2.9
Previous grades (N=26)	C- D+ D	12 9 5	46 35 19
Teaching approach in previous class (N=26)	Traditional Kahoot application	18 8	69 31

Table 2. Interval values

Degree of Agreement (DA)	Interval Values
Strongly Disagree (SD)	1.00 – 1.86
Disagree (D)	1.87 – 2.72
Somewhat Disagree (SOD)	2.73 – 3.58
Neutral (N)	3.59 – 4.44
Somewhat Agree (SOA)	4.45 - 5.30
Agree (A)	5.31 - 6.16
Strongly Agree (SA)	6.17 - 7.00

Table 3. Opinion on programming 1 class

Opinion on Programming 1 Class	Mean	DA
1. I enjoyed this semester's Programming 1 class	6.09	A
2. I often feel confused when I look at programming code	4.83	SOA
3. I enjoy doing programming	5.63	A
4. I am confident that I can solve real world problems	4.91	SOA
5. I would consider a job which involves programming	5.06	SOA
6. Towards the end of semester, I feel more confident with my programming knowledge	5.34	A
7. Now I feel programming is very easy and can be managed	4.91	SOA

Pair Programming

More specifically, the students were asked about the pair programming technique used in the class. Majority of them agreed that pair programming has a lot of benefits such as it helped them to enhance their programming skill, since they can learn from their pairs. Additionally, they responded that through pair programming, their communication skill can be improved and it facilitated them in producing better code quality. Notably, they also agreed that this technique helped them in getting solutions correctly and faster by following the proper steps. Apart from that, this technique also increased their interest in programming and enabled them to complete java programs more precisely. The results are exhibited in Table 4. The outcomes are consistent with prior researches conducted. For example, Maguire et al., (2014) indicated that through pair programming, the students are able to mingle around with other students in the class. Moreover, Corney, Teague and Thomson (2010) disclosed that the students' engagement can be improved through collaborative approach. Keefe, Sheard and Dick (2006) earlier highlighted that the use of pair programming makes the class more enjoyable. Interestingly, Salleh et al. (2011) examined the effects of pair programming and concluded that it can improve the achievement and interest to the programming class.

Table 4. Opinion on pair programming

Pair Programming	Mean	DA
1. I liked pair programming	5.46	A
2. I had to do all work when doing pair programming	4.23	N
3. I learned a lot from my pair	5.40	A
4. Pair programming helped me to enhance my programming skill	5.71	A
5. Pair programming helped me to enhance my communication skill	5.74	A
6. Pair programming helped me to produce better quality of code	5.69	A
7. Pair programming helped me to get solutions faster	5.77	A
8. Pair programming increases my interest to Programming 1 subject	5.69	A
9. I understand better what the question wants when I discuss with my pair	5.89	A
10. I understand better the steps to solve the problem given	5.71	A
11. I know better how to complete a Java program	5.54	A

Problem Based Learning (PBL)

Another collaborative learning technique used is the PBL. Overall, the students enjoyed the PBL activity. It is able to help the students think critically and support them in solving programming problem as a team. By being in team, they can learn and discuss and these enable them to produce better design. Besides, it helped the students to solve problems easier, improved their communication skills and gave them chance to discuss with the team members. On top of that, PBL is found to increase the students' interest on the subject. The outcomes are in line with earlier studies such as Soares et al. (2015), Hong and Seyal (2014) and Nilson (2010). On top of that, as claimed by Nuutila et al. (2005), programming students should not only possess programming skills, but also collaborative work, independent studying and communication. Accordingly, PBL is found to be as one of the ways to achieve these. Table 5 demonstrates the results.

Table 5. Opinion on PBL

Problem Based Learning (PBL)	Mean	DA
1. I enjoyed PBL activity	5.77	A
2. PBL increases my interest to Programming 1 subject	5.69	A
3. PBL helped me to think critically	5.77	A
4. PBL helped me to manage work as a team	5.69	A
5. PBL helped me to solve the problem easier rather than doing alone	5.86	A
6. PBL helped me to enhance my communication skill	5.71	A
7. PBL gave me more chance to ask others about programming	5.97	A
8. I learned a lot from my group mates	5.66	A
9. PBL helped me to produce better design	5.80	A

Buzz Group

Overall, the students enjoyed the buzz group activity and obtained benefits from it. This activity is found helpful for them in collaborating with team members. It also enhances the communication skills and enables students to get to know more people in class. More importantly, this activity is found to increase the students' interest in programming. The results are constant with previous studies such as Young (2010) and Selly (2013). Table 6 depicts the result obtained for the buzz group activity.

Focus Group Discussion Results

The focus group discussion which took around 45 minutes was joined by seven Programming 1 students. The questions raised to them are as follows:

1. Are you interested in learning programming? Why?
2. Do you like to attend Programming 1 class? Why?
3. Do you feel bored when the lecturer only gives lecture?
4. What do you think about pair programming activity?
5. What do you think about buzz group activity?
6. What do you think about problem based learning activity?
7. Do you feel stressful when these activities are conducted?
8. Do these activities able to increase your thinking and communication skills, teamwork, problem solving and programming?
9. Do these activities enhance your interest to the subject?
10. What are the improvements that can be made to enhance teaching and learning activities for Programming 1 subject?

Basically, the students feel that they are interested in learning programming since they are satisfied once they could solve the programming problems. Student A mentioned that she hopes to acquire more knowledge of Programming 2 subject and anticipate more challenging programming problems. Student B who is a repeater for this subject believed that he acquires more understanding in this semester and expects to perform better in the future. Subsequently, majority of the students concurred that they like to attend Programming 1 class. They are generally anticipated in learning new topics and eager to do the programs. They will still confront the difficulties despite everything, for example in understanding

Table 6. Opinion on buzz group activity

Buzz Group Activities	Mean	DA
1. I enjoyed buzz group activity	5.80	A
2. Buzz group activity increases my interest to Programming 1 subject	6.00	A
3. Buzz group activity helped me to collaborate with team members	6.00	A
4. Buzz group activity helped me to know more people in my class	5.91	A
5. Buzz group activity helped me to enhance my communication skill	5.94	A

the programming problem, designing the solutions and remembering the syntax. Hence, they trust that they can enhance their skills by time.

All of the students have the same insights about the traditional teaching approach; they feel bored if a class is only enclosed with lecture, without any activities. The repeaters shared their experience in the previous class and compared with the present class. Student B mentioned that by having collaborative activities, he does not feel bored and can be active in the class. In previous semester, he could not understand much because the lecturer only gave lecture, without considering whether the students understood the lesson or not. Student C added that because she could not understand the lesson, she ended up copying her friend's answer. However, in this semester, through the collaborative activities performed, she could improve herself by discussing and not copying.

The students are satisfied with the pair programming technique used. Student C pointed out that she can discuss with her pair and get better solutions through discussion, rather than finding the solution alone. Nonetheless, Student D highlighted that the pair plays an important role for the success of pair programming activity. There was once that he chose the pair wrongly and he had to do all of the work while his pair was relaxing and just relied on him. He highlighted the important factors that need to be considered in implementing a successful pair programming technique which are the compatibility with the partner, the programming ability and similar motivation levels. These factors are also consistent with the existing study by Eierman and Iversen (2018) and Keefe et al. (2006).

Regarding the buzz group and PBL, all of them agreed that they had benefited through those activities and hope their lecturers for other subjects can implement the same in their classes. Student E emphasized that he enjoyed the activities rather than feeling stressed. However, Student F said that he felt quite stressed when he could not find the solution and the lecturer would ask him to share the answer. The students totally agreed that these activities can enhance their thinking and communication skills, teamwork, problem solving and programming. On top of that, those activities are able to enhance their interest in programming. Student F suggested that these activities can be carried out in programming classes more frequently so that they will get more benefits. Overall, the students gave positive feedbacks on the techniques of collaborative learning approach used. These techniques were able to increase their interest in the class and they are looking forward for such activities conducted in other courses as well.

Comparison of Examination Results

Table 7 reveals the grades obtained by the 26 repeaters in the class for the current and previous semester. Majority of them have improved from their previous achievements despite two students who attained decreased grade. Another two students achieved the same grade. When traced back, these students are the ones who did not do their assignments and projects properly. They also did not perform well for their final exam. Thus, they could not improve their grades. On the contrary, four students (Student 6, 10, 20 and 25) improved intensely from failed grade to A or A-. These students are found to do their assignments, projects and final exam properly, thus they achieved good result. Additionally, they were found to practice programming more frequently compared to those who did not improve their grades. This shows that it is always possible for the students to improve their programming skills, if they change their learning style.

Outcomes from the survey and focus group indicate that the students enjoyed and obtained benefits from the activities. Other than increasing their interest in learning programming, students also experienced the following benefits such as critical thinking and communication skills, teamwork, problem

Table 7. Comparison of grades between previous semester with current semester

ID	Previous Grades	Current Grades	Status
Student 1	C-	D	Decreased
Student 2	C-	F	Decreased
Student 3	D+	D+	Same grade
Student 4	D+	D+	Same grade
Student 5	D	C	Improved
Student 6	C-	A-	Improved
Student 7	C-	B-	Improved
Student 8	C-	B-	Improved
Student 9	D+	B-	Improved
Student 10	D+	A-	Improved
Student 11	C-	B	Improved
Student 12	D+	C	Improved
Student 13	C-	B	Improved
Student 14	C-	C+	Improved
Student 15	D+	C	Improved
Student 16	C-	B+	Improved
Student 17	D	C	Improved
Student 18	D	B+	Improved
Student 19	D+	B+	Improved
Student 20	C-	A	Improved
Student 21	D	B-	Improved
Student 22	C-	B-	Improved
Student 23	D	C+	Improved
Student 24	C-	B-	Improved
Student 25	D+	A-	Improved
Student 26	D+	C	Improved

solving, programming teamwork skill and higher confidence level. As their interest is built up, they will automatically strive hard to understand the subject and try to learn deeply. Besides, by working in teams, they can exchange their knowledge and improve themselves. Over time, the students will be able to improve the quality of code and design that they produce.

Furthermore, the students' achievements can be seen in the examination results. The examination results of the 26 repeaters are examined to understand their improvement. Majority of them have upgraded their grades in this semester with only a few remained with the same grade or lower. Four students have improved dramatically when they achieved grade A or A-. This indicates that even though they failed the subject in the previous semester, they are able to achieve better results by gaining interest in the subject and improving their learning style. This is supported by another study by Hook and Eckerdal (2015). In their study, they found that students who achieved higher marks are those who spent more time in practicing programming, meanwhile, those students who just read books without sufficient programming practice obtained lower marks. Additionally, Maguire et al. (2014) found that female students achieved higher grades with the implementation of pair programming.

In a nutshell, this study has provided the initial evidence on the suitability of the collaborative learning approach in programming classes. The findings are aligned with the existing studies. The ability of collaborative learning techniques to enhance students' interest and achievement can be seen. Thus, this study suggests other lecturers to use the same approach in their programming classes.

CONCLUSION

Teacher centered learning has been widely used previously for teaching programming classes, whereby lecturers give lecture and students will have to understand the lecture. However, this approach seems to be insufficient in today's learning environment because students do not have the opportunity to interact and share ideas or work in teams collaboratively. Moreover, today's working environment requires graduates with necessary soft skills such as communication, teamwork and critical thinking. Thus, students need to be prepared with those skills before graduating and applying for future jobs. Similar situation adheres for BSc. IT students, where they face problem in solving the programming problems, producing low quality code and lack of interest to the subject. Their achievement is also affected as the failure rate for this subject is high. In order to increase their interest and achievement, collaborative learning techniques which are the pair programming, buzz group and PBL were adapted in Programming 1 class. Feedbacks on the implementation of those techniques were gathered from the students through an online survey and the focus group discussion. The result shows that the pair programming, buzz group and PBL enhanced students' interest in programming subject, as well as improving their performance. This indicates that the collaborative learning approach is suitable to be implemented in programming classes.

REFERENCES

Bashir, G. M. M., & Hoque, A. S. M. L. (2016). An effective learning and teaching model for programming languages. *Journal of Computers in Education*, *3*(4), 413–437. doi:10.100740692-016-0073-2

Braught, G., Eby, L. M., & Wahls, T. (2008). The effects of pair-programming on individual programming skill. *39th SIGCSE Technical Symposium on Computer Science Education (SIGCSE '08)*, 200-204. 10.1145/1352135.1352207

Carvalho, A. (2016). The impact of PBL on transferable skills development in management education. *Innovations in Education and Teaching International*, *53*(1), 35–47. doi:10.1080/14703297.2015.1020327

Chase, J. D., & Okie, E. G. (2000). Combining cooperative learning and peer instruction in introductory computer science. *ACM SIGCSE Bulletin*, *32*(1), 372–376. doi:10.1145/331795.331888

Cockburn, A., & Williams, L. (2000). The costs and benefits of pair programming. *Extreme programming examined*, 223-247.

Corney, M., Teague, D., & Thomas, R. N. (2010). Engaging students in programming. *Proceedings of the Twelfth Australasian Conference on Computing Education, 103*, 63-72.

Costa, E. B., Fonseca, B., Santana, M. A., de Araújo, F. F., & Rego, J. (2017). Evaluating the effectiveness of educational data mining techniques for early prediction of students' academic failure in introductory programming courses. *Computers in Human Behavior, 73*, 247–256. doi:10.1016/j.chb.2017.01.047

Denner, J., Werner, L., Campe, S., & Ortiz, E. (2014). Pair programming: Under what conditions is it advantageous for middle school students? *Journal of Research on Technology in Education, 46*(3), 277–296. doi:10.1080/15391523.2014.888272

Eierman, M. A., & Iversen, J. (2018). Comparing Test-Driven Development and Pair Programming to Improve the Learning of Programming Languages. *Journal of the Midwest Association for Information Systems, 1*(1), 23.

Fidge, C., Hogan, J., & Lister, R. (2013). What vs. how: comparing students' testing and coding skills. *Proceedings of the Fifteenth Australasian Computing Education Conference, 136*, 97-106.

Furberg, A., Kluge, A., & Ludvigsen, S. (2013). Student sense making with science diagrams in a computer-based setting. *International Journal of Computer-Supported Collaborative Learning, 8*(1), 41–64. doi:10.100711412-013-9165-4

Gallardo-Virgen, J. A., & DeVillar, R. A. (2011). Sharing, talking, and learning in the elementary school science classroom: Benefits of innovative design and collaborative learning in computer-integrated settings. *Computers in the Schools, 28*(4), 278–290. doi:10.1080/07380569.2011.621803

Gomes, A., & Mendes, A. J. (2007). An environment to improve programming education. *Proceedings of the 2007 International Conference on Computer Systems and Technologies, 88*, 1-6.

Hanapi, Z., & Nordin, M. S. (2014). Unemployment among Malaysia Graduates: Graduates' attributes, lecturers' competency and quality of education. *Procedia: Social and Behavioral Sciences, 112*, 1056–1063. doi:10.1016/j.sbspro.2014.01.1269

Hanapi, Z., Nordin, M. S., & Khamis, A. (2015). Challenges faced by engineering lecturers in integrating technical and employability skills in the curriculum: A case study in Community College, Malaysia. *International Journal of Social Science and Humanity, 5*(5), 483. doi:10.7763/IJSSH.2015.V5.504

Hong, C. L., & Seyal, A. H. (2014). Problem-based learning: An analysis of its application to the teaching of programming. *International Proceedings of Economics Development and Research*, 68–75.

Hook, L. J., & Eckerdal, A. (2015). On the bimodality in an introductory programming course: An analysis of student performance factors. *International Conference on Learning and Teaching in Computing and Engineering*, 79-86. 10.1109/LaTiCE.2015.25

Hung, W., Jonassen, D. H., & Liu, R. (2008). Problem-based learning. Handbook of research on educational communications and technology, 3, 485-506.

Iqbal, S., & Harsh, O. K. (2013). A self-review and external review model for teaching and assessing novice programmers. *International Journal of Information and Education Technology (IJIET), 2*(3), 120–123. doi:10.7763/IJIET.2013.V3.247

Ismail, R., Yusoff, I., & Sieng, L. W. (2011). Employer Perceptions on Graduates in Malaysian Services Sector. *International Business Management*, *5*(3), 184–193. doi:10.3923/ibm.2011.184.193

Ismail, W., Abedlazeez, N., & Hussin, Z. (2011). Epistemological beliefs of students at high schools: A survey study in Malaysia. *International Journal of Sustainable Development*, *2*(08), 39–46.

Isong, B. (2014). A Methodology for Teaching Computer Programming: First year students' perspective. *International Journal of Modern Education and Computer Science*, *6*(9), 15–21. doi:10.5815/ijmecs.2014.09.03

Jenkins, T. (2002). On the difficulty of learning to program. *Proceedings of the 3rd Annual Conference of the LTSN Centre for Information and Computer Sciences, 4*, 53-58.

Keefe, K., Sheard, J., & Dick, M. (2006). Adopting XP practices for teaching object oriented programming. *Proceedings of the 8th Australasian Conference on Computing Education, 52,* 91-100.

Law, K. M., Lee, V. C., & Yu, Y. T. (2010). Learning motivation in e-learning facilitated computer programming courses. *Computers & Education*, *55*(1), 218–228. doi:10.1016/j.compedu.2010.01.007

Lykke, M., Coto, M., Jantzen, C., Mora, S., & Vandel, N. (2015). Motivating Students through Positive Learning Experiences: A Comparison of Three Learning Designs for Computer Programming Courses. *Journal of Problem Based Learning in Higher Education*, *3*(2), 80–108.

Maguire, P., Maguire, R., Hyland, P., & Marshall, P. (2014). Enhancing collaborative learning using paired-programming: Who benefits? *AISHE-J: The All Ireland Journal of Teaching and Learning in Higher Education*, *6*(2), 1411–14125.

Malik, S. I. (2018). Improvements in introductory programming course: Action research insights and outcomes. *Systemic Practice and Action Research*, 1–20.

Malik, S. I., & Coldwell-Neilson, J. (2016). A model for teaching an introductory programming course using ADRI. *Education and Information Technologies*, 1–32.

Martakis, A., & Daneva, M. (2013). Handling requirements dependencies in Agile projects: A focus group with Agile software development practitioners. *7th International Conference on Research Challenges in IS*, 1-11. 10.1109/RCIS.2013.6577679

McDowell, C., Hanks, B., & Werner, L. (2003). Experimenting with pair programming in the classroom. *SIGCSE Conference on Innovation and Technology in Computer Science Education*, 60-64. 10.1145/961511.961531

McDowell, C., Werner, L., Bullock, H. E., & Fernald, J. (2006). Pair programming improves student retention, confidence, and program quality. *Communications of the ACM*, *49*(8), 90–95. doi:10.1145/1145287.1145293

MIDF Research. (2018, May 3). *Youth unemployment hit record high in 2017: MIDF Research.* Retrieved from http://www.thesundaily.my/news/2018/05/03/youth-unemployment-hit-record-high-2017-midf-research

Nilson, L. B. (2010). *Teaching at its best: A research-based resource for college instructors* (2nd ed.). San Francisco: Jossey-Bass.

Nuutila, E., Torma, S., & Malmi, L. (2005). PBL and computer programming—The seven steps method with adaptations. *Computer Science Education, 15*(2), 123–142. doi:10.1080/08993400500150788

Preston, D. (2005). Pair programming as a model of collaborative learning: A review of the research. *Journal of Computing Sciences in Colleges, 20*(4), 39–45.

Rogerson, C., & Scott, E. (2010). The fear factor: How it affects students learning to program in a tertiary environment. *Journal of Information Technology Education: Research, 9*, 147–171. doi:10.28945/1183

Salleh, N., Mendes, E., & Grundy, J. (2011). Empirical studies of pair programming for CS/SE teaching in higher education: A systematic literature review. *IEEE Transactions on Software Engineering, 37*(4), 509–525. doi:10.1109/TSE.2010.59

Sarker, F., Davis, H., & Tiropanis, T. (2010). *A review of higher education challenges and data infrastructure responses*. Retrieved from http://eprints.soton.ac.uk/271695/1/A_Review_of_HE_challenges_and_data_infrastructure_responses.pdf

Scannell, E. E., & Donaldson, L. (2000). *Human Resource Development*. New York: Perseus Publishing.

Selly, W. N. R. (2013). *Study Komparasi Hasil Belajar Pendidikan Kewarganegaraan Antara Metode Konvensional Dengan Metode Diskusi Kelas Teknik Buzz Groups Pada Siswa Kelas Viii Smp N 2 Kalasan* (Master's Thesis). Universitas Negeri Yogyakarta.

Serrano-Cámara, L. M., Paredes-Velasco, M., Alcover, C. M., & Velazquez-Iturbide, J. Á. (2014). An evaluation of students' motivation in computer-supported collaborative learning of programming concepts. *Computers in Human Behavior, 31*, 499–508. doi:10.1016/j.chb.2013.04.030

Shuhidan, S. M. (2012). *Probing the minds of novice programmers through guided learning (PhD thesis)*. RMIT University. retrieved from https://researchbank.rmit.edu.au/eserv/rmit:160247/MohamedShuhidan.pdf

Soares, A., Fonseca, F., & Martin, N. L. (2015). Teaching introductory programming with game design and problem-based learning. *Issues in Information Systems, 16*(3), 128–137.

Tarmudi, Muhiddin, Rossdy, & Tamsin. (2016). Fuzzy delphi method for evaluating effective teaching based on students' perspective. *E-Academic Journal UiTMT, 5*, 1-10.

Teague, D., & Roe, P. (2008). Collaborative learning: towards a solution for novice programmers. *Proceedings of the Tenth Conference On Australasian Computing Education, 78*,147-153.

Teague, D. M., & Roe, P. (2009). Learning to program: from pear-shaped to pairs. *Proceedings of the First International Conference on Computer Supported Education, 2*, 151-158.

Thompson, B., & Stapleton, J. C. (1979). A method for validating semantic differential referents. *Journal of Experimental Education, 48*(2), 110–113. doi:10.1080/00220973.1979.11011721

Tsai, W. T., Li, W., Elston, J., & Chen, Y. (2011). Collaborative learning using wiki web sites for computer science undergraduate education: A case study. *IEEE Transactions on Education*, *54*(1), 114–124. doi:10.1109/TE.2010.2046491

Umapathy, K., & Ritzhaupt, A. D. (2017). A Meta-Analysis of Pair-Programming in Computer Programming Courses: Implications for Educational Practice. *ACM Trans. Comput. Educ.*, *17*(4), 16:1–16:13. doi:10.1145/2996201

Wang, S.-L., & Lin, S. S. J. (2007). The effects of group composition of self-efficacy and collective efficacy on computer-supported collaborative learning. *Computers in Human Behavior*, *23*(5), 2256–2268. doi:10.1016/j.chb.2006.03.005

Wasserman, E. (2013). How do Lecturers Evaluate their Own Work? *Procedia: Social and Behavioral Sciences*, *89*, 144–147. doi:10.1016/j.sbspro.2013.08.824

Williams, L., McDowell, C., Nagappan, N., Fernald, J., & Werner, L. (2003). Building pair programming knowledge through a family of experiments. *Proceedings. International Symposium on Empirical Software Engineering*, 143-152. 10.1109/ISESE.2003.1237973

Williams, L., & Upchurch, R. L. (2001). In support of student pair-programming. *ACM SIGCSE Bulletin*, *33*(1), 327–331. doi:10.1145/366413.364614

Williams, L. A., & Kessler, R. R. (2000). All I really need to know about pair programming I learned in kindergarten. *Communications of the ACM*, *43*(5), 108–114. doi:10.1145/332833.332848

Yadin, A. (2011). Reducing the dropout rate in an introductory programming course. *ACM Inroads*, *2*(4), 71–76. doi:10.1145/2038876.2038894

Young, P. M. (2010). Teaching the Ethical Values Governing Mediator Impartiality Using Short Lectures, Buzz Group Discussions, Video Clips, a Defining Features Matrix, Games, and an Exercise Based on Grievances Filed Against Florida Mediators. *Pepp. Disp. Resol. LJ*, *11*, 309.

Zikmund, W. G., Babin, B. J., Carr, J. C., & Griffin, M. (2010). Business research methods (8th ed.). South-Western: Cengage Learning.

Chapter 17
IR4.0 Ready and Aware Academics at the National Defence University of Malaysia

Jowati Juhary
https://orcid.org/0000-0002-5314-1195
National Defence University of Malaysia, Malaysia

ABSTRACT

This chapter argues for the importance of equipping academics with the knowledge and impact of IR4.0 and beyond for future graduates. Particularly to prepare for military leaders for Malaysia at the National Defense University of Malaysia (NDUM), a new learning model may be apt and urgently needed. The objectives of the chapter are twofold: firstly, to gauge the readiness and awareness level of academics on implementing IR4.0 during their classroom teaching, and secondly, to investigate academics' views on their own teaching practices. The methodology for this chapter is quantitative in nature, where an online survey was used to collect data from the academics. It is found that, first, there are mixed reactions from academics on their readiness and awareness of IR4.0 and its impacts, and second, should the existing curriculum be revamped to serve students' needs on IR4.0, academics opined that the delivery methods must change too.

INTRODUCTION

Education is dynamic since it evolves more often than societies realize. With multifaceted digital technologies today, higher learning institutions are faced with the demands and needs of the students, who are digital natives. Teaching and learning with technology is inevitable. Today, academics are challenged with the ideas on implementing Industrial Revolution 4.0 (IR4.0) in their classroom practices. Malaysia is also now moving toward fulfilling this need through various initiatives especially at its higher learning institutions. This chapter argues for the importance of equipping academics with the knowledge and impact of IR4.0 and beyond for future graduates. Therefore, the objectives of the chapter are twofold; to gauge the readiness and awareness level of academics on implementing IR4.0 during their classroom

DOI: 10.4018/978-1-5225-7832-1.ch017

Copyright © 2019, IGI Global. Copying or distributing in print or electronic forms without written permission of IGI Global is prohibited.

teaching, and secondly, to investigate academics' views on their own teaching practices. The author opines that in order to prepare the new breed of leaders, the students today must be trained to face the unknown challenges. Particularly to prepare for military leaders for Malaysia at the National Defense University of Malaysia (NDUM), a new learning model may be apt, and urgently needed.

As a new topic of debates and discussions, there are various questions on IR4.0 and its implementation, including how best to cater to the learning needs of the military and civilian students in a military learning environment. Especially for military cadets, they are undergoing their academic and military training concurrently. Will they be affected academically by the rigorous military training? Will they have issues to excel academically? Similarly, will the civilian students, who are also required to be in the Reserved Officer Training Unit (ROTU), be able to perform academically since they also have military training every Wednesday and sometimes, over the weekends? On the other hand, critical questions such as the awareness and readiness of the academics to serve the students for IR4.0 must be checked as well. Are they equipped with relevant teaching skills to assist students to learn? Are they able to deliver curricula that are tailored for IR4.0? What types of assessments are involved to measure students' competency and skills? This chapter, nonetheless, focuses only on academics' perceptions about all these issues.

To begin with, an overview of the NDUM will be offered. The Defense University was initially known as the Military Academy of Malaysia (MAM), which started its academic and military training in June 1995 with 172 military cadets (Ahmad Azan Ridzuan, 2011). Although it was a military academy, the academics were staff of a public university, Universiti Teknologi Malaysia, who were seconded to the MAM. This suggests that there were two components to teaching and learning back then. One, the academic component was served by another public university, and two, the military component was provided by the Malaysian Armed Forces (MAF). It was a smart collaboration between these two institutions until the government decided that it is best for the nation to have its own defense university. Consequently, in November 2006, the NDUM was gazetted in the parliament, and its first batch of students reported for studies was about 450 in number including military cadets and civilian students. To date, the Defense University has 3,856 graduates over a period of 11 years, and about 65 percent of them are military officers now serving all three branches in the MAF, namely the Malaysian Army, Royal Malaysian Navy and Royal Malaysian Air Force.

After 11 years of operation, the NDUM is still the youngest public university in Malaysia. Being the youngest has its advantages since the university is able to learn from the sibling universities in Malaysia. Despite its age, the most critical element for the graduates of the NDUM is its Intellectual Leaders of Character program, where all graduates must be inspired to possess. There are six attributes in this program, specifically targeted at military cadets explained briefly next. Firstly, graduates must have bachelor degrees in any relevant fields offered by the Defense University. Secondly, graduates must be commissioned military officers, and to achieve this, military cadets must achieve a Cumulative Grade Point Average (CGPA) of 2.75 and above. Thirdly, graduates must be qualified Imams (leaders for prayers), as well as able to lead in all religious activities. Fourthly, graduates must acquire black belts in an unarmed combat; in the case of the NDUM, taekwondo. Fifthly, graduates must be qualified players and coaches for at least one sports. Lastly, graduates must possess the skills and competency as officers and gentlemen/ladies. This last attribute will only be reflected when graduates join the armed forces and civilian workforce.

Given the new scenario in the workforce today, where IR4.0 becomes significant in manufacturing and business, what and how could higher education in Malaysia produce suitable workforce to face the challenges of the new world? This chapter argues that there are at least two components that higher education could focus on, curriculum and its delivery. This suggests that higher education providers must be able to offer courses and academic programs that are relevant for the IR4.0 scenario and beyond because by the time the graduates join the workforce, perhaps they are no longer facing IR4.0. In addition, academics too must be able to offer a classroom learning experience that encourages students to critically think and solve real world issues, and challenges and engages meaningful interactions between and among students and academics.

To fulfill the objectives and address the issues mentioned earlier, this chapter will answer these five research questions.

- What is the percentage of academics' (i) readiness and (ii) awareness of IR4.0?
- What is the percentage of academics' understanding of the term Teaching and Learning 4.0 (T&L4.0)?
- Is there any relation between teaching experience of academics and the need to change curriculum delivery methods?
- Is there any relation between age and the need to change curriculum delivery methods?
- Is there any relation between curriculum revamp and changing curriculum delivery methods?

Before this chapter ventures further, it is necessary to highlight the outline of it. There are six main sections including the introduction. The second section examines the literature on the history of pedagogy and IR4.0 in higher education especially in Malaysia. The third section explores the research methodology adopted and the findings of the online survey. It then further analyses and discusses the findings in detail. The fourth section offers solutions and recommendations on how the NDUM could prepare its academics to face IR4.0 and T&L4.0, and the fifth section provides future research directions on IR4.0 and T&L4.0. The last section concludes the chapter.

BACKGROUND

This section focuses on two areas of interest of the chapter, namely the history of pedagogy in general and IR4.0. These two are critical since the former is the basis for an effective teaching and learning environment at the NDUM, and the latter is the topic of debate in the education setting today.

The Beginning of Pedagogy

Pedagogy is a small word with a massive impact in education. What constitutes an effective pedagogy? This sub section revolves around the historical account of pedagogy. It is argued that the founder of pedagogy is Herbart, a German philosopher and a Mathematician. Herbart's educational philosophy concentrated on the relationship between personal development and its influence on society. This suggests that learners realize their potential only when they are able to be productive citizens of the nation. According to Somr and Hruskova (2014), Herbat argued for a dominant role in the area of the progressive trend of the education, because without strong recommendations, the education reform in the United States would not possibly be developed.

Five formal steps in teaching, which were later translated into a practical teaching methodology are suggested by Herbert. These five steps include,

- **Preparation:** Relating new material to be learned to relevant existing ideas (memories) to stimulate the learner's interest (prepare learners to be ready for new lessons)
- **Presentation:** Presenting new material in the form of actual experience of concrete objects (present new lessons)
- **Association:** Comparison of the new idea with existing ideas to find similarities and differences, and thus implant new ideas in the mind of the learner (associate new lessons with ideas studied earlier)
- **Generalization:** Procedures designed to take learning beyond perception and experience of the concrete into the realm of abstract concepts (use examples to illustrate the lesson's major points)
- **Application:** Using the newly acquired knowledge so that it becomes an integral part of the life of the learner (test learners to ensure they learned these new lessons).

Herbart's ideas were widely adopted in Germany and also the United States, translated into the simple five-step teaching method that became the basic pedagogical practice in the 19th century. By the 20th century, however, the steps had become mechanical, and Herbart's underlying ideas on ethics, psychology and aesthetics had been forgotten. In their place, new pedagogical theories, such as those of John Dewey in the United States, which freed the child from what had become a controlled learning environment, grew in popularity. Although his teaching methodology was overtaken by new ideas, Herbart's institution of pedagogy as an academic field has remained. The idea of a science of education, including psychology as a source of information about the nature of the learner as well as the learning process, has continued to advance teaching methods. Refining Herbart's ideas, according to Dewey, there were four key educational issues on a reformed educational system. These continue to provide the basis of critiques of and discussion about contemporary education.

The first key issue proposed by Dewey was the value of learners' experiences. Dewey (1933) believed that the way learning occurred inside the formal educational setting should not be isolated from learning achieved elsewhere. He advocated a continuity of the process of learning between settings and viewed each learner's greatest asset as being his/her direct, personal experience that must not be ignored or thrown away in the process of learning. Dewey further proposed the principle of learning through personal experience that rested on two factors – internal and external experiences. For Dewey, education was a re-construction of experience that continuously unfolded learners' potential. The second key issue that Dewey (1916) argued for was that learning should be by doing. This meant giving more independence and active roles to learners in their learning. Learners were to describe the construction of their own learning by using their previous knowledge or experience. In other words, the role of active engagement was being advocated. Dewey (1916; 1938) noted that traditional schools kept insisting on telling learners what they needed to learn, despite research clearly demonstrating that learning by telling did not work and learning by doing did. Following Dewey, education should be democratic – it should be experiential, in the sense of engaging the interests of the learners; and it should be reflective, in encouraging learners not just to gather facts but to make connections and to critique knowledge. Thus, Dewey believed in the engagement of learners in the learning process through 'doing' and not just listening.

The third key educational issue promoted by Dewey was purposeful learning. For what purpose was the content to be learned? Dewey (1938) forcefully stressed the need for activities to be linked cumulatively, defining educative experiences as those that gave rise to the learners' need to gather more facts, become more skilled and use lessons learned in one experience as the basis for future experiences. He termed this a 'continuous spiral.' Also, Dewey proposed that when learners had a clear and strong purpose for learning, they became more committed and able to participate actively in the learning sessions. The last key educational issue stressed by Dewey was the importance of critical thinking in the education system. Dewey (1933) proposed that in order for learning to be truly effective, it must inculcate reflective thought, or what was referred to in contemporary education today as critical thinking. He further defined reflective thought as an active and persistent process that was able to support individuals' opinions. As this was an individual process, each learner would construct his/her own opinions and critically reflect on those opinions. Furthermore, Dewey elaborated that there were two important elements of reflective thinking. The first element was "a state of perplexity, hesitation and doubt," and the second element was "an act of search or investigation directed toward bringing to light further facts which serve to corroborate or to nullify the suggested belief" (Dewey, 1933). By these two elements, Dewey suggested that in order to be reflective or critical thinkers, learners should have doubts or question ideas that they just encountered. They must not believe or merely accept the ideas without trying to challenge their validity. The next step was to analyze the ideas and try to find support to accept or reject them. Thus, for Dewey, critical thinking or reflective thought helped learners to construct and build new knowledge, making them aware that they must be active in questioning and challenging new information.

Dewey's interpretations of learning have proven to be useful in the modern world of education. It is in the 21st century that scholars such as Hickman (1990; 2001) and Phillips (2002) have realized that what Dewey stated as the critical issues in education are indeed valid, and these issues continue to form the basis of effective pedagogy. This is because when Dewey began his philosophizing on education, many misunderstood his ideas of progressive and reflective education. Some even understood Dewey's ideas as only having relevance to children and their education. Nonetheless today, it has become increasingly obvious that Dewey's writings were not limited to children's education only.

Given the above analysis of pedagogy, there now exist several branches of pedagogy; some can be argued as the extension of the other, and some are viewed as something totally different. Despite this, all of these 'gogies' have something critical in common. It is the fact that these 'gogies' emerge as ways to facilitate teaching and learning, especially when learning itself is a dynamic process, and they are the platforms to ensure learning is achieved. Table 1 illustrates the summary of the other 'gogies.'

For teaching and learning at higher education, most scholars agree that the best 'gogy' to use are the ones that promote critical thinking and active learning. This is can be better explained in the next sub section in the context of IR4.0.

Industrial Revolution and Higher Education

Higher education in the IR4.0 environment is a complex, yet exciting opportunity which can be a paradigm shift for society. IR4.0 relies on artificial intelligence, and according to Xing and Marwala (2017), IR4.0 will be able to transform working environment from tasks-based characteristics to the human centered characteristics. Consequently, the Ministry of Education Malaysia (Higher Education) has outlined what T&L4.0 consists of. It must be emphasized that some of this T&L4.0 has very little references and sources. Nonetheless, the author argues that as it is in its infancy, more scholars will be investigating T&L4.0 in the immediate future. Basically, T&L4.0 has several key aspects, including learning spaces, pedagogy, fluid and organic curriculum and technologies, and all these aspects are supported by its various pedagogies: heutagogy, paragogy and cybergogy (see Table 1; no mention of military pedagogy as it only applies in the NDUM). T&L4.0 emphasizes learning without lectures, evaluation without examinations,

Table 1. Comparison between heutagogy, paragogy, cybergogy and military pedagogy

Critical Aspects	Heutagogy: Self-Determined Learning	Paragogy: Peer Assisted Learning	Cybergogy: Cognitive, Emotive and Social Learning	Military Pedagogy: Independent, Directed Learning
Supporters	Blaschke (2012); Dick (2013); Hase and Kenyon (2013)	Corneli and Danoff (2011); Lee and Rofe (2016); Boulos, Giustini and Wheeler (2016)	Wang and Kang (2006); Muresan (2013)	Toiskallio (2003); Falk (2008)
Dependence	Learners are 'problem finders.' They know their destination and become interdependent on those who can help them determine the route	Learners are independent but rely upon peers for learning support. They explore and share their learning experience. Technology plays a critical factor too	Learners are engaged in learning through three domains, cognitive, emotive and social aspects	Learners are independent and 'problem solvers.' They listen and follow instructions well but are able to question and probe further for more
Reasons for Learning	Learning is not necessarily sequential or linear. Learners accept full responsibility for their learning, welcoming challenges and serendipity	Learning is distributed and non-linear. Learners seek feedback more often from peers	Learners seek guidance/ mentorship but aspire to increase responsibility for the direction of their learning	Learning is sequential in order to allow for progress. Learners aspire to increase responsibility as they become more familiar with learning and working environment
Role of Academics	Coach – brings together opportunity, context, external relevance and extended complexity. Fosters a culture of collaboration and curiosity	Facilitator – sets tasks but encourages diverse routes to solutions. Allows peers to offer feedback and monitors the feedback given	Facilitator – sets tasks but encourages diverse routes to solutions. Pursues meta-cognition in learners	Mentor (Murabbi) – becomes a role model for knowledge and personality. Designs the learning process, set tasks and encourages collaboration and sharing of ideas

Sources: Gerstein, 2016; the author, 2018

ability of graduates to be employed, the need for industry and academia to become one and an education that focuses on humanity (Idris Jusoh, 2018).

What exactly is IR4.0? It deals with the fourth industrial revolution in manufacturing and industry. IR4.0 places emphasis on "the current industrial transformation with automation, data exchanges, cloud, cyber-physical systems, robots, Big Data, artificial intelligence, Internet of Things and semi-autonomous industrial techniques to realize smart industry and manufacturing goals in the intersection of people, new technologies and innovation" (*Industry 4.0: The Fourth Industrial Revolution – Guide to Industrie 4.0*, 2016). Accordingly, as societies face a technological revolution, the way people live, work and relate to one another changes, and it will be a transformation that humankind has never experienced before.

In Malaysia, the manifestation of IR4.0 in T&L4.0 can be exemplified as such (see Figure 1). Teaching and learning was once teacher centered (MyHE 1.0), followed by how students were able to actively respond during classroom sessions (MyHE2.0). In MyHE 3.0, instructors became facilitators, and students were expected to collaborate through the use of digital technologies. To cope with IR4.0, students are now required to connect, create and construct new ideas from their existing and new-found knowledge. For Malaysian higher learning institutions, the challenges, now, among others, are to prepare appropriate curriculum to cater to this demand, and to find the best approaches to deliver the curriculum.

Figure 1. The transformation of higher education delivery in Malaysia
(Gerstein, 2016; Siti Hamisah Tapsir, 2017)

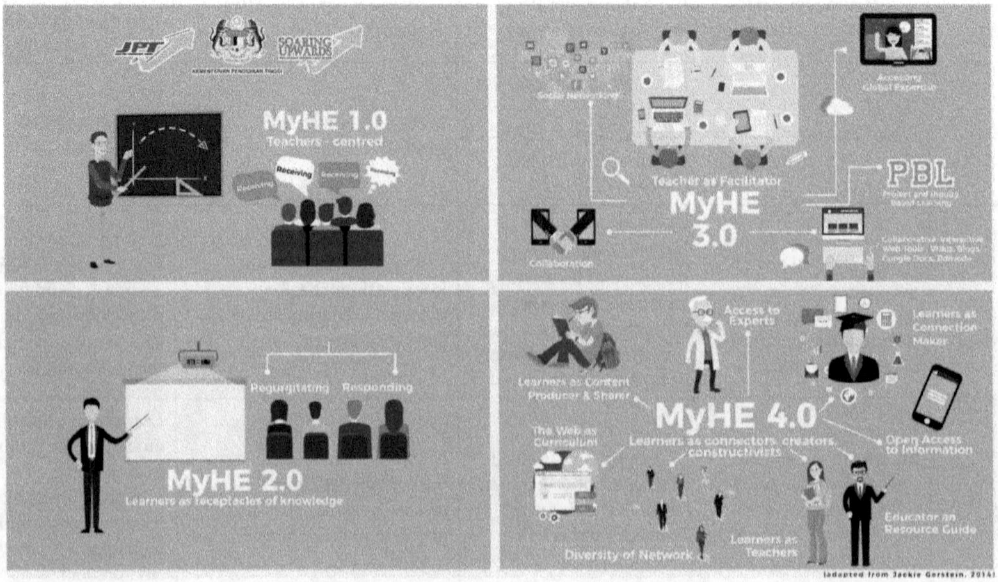

Figure 2. Malaysian higher education framework
(Siti Hamisah Tapsir & Marlia Puteh, 2018)

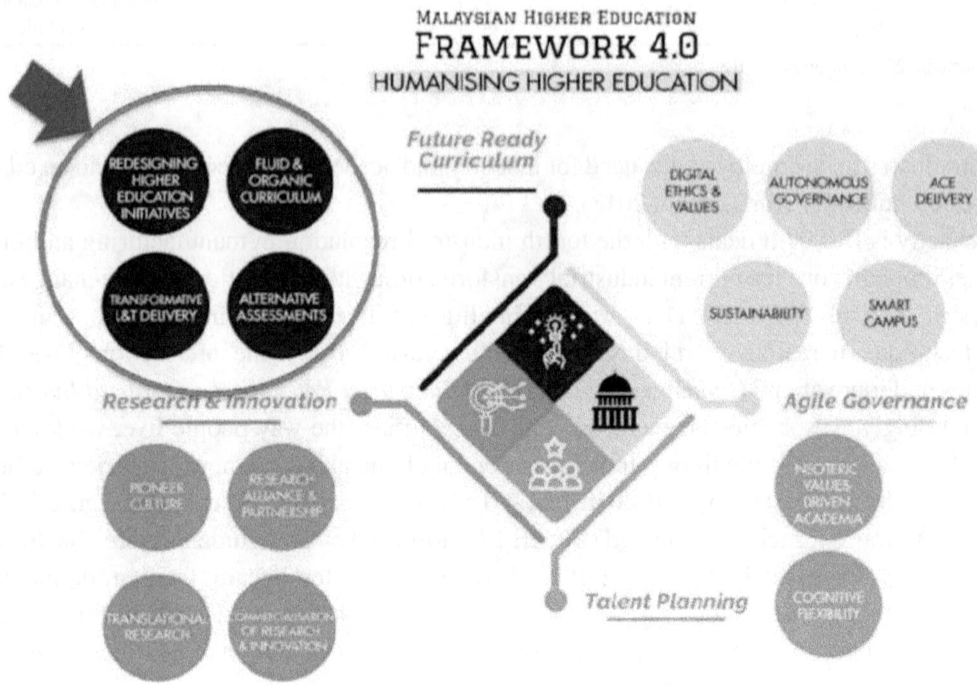

For Malaysia, the aspiration to prepare the graduates for IR4.0 can be seen in its initiatives and current policies. The latest one, which includes T&L4.0 is Malaysian Higher Education Framework 4.0, and its theme is 'Humanizing Higher Education,' can be exemplified in Figure 2 above. If the framework is analyzed carefully, one would notice that the focus of this framework is on producing future graduates who are able to sustain their own existence in the workforce by mainly creating job opportunities for others. This translates the importance of suitable curriculum and delivery approaches for the curriculum to be provided by higher education providers in the country. Both of these aspects fall under the Future Ready Curriculum in the framework.

In concluding this section, it is pertinent to emphasize that discussions on IR4.0 in this chapter must be placed into the context of teaching and learning in Malaysian higher learning institutions. This is because preparing future workforce for IR4.0 requires critical understanding on what constitutes suitable teaching and learning approaches in order to achieve the desired results: For the Defense University, military pedagogy caters to the needs of the graduates in ways that could not be served by other 'gogies.' Nevertheless, is it enough to prepare future defenders of the nation to face IR4.0 on the battlefield?

MAIN FOCUS OF THE CHAPTER

Research Methodology

As IR4.0 may be unfamiliar to some academics especially society at large, the author argues that this chapter could become a starting point for understanding whether academics especially at the NDUM are aware and ready to deliver their lessons accordingly and prepare students to face IR4.0. This chapter then adopts a quantitative approach to research where academics at the Defense University would have to volunteer to participate in an online survey. There are currently about 399 active academics (this number is true as at 29th June 2018) at the NDUM, and using Google Forms, the links to the survey would be sent through emails to all academics.

As IR4.0 is still at its infancy, not much empirical data could be found on it as very little (if any) research has been conducted. Therefore, the items in the survey originated from the existing literature on the discussions on pedagogy and industrial revolution. The survey was prepared in Bahasa Melayu and English language to cater to local and international academics at the Defense University. There are basically 13 items in the survey with only two sections, the demographic section with three items, and the main section, where 10 items are located (see http://bit.ly/2LN08Qe for Bahasa Melayu version of the survey and http://bit.ly/2LyAoKl for the English version). The main section has three different types of questions, including five multiple choices items; three short answers items; and two checkboxes items.

Emails were sent to all academics on 26th July 2018, and the author allowed for about two weeks for the responses to come in. Nonetheless, for the first invitation, not many academics responded. The author sent another invitation and allowed for another week for the responses to come in. The response button on Google Forms was closed on 13th August 2018. The final number of respondents amounts to 38 only. As the survey is voluntary in nature, the author did not force academics to complete the survey. Despite this small number of respondents, the author has to accept the fact that most of the academics were either too busy or were not interested in the topic under investigation.

Prior to the data collection, the survey was sent to a group of academics consisting of various age groups and gender. The reason for this is to validate the items in the survey. Since there is not much information and research in Malaysia on this topic done quantitatively, the author argues that validation by five academics ranging from professors to language instructors is sufficient at this preliminary stage. Despite this, the author is cautious in framing the items to address the issue of validity since there is no validity nor reliability tests are conducted. The items asked in the main section serve their purposes. They become critical indicators and variables on understanding academics' perceptions toward IR4.0 and its importance. The final data collected from 38 respondents are analyzed using Statistical Package for Social Sciences (SPSS) Version 25, and they are presented in mostly descriptive statistics. Nonetheless, both descriptive and inferential data are used to answer the research questions posed earlier in the chapter.

Findings

This sub section is presented based on the items asked in the survey. Explanation will not be offered in detail in this sub section as analysis is offered in the next.

Demographic Information

Based on Figure 3, there were more male respondents compared to female respondents. In addition, the majority of the respondents aged between 31 and 40 years old. In terms of teaching experiences, the majority of the respondents had between 11 and 15 years of experiences, followed by those with six to 10 years of teaching experiences.

Main Items

Based on these preliminary findings, there are mixed responses on the awareness and understanding of IR4.0 and T&L4.0 among the academics at the NDUM. At the same time, some argued that if they were not sure of IR4.0 and T&L4.0, it is highly unlikely that they were going to apply changes in their approaches to classroom teaching. To further analyze the findings in order to understand the scenario at the Defense University, the next sub section will answer the research questions accordingly.

Figure 3. Gender of respondents

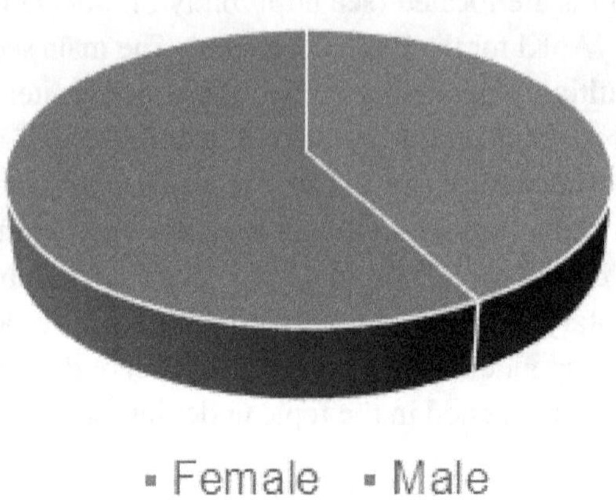

Figure 4. Age groups of respondents

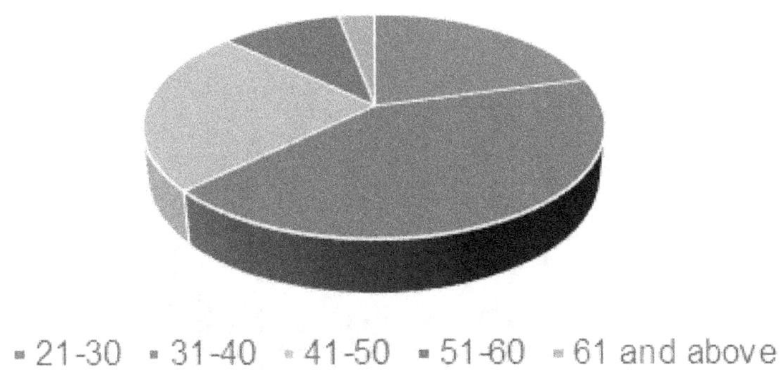

Figure 5. Teaching experience of respondents

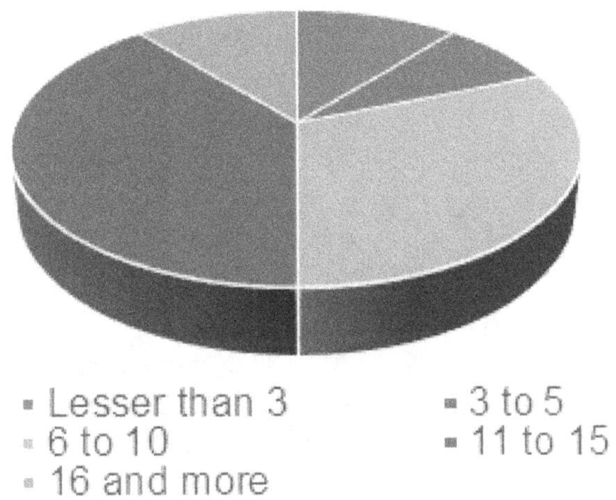

Discussions and Analysis

This sub section is divided into two parts. The first one uses the research questions in order to discuss and further analyze the findings. Research Questions 1 and 2 are answered descriptively, and Research Questions 3, 4 and 5 are answered inferentially using Pearson Correlations. This is followed by the second part that consolidates the descriptive and inferential data.

Answering Research Questions

Research Question 1: *What is the percentage of academics' (i) readiness and (ii) awareness of IR4.0?*

Based on Figures 6 and 8, it can be concluded that the majority of academics at the Defense University was ready and aware of IR4.0. This is because 53 percent of them would apply the knowledge on IR4.0 in their teaching and learning, and 71 percent of the respondents was aware about IR4.0. As IR4.0 requires understanding and acceptance of new technologies, it can be argued that the results are as such since the majority of the respondents fell under the age groups of 40 years old and below, those who are

Figure 6. Understand the term IR4.0

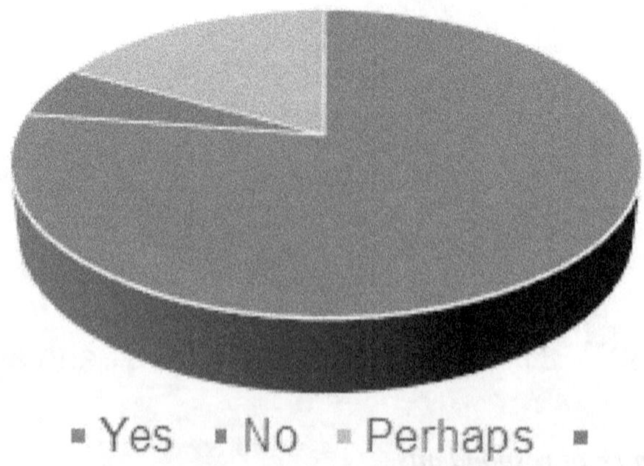

Figure 7. Understand the term T&L4.0

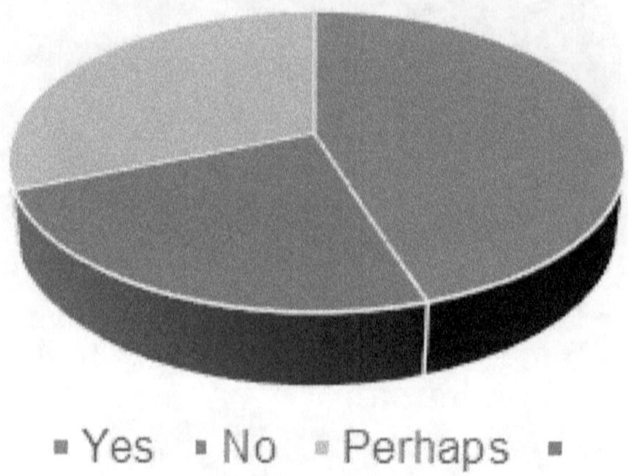

Figure 8. Apply knowledge of IR4.0 for T&L4.0

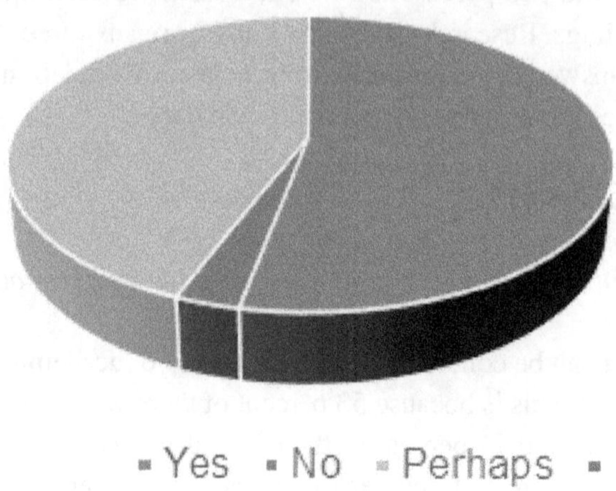

Figure 9. Curriculum revamp

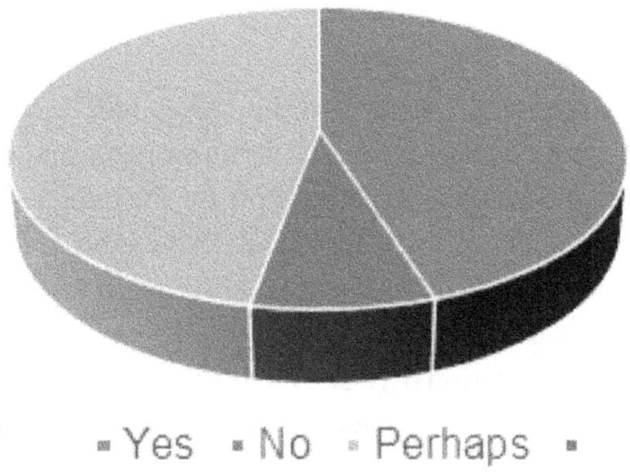

Figure 10. Curriculum delivery must change

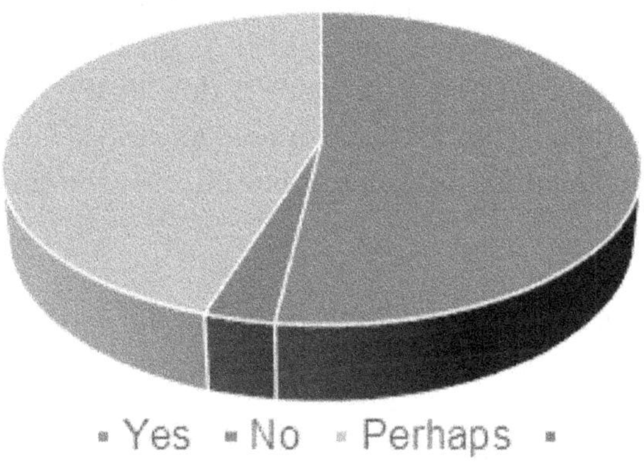

considered Generation X (born between the mid-1960s and 1980s) and Generation Y (born between 1984 and 1996). Both generations, Generation X (42 percent respondents) and Generation Y (21 percent respondents) are generally more susceptible toward the use of technology, thus making them more technology ready and aware. This question signals the impact of IR4.0 in teaching and learning in higher education. Without understanding IR4.0 and its impact, academics may not view IR4.0 as an important aspect in preparing the workforce for tomorrow.

Research Question 2: *What is the percentage of academics' understanding of the term T&L4.0?*

T&L4.0 requires academics to incorporate various teaching approaches so that students are able to connect, create and construct new knowledge. In this sense, students, in this T&L4.0 environment should connect new and old knowledge, and should make connections with peers and experts on their field. At the same time, they should also create and construct new knowledge based on their old knowledge. They should be able to relate and ultimately make meaning of their knowledge. Based on Figure 7, 45 percent of the respondents agreed that they understood the term 'T&L4.0;' 24 percent did not understand the term, while 31 percent respondents were unsure of the term. This may lead to resistance to adopt T&L4.0 in classroom, and may hinder students from connecting, creating and constructing knowledge on their own.

Research Question 3: *Is there any relation between teaching experience of academics and the need to change curriculum delivery methods?*

For the next three questions, the indicators in the table below will be used. It explains the strength of the relation between variables being tested. There are various ranges and explanations offered by reputable scholars on the correlation indicators; the author opts for the indicators recommended by Sarwono (2012) simply because the indicators are more representative of the author's data. The indicators are divided into six ranges; it begins with '0' to indicate no correlation of the variables, and ends with '1' to depict a perfect correlation between two variables.

In looking at whether there is any relation between the respondents' teaching experiences and whether they wanted to change the delivery method of the curriculum, Table 3 illustrates that there is a negative correlation between these two variables (-0.95). This is a very strong (negative) correlation, whereby as years of teaching increase, the need to change delivery method decreases. The author argues that because academics with more years of experience may be comfortable with their teaching methods, they could find it difficult to change their approaches. Based on their experiences, they may believe that what they are doing is the best method of teaching their students. The author does not deny these academics' experience; perhaps changes may disturb the flow of teaching. Notwithstanding this, the current students in higher learning institutions are exposed to various facets of life differently than the academics, and this may hinder the students from learning effectively if their learning environment does not support the learning needs.

Research Question 4: *Is there any relation between age and the need to change curriculum delivery methods?*

As for this research question, it is found that there is a sufficient (negative) correlation between age and changing curriculum delivery. While this comes as a surprise, since age is always associated with years of experience, it is also possible since prior to teaching, the academics may have had other jobs. This suggests that although they are older in terms of age, their years of teaching experience may be little. Table 4 illustrates a sufficient (negative) correlation at -0.273. Thus, the data suggest that as the age increases, the need to change delivery methods decreases. This has a similar pattern to the finding of Research Question 3 explained previously.

Table 2. Correlation indicators (positive and negative)

Range	Explanation
0	No correlation
0.00-0.25	Very weak correlation
0.25-0.50	Sufficient correlation
0.50-0.75	Strong correlation
0.75-0.99	Very strong correlation
1	Perfect correlation

Source: Sarwono, 2012

Table 3. Correlation between teaching experiences and changing delivery methods

Correlations			
		Teaching Experience	**Delivery Method**
Teaching Experience	Pearson Correlation	1	-.095
	Sig. (2-tailed)		.571
	N	38	38
Delivery Method	Pearson Correlation	-.095	1
	Sig. (2-tailed)	.571	
	N	38	38

Table 4. Correlation between age group and changing delivery methods

Correlations			
		Delivery Method	**Age Group**
Delivery Method	Pearson Correlation	1	-.273
	Sig. (2-tailed)		.097
	N	38	38
Age Group	Pearson Correlation	-.273	1
	Sig. (2-tailed)	.097	
	N	38	38

Research Question 5: *Is there any relation between curriculum revamp and changing curriculum delivery methods?*

Table 5 illustrates the relation between curriculum revamp and the need to change curriculum delivery method. There is a sufficient positive correlation between these two variables (0.470). This suggests that as the need to revamp curriculum increases, so does the need to change curriculum delivery methods. This finding is in line with data in Figures 9 and 10, where the number of respondents, who believed that

Table 5. Correlation between curriculum revamp and changing delivery methods

Correlations			
		Delivery Method	**Curriculum Revamp**
Delivery Method	Pearson Correlation	1	.470**
	Sig. (2-tailed)		.003
	N	38	38
Curriculum Revamp	Pearson Correlation	.470**	1
	Sig. (2-tailed)	.003	
	N	38	38

**. Correlation is significant at the 0.01 level (2-tailed).

the current curriculum should change is almost similar to the number of respondents, who opined that the curriculum delivery should change too. It appears that respondents were aware that once curriculum revamp happens to suit students' needs for IR4.0, they too must apply appropriate teaching strategies to match the curriculum.

Consolidating All Data

In consolidating the data, it is pertinent to look at the reasons given by respondents of three items asked in the survey. Those are the reasons for using IR4.0 knowledge in teaching and learning; reasons for curriculum revamp; and reasons for changing the curriculum delivery methods. Further, academics were also asked to provide opinions about the most effective curriculum delivery methods and their understanding on branches of pedagogy.

Using IR4.0 Knowledge

Firstly, respondents gave their reasons for using IR4.0 knowledge in T&L4.0, and these reasons are briefly explained below. It cannot be assumed that only those who chose 'Yes' (see Figure 8) had given their reasons since all respondents were required to answer all items. The responses for using/or not using the IR4.0 knowledge are categorized into several themes. Firstly, five respondents opined that the knowledge on IR4.0, when shared with the students, would make their classroom sessions interactive and engaging. The author argues that since the knowledge allows students to be ready for their future vocation, perhaps the students find IR4.0 to be highly relevant for them, thus making their own learning interactive and engaging. Secondly, 16 respondents believed that their knowledge on IR4.0 will allow their students to keep up with the dynamic of technology, as IR4.0 requires the sophistication of utilizing technology in the workforce.

Further, seven respondents claimed that they have had no prior exposure to IR4.0. This is somehow supported by the fact that 13 percent of the respondents did not understand the term 'IR4.0' while 16 percent was unsure of the term (see Figure 6). Meanwhile, three respondents wanted an awareness program on IR4.0; perhaps this is also a sign for not being exposed to the implication and functions of IR4.0. The last theme is offered by six respondents, who felt that knowledge on IR4.0 should be used only when it is needed. This suggests that not all courses and possibly academic programs require the knowledge on IR4.0.

Curriculum Revamp

Similar to reasons for using IR4.0 knowledge in T&L4.0, reasons for curriculum revamp exercise (see Figure 9) are grouped into several themes. Firstly, 10 respondents opined that the current curriculum must be revamped in order to ensure students are able to face the changes in the industry. Secondly, 14 respondents believed that the current curriculum must be revisited in order to fit the new requirement in the industry. Both reasons are inter related since both are concerned with the ability of the graduates to effectively function in the workplace environment. Next, two respondents wanted the curriculum to allow for a blended approach to teaching and learning. The author argues that this reason is highlighting the fact that in today's classroom learning, the curriculum must be set to engage students in various learning approaches so that they become more responsive, and ultimately able to make meaning of their learning.

One respondent argued that curriculum revamp must abide by the requirement of the accreditation body of the academic program. What this implies is that changes in the curriculum must be granted approval by the relevant accreditation body. Similarly, only one respondent, who felt that whatever changes made in the curriculum must be followed by an upgrade of the teaching and learning facilities at the NDUM. Meanwhile, surprisingly 10 respondents confirmed that curriculum revamp is not needed. This suggests that the current curriculum is effective in producing Intellectual Leaders of Character. The author argues that this is a possible phenomenon if the respondents were unsure of IR4.0 and T&L4.0, and how these two concepts could help future graduates of the Defense University to face the unknown challenges of the 21st century.

Approaches to Curriculum Delivery

The survey respondents were also asked to offer reasons on the need to have various approaches to deliver the curriculum. Respondents' responses are grouped into a few themes too (see Figure 10). The majority of the respondents (25 respondents) believed that they had to prepare the students with current technologies. As such, the use of various delivery methods is significant. To complement this, one respondent stated that the traditional method of curriculum delivery is no longer suitable. Perhaps, this respondent was referring to the board-and-chalk method of teaching where it is no longer appropriate for the current generation of learners, who are digital natives.

On the other hand, six respondents felt that when necessary, academics can change their delivery methods to cater to IR4.0. This further implies that not all courses should change the delivery methods. Next, one respondent claimed that he/she needed more exposure on applying changes to curriculum delivery to serve IR4.0, and lastly, five respondents were not sure about changing the delivery methods altogether.

When asked about the most effective curriculum delivery methods in order to prepare the future graduates, respondents, who were allowed to choose more than one option listed, illustrated that they would like for their students to experience various approaches of learning as illustrated in Table 6.

Given the findings in Table 6, it could be concluded that the respondents at the Defense University were aware of various effective delivery methods. It is further strengthened in Table 6 that what the

Table 6. Responses on effective curriculum delivery methods

No.	Options Listed	Number of Responses
1	Lecture	22 responses
2	Tutorial	24 responses
3	Fieldtrip	28 responses
4	Blended Learning (conventional and technology; flipped classroom)	23 responses
5	e-Learning (MOOCs, OER, OCW)	25 responses
6	Seminar/Workshop	21 responses
7	Others: *Industrial Training* *A combination of all listed* *Group Project* *Reflective Learning* *Online application such as 'Kahoot'*	 1 response 2 responses 1 response 1 response 1 response

respondents were suggesting is a form of blended learning for the students at the NDUM. The question is whether they are able to apply all these or whether they have applied all these methods in their teaching. Alas, this question could not be answered in this chapter.

Curriculum delivery methods too heavily rely on the understanding of various branches of pedagogy. Table 7 illustrates the responses of respondents on their understanding of these branches.

Almost all respondents (37 out of 38 respondents) understood what pedagogy is as expected since this is the basis for teaching. It is no surprise too that 14 respondents understood what makes up military pedagogy since the Defense University is operating in a military learning environment. Cybergogy, heutagogy and paragogy receive among the least responses with 12, eight and five responses respectively. Although these three branches of pedagogy are critical for T&L4.0, it shows that respondents need to be trained in understanding and applying these branches of pedagogy in their teaching.

SOLUTIONS AND RECOMMENDATIONS

What could be determined from all these discussions and analyses are twofold. First, despite the acceptance of the majority of the respondents on their responsibility to share their knowledge on IR4.0 to their students, some academics must be first given appropriate amount of exposure to ensure that they are IR4.0 ready and aware. The Defense University could not expect the graduates to be able to face IR4.0 when the academics themselves are uncertain about what IR4.0 is all about. Second, some academics too are unsure about T&L4.0, when they are supposed to have first-hand knowledge on this. Ignorance or not, academics who are preparing the future leaders of Malaysia must be fully equipped with various strategies and approaches to engage students in their learning. Providing students with the best possible learning environment is in the hand of the academics. The author opines that among possible yet immediate interventions that can be taken are, firstly, engaging relevant industries for collaborations such as industry placements of academics and block release of students at the industry; and secondly, continual awareness programs on IR4.0 and T&L4.0, where academics have roundtable sessions to exchange and share current practices and knowledge in teaching and learning.

Table 7. Responses on understanding branches of pedagogy

No.	Options Listed	Number of Responses
1	Pedagogy	37 responses
2	Cybergogy	12 responses
3	Andragogy	12 responses
4	Heutagogy	8 responses
5	Paragogy	5 responses
6	Military Pedagogy	14 responses

FUTURE RESEARCH DIRECTIONS

Much is yet to be discovered and investigated on IR4.0 and T&L4.0 in the Malaysian context. Future research should look at how IR4.0 could be seamlessly integrated in the curriculum, and how best to deliver the curriculum. Other scholars should also look at firstly, the curriculum for IR4.0 and beyond, and secondly, identify courses and academic programs that must change. As for T&L4.0, future research must investigate the most effective teaching and learning models that can prepare future graduates in facing IR4.0 and beyond. At the same time, the roles of the industry too must be investigated and further enhanced as graduates would eventually become employees of the industry.

CONCLUSION

In concluding this chapter, it is pertinent to revisit the reasons why IR4.0 and T&L4.0 are incorporated in the Malaysian higher education scenario. To ensure that the graduates are industry ready, they must be trained in an environment that allows them to develop their cognitive and affective domains appropriately. Based on the data analysis and discussions, it is inevitable that the vast range of age groups, teaching experiences and exposure on IR4.0 and T&L4.0 call for an intervention by the top management in facilitating the needs of the academics for various relevant programs on humanizing higher education in Malaysia. This intervention comes in the forms of various programs such as ones mentioned in the solutions and recommendations section.

ACKNOWLEDGMENT

This research received no specific grant from any funding agency in the public, commercial, or not-for-profit sectors.

REFERENCES

Blaschke, L. M. (2012). Heutagogy and Lifelong Learning: A Review of Heutagogical Practice and Self-Determined Learning. *International Review of Research in Open and Distance Learning*, *13*(1), 56–71. doi:10.19173/irrodl.v13i1.1076

Boulos, M. N. K., Giustini, D. M., & Wheeler, S. (2016). Instagram and WhatsApp in Health and Healthcare: An Overview. *Future Internet, 8*(3). Retrieved from http://www.mdpi.com/1999-5903/8/3/37/htm

Corneli, J., & Danoff, C. J. (2011). *Paragogy: Synergizing Individual and Organizational Learning.* Retrieved from https://pdfs.semanticscholar.org/1b4b/4c5b2a53705e161ca797be3562f4eeda1fc1.pdf

Dewey, J. (1916). *The Child & The Curriculum: The School & Society.* Chicago: University of Chicago Press.

Dewey, J. (1933). *How We Think*. Boston: Heath.

Dewey, J. (1934). *Art as Experience*. New York: Capricorn Books.

Dewey, J. (1938). *Experience & Education*. New York: Macmillan Publishing Co. Inc.

Dewey, J. (1968). *Democracy & Education*. New York: Macmillan.

Dick, B. (2013). Crafting learner-centered processes using action research and action learning. In S. Hase & C. Kenyon (Eds.), *Self-determined Learning: Heutagogy in Action*. London: Bloomsbury Academic.

Falk, C. (2008). All Pedagogy is Military. In T. Kvernbekk, H. Simpson, & M. A. Peters (Eds.), *Military Pedagogies: And Why They Matter*. Rotterdam: Sense Publishers.

Gerstein, J. (2016). *Maker Education: Pedagogy, Andragogy and Heutagogy*. Retrieved from https://usergeneratededucation.wordpress.com/2016/06/14/maker-education-pedagogy-andragogy-heutagogy/

Herbart, J. F. (1906). *A Study in Pedagogics (A. M. Williams, Trans.)*. Kessinger Publishing.

Hickman, L. A. (1990). *John Dewey's Pragmatic Technology*. Bloomington, IN: Indiana University Press.

Hickman, L. A. (2001). *Philosophical Tools for Technological Culture*. Bloomington, IN: Indiana University Press.

Industry 4.0: The Fourth Industrial Revolution – Guide to Industrie 4.0. (2016). Retrieved from https://www.i-scoop.eu/industry-4-0/

Jusoh. (2018). *Annual Speech by the Minister of Higher Education*. Retrieved from https://www.mohe.gov.my/speech-text/483-teks-ucapan-amanat-tahun-baharu-2018-yb-menteri-pendidikan-tinggi

Kenyon, C., & Hase, S. (2013). Heutagogy Fundamentals. In S. Hase & C. Kenyon (Eds.), *Self-determined Learning: Heutagogy in Action*. London: Bloomsbury Academic.

Lee, Y., & Rofe, J. S. (2016). Paragogy and Flipped Assessment: Experience of Designing and Running a MOOC on Research Methods. *Open Learning*, *31*(2), 116–129. doi:10.1080/02680513.2016.1188690

Muresan, M. (2013). A Blended Learning System within the Cybergogy Paradigm. *Procedia: Social and Behavioral Sciences*, *89*, 193–198. doi:10.1016/j.sbspro.2013.08.833

Phillips, A. G. (2002). John Dewey and His Religious Critics. *Religious Education (Chicago, Ill.)*, *29*(1), 31–48.

Ridzuan. (2011). *Sedekad Akademi Tentera Malaysia (1995-2005): Dari Kontena ke Kelas Pintar*. Kuala Lumpur: Penerbit UPNM.

Sarwono, J. (2012). *Path Analysis Dengan SPSS*. Jakarta: PT Alex Media Computindo.

Somr, M., & Hruskova, L. (2014). Herbart's Philosophy of Pedagogy and Educational Teaching (The Views and Differences of Opinion). *Studia Edukacyjne*, *33*, 413–429. doi:10.14746e.2014.33.21

Tapsir. (2017). *9th UTM University Presidents Forum 2017*. Universiti Teknologi Malaysia.

Tapsir & Puteh. (2018). *Framing Malaysian Higher Education 4.0: Future-Proof Talent*. Putrajaya: Department of Higher Education, Ministry of Higher Education Malaysia.

Toiskallio, J. (2003). Ethics, Military Pedagogy and Action Competence. In E. R. Micewski (Ed.), *Civil-Military Aspects of Military Ethics*. Frankfurt, Germany: Verlag.

Wang, M., & Kang, M. (2006). Cybergogy for engaged teaming: A Framework for Creating Learner Engagement through Information and Communication Technology. In D. Hung & V. I. S. Khine (Eds.), *Engaged Learning with Emerging Technologies*. Dordrecht, The Netherlands: Springer. doi:10.1007/1-4020-3669-8_11

Xing, B., & Marwala, T. (2017). *Implications of the Fourth Industrial Age on Higher Education*. Retrieved from https://arxiv.org/ftp/arxiv/papers/1703/1703.09643.pdf

KEY TERMS AND DEFINITIONS

Cybergogy: A branch of pedagogy that allows learners to engage in learning through three main domains, including cognitive, emotive, and social aspects.

Heutagogy: A branch of pedagogy that relies on learners' ability to be independent and promotes self-determined learning and is best promoted by adopting digital technologies.

Industrial Revolution: A period where automation becomes second nature to workforce in the nation.

Military Pedagogy: A branch of pedagogy that requires the conditions of learning to be set in a military learning environment and for the purposes of defense.

National Defence University of Malaysia: The only defense higher learning institution in Malaysia that offers military and academic training concurrently to the learners, who are divided into military cadets and civilians.

Paragogy: A branch of pedagogy that is also known as peer assisted learning, where learners rely on the support of peers during the learning process.

Pedagogy: The act of educating learners in order to promote change of behaviors.

Teaching and Learning: The process that allows for transformation of thoughts and behaviors, and may happen formally or informally, in and out of classroom.

Chapter 18
Redesigning Continuing Professional Development Training (CPDT) in Higher Education:
Enhancing ICT Integration Among University Lectures/Teachers in Classrooms

Abdul Halim Mohamed
Universiti Utara Malaysia, Malaysia

Azlida Ahmad
Tuanku Syed Sirajuddin Polytechnic Perlis, Malaysia

ABSTRACT

This chapter discusses the impacts of Industrial Revolution (IR) 4.0 on the continuing professional development training (CPDT), demanding lecturers/teachers to acquire knowledge of instructional design and information and communications technology (ICT). IR 4.0, often referred to as Education 4.0, requires universities to upgrade their CPDT that can enhance ICT skills of lecturers/teachers as well as students. The Ministries of Education (MOEs) and Ministries of Higher Education (MOHE) among ASEAN countries need to share their initiatives to help produce effective digital citizens. The quality of students depends on the quality of lecturers/teachers trained in CPDT. CPDT should be carefully redesigned to encourage more lecturers to enhance their student learning as well as engagement through ICT integration. Guidelines of redesigning CPDT are provided to redesign better and effective CPDT in the future. It is hoped that lecturers can increase their teaching effectiveness, efficiency, and productivity through continuous successful exposure of ID knowledge, skills, and ICT integration.

DOI: 10.4018/978-1-5225-7832-1.ch018

Copyright © 2019, IGI Global. Copying or distributing in print or electronic forms without written permission of IGI Global is prohibited.

INTRODUCTION

All ASEAN countries especially the Ministries of Education (MOE) and Ministries of Higher Education (MOHE) collaborate among themselves to share and promote effective Information and Communications Technology (ICT) policies, initiatives and platforms through the Southeast Asian Ministers of Education Organization (SEAMEO) and The United Nations Educational, Scientific and Cultural Organization (UNESCO). A variety of instructional reforms and views from research conducted by countries participating in the Organisation for Economic Co-operation and Development (OECD) have also contributed to the redesign and restructure of Continuing Professional Development Training (CPDT) for lecturers and teachers in Malaysia. Lecturers in universities especially those involved in teacher training departments need to improve their knowledge, attitude and skills in order to prepare prospective (future) teachers and in-service teachers undergoing training in universities. These teachers can deliver quality instruction if the lecturers are well-equipped with necessary skills and knowledge of Instructional Design (ID). The knowledge of ID can certainly enhance the development and production of better, effective and more interactive teaching and learning materials used for their students. Teachers need to shift from teacher-centred learning to more learner-centred learning approaches, individualistic to collaborative learning, shallow learning to deeper learning, single- sense sensory learning to multiple sense sensory learning, mundane to more interactive learning and low skills to higher skills' learning (Roblyer and Hughes, 2018). Just like any other countries, Malaysia is preparing its teachers to be ICT-competent leaders in schools. Focus of CPDT is on training teachers to not only teaching with ICT such as computers and mobile devices but also integrating ICT into a variety of subjects taught in schools. Eventually, CPDT needs to produce excellent teachers who focus more on producing life-long learners (Su & Wood, 2012; Carr, 2003 & Ramsden, 2003).

Besides SEAMEO, UNESCO and OECD, there are also many world's ICT organizations such as International Telecommunication Union (ITU), The United Nations Information and Communication Technologies Task Force (UNICTTF), The World Health Organization (WHO) and the International Society for Technology in Education (ISTE) which organize meetings consistently to maximize the impact and the potential of ICT in classrooms at universities and schools. In ensuring the achievement of Malaysia's Vision 2020 and development of holistic and balanced human beings, MOE (MOHE merged with MOE in 2018) and its agencies has to design and structure CPDT to align with expectations of Industrial Revolution (IR) 4.0 and referred to as Education 4.0 as educators at higher institutions of learning. The Internet of Things (IOT) and Web 2.0 have revolutionized and transformed the way lecturers and teachers present good pedagogy of learning to their students (Seale, 2010 & Carr, 2003). Undoubtedly, "[t]he regulation of quality is central to the credibility of the higher education system" (Brown, 2010, p. 47). Brown cites that the quality of lecturers needs to be seen from students' lenses or views.

To ensure teaching excellence at universities or schools, collaboration and interaction are expected to be integrated into teaching and learning processes (Luppicini, 2012; Kanter & Litow, 2009). Education 4.0 along with IR 4.0 are *"expected to bring immense benefits and many challenges"* (Morar, Arman, & Mousa, 2017, p. 2). With the help of Education 4.0, lecturers can increase their productivity and teach their students to be creative and innovative with technology and its tools (Waters, 2016; Drucker, 2014; Cordes & Stacey, 2017). Students are required to be effective and competent users of ICT (Roblyer & Hughes, 2018; Roblyer & Doering, 2012) as well as digital citizens (McAlister& Crappell, 2014; Winn, 2012). Universities need to train their lecturers to be skillful in dealing with a variety of digital resources that can enhance students' skills in ICT-rich environment. They need to educate their students to be responsible digital citizens who can safely use a variety of ICT tools available online and appropriately utilise these new online tools.

Lecturers of universities are expected to attend CPDT to improve their attitudes, knowledge and skills through ID and ICT. The successful use of ICT in higher education depends on the effective CPDT of universities (Kheswa, 2014; Whitaker, 1993, 1997). Lecturers attending any CPDT need to have a sense of ownership before they can bring about commitment, positive change in their teaching professions and adopt or adapt ICT to their teaching styles (Lai & Pratt, 2004; Roblyer & Doering, 2018). As universities are *"dynamic socio-technical systems"* (Kheswa, 2014, p. 1043), change and transformation are necessary to ensure acquisition of new knowledge and skills for sustainability and growth (Kheswa, 2014). Each lecturer plays a variety of roles in implementing ICT integration in their classrooms. With rapid and advancement of ICT, lecturers are expected to carry out face-to-face and online instruction effectively. ICT is expected to sustain and transform learning and facilitation in higher education. Regardless of courses they teach at universities, lecturers as ICT leaders need to be equipped with sufficient training to prepare their students (in-service teachers) for challenging work places. Besides, they need to be continuously supported and recognized in efforts to integrate ICT integration in classrooms. Lecturers and teachers need to learn how to effectively integrate ICT in order to *"engage students in collaborative communities of inquiry that facilitate deep and meaningful approaches to learning"* (Garrison & Aykol, 2009, p. 26). They stress that it is crucial to instill the commitment of lecturers and mastery of learning contents while integrating ICT into their subjects (Tearle, 2003).

With Industry Revolution 4.0, lecturers as well as teachers need to take advantage of the advancement of ICT and integrate new tools that promote communication, collaboration and critical thinking skills among their in-service teachers. ICT has created many possibilities for lecturers to educate their trainees to spread the ICT integration in schools through Professional Learning Community (PLC) and Action Research (AR) (Phillips & Carr, 2014; Foord & Haar, 2013; Thessin & Starr, 2011; Gummesson, 2000). It is hoped that in-service teachers through PLCs, new and old ICT tools are explored in providing effective instruction to students. These teachers are expected to encourage their colleagues to integrate ICT in their lessons effectively and efficiently. According to Garrison and Aykol (2009), ICT *"has provided new possibilities where educators can create and sustain collaborative learning communities not constrained by time and space"* (p. 20). Besides IT integration, English language proficiency among teachers can also be upgraded through ICT and its advancement. In order to sustain ICT integration among lecturers, AR should be encouraged in CPDT as it encourages positive changes in classrooms (Gummesson, 2000) and promote critical reflection and democratic practices (Smaldino, Donalson & Herring, 2012; Walker, 1997). PLCs can be easily nurtured through effective CPDT (DuFour, DuFour, Eaker, & Karhanek, 2004).

Ministries of Education and Ministries of Higher Education (MOHE) around the globe always strive to produce effective lecturers who can effectively train their students to maximize Information and Communications Technology (ICT) integration at universities and polytechnics. Lecturers of universities are expected to attend CPDT to improve their attitudes, knowledge and skills through Instructional Technology (ICT). Malaysian MOHE has always tried to improve CPDT programmes by redesigning its staff development to produce quality graduates who will be at par with other developed nations and to become competent and competent employees. MOHE has never failed to come up with many strategic policies such as Modular and Stackable (TVET), Accreditation of Prior Experiential Learning (APEL), Massive Open Online Courses (MOOCs), 2u2i (2 years in a university, 2 years in industrial attachments),

Program CEO @ Faculty and Integrated Cumulative Grade Point Average (iCGPA) in order to produce holistic and balanced digital citizens. In ensuring quality prospective employees, the quality of graduate or students depends on the quality of lecturers who have been continuously trained at their universities.

The successful use of ICT in higher education depends on the effective CPDT of universities (Subedi, 2015; Vatanartiran, & Karadeniz, 2014; Kheswa, 2014; Whitaker, 1993, 1997). Lecturers attending any CPDT need to have a sense of ownership before they can bring about commitment, positive change in their teaching professions and adopt or adapt ICT to their teaching styles (Roblyer & Hughes, 2018; Roblyer & Doering, 2013; Cook-Sather, 2011). As universities are *"dynamic socio-technical systems"* (Kheswa, 2014, p. 1043), change and transformation are necessary to ensure acquisition of new knowledge and skills for sustainability and growth (Kheswa, 2014). Each lecturer plays a variety of roles in implementing ICT integration in their classrooms. With rapid and advancement of ICT, lecturers are expected to carry out face-to-face and online instruction effectively. ICT is expected to sustain and transform learning and facilitation in higher education. Regardless of courses they teach at universities, lecturers as ICT leaders need to be equipped with sufficient training to prepare their students (in-service teachers) for challenging work places. Besides, they need to be continuously supported and recognized in efforts to integrate ICT. They need to learn how to effectively integrate ICT in order to *"engage students in collaborative communities of inquiry that facilitate deep and meaningful approaches to learning"* (Garrison & Aykol, 2009, p. 26).

COLLABORATION AMONG ASEAN MEMBER COUNTRIES

All 11-member ASEAN countries (Malaysia, Singapore, Indonesia, Thailand, Indonesia, Philippines, Socialist Republic of Vietnam, Lao, Cambodia, Republic of the Union of Myanmar and Timor-Leste) have collaborated and established the Southeast Asian Ministers of Education Organization (SEAMEO) Council in order in order to make lives better in quality and equity in education, preventive health education, culture and tradition, Information and Communication Technology (ICT), languages, poverty alleviation and agriculture and natural resources. They work together to nurture human capacities and explored the peoples' fullest potential. There are many SEAMEO centres located around ASEAN countries such as Regional Centre for Education in Science and Mathematics (RECSAM) (Malaysia), Regional Language Centre (RELC) (Singapore), Regional Centre for Tropical Medicine (Thailand), Regional Centre for Vocational and Technical Education and Training (VOCTECH) (Brunei), Regional Centre for Quality Improvement of Lecturers and Education Personnel (QITEP) (Indonesia), Regional Centre for Community Education Development (CED) (Laos) and Regional Centre for History and Tradition (CHAT) (Myanmar). With IR 4.0, these centres work together to training lecturers and teachers to become effective ICT leaders in their respect organisations.

Malaysian MOHE through its Higher Education Leadership Academy (AKEPT) and other agencies such SEAMEO design and develop CPDT to improve ICT integration among lecturers (Fullan & Hargreaves, 2013; Clegg, 2003; Becher, 2000 & Eraut, 2000). Researchers such as Friedman et al. (2001) feel that CPDT should encourage *"a continuous process of learning through reflection on practice"* (p. 13). Furze and Pearcey (1999) defines CPDT as *"the lifelong process of active participation in learning activities to enhance professional practice."* It is expected through effective CPDT, scholarship of teaching can be enhanced (Lueddeke, 2003; Friedman et al., 2001; Bercher, 2000). CPDT refers to *"the systematic maintenance, improvement and broadening of knowledge and skills, and the development of personal qualities necessary for execution of professional and technical duties throughout the individual's working life"* (Construction Industry Council, 1986, p. 3, cited Friedman & Phillips, 2004). It is hoped CPDT organized can sustain positive learning and provide continuous improvement of lecturers and students as well as build a career (Friedman & Phillips, 2004; Friedman et al., 2001). CPDT in Malaysian universities need to focus not only on ICT integration but also the use of medium of English as a Second Language (ESL). ID and ICT can be enhanced through the integration of ICT and Web 2.0 tools in enhancing student learning and increasing teacher-student interactions (Roblyer & Hughes, 2018; Smith & Ragan, 2005; Savery & Duffy, 1996).

MOHE needs to work closely with AKEPT, SEAMEO RECSAM and other agencies to develop quality CPDT at higher learning institutions. CPDT in universities or polytechnics need to be redesigned and restructured to produce quality lecturers for the 21st century in line with rapid ICT development in education. There is always a need to consolidate and enhance present CPDT organized and designed by MOHE and various training agencies so as to empower lecturers with the power of ICT and ICT integration. It is important to ensure CPDT prepare lecturers for the 21st century where ICT is maximised to make teaching more enjoyable and exciting. MOHE needs to develop better CPDT in order to produce successful instructors who can take advantages of ICT and the Internet. Higher institutions of learning will be good at serving their students especially to prepare them as marketable future employees for the Information Age. Upon completing their degrees, they are well prepared to deal with ICT tools at their workplaces.

ICT AND CPDT IN MALAYSIA

Malaysian government has come up with many initiatives such as New Economic Model, Economic Transformation Plan and Government to ensure that Malaysian education system continues to progress in tandem (MOE, 2013). Our roles as educators are to ensure that our efforts are in line with our National Education Philosophy that is to *"produce Malaysian citizens who are knowledgeable and competent, who possess high moral standards, and who are responsible and capable of achieving high level of personal well-being as well as being able to contribute to the harmony and betterment of the family, the society and the nation at large"* (MOE, 1988). MOHE and MOE are merged in Malaysia to form MOHE. The ministry and its agencies such as State Education Department, District Education Office, school administrators, teachers and parents must collaborate to further develop *"the potential of individuals in a holistic and integrated manner, so as to produce individuals who are intellectually, spiritually, emotionally and physically balanced and harmonious, based on a firm belief in and devotion to God"* (Education Act, 1996).

Universities and schools must ensure that their graduates and school leavers have the ability to compete in global market nationally and internationally where ICT is integrated at workplaces. Together, MOHE has to ensure that all students regardless of educational institutions to have six aspirations: knowledge, thinking skills, leadership skills, bilingual proficiency, ethics and spirituality and national identity (MOE, 2013). In bilingual proficiency, every student will be proficient in Bahasa Malaysia as the national language and in English language as a second language and the international language of communication (MOE, 2013). According to Malaysia Educational Statistics, there are 10,000 schools and 410,000 teachers (MOE, 2013). MOHE (2013) wants to equip our students holistically to allow them to succeed in the 21st century, with all of the opportunities and challenges that this new era presents (p. 8). Undoubtedly, quality teachers and lecturers must be trained to develop young Malaysians who are knowledgeable, think critically and creatively, have leadership skills and are able to communicate with the rest of the world.

CPDT IN UNIVERSITIES

Universities need to design and structure effective CPDT to develop skillful lecturers who can teach their students to integrate ICT into their activities where communication, collaboration, creativity and critical things skills are enhanced. CPDT is deemed useful in successfully ICT integration in Malaysia and in helping lecturers as agents of change to encourage or discourage the adoption of innovations and further diffusion of innovation (Rogers, 1995). Through an effective CPDT, instructors can help the MOHE attain the vision and mission of the educational needs. CPDT needs to be effective since the ultimate goal of professional development is to enhance and improve student learning (Speck, 1996).

The advancement of Information and Communications Technology (ICT) provide numerous Web 2.0 tools that lecturers can use to encourage the 4 Cs (Communication, Collaboration, Creativity and Critical thinking skills). Malaysian MOHE has spent lots of expenditures to ensure that lecturers are taking advantage of ICT tools in classrooms. Besides Malaysian Government's Transformation Plan, Malaysian Educational Blueprint (2013-2025) outlines many strategic initiatives that can not only encourage ICT implementation and integration in schools but also ensure the return of investment through well planned efforts by MOE and MOHE. However, some lecturers are still struggling to integrate ICT into their teaching and learning. Based on various studies conducted by Malaysian researchers, there are some challenges in ICT training conducted by the MOHE. There are many reasons lecturers fail to integrate ICT into their classrooms, Some of the reasons lack of time, lack of knowledge and abilities, lack of good ICT integration ways and lack of ICT skills (Hu & Garimella, 2014; Heyberi, 2013; Ageel & Woollard, 2012; Eristi, Kurt & Dindar, 2012; Richard, 2008; Abdul Halim, 2006; Pelgrus, 2001).

There are three types of CPD. Lieberman (1996) classifies CPD into three types: direct teaching (such as courses, workshops and so on); learning in school (such as peer coaching, critical friendships, mentoring, action research, and task-related planning teams); and out of school learning (such as learning networks, visits to other schools, school-university partnerships and so on). Lecturers who are successfully trained would surely able to design and develop digital age learning experiences and assessments, facilitate student learning with ICT, increase students' productivity, promote and model digital citizenship and responsibility and engage in professional growth and leadership. It is hoped lecturers can increase their teaching effectiveness and efficiency through ICT integration. Organisation for Economic Co-operation and Development (OECD) (2009) states that *"CPD cannot be expected to prepare lecturers for all the challenges they will face throughout their careers"* (p. 49). OECD (2005) states that "[s]uccessful programmes involve lecturers in learning activities that are similar to ones they will use with their students, and encourage the development of lecturers' learning communities. There is growing interest in developing schools as learning organisations, and in ways for lecturers to share their expertise and experience more systematically" (p. 49).

Furthermore, CPDT should provide a platform for instructors to mingle with their colleagues professionally. It should also give an opportunity for them to improve their Subject Knowledge (SK) as well as Pedagogical Content Knowledge (PCK). Lecturers as trainees must not be haphazardly sent for training programmes without analysing their needs to attend it! It is essential to maintain the reputation and professionalism of lecturers so that they remain a respected profession in society. In line with educational changes, the MOHE's CPDT should aim at improving lecturers' pedagogical content knowledge, including general pedagogical knowledge; knowledge of the subjects taught, the curriculum, educational contexts, goals and values; and knowledge about learners (Shulman, 1987). When dealing with any substantial change, people go through three reactive stages: *"shock, withdrawal, acknowledgement and adaptation"* (p. 64). Therefore, CPDT has to be carefully designed in order to produce effective results and high quality lecturers (Adu & Okeke, 2017; Kemp, Preston, Page, Harper, Dillard, Flynn, & Yamaguchi, 2014; Bierama & Eraut, 2004; Harwell, 2003; Kennedy, 2005; Weindling, 2001; Friedman, Durkin & Phillips, 2000; Becher, 2000).

GUIDELINES OF REDESIGNING CPDT

There are some suggestions for organizing CPDT for lecturers and teachers.

Some of the recommendations made are as follows:

1. *Before sending lecturers out for training, it is crucial to perform a basic needs' assessment of their general, specialised and technical skills.*

Needs assessment or analysis of attendees would ensure that all lecturers in a training session are of the same ability and do not feel threatened by their colleagues who attend the same session (Farrell, Shafiei & Salmon, 2010; Arthur, Benner, Eden & Bell, 2003; Brown, 2002).

It is important to conduct assessments periodically as an effective way of discovering emerging or immediate needs after lecturers have undergone CPDT. Lecturers could easily resolve problems faced by students or the lecturers themselves if they were able to discuss them with their administrators when they first occur. Lecturers lacking in confidence could be allowed to undertake more courses when they were not teaching. Needs assessment ensures that lecturers should be continuously assessed for professionalism and for the development of their professionalism to maintain high standards of teaching and learning.

There should be a non-threatening climate in which lecturers do not hesitate to learn from lecturers and colleagues. Lecturers would be encouraged to share and transfer their skills if the knowledge gaps between them were not excessive. It is important to keep the gap minimal in order to improve working relationships. The lecturers need to build good mutual relationships, as collegiality affects lecturers' morale, happiness and satisfaction (Talbert, 2010; McLaughlin & Talbert, 2006; McLaughlin & Talbert, 2001). A good relationship among lecturers influences their motivation and career commitment, and the extent to which they are willing to modify classroom practice. It is important to keep in mind that implementing curriculum innovation is not an easy task. It cannot be accomplished in a day of training. CPD may take several years to accomplish educational changes such as curriculum innovations (Speck, 1996). Fullan (1996) advocates the use of multilevel training to address the needs of lecturers who may be at different knowledge and skills levels.

2. *The CPDT organised should be attended by lecturers with similar levels of ICT skills.*

Even though lecturers of the same subject are involved together in the training programme, the training would be more effective if they were also grouped based on their levels of ICT ability. The selection of participants based on certain abilities must be the most important criteria in determining which lecturers should attend the training programme. It is advisable that administrators to select trainees at similar ICT levels in order to maximise the use of ICT during the training. Lecturers with widely differing ICT skills affect the performance of the group psychologically, as those who are not skilled may mistakenly be considered as weak lecturers. It is important that lecturers work collaboratively with other lecturers of similar capabilities to produce effective results based on the targeted learning outcomes (Drucker, 2014; Van Eerde, Tang, & Talbot, 2008).

3. *Lecturers should be divided into groups based on their levels of ICT skills.*

Specific general and technical skills training can be addressed in groups of similar ability. In training lecturers to use computers or ICT, for instance, it is best to divide them into beginner, intermediate and advanced capabilities. This ensures that more skilled lecturers are not slowed down by those with less ability, and beginners do not feel that their skills are being tested. Although heterogeneous grouping provides a more conducive learning atmosphere, multilevel training is still considered to be effective (Winn, 2012; Ramsden, 2003; Harwell, 2003; Fullan, 1996).

4. *CPDT should be conducted as an ongoing series of CPDT courses.*

Lecturers should not have to attend long sessions of in-service training, as these do not guarantee or produce effective results. It is suggested that the programme would be more effective broken into a series of shorter sessions. Moreover, CPDT should be continuous and multilevel (Fullan, 1996; Speck, 1996). Moreover, it should be backed up with sufficient technical and social support (Harwel, 2003; Fullan, 1996; Speck, 1996). CPDT could be divided into several sessions over a number of years. More than one of each session could be provided at different times to suit lecturers with varying commitments. Lecturers should be given choices, not ultimatums.

5. *CPDT does not have to be a residential programme.*

The training programme could be conducted in teacher training colleges or universities rather than in hotels, and the money saved spent on educational resources such as CDs and other programmed instructions; on research and development; high quality reference and programming books; CD-ROMs, videos and cassettes; audio-visual materials and teaching kits. Lecturers could improve their skills in using and evaluating instructional computer-based programmes in their schools. With effective CPDT, those lecturers who are motivated and enthusiastic would motivate students in their learning process (Baker, 1994).

6. *CPDT should be brief and well-focused.*

Those who attend a very long CPDT feel that attending long training sessions was ineffective and unproductive, as they became too tired to learn. They complained that their minds were unable to focus on some lectures, and were distracted by thoughts about their domestic affairs as the training was held far from their homes (Abdul Halim, 2006; Harwell, 2003; Kraiger, 2002). The CPDT should be brief and well-focused. Lecturers should be given time to comprehend and digest information given during the training. CPDT can be short, but should be continuous and effective, allowing lecturers to find a suitable time to attend as long as they complete the requirements within a certain limited time.

7. *CPDT offered should be not only relevant but also continuous.*

It is easy to frustrate lecturers by forcing them to attend training which they do not feel they want or need. Therefore they need to be allowed to make choices about attending, thus creating more motivated and enthusiastic agents of change. Quality teaching is needed to foster student learning (Harwell, 2003; Darling-Hammond and Berry, 1998).

8. *CPDT trainees should be monitored for a period of time after the training.*

The effectiveness and efficiency of any CPDT could be determined by monitoring lecturers to see whether they are applying the elements that they have learnt. Their approach to students should be evaluated for a considerable number of months after the CPDT. Lecturers should be appraised based on the elements taught in the CPDT to encourage their use. With the help of trainers, systematic evaluation could be used to observe lecturers. The trainers could be asked to provide principals with reports on lecturers' levels of ICT use and integration (Tsai & Chai, 2012; Riasati, Alllahyar & Tan, 2012; Pelgrum, 2001).

9. *CPDT, if focused on ICT skills, should be presented as three separate ICT courses.*

Lecturers could be exposed to three main ICT courses: computer as a tutee, computer as a tutor and computer as a tool. A prerequisite course such as keyboard skills could be taught by coordinators. Keyboard skills training could avoid lecturers with low ICT capability feeling inferior or dejected. This low-level course could be conducted in schools, presented by the schools' ICT coordinators. As ICT skills encompass many skills, the skills could be broken into manageable chunks, for instance, by having a series of ICT training programmes (Abdul Halim, 2006). Lecturers who have attended the SSTP mentioned that the teaching of ICT had not been considered effective as they were exposed to too many concepts and needed more time to grasp them. CPDT involving ICT integration could be conducted in schools or regionally. Lecturers in the northern states of Malaysia, for instance, would attend CPD in the northern region.

10. *CPDT should focus on selected skills.*

The MOHE has been ambitious in conducting a training programme which incorporates so many skills at once. Instead, CPDT should be limited to certain skills such as critical and creative thinking and ICT skills. It is important to listen to the voices of academics (Crawford, 2007). A short course in learning skills could be easily conducted by faculty members as in-house training. Some teaching and learning skills are sold as videotapes or CDs and can be purchased by schools.

11. CPDT should be taught by competent, up-to-date trainers.

Trainers appointed to train lecturers in CPDT should be not only knowledgeable in pedagogy but also familiar with present teaching in schools. Based on a study by Abdul Halim (2006), those who attended the CPDT felt that some of the trainers, who were lecturers at lecturers' colleges, had not been involved in teaching students for some time and could not relate to lecturers' experiences. It is crucial that trainers know about current teaching scenarios. Thus, the credibility of trainers could be questioned by lecturers and teachers.

The MOHE should ensure that CPDT consists of trainers who are competent as well as knowledgeable, and all elements of the training are useful and relevant to improving student learning. An in-service training programme would be useful to increase lecturers' productivity as well as to enhance their skills, knowledge and competence (Crawford, 2007; Harwell, 2003;).

12. Provide practical teaching strategies and skills during CPDT.

Lecturers should be exposed to effective instruction. The MOHE especially the trainers should be aware of and document all effective teaching happening in schools. This would help lecturers to review their teaching approaches and beliefs as well the instructional materials. It is crucial for lecturers to adopt or adapt teaching strategies or skills that are proven to be effective. Trainers need to provide more practical approaches to teaching and learning. Lecturers of science, for instance, could be exposed to Cognitive Acceleration in Science Education (CASE) to enhance students' thinking skills. CASE is based on the principle that children can be developed using challenging activities. Perhaps, the SSTP could use CASE materials and Thinking Science (TS) to train these Malaysian lecturers as these teaching materials involve concrete preparation, cognitive conflict, social construction and metacognition (Adey and Shayer, 1994; Adey et al., 2001). These materials can also be used to promote children's procedural understanding and thus, help lecturers to present more coherent picture of science (Jones and Gott, 1998).

13. Head of Departments (HODs) should be the main drivers of educational change in schools.

The skills taught on the CPDT should be emphasised by HODs, administrators at staff meetings and in speeches. HODs, as leaders, should be the drivers of the initiative, and should encourage lecturers' use of the skills taught during the programme. The examples of good leadership would be emulated by others. According to Fullan (2001), *"[g]ood leaders foster good leadership at other levels.* Staff expertise *"is the most important asset in a university; without it literally nothing can be achieved"* (Blackmore & Blackwell 2003, p. 23). Efforts should be on developing their knowledge and skills to integrate new methods of teaching and ICT integration. *Leadership at other levels produces a steady stream of future leaders for the system as a whole" (p. 10).* Crawford (2009) states that "*it is necessary to start with an exploration of what academics understand by continuing professional development, what they do, and why, taking account of the context within which it happens*" (p. 112).

14. There should be a follow-up training after CPDT.

To ensure the effectiveness of the programme, it is advisable to monitor the extent to which lecturers are using the skills taught during CPDT. Failures to organise follow-up training will certainly wastes time and efforts. Lecturers who have undergone the training are valuable resources and informants in finding out the effectiveness of the present CPDT they are attending (Crawford, 2009). They can also be asked to give feedbacks pertaining to the teaching approaches and effectiveness. With new and modern teaching strategies to be integrated in CPDT, the teachers' roles of engaging with their students are crucial (Deem et al., 2008; Dexter, 2007; Abdul Halim, 2006; Blackmore & Blackwell, 2003; Roscoe, 2002).

15. In-house training CPDT should be a frequent event in universities.

Lecturers who attend CPDT should collaborate with lecturers who have not, providing a collegial teaching and learning platform where lecturers can discuss their successes or failures with certain teaching strategies and skills. Collegial sharing is vital for systemic changes (Dexter, 2007; Abdul Halim 2006; Fullan, 1996). Involving lecturers in conducting in-house in-service training builds teacher leadership (Harris and Lambert, 2003). CPDT changes along with the expectations of world's ICT rapid development in higher education (Woodward, 1996). According to Woodward (1996), there is *"accelerating pace of economic, social and technological change"* (p. 1).

16. CPDT should be the main agenda for lecturers' development since ICT is considered crucial.

There is no doubt that even after CPDT, lecturers need to learn more on teaching with ICT than on teaching about ICT. If schools are to be equipped with ICT and lecturers are meant to integrate ICT, there should be in-service training courses related to ICT skills. Professional development in ICT should address the attitudes of lecturers and towards ICT tools. Friedman et al. (2008) define CPDT as "[t]he systematic maintenance, improvement and broadening of knowledge and skills, and the development of personal qualities necessary for execution of professional and technical duties throughout the individual's working life" (p. 18). If lecturers' anxiety level is high, trainers should use a more clinical approach and make the trainees more comfortable before immersing them in the actual and robust programmes that deal with ICT professional development (Rothwell & Arnold, 2005). It is also important for in-house technical assistance and support to help lecturers increase their ability in troubleshooting when any problem pertaining to ICT rises and becoming competent users of ICT.

17. Encourage CPDT trainees to conduct action research into the skills taught.

It is crucial to cite that CPDT is *"a process by which individuals take control of their own learning and development, by engaging in an on-going process of reflection and action"* (Megginson & Whitaker 2007, p. 3). Lecturers can improve and enhance certain transferable skills as well as skills of ICT integration by conducting action research in their classrooms (Kheswa, 2014; Friedman & Phillips, 2004; Jenkins et al., 2003; Friedman, Durkin & Phillips, 2000; Roberts, 2002). They can learn to "acquire new knowledge, skills and ability to deal with more small groups, more problem – based tutorials and new computer- based instructional programme" (ibid, p. 1). Administrators should encourage lecturers to conduct research involving their students using the teaching and learning strategies and styles they gained from the training. They could use such research as a platform to test their teaching ideas and to incorporate new teaching innovations into their research. Kheswa (2014) recommend that (1) lecturers should introduce various forms of teaching such as participatory approach and (2) educational resources be available for students to access computer centres.

18. Promote self-assessment to engage students in classrooms.

Students can be taught to be critical readers. They can be guided to help assess their peers' work. Lecturers can use peer assessment even in primary schools as an effective way of encouraging students to think aloud about certain concepts, help them detect errors made by their peers, find alternative ways of remembering facts and develop awareness of good quality science work. Engaging students in assessing their peers' work will encourage them to develop an ethos where producing quality work is imperative.

19. Malaysian lecturers should be guided by English lecturers on the use of English as a medium of instruction.

Lecturers of science and other subjects who lack confidence in delivering the content of science in English should be helped through peer learning or peer coaching (Joyce and Showers, 1995). Peer coaching may involve observation of lecturers, analysis of their performance and providing feedback; identification of strategies for improvement; monitoring of change in performance and consolidation of improvements (West-Burnham and O' Sullivan, 1998). A collaborative culture is crucial to improve the professional standards of lecturers as members of the professional learning community (Smith, 2008; DfES, 2003; Jones & Jones, 2007; Jenkins et al., 2003; Zuber-Skerritt, 1992). For instance, English language lecturers can help science lecturers improve their interactions with their students by teaching them to use different questioning techniques and dialogue and language games. Classroom dialogues can be divided into four different types: *"learning the official language of science; questioning; explaining; and focusing and shaping in interactions"* (Wellington and Osborne, 2001, p. 28).

CONCLUSION

It is crucial that HODs and administrators play a more effective role in managing educational change especially the influence of IR 4.0 (Cordes & Stacey, 2017). In order to implement curriculum innovations, lecturers should be trained in understanding the concepts of curriculum implementation process; the development of curriculum; ICT integration; assessing skills; adopting and adapting to new teaching approaches to be used together with old ones; and understand that their teaching is influenced by their beliefs, pedagogy and teaching materials as well as by their students. Kreber (2001) states that scholarship of teaching entails three different knowledge- domains, namely; instructional knowledge (i.e. knowledge acquisition of the course by lecturers), pedagogical ID knowledge (i.e. knowledge that lecturers have about how students learn), and curricular knowledge (i.e. teaching objectives and the purpose or future prospects of the course). CPDT needs to be designed and developed to improve the ID and ICT knowledge and produce good quality instruction for lecturers at higher institutions of learning. CPDT is vital for teacher preparation programme for the twenty-first century (Carr-Chellman & Rowland, 2017; Fullan & Hargreaves, 2013; Talbert, 2010; Smith, 2008).

In line with expectations of IR 4.0 and education 4.0, lecturers needs to be skillful in showing teachers (their trainees) to integrate ICT tools that can promote interactions among students and teachers in schools. There is no doubt that even after CPDT, lecturers need to learn more on teaching *with* ICT than on teaching *about* ICT. As stated in the guidelines above, CPDT is directed towards producing effective and quality lecturers and teachers for their learners. If schools are to be equipped with ICT and lecturers are meant to encourage ICT and its integration, there should be in-service training courses related to ICT. CPDT in ICT should address the attitudes of lecturers and towards computers (Dexter,

2007; Abdul Halim, 2006; Day & Sachs, 2004; Nixon, Marks, Rowland & Walker, 2001). If lecturers' anxiety level is high, trainers should use a more clinical approach and make the trainees more comfortable before immersing them in the actual programmes that deal with ICT professional development. It is also important for carrying out needs analysis as well as in-house technical assistance and support to help lecturers increase their ability to engage with their students (Cho, Park, Jo & Suh, 2013; Van Eerde, Tang & Talbot, 2008).

In training lecturers about ICT integration, lecturers who are not competent in using English language can develop their language competence through collaboration with their colleagues who has good English language proficiency. They can learn three skills in the art of good communication in science teaching: *(1) the recognition that teaching involves a range of modes of communication; (2) the awareness of these different modes and the recognition that different modes suit different learners; and (3) the ability to switch from one mode to another when teaching* (Wellington, 2000, p. 189).

Lecturers should be taught to be sensitive to their proficiency of English language in any subjects taught, dividing students according to their proficiency levels in order to teach lessons effectively. With ICT, lecturers can use ICT to reduce language barriers and increase interactions among students. They can use IT tools to facilitate instruction. They can engage their students to use practical classroom tactics that involve '*giving instructions'* (p. 122), '*adopting a multisensory approach*' (p. 122), '*worksheets'* (p. 123), '*variety in submitted work*' (p. 123), '*support with writing and spelling* ' (p. 124) '*using computers (for task and outcome'* (p. 124), '*using laminated cards to help to enrich reading'* (p. 126) and '*using the science word banks'* (p. 131) (Wellington and Osborne, 2001).

CPDT or any effective staff development programme should provide lecturers with opportunities to re-examine their teaching beliefs, values, teaching approaches and teaching materials (. CPDT should provide a platform for lecturers to mingle with their colleagues professionally. It should also give an opportunity for them to improve their knowledge of ID and ICT integration. Through effective and efficient CPDT, they should be able to learn to integrate ICT tools into their teaching practices. Regardless of their age and circumstances, trainees of lecturers who are school teachers or pre-service teaching can learn as much as they can so that they can return to schools as confident ICT leaders or trainers. They will also, then, conduct in-house training for those who have not had the opportunity to attend any CPDT.

REFERENCES

Abdul Halim, M. (2006). Smart schools' training programme (SSTP): Teachers perspectives (Doctoral Dissertation). Universiti of East Anglia, UK.

Adey, P. S., & Shayer, M. (1994). *Really Raising Standards: Cognitive Intervention and Academic Achievement*. London: Routledge.

Adey, P. S., Shayer, M., & Yates, C. (2001). *Thinking Science: The curriculum materials of the cognitive acceleration through Science (CASE) Project* (3rd ed.). Cheltenham, UK: Nelson Thornes.

Adu, E. O., & Okeke, C. I. O. (2014). Factors Affecting Lecturers' Participation in Continuing Professional Development (CPD). *Journal of Sociology and Social Anthropology*, *5*(3), 271–281. doi:10.1080/09766634.2014.11885631

Ageel, M., & Woollard, J. (2012). Enhancing university teachers' information and communication technology usage by using a virtual learning environment training course. In *International Technology, Education and Development Conference* (pp. 5599–5606). Valencia, Spain: Academic Press.

Ahmed, S. (2011). Improving cognitive development in secondary chemistry through Gagne's events of instruction. *Journal of Education and Practice*, *2*(4), 2011.

Arthur, W. J., Bennett, W. J., Edens, P., & Bell, S. T. (2003). Effectiveness of training in organizations: A metaanalysis of design and evaluation features. *The Journal of Applied Psychology*, *88*(2), 234–245. doi:10.1037/0021-9010.88.2.234 PMID:12731707

Baker, F. G. (1994). *Integrating Computer Usage in the Classroom Curriculum through Teacher Training*. Nova Southeastern University.

Ball, D. (2000). Bridging practices: Interwining content and pedagogy in teaching and learning to teach. *Journal of Teacher Education*, *51*(3), 241–247. doi:10.1177/0022487100051003013

Becher, T. (2000). *Professional practices: commitment and capability in a changing environment.* Somerset, NJ: Transaction.

Bierama, L. L., & Eraut, M. (2004). Workplace-focused learning: Perspective on continuing professional education and human resource development. *Advances in Developing Human Resources*, *6*(1), 52–68. doi:10.1177/1523422303260859

Blackmore, P., & Blackwell, R. (2003). Academic roles and relationships. In R. Blackwell & P. Blackmore (Eds.), *Towards Strategic Staff Development in Higher Education* (pp. 16–28). Berkshire: SRHE and Open University Press.

Boliton, K. (1985). *A study on the secondary school English training programs.* Seoul: Hong-Ik University.

Brown, J. (2002). Training needs assessment: A must for developing an effective training programme. *Public Personnel Management*, *31*(4), 569–578. doi:10.1177/009102600203100412

Browne, J. (2010). *Securing a sustainable future for higher education.* Available at: www. independent. gov.uk/browne-report

Capperucci, D. (2017). English language teaching and learning in Primary School. theoretical and Methodological Perspectives. *Studi sulla Formazione, 20*(2), 203-217.

Carr, D. (2003). *Making sense of education.* London: Routledge Falmer.

Carr-Chellman, A. A., & Rowland, G. (Eds.). (2017). *Issues in technology, learning, and instructional design: Classic and contemporary issues.* Routledge.

Cho, Y., Park, S., Jo, S. J., & Suh, S. (2013). The landscape of educational technology viewed from the *ETR&D* journal. *British Journal of Educational Technology*, *44*(5), 677–694. doi:10.1111/j.1467-8535.2012.01338.x

Clegg, S. (2003). Problematising ourselves: Continuing professional development in higher education. *The International Journal for Academic Development*, *8*(1/2), 37–50. doi:10.1080/1360144042000277928

Cook-Sather, A. (2011). Layered learning: Student consultants deepening classroom and life lessons. *Educational Action Research, 19*(1), 41–57. doi:10.1080/09650792.2011.547680

Cordes, F., & Stacey, N. (2017). *Is UK Industry Ready for the Fourth Industrial Revolution?* Boston, MA: The Boston Consulting Group.

Crawford, K. (2007). Continuing professional development in higher education: Debating the academic perspective. *The International Journal for Academic Development, 13*(2), 141–146. doi:10.1080/13601440802076657

Crawford, K. (2009). Continuing Professional Development in Higher Education – tensions and debates in a changing environment. In L. Bell, H. Stevenson, & M. Neary (Eds.), *The future of higher education: Policy, pedagogy and the student experience* (pp. 69–83). London: Continuum.

Cummins, J. (1994). The acquisition of English as a Second Language. In K. Spangenberg-Urbschat & R. Pritchard (Eds.), *Reading Instruction for ESL Students*. International Reading Association.

Dabbagh, N. (2002). Scaffolding: An Important Teacher Competency in Online Learning. *Tech-Trends, 47(*2), 39-44.

Darling-Hammond, L., Chung, R., & Frelow, F. (2002). Variation in teacher preparation: How well do different pathways prepare teacherst o teach? *Journal of Teacher Education, 53*(4), 286–302. doi:10.1177/0022487102053004002

Day, C. (1999). *Developing teachers: The challenges of lifelong learning*. London: Falmer Press.

Day, C., Harris, A., Hadfield, M., Tolley, H., & Beresford, J. (2000). *Leading schools in times of change*. Buckingham, UK: Open University Press.

Day, C., & Sachs, J. (2004). *International handbook on the continuing professional development of teachers*. Berkshire, UK: Open University Press.

Department for Education and Employment (DfEE). (2001). *Good value CPD: A code of practice for providers of professional development for teachers*. London: DfEE.

Department for Education and Skills (DfES). (2001). *Helping you develop: guidance on producing a professional development record*. London: DfES.

Department for Education and Skills (DfES). (2003). *The future of higher education*. London: The Stationery Office.

Dexter, B. (2007). Continuous professional development for dual professionals: Introducing CPD in a university setting. *The International Journal of Interdisciplinary Social Sciences: Annual Review*, 2(4), 21–327. doi:10.18848/1833-1882/CGP/v02i04/52363

Drucker, P. (2014). *Innovation and entrepreneurship*. New York: Routledge. doi:10.4324/9781315747453

DuFour, R., DuFour, R., & Eaker, R. (2008). *Revisiting professional learning communities at work. New insights for improving schools*. Bloomington, IN: Solution Tree.

DuFour, R., DuFour, R., Eaker, R., & Karhanek, G. (2004). *Whatever It Takes: How Professional LearningCommunities Respond When Kids Don't Learn*. Bloomington, IN: National Educational Service.

DuFour, R., DuFour, R., Eaker, R., & Many, T. (2006). *Learning by doing: A handbook for professional learning communities at work* (2nd ed.). Bloomington, IN: Solution Tree.

Eraut, M. (1994). *Developing Professional Knowledge and Competence*. London: The Falmer Press.

Eristi, S. D., Kurt, A. A., & Dindar, M. (2012). Teachers ' views about effective use of technology in classrooms. *Research on Computing in Education, 33*(5), 1–27.

Farrell, G. A., Shafiei, T., & Salmon, P. (2010). Facing up to 'challenging behaviour': A model for training in staff–client interaction. *Journal of Advanced Nursing, 66*(7), 1422–1434. doi:10.1111/j.1365-2648.2010.05340.x PMID:20497270

Foord, K. A., & Haar, J. M. (2013). *Professional learning communities: An implementation guide and toolkit*. New York: Routledge.

Freeman, D. (1991, December). Three views to teachers' knowledge. *IATEFL Teacher Development Newsletter*, 1-4.

Freeman, D. (1993). Renaming experience/reconstructing practice: Developing new understandings of teaching. *Teaching and Teacher Education, 94*(5-6), 485–497. doi:10.1016/0742-051X(93)90032-C

Freeman, D., & Johnson, K. E. (1998). Reconceptualizing the knowledge-base of language teacher education. *TESOL Quarterly, 32*(3), 397–418. doi:10.2307/3588114

Frey, N., Fisher, D., & Allen, A. (2009). Productive group work in middle and high school classrooms. In S. Parris, D. Fisher, & K. Headley (Eds.), *Adolescent literacy: Effective solutions for every classroom* (pp. 70–81). Newark, DE: International Reading Association. doi:10.1598/0695.06

Friedman, A., Davis, K., & Phillips, M. (2001). *Continuing professional development in the UK: Attitudes and experiences of practitioners*. Bristol: PARN.

Friedman, A., Durkin, C., & Phillips, M. (2000). CPD: What are the true costs of continuing professional development? *Continuing Professional Development, 3*(3), 78–87.

Friedman, A., & Phillips, M. (2004). Continuing professional development: Developing a vision. *Journal of Education and Work, 17*(3), 361–376. doi:10.1080/1363908042000267432

Friedman, A., Williams, C., Hopkins, S., & Jackson, L. (2008). *Linking professional associations with higher education institutions in relation to the provision of continuing professional development*. Bristol: Professional Associations Research Network.

Fullan, M. (1996). Turning Systemic Thinking on its Head. *Phi Delta Kappan, 77*(6), 420–423.

Fullan, M. (2001). *Leading in a Culture of Change: Being Effective in Complex Times*. San Francisco: Jossey-Bass Wiley.

Fullan, M., & Hargreaves, A. (Eds.). (2013). *Teacher development and educational change*. London: Routledge.

Furze, G., & Pearcey, P. (1999). Continuing education in nursing: A review of the literature. *Journal of Advanced Nursing*, *29*(2), 355–363. doi:10.1046/j.1365-2648.1999.00896.x PMID:10197935

Gagné, R. M., Briggs, L. J., & Wager, W. W. (1992). *Principles of instructional design* (4th ed.). Fort Worth, TX: Harcourt Brace Jovanovich College Publishers.

Gagne, R. M., Wager, W. W., Golas, K. C., & Keller, J. M. (2005). *Principles of instructional design* (5th ed.). CA: Wadsworth.

Gagne, R. M., Wager, W. W., Golas, K. C., & Keller, J. M. (2005). *Principles of instructional design* (5th ed.). Wadsworth.

Garrison, D. R., & Aykol, Z. (2009). Role of instructional technology in the transformation of higher education. *Journal of Computing in Higher Education*, *21*(1), 19–30. doi:10.100712528-009-9014-7

Goh, R., & Fang, Y. (2017). Improving English language teaching through lesson study: Case study of teacher learning in a Singapore primary school grade level team. *International Journal for Lesson and Learning Studies*, *6*(2), 135–150. doi:10.1108/IJLLS-11-2015-0037

Gummesson, E. (2000). *Qualitative methods in management* (2nd ed.). Thousand Oaks, CA: Sage.

Harmer, J. (2003). *The practice of English language teaching* (3rd ed.). Harlow, UK: Pearson.

Harwell, S. H. (2003). *Teacher professional development: It's not an event, it's a process*. Waco, TX: Cord.

Heyberi, E. (2013). *Integrating Technology in the Curriculum for Enhanced Learning : A Comparative Study in England and North Cyprus* (Doctoral dissertation). University of Birmingham.

Hodell, C. (2011). *ISD from the ground up: A no-nonsense approach to instructional design* (3rd ed.). Alexandria, VA: ASTD Press.

Hu, H., & Garimella, U. (2014). iPads for STEM teachers: A case study on perceived usefulness, perceived proficiency, intention to adopt, and integration in K-12 instruction. *Journal of Educational Technology Development and Exchange*, *7*(1), 49–66. doi:10.18785/jetde.0701.04

Hurst, B., Wallace, R., & Nixon, S. B. (2013). The impact of social interaction on student learning. *Reading Horizons*, *52*(4). Retrieved from https://scholarworks.wmich.edu/reading_horizons/vol52/iss4/5

Jenkins, A., Breen, R., Lindsay, R., & Brew, A. (2003). *Reshaping teaching in higher education: Linking teaching with research*. London: Kogan Page. doi:10.4324/9780203416884

Jones, C., & Jones, K. (2007). Language of Professional Learning. *Teaching Expertise*. Available from: www.teachingexpertise.com/articles/language-of-professional-learning-1791

Jones, M., & Gott, R. (1998). Cognitive acceleration through science education: Alternative perspectives. *International Journal of Science Education*, *20*(7), 755–768. doi:10.1080/0950069980200701

Joyce, B. R., & Showers, B. (1995). *Student achievement through staff development* (2nd ed.). White Plains, NY: Longman.

Kanter, M. R., & Litow, S. S. (2009). *Informed and Interconnected: A Manifesto for Smarter Cities*. Harvard Business School Working Paper 09-141. Boston, MA: Harvard Business School.

Kasten, W. C. (1997). Learning is noisy; the myth of silence in the reading-writing classroom. In J. R. Paratore & R. L. McCormack (Eds.), *Peer talk in the classroom: Learning from research* (pp. 88–101). Newark, DE: International Reading Association.

Kemp, A. T., Preston, J., Page, C. S., Harper, R., Dillard, B., Flynn, J., & Yamaguchi, M. (2014). Technology and teaching :A conversation among faculty regarding the pros and cons of technology. *Qualitative Report*, *19*(3), 1–23.

Kennedy, A. (2005). Models of continuing professional development: A framework for analysis. *Journal of In-service Education*, *31*(2), 235–250. doi:10.1080/13674580500200277

Kheswa, J. G. (2014). Action Research: A practice to improve academic performance among undergraduate psychology students at the South African university campus. *Mediterranean Journal of Social Sciences*, *5*(2), 1043–1049.

Kissock, C., & Iyortsuun, P. T. (1982). *A guide to questioning: Classroom procedures for teachers.* London: Macmillan.

Kraiger, K. (2002). Decision-based evaluation. In K. Kraiger (Ed.), *Creating, implementing, and maintaining effective training and development: State-of-the-Art Lessons for Practice* (pp. 331–375). San Francisco, CA: Jossey-Bass.

Kreber, C. (2001). *Scholarship revisited: perspectives on the scholarship of teaching*. San Francisco, CA: Jossey-Bass.

Lai, K. W., & Pratt, K. (2004). Information and communication technology (ICT) in secondary schools: The role of the computer coordinator. *British Journal of Educational Technology*, *35*(4), 461–475. doi:10.1111/j.0007-1013.2004.00404.x

Lever-Duffy, J., & McDonald, J. B. (2011). *Teaching and learning with technology* (4th ed.). Boston, MA: Pearson.

Likitrattanaporn, W. (2017). The development of English language teaching skills for graduate students through the process of learning by doing. *English Language Teaching*, *10*(7), 96. doi:10.5539/elt.v10n7p96

Lueddeke, G. R. (2003). Professionalizing teaching practice in higher education: A study of disciplinary variation and 'teaching scholarship'. *Studies in Higher Education*, *28*(2), 213–228. doi:10.1080/0307507032000058082

Luppicini, R. (2012). *Ethical Impact of Technological Advancements and Applications in Society*. Hershey, PA: Information Science Reference. doi:10.4018/978-1-4666-1773-5

McAlister, A., & Crappell, C. (2014). Twenty-first century pedagogy: A whole new world again. *Clavier Companion*, *6*(1), 56.

McLaughlin, M. W., & Talbert, J. (1993). *Contexts that Matter for Teaching and Learning. Stanford, CA Stanford University*. Stanford, CA: Centre for Research on the Context of Secondary School Teaching.

McLaughlin, M. W., & Talbert, J. E. (2006). *Building school-based teacher learning communities: Professional strategies to improve student achievement.* Chicago: University of Chicago Press.

Megginson, D., & Whitaker, V. (2007). *Continuing Professional Development* (2nd ed.). London: Chartered Institute of Personnel and Development.

MOE. (2013). *Education Blueprint 2013–2025 (Preschool to Post-secondary education)*. Putrajaya: MOE.

Morrar, R., Arman, H., & Mousa, S. (2017). The Fourth Industrial Revolution (Industry 4.0): A Social Innovation Perspective. *Technology Innovation Management Review*, *7*(11), 12–20. doi:10.22215/timreview/1117

Nixon, J., Marks, A., Rowland, S., & Walker, M. (2001). 'Towards a New Academic Professionalism: A manifesto of hope'. *British Journal of Sociology of Education*, *22*(2), 227–244. doi:10.1080/01425690124202

Page, B. (2010). 12 things teachers must know about learning. *Education Digest*, *75*(8), 54–56.

Pelgrum, W. J. (2001). Obstacles to the integration of ICT in education : Results from a worldwide educational assessment. *Ournal of Science Education and Technology*, *37*(2), 163–178.

Phillips, D. K., & Carr, K. (2014). *Becoming a teacher through action research: Process, context, and self-study* (3rd ed.). New York: Routledge. doi:10.4324/9781315867496

Ramsden, P. (2003). *Learning to teach in higher education*. London: Routledge Falmer. doi:10.4324/9780203507711

Reiser, R. A. (2012). Getting an instructional design position: Lessons from a personal history. In R. A. Reiser & J. V. Dempsey (Eds.), *Trends and issues in instructional design and technology* (3rd ed.; pp. 256–262). Boston, MA: Pearson Education, Inc.

Reiser, R. A., & Dempsey, J. V. (2007). *Trends and issues i0n instructional design and technology* (2nd ed.). Pearson Education, Inc.

Reiser, R. A., & Dempsey, J. V. (Eds.). (2018). *Trends and issues in instructional design and technology* (4th ed.). Boston, MA: Pearson Education, Inc.

Riasati, J. M., Alllahyar, N., & Tan, K. A. (2012). Technology in Language Education: Benefits and Barriers. *Journal of Education and Practice*, *3*(5), 25–31.

Richards, J. (2008). Second language teacher education today. *RELC Journal: A Journal of Language and Teaching Research, 39*(2), 158-177.

Richards, J., & Rodgers, T. (1986). *Approaches and Methods in Language Teaching*. Cambridge, UK: Cambridge University Press.

Rieber, L. (2018). *The Proper Way to Become an Instructional Technologist. In R. E. West, Foundations of Learning and Instructional Design Technology: The Past, Present, and Future of Learning and Instructional Design Technology*. Retrieved from http://edtechbooks.org/lidtfoundations/proper_way

Roblyer, M. D., & Hughes, J. E. (2018). *Integrating educational technology into teaching: Transforming learning across disciplines* (8th ed.). Pearson.

Rogers, E. M. (2003). *Diffusion of innovations*. New York: Free Press.

Roscoe, J. (2002). Continuing professional development in higher education. *Human Resource Development International, 5*(1), 3–9. doi:10.1080/13678860110076006

Rothwell, A., & Arnold, J. (2005). How HR professionals rate "continuing professional development. *Human Resource Management Journal, 15*(3), 18–32. doi:10.1111/j.1748-8583.2005.tb00151.x

Sampson, R. (2012). The language-learning self, self-enhancement activities, and self-perceptual image. *Language Teaching Research, 16*(3), 317–335. doi:10.1177/1362168812436898

Savery, J. R., & Duffy, T. M. (1996). Problem based learning: An instructional model and its constructivist framework. In B. Wilson (Ed.), *Constructivist learning environments: Case studies in instructional design* (pp. 135–148). Englewood Cliffs, NJ: Educational Technology Publications.

Seale, J. (2010). Doing student voice work in higher education: An exploration of the value of participatory methods. *British Educational Research Journal, 36*(6), 995–1015. doi:10.1080/01411920903342038

Shamsutdinova, E. K., Khakimzyanova, D. F., & Melnikova, O. K. (2017). Implementing interactive methods in teaching General English for future programmers. *International Journal of Scientific Study, 5*(5), 362–366.

Shulman, L. (1987). Knowledge and teaching: Foundations of the new reform. *Harvard Educational Reform, 57*(1), 1–22. doi:10.17763/haer.57.1.j463w79r56455411

Skelton, A. (2005). *Understanding Teaching Excellence in Higher Education. Towards a Critical Approach*. Abingdon, UK: Routledge. doi:10.4324/9780203412947

Smaldino, S. E., Donaldson, J. A., & Herring, M. (2012). Professional ethics: Rules applied to practice. In R. A. Reiser & J. V. Dempsey (Eds.), *Trends and issues in instructional design and technology* (3rd ed.; pp. 342–347). Boston, MA: Pearson Education, Inc.

Smith, K. (2008). "Who do you think you're talking to?" – the discourse of learning and teaching strategies. *Higher Education, 56*(4), 395–406. doi:10.100710734-007-9100-6

Smith, P. L., & Ragan, T. J. (2005). Instructional Design (3rd ed.). Hoboken, NJ: John Wiley Sons, Inc.

Speck, M. (1996). Best Practice in Professional Development for Sustained Educational Change. *ERS Spectrum*, 33–41.

Subedi, B. S. (2015). Assessing the effectiveness of teacher training programs to improve the quality of school education in Nepal. *Journal of Training and Development, 1*(1), 9–14. doi:10.3126/jtd.v1i0.13084

Talbert, J. (2010). Professional learning communities at the crossroads: How systems hinder or engender change. In A. Hargreaves, A. Lieberman, M. Fullan, & D. Hopkins (Eds.), *Second international handbook of educational change* (pp. 555–571). New York: Springer. doi:10.1007/978-90-481-2660-6_32

Taylor, P. J., O'Driscoll, M. P., & Binning, J. F. (1998). A new integrated framework for training needs analysis. *Human Resource Management Journal, 8*(2), 29–50. doi:10.1111/j.1748-8583.1998.tb00165.x

Tearle, P. (2003). ICT implementation: What makes the difference? *British Journal of Educational Technology, 34*(5), 567–583. doi:10.1046/j.0007-1013.2003.00351.x

Thessin, R. A., & Starr, J. P. (2011). Supporting the GROWTH of effective professional learning communities districtwide. *Phi Delta Kappan*, *92*(6), 48–54. doi:10.1177/003172171109200611

Thomas, G. (2009). *How to do Your Research Project*. London: Sage.

Tsai, S., & Chai, C. S. (2012). The "Third " order barrier for technology-integration instruction : Implications for teacher education. *Australasian Journal of Educational Technology*, *28*(6), 1057–1060. doi:10.14742/ajet.810

Vacca, R. T., & Vacca, J. L. (2002). *Content area reading: Literacy and learning across the curriculum* (7th ed.). Boston: Allyn and Bacon.

Vacca, R. T., Vacca, J. L., & Mraz, M. (2011). *Content area reading: Literacy and learning across the curriculum* (10th ed.). Boston: Pearson.

Van Eerde, W., Tang, K., & Talbot, G. (2008). The mediating role of training utility in the relationship between training needs assessment and organizational effectiveness. *International Journal of Human Resource Management*, *19*(1), 63–73.

Vatanartiran, S., & Karadeniz, S. (2015). A needs analysis for technology integration plan: Challenges and needs of teachers. *Contemporary Educational Technology*, *6*(3), 206–220.

Walker, M. (1993). Developing the theory and practice of action research: A South African case. *Educational Action Research*, *1*(1), 95–109. doi:10.1080/0965079930010106

Waters, R. (2016, January 5). Artificial Intelligence: Can Watson Save IBM? *Financial Times*. Accessed November 1, 2017: https://www.ft.com/content/dced8150-b300-11e5-83589a82b43f6b2f

Weindling, A. M. (2001). Education and training: Continuing professional development. *Current Paediatrics*, *11*(5), 369–374. doi:10.1054/cupe.2001.0217

Wellington, J. (2000). *Teaching and learning secondary Science: Contemporary Issues and Practical Approaches*. London: Routledge. doi:10.4324/9780203059876

Wellington, J., & Osborne, J. (2001). *Language and Literacy in Science Education*. Buckingham, UK: Open University Press.

Wenglinsky, H. (2000). *How teaching matters: Bringing the classroom back into discussions of teacher quality*. Princeton, NJ: Educational Testing Service.

West-Burnham, J., & O'Sullivan, F. (1998). *Leadership and professional development in schools*. London: Pitman.

Whitaker, P. (1993). *Managing change in schools*. Buckingham, UK: Open University Press.

Whitaker, P. (1997). Changes in professional development: The personal dimension. In L. Kydd, M. Crawford, & C. Riches (Eds.), *Professional development and educational management*. Buckingham, UK: Open University Press.

Wilkinson, I. A. G., Soter, A. O., & Murphy, P. K. (2010). Developing a model of quality talk about literacy text. In M. G. McKeown & L. Kucan (Eds.), *Bringing reading research to life* (pp. 142–169). New York: The Guilford Press.

Winn, M. R. (2012). Promote digital citizenship through school-based social networking. *Learning and Leading with Technology*, *39*(4), 10–13.

Wiske, M. S., Franz, K. R., & Breit, L. (2005). *Teaching for understanding with technology*. San Francisco: Jossey-Bass Inc.

Woodward, I. (1996). Dimensions of Continuing Professional Development. In I. Woodward (Ed.), *Continuing professional development* (pp. 1–18). London: Cassell.

Zuber-Skerritt, O. (1992). *Professional development in higher education*. London: Kogan Page.

Chapter 19
The New Challenges in the Training of the Engineer for the Industry 4.0:
A Case Study of a Brazilian University Center

Sergio Ricardo Mazini
University Center Toledo Araçatuba, Brazil

Márcia Maria Teresa Baptistella
University Center Toledo Araçatuba, Brazil

ABSTRACT

The growing changes brought about by new concepts and new technologies, such as Industry 4.0, have demanded that educational institutions seek new teaching and learning methodologies, as well as new resources that can contribute to the training of the future engineer. This chapter demonstrates some practices adopted in the process of training the future engineer in a university center in the interior of the state of São Paulo through the use of the CDIO initiative. The results presented confirm the importance and necessity of changes in the teaching and learning process in higher education institutions.

INTRODUCTION

Since the beginning of time, evolution has been a constant in all areas, there is always the need to adapt and innovate to follow the various changes that have been taking place in the world, both for companies, industries, services, as well as people. Industry 4.0 or the fourth industrial revolution, as many call it, reflects these changes and the various adaptations that are necessary for business development and continuity to be possible, where people need to innovate and adapt to new realities and technological and market trends, as new professions emerge while others disappear, new business segments thrive while others decline, and so on.

DOI: 10.4018/978-1-5225-7832-1.ch019

Copyright © 2019, IGI Global. Copying or distributing in print or electronic forms without written permission of IGI Global is prohibited.

While executives are aware of the imperative need to embrace innovation as a business strategy, they recognize that they lack the knowledge of the tools and tools that make it possible to put theory into practice. "Companies do not adopt consistent practices of innovation, do not prepare their leaders, and do not have ways to measure the innovation process" (Scherer & Carlomagno, 2016, p. 4).

Still according to Scherer & Carlomagno (2016), few companies are considered innovative because innovating demands time, needs to value people, means tolerating mistakes, taking risks and changing, but above all must lead to results. However, in the world where total quality and zero defect prevail, there is not much room for creativity and innovation, as it is cheaper to continue doing the usual and routine than to devote time and resources in the search for new and uncertain alternatives, where the development of a new idea, a new product or a new process can take months or even years, because to innovate it is necessary to incorporate the new knowledge obtained for the elaboration of new products and services that add value to the company. The process of innovation is based on the sharing of tacit knowledge, which is in people's minds, and encouraging the sharing of this knowledge means valuing people.

Knowledge is one of the greatest human assets because it is the result of a complex combination of skills, competencies and attitudes that allow a person to understand, critique, analyze and interpret facts and data, developing the ability to master techniques, processes, ideas, concepts and evolutions (PÁDUA FILHO, 2016). For Alvarenga Neto (2008), knowledge companies are organizations that have information and knowledge that make them differentiated, with capacities of perception and discernment, capable of generating value for themselves. However, "successful organizations will increasingly move from hierarchical structures to more collaborative and networked models" (SCHWAB, 2016, p.65).

In this context the approaches in this work are made, with the objective of presenting the industry scenario 4.0, as well as the skills and competences for the production engineering professional that will act in this new world scenario.

THEORETICAL BACKGROUND

Effectively started in the second decade of the 21st century, more precisely between the years 2013 and 2016, the fourth industrial revolution is characterized by the convergence of important levels of sensing, control and artificial intelligence driven by globally established mass communication and intercommunication requirements (STEVAN JUNIOR; LEME & SANTOS, 2018).

However, according to Schwab (2016), the fourth industrial revolution is not just about connected systems and intelligent machines, because waves of new discoveries occur simultaneously in areas ranging from gene sequencing to nanotechnology, from renewable energies to quantum computing. "What makes the fourth industrial revolution fundamentally different from the previous ones is the merging of these technologies into the interaction between physical, digital, and biological domains" (SCHWAB, 2016, p.16).

Industrial Revolutions: Historical Evolution

To arrive at the fourth industrial revolution, radical and abrupt changes occurred throughout history. According to Assad Neto et al. (2017) and Schwab (2016), the first industrial revolution occurred between 1760 and 1840, provoked by the construction of highways and the advent of the steam engine, beginning the mechanical production; the second industrial revolution occurred between the late nineteenth and early twentieth century, with the invention of electricity and the creation of the assembly line, enabling mass production; the third industrial revolution began in the 1960s, characterized by the use of electronics and information technology to obtain greater automation in manufacturing processes, it is also often called the digital revolution, because it is driven by the development of semiconductors, mainframe computing (1960s), personal computing (1970s and 1980s) and the internet (1990s), as shown in Table 1.

According to Schwab (2016: 114), "the fourth industrial revolution can robotize mankind and thus compromise our traditional sources of meaning - work, community, family and identity. Or, then, we can use the fourth industrial revolution to elevate humanity to a new collective and moral consciousness based on a common sense of destiny. It is up to all of us to ensure that this latter scenario occurs. "

For Assad Neto et al. (2017), the High-Tech Strategy 2020 technological development plan was launched in 2010, aimed at strengthening the partnership between industry and science, as well as improving the conditions for technological innovation in the country. Germany has been at the forefront of the development of Industry 4.0 (I4.0), being the first country to introduce it, transforming its manufacturing processes to increase efficiency, ensure resource conservation, flexibility and competitiveness (YANAI et al., 2017).

Industry 4.0: Fourth Industrial Revolution

According to Stevan Junior, Leme & Santos (2018), an industry is considered to be extremely efficient when all processes occur without failure or waste of materials, focusing on the constant reduction of production costs, as well as when processes involve development of communication between sensors and equipment, making them more agile and reliable. For Schwab (2016), society is aboard a technological

Table 1. The 4th industrial revolutions

Period	Revolution	Description
Late 18th century	1st Industrial Revolution (Industry 1.0)	The first Industrial Revolution began in Britain in the late eighteenth century and ended in the mid-nineteenth century. It represented the shift from an agrarian, craft-based economy to an industry-led and machine-manufacturing economy with the introduction of mechanical production methods and the application of steam energy.
Early 20th century	2nd Industrial Revolution (Industry 2.0)	The second period began in the early twentieth century. It was marked by the era of mass industrial production, in which principles of the assembly line were focused on the creation of mass consumer products. The introduction of electric power aided the set of changes.
From the 1960s	3rd Industrial Revolution (Industry 3.0)	The third revolution started in the 60's is marked by the automation of production processes with the deployment of electronic products and IT in industrial processes.
Nowadays	4th Industrial Revolution (Industry 4.0)	It began at the turn of the century, and is based on the digital revolution, characterized by a more ubiquitous and mobile Internet, smaller and more powerful sensors, artificial intelligence and machine learning. Over the next decade, the fourth industrial revolution will usher in the era of decentralized production. The use of sensor technology, interconnectivity, and data analysis will allow the merging of real and virtual worlds into production.

Source: Prepared by the authors, adapted from Pinheiro & Gargaglione & Gonçalves (2017), Assad Neto et al. (2017) e Schwab (2016)

revolution that will fundamentally transform the way people are living, working, or relating, for to him, transformation will be different from anything that the human being has experienced before. According to Stevan Junior, Leme & Santos (2018, p.137):

For many, Industry 4.0 is a new model of management and organization of the manufacturing industry, possible from the development of automation and data exchange technologies. It is considered the manufacturing system of the future.Connectivity between people and systems provides a more expressive and functional level of integration. Thus, Industry 4.0 facilitates the vision and possibility of an intelligent factory, with a high level of modularity, flexibility and decentralized decision-making. Integration and communication between devices promote better operation, virtualization, decentralization, service orientation, modularity, real-time monitoring, flexibility and optimization of the manufacturing system.

Thus, Industry 4.0 represents a set of technological advances that provide intelligent networks, where machines and products in process interact without the need for human intervention, allowing the exchange of information instantly between the company's units, being able to optimize and improve the decisions along the supply chain, as well as to enable a combination of custom and mass production (FUKUDA; MARIZ & MESQUITA, 2017).

According to Stevan Junior, Leme and Santos (2018: 81), "The Fourth Industrial Revolution is driven by connectivity trends, advanced materials that enable sensor development, faster processing technology, advanced production networks, device networks of manufacturing and controlled by computers, allowing an interaction between the real and the virtual in a much more integrated way. "However, according to Pinheiro & Gargaglione & Gonçalves (2017) and Yanai et al. (2017) for the formation of Industry 4.0, four key components are required:

- **Cyber Physical Systems (CPS):** These are systems that allow the integration of the digital and physical world. They are composed of virtual units for the purpose of controlling physical units;
- **Internet of Things (IoT):** Network of objects, systems and platforms. It can be described as the relationship between things (products, services, places) and people, through various platforms and connected technologies;
- **Internet of Services (IoS):** When Internet of Things works properly, it is possible to identify opportunities and introduce new services. Is the provision of services offered via the internet;
- **Smart Factories:** In these factories CPS is used, where consumers, devices and systems form a dynamic production network, intelligently organized, generating tangible gains in efficiency, time and cost when compared to a traditional factory. There is communication between all sectors, working together, exchanging data at all times through the network.

According to CNI (2016), the key enabling technologies behind I4.0 include the Internet of Things, Big Data, Cloud Computing, Advanced Robotics, Artificial Intelligence, New Materials, and New Technology additive manufacturing (3D printing) and hybrid manufacturing (additive and machining functions on the same machine).

However, in Brazil there is a certain delay in relation to Industry 4.0, since experts say that the national industry is in transit between industry 2.0 and industry 3.0, differing only in the automotive sector, which has strong tendencies to be the pioneer to receive Industry 4.0 due to the high number of qualified professionals and the possibility to work in key sectors in the country. The major challenge for

Brazil will be to focus forces on key factors for the development of this new level. Innovative strategic policies are needed, with incentives, availability of new technologies and the development of qualified professionals to work in this new scenario (PINHEIRO; GARGAGLIONE & GONÇALVES, 2017). Figure 1 shows the transition of Brazilian industry.

For Schwab (2016), the main effects that the fourth industrial revolution has on the business of all industries are:

- Changes in customer expectations;
- Product enhancements are being made possible through easy access to data, improving productivity;
- Formation of new partnerships, as companies learn the importance of new forms of collaboration; and
- Transformation of operational models into new digital models.

In this new scenario, through the cyber-physical systems, the internet of things and the internet of services, the productive processes tend to become increasingly efficient, autonomous and customizable, with better synchronization of demands, enabling direct contact and in time with the chain of suppliers and customers. With this, production increases or decreases from the set of data and information received / transmitted, causing productivity to increase and operating costs gradually decrease (PINHEIRO; GARGAGLIONE & GONÇALVES, 2017).

Figure 1. Transition of Brazilian industry
Source: Prepared by the authors, adapted from Stevan Junior, Leme & Santos (2018, p. 137)

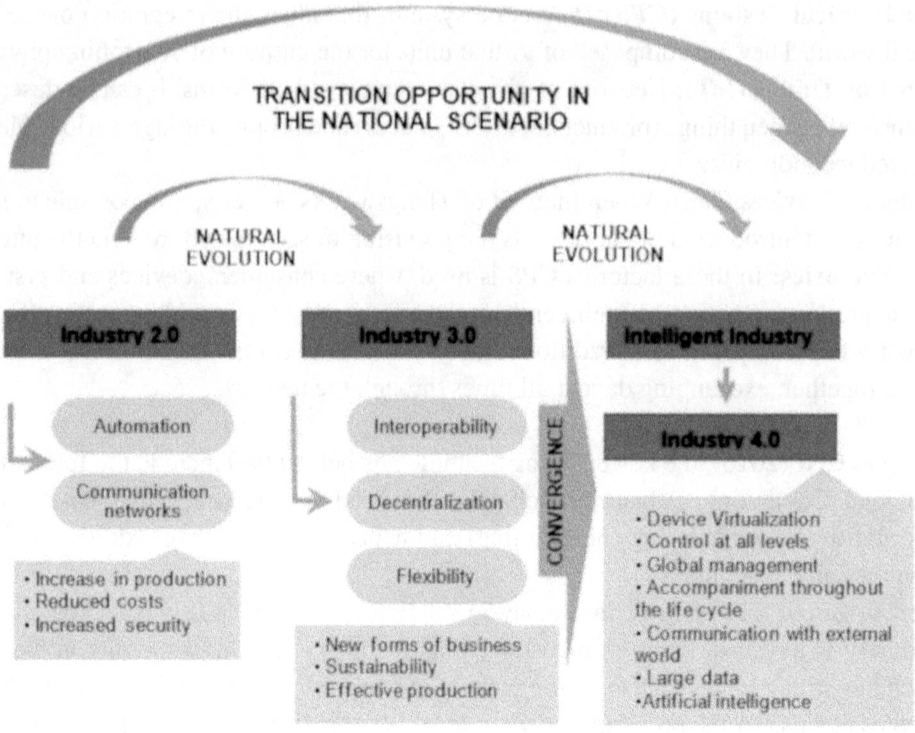

Still according to the authors, with this process of Industry 4.0, automation brings several benefits in the execution of the works, such as:

- Replacement of human capital in tasks that can put the physical integrity of people at risk, such as handling heavy or hot loads, nuclear power plants, furnaces;
- Increased economic development, since there is an increase in productivity and efficiency in the processes;
- Faster performance of tasks, with lower incidence and probability of errors.

According to CNI (2016: 17), "McKinsey estimates that, by 2025, processes related to Industry 4.0 could reduce equipment maintenance costs by 10-40%, reduce energy consumption by 10-20% and increase work efficiency by between 10% and 25% ".

In this new scenario, the constant exchange of information between different people and companies will be increasingly common, generating new impacts and new issues regarding reliability, risks, sustainability, safety and know-how (STEVAN JUNIOR; LEME & SANTOS, 2018).

Thus, in order to effectively implement and implement the I4.0 concepts, workers will have to combine know-how related to robotics and information technology, be open to changes and adaptations, as well as having multiple skills, flexibility and adaptability. new rules and work environments, and continuous interdisciplinary learning (VOLPE et al., 2017).

Skills and Competencies for Industry 4.0

According to Perasso (2016), David Ritter, CEO of Greenpeace Australia / Pacific said in a column on the fourth industrial revolution for the British newspaper The Guardian "The future of employment will be made by vacancies that do not exist in industries that use technologies new, in planetary conditions that no human being has ever experienced. "

For Volpe et al. (2017), based on a research carried out, the profile of the I4.0 professional should be developed and guided by the requirements of technical, multidisciplinary vision, sense of collaboration, language proficiency, critical sense and flexibility. However, to leverage the technologies adopted in I4.0, it is important to select employees with a high cognitive capacity to understand and use the adopted systems, as well as the ability to follow instructions to help learn the technology about the systems adopted.

Thus, the concepts introduced by Industry 4.0 in production management processes have triggered changes in the skills and competencies required of employees as well as of people managers in the manufacturing processes, highlighting five competencies as essential for educational activities of employers and employees: communication and collaboration, creativity, self-management, problem-solving thinking and constant learning (VOLPE et al., 2017).

The Industry 4.0 professional needs to improve on platforms essential to the industrial model and should learn how to handle software and specific programs such as 3D modeling software, information management and system optimization (VOLPE et al., 2017).

For Schwab (2016), based on a survey of human resources managers of the largest employers in 10 industries and 15 countries on the impact on employment, work and skills by the year 2020, it was noted that the greater demand is for complex problem solving skills, process skills, social and system skills, and less for physical skills or more specific technical skills.

Therefore, according to Volpe et al. (2017), the profile of the new professional should prioritize characteristics such as: critical sense, use of new media, social intelligence, flexibility, abstraction ability (understanding and interpreting concepts and data), cross-cultural competence (knowing how to relate to people of different cultures and countries), interdisciplinarity (knowing how to work in multidisciplinary and global teams), collaboration at a distance (creating both online and virtual connections) and prioritization (ability to filter, retain and take advantage of only what is important).

In this way, it is necessary to have knowledge and talent, as well as to develop personal skills and competences so that the market is very promising for the professional in Industry 4.0, aiming at clear and integrated communication so that transparency and accuracy are fundamental throughout the process.

Concept, Design, Implement, Operate: CDIO

The CDIO™ Initiative (www.cdio.org) is an innovative educational framework for producing the next generation of engineers and consists of more than 120 schools around the world organized into 7 regions; Europe, North America, Asia, UK-Ireland, Latin America, Australia, New Zealand and Africa. These are led by one or several regional leaders appointed by each region. The CDIO™ Initiative adopted 12 standards to describe CDIO programs. These guiding principles were developed in response to program leaders, alumni, and industry partners who wanted to know how they would recognize CDIO programs and their graduates.

The main role of these 12 CDIO Standards is to serve as a guideline for educational program reform and evaluation, create benchmarks and goals with worldwide application, and provide a framework for continuous improvement. There is no formal certification as a CDIO Program; each institution/institutional department self-certifies using the CDIO Standards and demonstrates certification to its normal accrediting agency or organization. The CDIO Standards allow other academics and industry to identify clearly CDIO Programs and their graduates.

The 12 CDIO Standards:

- Define the distinguishing features of a CDIO program;
- Serve as guidelines for educational program reform and evaluation;
- Create benchmarks and goals with worldwide application; and
- Provide a framework for continuous improvement.

The 12 CDIO Standards address program philosophy (Standard 1), curriculum development (Standards 2, 3 and 4), design-build experiences and workspaces (Standards 5 and 6), new methods of teaching and learning (Standards 7 and 8), faculty development (Standards 9 and 10), and assessment and evaluation (Standards 11 and 12).

Methodology

As the term Industry 4.0 is quite recent, there is a certain difficulty in obtaining many references on the subject. However, this work also aims to contribute to make the subject more and more widespread, understood and shared by society in general.

Thus, the search for knowledge for the elaboration of this document was done using the bibliographic research, as well as the case study to present the results obtained. For Cervo, Bervian & Silva (2007: 60), "the bibliographical research seeks to explain a problem from theoretical references published in articles, books, dissertations and theses," and the objective is to know and analyze cultural contributions, scientific and technical information on a particular subject, theme or problem, so that a new work can be developed, regardless of how it is used.

According to Martins (2014: 141), a case study is defined as "an empirical investigation that investigates a contemporary phenomenon within its real-life context, especially when the boundaries between phenomenon and context are not clearly defined."

Thus, the context and the objectives of this research were initially defined, to carry out the bibliographical survey and the case study, in order to do the work elaboration. For that, the researches were carried out in books and articles related to the subject in question so that the theoretical reference was based on studies and research previously carried out by other authors.

Case Study

The Centro Universitário Toledo Araçatuba - SP, UNITOLEDO, has more than 50 years of tradition in higher education, working in several areas such as education, health, business and engineering, both undergraduate and postgraduate. In the area of engineering the institution has the courses of Civil Engineering, Mechanical Engineering, Production Engineering, Electrical Engineering, Chemical Engineering

With the objective of aligning the growing needs of the labor market, among them, the concept of Industry 4.0, the institution has been working some actions in the scope of engineering teaching, based on the CDIO initiative:

- **Integrating Projects:** it was instituted the realization of an integrative project from the first semester of the engineering courses until the eighth semester, with the objective of integrating concepts and practices of the disciplines of each semester. The CDIO standards developed in the integrative projects can be seen in Table 2.
- **Laboratory of Modeling and Simulation of Productive Systems:** conceived with the objective of allowing the modeling and simulation of productive systems, aiming at the identification and solution of real problems found in the industrial environment. In this lab the available resources are ProModel software, the Lean Board Game and business management software.
- **Laboratory of Robotics and Industrial Automation:** designed with the objective of allowing the use of robotics in real issues related to industrial automation.
- **Groups and Lines of Research:** scientific research is being organized in undergraduate courses with topics related to the real needs of companies. Through research projects and scientific initiation projects.
- **Active Learning Methodologies:** use of active learning methodologies such as Project Based Learning, Problem Based Learning in disciplines and also in integrative projects.
- **Introduction to Engineering:** in the first two semesters of engineering courses the Introduction to Engineering I and II discipline is an introductory course that provides a framework for the practice of engineering in the elaboration of products, processes and systems construction, in addition to introduce the personal and interpersonal skills essential to professional practice.

- **Partnerships with Institutions:** some activities are carried out in partnership with institutions that work directly with industry issues, such as SENAI, where the exchange of experiences and activities together complements the student's training. Another partnership is with CIESP aimed at aligning the skills needed for the future professional.

CONCLUSION

The changes brought about by new technologies increasingly demand that higher education institutions seek alternatives for a complete training of the future engineer, aligning the expectations and needs of the labor market. The development of personal and interpersonal skills is also necessary in this scenario, in addition to the knowledge and technical skills required for the professional in training.

The actions carried out by the Centro Universitário Toledo Araçatuba - UNITOLEDO demonstrate an alignment with the new training needs of the engineer and that Industry 4.0 requires the professional that will work. The requirements of Industry 4.0 presented in the text allow the identification of technical and behavioral competencies that the actions carried out by the institution seek to apply in the process of teaching and learning.

Table 2. CDIO standards developed in the integrator project

CDIO Standards	Description	Note
2. Learning Outcomes	Specific, detailed learning outcomes for personal and interpersonal skills, and product, process, and system building skills, as well as disciplinary knowledge, consistent with program goals and validated by program stakeholders	We identified the personal and interpersonal skills to be developed throughout the project, as well as the skills of building products, processes and systems.
3.Integrated Curriculum	A curriculum designed with mutually supporting disciplinary courses, with an explicit plan to integrate personal and interpersonal skills, and product, process, and system building skills	The course subjects allow integration for the development of multidisciplinary projects.
5.Design-Implement Experiences	A curriculum that includes two or more design-implement experiences, including one at a basic level and one at an advanced level	The disciplines allow each semester to elaborate integrative projects with the theme of development of new products, processes or systems.
6.Engineering Workspaces	Engineering workspaces and laboratories that support and encourage hands-on learning of product, process, and system building, disciplinary knowledge, and social learning	The institution's laboratories used in the project allow the production of products from several industrial segments.
7. Integrated Learning Experiences	Integrated learning experiences that lead to the acquisition of disciplinary knowledge, as well as personal and interpersonal skills, and product, process, and system building skills	The integration with another course of the institution throughout the integrating project made possible the exchange of information and experiences that contributed to the development of established competencies.
8.Active Learning	Teaching and learning based on active experiential learning methods	The use of Project Based Learning enabled the use of active learning methodologies including students at the center of the teaching and learning process.
11.Learning Assessment	Assessment of student learning in personal and interpersonal skills, and product, process, and system building skills, as well as in disciplinary knowledge	Data collection before and after the project made it possible to evaluate the evolution of the students and also of the learning process.

REFERENCES

ACATECH, National Academy of Science and Engineering. (2013). *Recommendations for implementing the strategic initiative INDUSTRIE 4.0.* Disponível em: http://www.acatech.de/fileadmin/user_upload/Baumstruktur_nach_Website/Acatech/root/de/Material_fuer_Sonderseiten/Industrie_4.0/Final_report__Industrie_4.0_accessible.pdf

Alvarenga, N. E. T. O. (2008). *Rivadávia Correa Drummond de. Gestão de conhecimento nas organizações: proposta de mapeamento conceitual integrativo.* São Paulo: Saraiva.

Assad, N. E. T. O. (2017). A busca de uma identidade para a indústria 4.0. In XXXVII Encontro Nacional de Engenharia de Produção, 2017. Joinville/SC: Enegep.

Cervo, Bervian, & Silva. (2007). Metodologia científica (6th ed.). São Paulo: Pearson Prentice Hall.

CNI. (2016). *Confederação Nacional da Indústria. Desafios para a Indústria 4.0 no Brasil.* Brasília: CNI.

Fukuda, Mariz, & Mesquita. (2017). Impactos da Indústria 4.0 na gestão de operações. In XXXVII Encontro Nacional de Engenharia de Produção, 2017. Joinville/SC: Enegep.

Junior, Leme, & Santos. (2018). Indústria 4.0: fundamentos, perspectivas e aplicações. São Paulo: Érica.

Martins. (2014). Guia para elaboração de monografia e TCC em engenharia de produção. São Paulo: Atlas.

Pádua, . (2016). *Inovação é tudo.* São Paulo: Atlas.

Perasso. (2018). O que é a 4ª revolução industrial - e como ela deve afetar nossas vidas. *BBC News Brasil.* Disponível em: https://www.bbc.com/portuguese/geral-37658309

Pinheiro. (2017). Indústria 4.0: uma análise conceitual do tema, seus impactos na economia e a colocação do profissional neste novo cenário. In XXXVII Encontro Nacional de Engenharia de Produção, 2017. Joinville/SC: Enegep.

Scherer. (2016). *Gestão da inovação na prática: como aplicar conceitos e ferramentas para alavancar a inovação. 2.* São Paulo: Atlas.

Volpe. (2017). Habilidades e competências do profissional para o ambiente da Indústria 4.0: uma revisão sistemática. In XXXVII Encontro Nacional de Engenharia de Produção, 2017. Joinville/SC: Enegep.

Yanai. (2017). O desenvolvimento da Indústria 4.0: um estudo bibliométrico. In XXXVII Encontro Nacional de Engenharia de Produção, 2017. Joinville/SC: Enegep.

Compilation of References

Abdo, A. A. M. (2016). *Factors affecting career choice among undergraduate students in Universitas Indonesia* (Doctoral dissertation). Universiti Utara Malaysia.

Abdul Halim, M. (2006). Smart schools' training programme (SSTP): Teachers perspectives (Doctoral Dissertation). Universiti of East Anglia, UK.

Abdullah, N. H., Ramli, N. H. L., & Rafek, M. (2017). Mass lecture in language learning: What do the boys and girls think? *Journal of Advances in Humanities and Social Sciences*, *3*(2), 115–123.

ABET. (2011). *Criteria for Accrediting Engineering Programs*. Accreditation Board for Engineering and Technology Inc.

Abeysekera, L., & Dawson, P. (2015). Motivation and cognitive load in the flipped classroom: Definition, rationale and a call for research. *Higher Education Research & Development*, *34*(1), 1–14. doi:10.1080/07294360.2014.934336

ACARA. (2014). *The Australian Curriculum V5.2*. Australian Curriculum and Reporting Authority. Retrieved from http://www.australiancurriculum.edu.au/

ACARA. (2017). *The Australian Curriculum: General Capabilities: Critical and Creative Thinking*. Retrieved 30/5/2017, from ACARA: http://www.australiancurriculum.edu.au/generalcapabilities/critical-and-creative-thinking/introduction/in-the-learning-areas

ACATECH, National Academy of Science and Engineering. (2013). *Recommendations for implementing the strategic initiative INDUSTRIE 4.0*. Disponível em: http://www.acatech.de/fileadmin/user_upload/Baumstruktur_nach_Website/Acatech/root/de/Material_fuer_Sonderseiten/Industrie_4.0/Final_report__Industrie_4.0_accessible.pdf

Adams, C., Yin, Y., Vargas Madriz, L., & Mullen, C. (2014). A phenomenology of learning large: The tutorial sphere of xMOOC video lectures. *Distance Education*, *35*(2), 202–216. doi:10.1080/01587919.2014.917701

Adey, P. S., & Shayer, M. (1994). *Really Raising Standards: Cognitive Intervention and Academic Achievement*. London: Routledge.

Adey, P. S., Shayer, M., & Yates, C. (2001). *Thinking Science: The curriculum materials of the cognitive acceleration through Science (CASE) Project* (3rd ed.). Cheltenham, UK: Nelson Thornes.

Adu, E. O., & Okeke, C. I. O. (2014). Factors Affecting Lecturers' Participation in Continuing Professional Development (CPD). *Journal of Sociology and Social Anthropology*, *5*(3), 271–281. doi:10.1080/09766634.2014.11885631

Ageel, M., & Woollard, J. (2012). Enhancing university teachers' information and communication technology usage by using a virtual learning environment training course. In *International Technology, Education and Development Conference* (pp. 5599–5606). Valencia, Spain: Academic Press.

Ahmed, S. (2011). Improving cognitive development in secondary chemistry through Gagne's events of instruction. *Journal of Education and Practice*, *2*(4), 2011.

Aldersey-Williams, H. (2011). *The new tin ear: Manufacturing, materials and the rise of the user-maker*. London, UK: RSA Projects.

Aldowah, H., Ghazal, S., & Muniandy, B. (2015). Issues and challenges of using E-learning in a Yemeni Public University. *Indian Journal of Science and Technology*, *8*(32). doi:10.17485/ijst/2015/v8i32/92160

Aldowah, H., Rehman, S. U., Ghazal, S., & Umar, I. N. (2017). Internet of Things in Higher Education: A Study on Future Learning. *Journal of Physics: Conference Series*, *892*(1), 012017. doi:10.1088/1742-6596/892/1/012017

Alexander, B. (2004). Going nomadic: mobile learning in higher education. *EDUCASE Review, 39*(5), 28-35. Retrieved from http://www.educause.edu/ero/article/going-nomadic-mobile-learning-higher-education

Alexander, R. (2006). *Towards dialogic teaching: Rethinking classroom talk*. Dialogos York.

Al-Fuqaha, A., Guizani, M., Mohammadi, M., Aledhari, M., & Ayyash, M. (2015). Internet of things: A survey on enabling technologies, protocols, and applications. *IEEE Communications Surveys and Tutorials*, *17*(4), 2347–2376. doi:10.1109/COMST.2015.2444095

Ali, Z. (2015). A case study of tertiary students' experiences using Edmodo in language learning. *International Journal of Language Education and Applied Linguistics.*, *2*, 39–48.

Allen, I., & Seaman, J. (2015). *Grade level: Tracking online education in the United States*. Babson Survey Research Group.

Almacen, B. (2010). *Teacher training program: The Malaysian perspective*. Retrieved on 3 July 2012 from http://independent.academia.edu/DrBayaniAlmacen/Books/408843/Teacher_Training_Program_The_Malaysian_Perspective

Al-Maqaleh, B. M., & Abdullah, A. M. G. (2017). Intelligent predictive system using classification techniques for heart disease diagnosis. *International Journal on Computer Science and Engineering*, *6*(6), 145–151.

Alraimi, K., Zo, H., & Ciganek, A. (2015). Understanding the MOOCs continuance: The role of openness and reputation. *Computers & Education*, *80*, 28–38. doi:10.1016/j.compedu.2014.08.006

Altemueller, L., & Lindquist, C. (2017). Flipped classroom instruction for inclusive learning. *British Journal of Special Education*, *44*(3), 341–358. doi:10.1111/1467-8578.12177

Alvarenga, N. E. T. O. (2008). *Rivadávia Correa Drummond de. Gestão de conhecimento nas organizações: proposta de mapeamento conceitual integrativo*. São Paulo: Saraiva.

Alvarez, B. (2012). Flipping the classroom: Homework in class, lessons at home. *Education Digest: Essential Readings Condensed for Quick Review*, *77*(8), 18–21.

Ambrose, S., Bridges, M. W., Lovett, M. C., DiPietro, M., & Norman, M. K. (2010). *7 Research-based principles for smart teaching*. San Francisco, CA: Jossey-Bass. Retrieved from https://firstliteracy.org/wp-content/uploads/2015/07/How-Learning-Works.pdf

Anderson, L. W., & Krathwohl, D. R. (Eds.). (2001). *A taxonomy for learning, teaching, and assessing: A revision of Bloom's taxonomy of educational objectives*. New York, NY: Longman.

Anderson, L., Krathwohl, D., & Airasian, P. (2001). *A taxonomy for learning, teaching, and assessing: A revision of Bloom's Taxonomy of educational objectives*. New York, NY: Pearson, Allyn & Bacon.

Anderson, T., & Dron, J. (2011). Three generations of distance education pedagogy. *The International Review of Research in Open and Distributed Learning, 12*(3), 80. doi:10.19173/irrodl.v12i3.890

Apostel, L., Berger, G., & Michaud, G. (1972). *Interdisciplinarity: problems of teaching and research in universities.* Paris: OECD.

Arbaugh, J. B. (2002). Managing the on-line classroom: A study of technological and behavioral characteristics of web-based MBA courses. *The Journal of High Technology Management Research, 13*(2), 203–223. doi:10.1016/S1047-8310(02)00049-4

Archbald, D. (2010). Breaking the mold in the dissertation. *Journal of Continuing Higher Education, 58*(2), 99-107.

Archbald, D. (2011). The emergence of the nontraditional doctorate: A historical overview. *New Directions for Adult and Continuing Education, 2011*(129), 7–19. doi:10.1002/ace.396

Archer, R., & Wood, L. (1992). Production tissue culture by robots. In R. Spier, J. Griffiths, & C. MacDonald (Eds.), *Animal Cell Technology: Developments, Processes and Products* (pp. 403–408). Oxford, UK: Butterworth-Heinemann. doi:10.1016/B978-0-7506-0421-5.50094-5

Arroyo, G. (2011). *On-Line Social Network: Innovative Ways towards the Boost of Collaborative Language Learning.* Paper presented at the International Conference "ICT for Language Learning", Florence, Italy. Retrieved from https://conference.pixel-online.net/conferences/ICT4LL2011/common/download/Paper_pdf/CLL16-428-FP-Gonzalez-ICT4LL2011.pdf

Arthur, W. J., Bennett, W. J., Edens, P., & Bell, S. T. (2003). Effectiveness of training in organizations: A metaanalysis of design and evaluation features. *The Journal of Applied Psychology, 88*(2), 234–245. doi:10.1037/0021-9010.88.2.234 PMID:12731707

Arumugam. (2012). The usage of technology among education students in University Utara Malaysia: An application of extended Technology Acceptance Model. *IJEDICT, 7*(3).

Assad, N. E. T. O. (2017). A busca de uma identidade para a indústria 4.0. In XXXVII Encontro Nacional de Engenharia de Produção, 2017. Joinville/SC: Enegep.

Auerback, S. (2011). It's not just going to collect dust on a shelf: Faculty perceptions of the applied dissertation in the new CSU. *Journal of Research on Leadership Education, 6*(3), 59-82.

Auster, E. R., & Wylie, K. K. (2006). Creating active learning in the classroom: A systematic approach. *Journal of Management Education, 30*(2), 333–353. doi:10.1177/1052562905283346

Azizan & Habib. (1995). *Pengajaran dalam bilik darjah: Kaedah dan strategi.* Masa Enterprise.

Azizan, F. Z. (2010). Blended Learning in Higher Education Institution in Malaysia. *Proceedings of Regional Conference on Knowledge Integration in ICT.*

Bacsich, P., Bristow, S. F., Camilleri, A., Op de Beeck, I., Pepler, G., & Phillips, B. (2013). *Virtual schools and colleges. Providing alternatives for successful learning, 2.* Roosbeek, Belgium: ATiT bvba. Retrieved from https://www.pedocs.de/volltexte/2013/8286/pdf/VISCED_Handbook_2013_2.pdf

Baepler, P., & Murdoch, C. J. (2010). Academic analytics and data mining in higher education. *International Journal for the Scholarship of Teaching and Learning, 4*(2), 17. doi:10.20429/ijsotl.2010.040217

Bainbridge, J., Melitski, J., Zahradnik, A., Lauría, E. J., Jayaprakash, S., & Baron, J. (2015). Using learning analytics to predict at-risk students in online graduate public affairs and administration education. *Journal of Public Affairs Education, 21*(2), 247–262. doi:10.1080/15236803.2015.12001831

Baker, F. G. (1994). *Integrating Computer Usage in the Classroom Curriculum through Teacher Training*. Nova Southeastern University.

Baker, J. W. (2000). The "classroom flip": Using web course man- agement tools to become the guide on the side. *11th International Conference on College Teaching and Learning.*

Baker, L. M., & Settle, Q. (2013, September). Flipping the classroom and furthering our careers. *NACTA Journal, 57*(3), 75.

Balasubramanian, K., Jaykumar, V., & Fukey, L. N. (2014). A study on "Student preference towards the use of Edmodo as a learning platform to create responsible learning environment". *Procedia: Social and Behavioral Sciences, 144*, 416–422. doi:10.1016/j.sbspro.2014.07.311

Ball, D. (2000). Bridging practices: Interwining content and pedagogy in teaching and learning to teach. *Journal of Teacher Education, 51*(3), 241–247. doi:10.1177/0022487100051003013

Banados, E. (2006). A blended-learning pedagogical model for teaching and learning EFL successfully through an online interactive multimedia environment. *CALICO Journal, 23*(3), 534.

Bandura, A. (1997). *Self-efficacy: The exercise of control*. Macmillan.

Banerjee, S., & Morley, C. (2013). Professional Doctorates in Management: Toward a Practice-Based Approach to Doctoral Education. *Academy of Management Learning & Education, 12*(2), 173–193. doi:10.5465/amle.2012.0159

Barak, M., & Shakhman, L. (2008). Teachers and Teaching. *Theory and Practice, 14*(3), 191-208.

Barakat, S. M. (2016). Internet of things: Ecosystem and applications. *Journal of Current Research in Science, 4*(1), 32.

Barata, G., Gama, S., Jorge, J., & Goncalves, D. (2013). Engaging Engineering Students with Gamification. *Proceedings of the 5th International Conference on Games and Virtual Worlds for Serious Applications*, 1-8.

Bareither, R., & Pollard, D. (2011). A review of advanced small-scale parallel bioreactor technology for accelerated process development: Current state and future need. *Biotechnology Progress, 27*(1), 2–14. doi:10.1002/btpr.522 PMID:21312350

Bashir, G. M. M., & Hoque, A. S. M. L. (2016). An effective learning and teaching model for programming languages. *Journal of Computers in Education, 3*(4), 413–437. doi:10.100740692-016-0073-2

Bates, A. W. (1984). Putting it together: Now and the future. In A. W. Bates (Ed.), *TheRole of Technology in Distance Education*. London: Croom Helm.

Beadle, M., & Santy, J. (2008). The early benefits of a problem-based approach to teaching social inclusion using an online virtual town. *Nurse Education in Practice, 8*(3), 190–196. doi:10.1016/j.nepr.2007.07.004 PMID:17855168

Becher, T. (2000). *Professional practices: commitment and capability in a changing environment*. Somerset, NJ: Transaction.

BECTA. (2004). *A review of the research literature on barriers to the uptake of ICT by teachers*. BECTA ICT Research. Retrieved from https://ezp.lib.unimelb.edu.au/login?url=https://search.ebscohost.com/login.aspx?direct=true&db=edsoai&AN=edsoai.771640667&site=eds-live&scope=site

Bergmann, J., & Sams, A. (2013). Flip your students' learning. *Educational Leadership, 70*(6), 16-20. Retrieved from ASCD website: http://www.ascd.org/publications/educational-leadership/mar13/vol70/num06/Flip-Your-Students%27-Learning.aspx

Bergmann, J., & Sams, A. (2012). *Flip your classroom: Reach every student in every class every day*. Eugene, OR: International Society for Technology in Education.

Berrett, D. (2012). How "flipping" the classroom can improve the traditional lecture. *Education Digest, 78*(1), 36–41.

Bicen, H., & Kocakoyun, S. (2018). Perceptions of Students for Gamification Approach: Kahoot as a Case Study. *International Journal of Emerging Technologies, 13*(2), 72–93. doi:10.3991/ijet.v13i02.7467

Bierama, L. L., & Eraut, M. (2004). Workplace-focused learning: Perspective on continuing professional education and human resource development. *Advances in Developing Human Resources, 6*(1), 52–68. doi:10.1177/1523422303260859

Bishop, J., & Verleger, M. (2013). The flipped classroom: A survey of the research. In *ASEE Conference Proceedings*. Available at http://www.asee.org/public/conferences/20/papers/6219/download

Blackboard. (2018). *Retention centre*. Available from: https://help.blackboard.com/Learn/Instructor/Performance/Retention_Center

Blackmore, P., & Blackwell, R. (2003). Academic roles and relationships. In R. Blackwell & P. Blackmore (Eds.), *Towards Strategic Staff Development in Higher Education* (pp. 16–28). Berkshire: SRHE and Open University Press.

Blair, N. (2012, January). Technology Integration for the New 21st Century Learner. *Principal*.

Blanchette, J. (2001). Questions in the online learning environment. *Journal of Distance Education, 16*(2), 13–49. Retrieved from http://cade.athabascau.ca/vol16.2/blanchette.html

Bland, L. (2006). *Applying flip/inverted classroom model in electrical engineering to establish lifelong learning*. Presented at the 2006 ASEE Annual Conference & Exposition, Chicago, IL. Retrieved from https://peer.asee.org/491

Blaschke, L. M. (2012). Heutagogy and Lifelong Learning: A Review of Heutagogical Practice and Self-Determined Learning. *International Review of Research in Open and Distance Learning, 13*(1), 56–71. doi:10.19173/irrodl.v13i1.1076

Blau, I., Mor, N., & Neuthal, T. (2013). Interacting for learning: Digital portfolios for a learning community in a university course. *Learning, Media and Technology, 38*(3), 241–255. doi:10.1080/17439884.2012.709864

Blended Learning Toolkit. (n.d.). *What is blended learning?* Retrieved from https://blended.online.ucf.edu/about/what-is-blended-learning/

Blessinger, P., & Stockley, D. (Eds.). (2016). *Emerging directions in doctoral education*. Bingley, UK: Emerald. doi:10.1108/S2055-364120166

Bloom, B. S., Engelhart, M. D., Furst, E. J., Hill, W. H., & Krathwohl, D. R. (1956). *Taxonomy of Educational Objectives: The Classification of Educational Goals. Handbook I: Cognitive Domain*. New York: Longmans.

Bloom, B. S. (1956). *Taxonomy of educational objectives, Handbook I: The dognitive domain*. New York, NY: David McKay Co Inc.

Bloom, B. S. (1956). *Taxonomy of educational objectives: The classification of educational goals*. New York: McKay.

Bodnar, C. A., & Clark, R. M. (2017). Can Game-Based Learning Enhance Engineering Communication Skills? *IEEE Transactions on Professional Communication, 13*(1), 24–41. doi:10.1109/TPC.2016.2632838

Boliton, K. (1985). *A study on the secondary school English training programs*. Seoul: Hong-Ik University.

Bolliger, D. U., & Martindale, T. (2004). Key factors for determining student satisfaction in online courses. *International Journal on E-Learning, 3*(1), 61–67.

Borrego, M., & Newswander, L. K. (2008). Characteristics of Successful Cross-disciplinary Engineering Education Collaborations. *Journal of Engineering Education, 97*(2), 123–134. doi:10.1002/j.2168-9830.2008.tb00962.x

Borrego, M., & Newswander, L. K. (2010). Definitions of interdisciplinary research: Toward graduate-level interdisciplinary learning outcomes. *The Review of Higher Education, 34*(1), 61–84. doi:10.1353/rhe.2010.0006

Boulos, M. N. K., Giustini, D. M., & Wheeler, S. (2016). Instagram and WhatsApp in Health and Healthcare: An Overview. *Future Internet, 8*(3). Retrieved from http://www.mdpi.com/1999-5903/8/3/37/htm

Bourner, T., Bowqden, R., & Laing, S. (1999). A National Profile of Research Degree Awards: Innovation, Clarity and Coherence. *Higher Education Quarterly*, *53*(3), 264–280. doi:10.1111/1468-2273.00130

Bowyer, J., & Chambers, L. (2017). Evaluating blended learning: Bringing the elements together. *Research Matters, 23*, 17-26.

Boyer, E. L. (1996). From scholarship reconsidered to scholarship assessed. *Quest*, *48*(2), 129–139. doi:10.1080/00336297.1996.10484184

Brace-Govan, J., & Clulow, V. (2001). Comparing face-to-face with online: Learners' perspective. *Academic Exchange Quarterly*, *5*(4), 112–118.

Brame, C. (2013). *Flipping the classroom*. Vanderbilt University Center for Teaching. Retrieved from http://cft.vanderbilt.edu/guides-sub-pages/flipping-the-classroom/

Braught, G., Eby, L. M., & Wahls, T. (2008). The effects of pair-programming on individual programming skill. *39th SIGCSE Technical Symposium on Computer Science Education (SIGCSE '08)*, 200-204. 10.1145/1352135.1352207

Braxton, J., Jones, W., Hirschy, A., & Hartley, H. III. (2008). The role of active learning in college student persistence. *New Directions for Teaching and Learning*, *2008*(115), 71–83. doi:10.1002/tl.326

Bregman, R. (2017). *Utopia for realists and how we can get there*. London, UK: Bloomsbury.

Brenneke, J. S. (1990). Case studies and other student based programs in entrepreneurship programs. In Entrepreneurship education: intergrating entrepreneurship in Secondary Curriculum (pp. 231-242). New York: Quorum Books.

Breslow, L. B., Pritchard, D. E., DeBoer, J., Stump, G. S., Ho, A. D., & Seaton, D. T. (2013). Studying learning in the worldwide classroom: Research into edX's first MOOC. *Research & Practice in Assessment*, *8*, 13–25.

Brint, S. G., Turk-Bicakci, L., Proctor, K., & Murphy, S. P. (2009). Expanding the social frame of knowledge: Interdisciplinary, degree-granting fields in American colleges and universities, 1975–2000. *The Review of Higher Education*, *32*(2), 155–183. doi:10.1353/rhe.0.0042

Brookhart, S. (2010). *How to assess higher-order thinking skills in your classroom*. Alexandria, VA: ASCD. Retrieved from http://www.ascd.org/Publications/Books/Overview/How-to-Assess-Higher-Order-Thinking-Skills-in-Your-Classroom.aspx

Brown, C., & Czerniewicz, L. (2010). Debunking the 'digital native': Beyond digital apartheid, towards digital democracy. *Journal of Computer Assisted Learning*, *26*(5), 357–369. doi:10.1111/j.1365-2729.2010.00369.x

Browne, J. (2010). *Securing a sustainable future for higher education*. Available at: www. independent.gov.uk/browne-report

Brown, J. (2002). Training needs assessment: A must for developing an effective training programme. *Public Personnel Management*, *31*(4), 569–578. doi:10.1177/009102600203100412

Brown, J., & Adler, R. (2008). Minds on fire: Open education, the long tail, and learning 2.0. *EDUCAUSE Review*, *43*(1), 16–32.

Buckley, K. M. (2003). Evaluation of classroom-based, web-enhanced, and web-based distance learning nutrition courses for undergraduate nursing. *The Journal of Nursing Education*, *42*(8), 367–370. PMID:12938900

Bullen, M. (1997). *A case study of participation and critical thinking in a university-level course delivered by computer conferencing*. University of British Columbia.

Burd, B., Elahi, A., Russell, I., Barker, L., Pérez, A. F., Siever, B., ... Guerra, J. G. (2017). The Internet of Things in CS education: current challenges and future potential. In *Proceedings of the 2017 ACM Conference on Innovation and Technology in Computer Science Education* (pp. 392-392). ACM. 10.1145/3059009.3081331

BusinessWire. (2014). *Edmodo releases app for Windows 8.1 at ISTE 2014*. Retrieved from https://www.businesswire.com/news/home/20140628005006/en/Edmodo-Releases-App-Windows-8.1-ISTE-2014

Byrne, J., Jørgensen, T., & Loukkola, T. (2013). *Quality assurance in doctoral education: Results of the ARDE Project*. EUA Publications. Retrieved from https://brjk.amu.edu.pl/__data/assets/pdf_file/0018/224262/EUA_ARDE_Publication.pdf

Byrne, R. (2013). *Free technology for teachers: Kahoot! - create quizzes and surveys your students can answer on any device*. Retrieved from http://www.freetech4teachers.com/2013/11/kahoot-create-quizzes-and-surveys-your.html#.VLnc78buzuU

CAE Computers Aided USA Corp. (2018). *What is an LMS and why do I need one?* Retrieved from https://www.cae.net/what-is-an-lms

Cameron, N. (2017). *Will robots take your job?* Cambridge, UK: Polity.

Capperucci, D. (2017). English language teaching and learning in Primary School. theoretical and Methodological Perspectives. *Studi sulla Formazione, 20*(2), 203-217.

Capterra. (2018a). *Best LMS software/ 2018 Review of the most popular systems*. Retrieved from https://www.capterra.com/learning-management-system-software/#infographic

Capterra. (2018b). *Edmodo reviews and pricing*. Retrieved from https://www.capterra.com/p/129886/Edmodo/

Carman, J. M. (2002). *Blended learning design: five key elements*. Retrieved from http://blended2010.pbworks.com/f/Carman.pdf

Carr-Chellman, A. A., & Rowland, G. (Eds.). (2017). *Issues in technology, learning, and instructional design: Classic and contemporary issues*. Routledge.

Carr, D. (2003). *Making sense of education*. London: Routledge Falmer.

Carrol. (1989). Critical thinking has a place in every classroom. *Journal of Reading, 33*(1). Retrieved from https://smallbiztrends.com

Carvalho, A. (2016). The impact of PBL on transferable skills development in management education. *Innovations in Education and Teaching International, 53*(1), 35–47. doi:10.1080/14703297.2015.1020327

Case, S. M., & Swanson, D. B. (2002). *Constructing written test questions for the basic and clinical sciences* (3rd ed.). Philadelphia: National Board of Medical Examiners.

Cashin, W. E. (1995). *Student ratings of teaching: The research revisited. IDEA Paper No. 32. Center for Faculty Evaluation and Development in Higher Education*. Manhattan, KS: Kansas State University. Available from https://eric.ed.gov/?id=ED402338

Cassidy, E. D., Colmenares, A., Jones, G., Manolovitz, T., Shen, L., & Vieira, S. (2014). Higher education and emerging technologies: Shifting trends in student usage. *Journal of Academic Librarianship, 40*(2), 124–133. doi:10.1016/j.acalib.2014.02.003

Caţă, M. (2015). Smart university, a new concept in the Internet of Things. In *RoEduNet International Conference-Networking in Education and Research (RoEduNet NER), 2015 14th* (pp. 195–197). IEEE. doi:10.1109/RoEduNet.2015.7311993

Cauley, P. (n.d.). *A guide to explain it all.* Retrieved from http://www.csub.edu/~tfernandez_ulloa/edmodo%20user%20 guide.pdf

Cervo, Bervian, & Silva. (2007). Metodologia científica (6th ed.). São Paulo: Pearson Prentice Hall.

Chadyiwa, M., & Mgutshini, T. (2015). Using mobile handheld devices as tools of learning and teaching for student EHPS: A blessing or a curse? *International Journal of Humanities, Arts and Social Sciences, 1*(2), 85–91.

Chan, Y., F. (2001). Cabaran Pengurusan Pendidikan di Era Baru. *Jurnal Pengurusan dan Kepimpinan Pendidikan, 11*, 68-79.

Chang, C. H. (2011). Smart classroom roll caller system with IOT architecture. In *Innovations in Bio-inspired Computing and Applications (IBICA), 2011 Second International Conference on* (pp. 356-360). IEEE. 10.1109/IBICA.2011.94

Charles, M. (2018). Teaching, in spite of excellence: Recovering a practice of teaching-led research. *Studies in Philosophy and Education, 37*(1), 15–29. doi:10.100711217-017-9568-1

Chase, J. D., & Okie, E. G. (2000). Combining cooperative learning and peer instruction in introductory computer science. *ACM SIGCSE Bulletin, 32*(1), 372–376. doi:10.1145/331795.331888

Chee-wei, T., Pei-duo, Y., Ling, L., Chung-kit, F., Chun-kiu, L., & Yanru, C. (2017). *Teaching computational thinking by gamification of k-12 mathematics: Mobile app math games in mathematics and computer science tournament. In Proceedings of Computational Thinking Education 2017* (pp. 55–59). Hong Kong, China: The Education University of Hong Kong.

Chen, C.-Y., Chang, R.-E., Hung, M.-C., & Lin, M.-H. (2009). Assessing the quality of a web-based learning system for nurses. *Journal of Medical Systems, 33*(4), 317–325. doi:10.100710916-008-9193-1 PMID:19697698

Cheng, H. C., & Liao, W. W. (2012). Establishing an lifelong learning environment using IOT and learning analytics. In *Advanced Communication Technology (ICACT), 2012 14th International Conference on* (pp. 1178-1183). IEEE.

Chew, E. (2009). *A blended learning model in higher education: A comparative study of blended learning in UK and Malaysia* (Doctoral dissertation). University of Glamorgan.

Chew, C. B., Mahinderjit-Singh, M., Wei, K. C., Sheng, T. W., Husin, M. H., & Malim, N. H. A. H. (2015). Sensors-enabled smart attendance systems using NFC and RFID technologies. *Int. J. New Comput. Archit. Appl, 5*, 19–29.

Chin, J., & Callaghan, V. (2013). Educational living labs: a novel internet-of-things based approach to teaching and research. In *Intelligent Environments (IE), 2013 9th International Conference on* (pp. 92-99). IEEE. 10.1109/IE.2013.48

Chingos, M. M., & Whitehurst, G. J. (2011). Choosing blindly: Instructional materials, teacher effectiveness, and the common core. The Brookings Institution.

Chinnappan, M. (1998). Schemas and mental models in geometry problem solving. *Educational Studies in Mathematics, 36*(3), 201–217. doi:10.1023/A:1003134323371

Cho, Y., Park, S., Jo, S. J., & Suh, S. (2013). The landscape of educational technology viewed from the *ETR&D* journal. *British Journal of Educational Technology, 44*(5), 677–694. doi:10.1111/j.1467-8535.2012.01338.x

Choy, C. S., Sedhu, D. S., Liew, Y. L., Lee, M. Y., Malanee, A., & Anua, N. (2015). Influence of culture on students' awareness of how and why they learn. *Malaysian Journal of Learning and Instruction, 12*, 49–67.

Chuesathuchon, C. (2008). *Computerized adaptive testing in mathematics for primary schools in Thailand.* Retrieved from http://ro.ecu.edu.au/theses/1591

Chunyu, L. I. U., Meng, P., Zhang, Z., & Pan, Y. (2010). Research on Computerized Adaptive Testing System Based on IRT. *2010 Seventh International Conference on Fuzzy Systems and Knowledge Discovery (FSKD 2010).* Retrieved from https://ieeexplore.ieee.org/stamp/stamp.jsp?tp=&arnumber=5569425

Ciolacu, M., & Beer, R. (2016). Adaptive user interface for higher education based on web technology. *2016 IEEE 22nd International Symposium for Design and Technology in Electronic Packaging (SIITME).* 10.1109/SIITME.2016.7777299

Clark, D. (2004). *Gains in critical thinking using online discussions.* International Forum of Educational Technology & Society. Retrieved Jun 16, 2018, from https://mail.fit.fraunhofer.de/pipermail/ifets/2004q1/000553.html

Clark, R. C., Nguyen, F., & Sweller, J. (2005). *Efficiency in learning: Evidence based guidelines to manage cognitive load.* San Francisco, CA: Pfeiffer & Co.

Clegg, S. (2003). Problematising ourselves: Continuing professional development in higher education. *The International Journal for Academic Development, 8*(1/2), 37–50. doi:10.1080/1360144042000277928

Clow, D. (2013). An overview of learning analytics. *Teaching in Higher Education, 18*(6), 683–695. doi:10.1080/13562517.2013.827653

CNI. (2016). *Confederação Nacional da Indústria. Desafios para a Indústria 4.0 no Brasil.* Brasília: CNI.

Cockburn, A., & Williams, L. (2000). The costs and benefits of pair programming. *Extreme programming examined,* 223-247.

Cockrum, T. (2014). *Flipping your English class to reach all learners: Strategies and lesson plans. Abingdon-on-Thames.* Routledge.

Collins, R. (2014). Skills for the 21st Century: Teaching higher-order thinking. *Curriculum & Leadership Journal, 12*(14). Retrieved from http://www.curriculum.edu.au/leader/teaching_higher_order_thinking,37431.html?issueID=12910

Conroy, G. (2016). *Kick-starting student start-ups.* Retrieved from http://sciencemeetsbusiness.com.au/kick-starting-student-startups/

Cook-Sather, A. (2011). Layered learning: Student consultants deepening classroom and life lessons. *Educational Action Research, 19*(1), 41–57. doi:10.1080/09650792.2011.547680

Corbeil, J. R., & Valdes-Corbeil, M. E. (2007). Are You Ready for Mobile Learning? *EDUCAUSE Quarterly, 30*(2).

Cordes, F., & Stacey, N. (2017). *Is UK Industry Ready for the Fourth Industrial Revolution?* Boston, MA: The Boston Consulting Group.

Cormier, D., & Siemens, G. (2010). The open course through the open door: open courses and research, learning and engagement. *Educase Review,* 31-39.

Corneli, J., & Danoff, C. J. (2011). *Paragogy: Synergizing Individual and Organizational Learning.* Retrieved from https://pdfs.semanticscholar.org/1b4b/4c5b2a53705e161ca797be3562f4eeda1fc1.pdf

Corney, M., Teague, D., & Thomas, R. N. (2010). Engaging students in programming. *Proceedings of the Twelfth Australasian Conference on Computing Education, 103*, 63-72.

Costa, E. B., Fonseca, B., Santana, M. A., de Araújo, F. F., & Rego, J. (2017). Evaluating the effectiveness of educational data mining techniques for early prediction of students' academic failure in introductory programming courses. *Computers in Human Behavior, 73*, 247–256. doi:10.1016/j.chb.2017.01.047

Council of Graduate Schools. (2007). *CGS Task Force Report on the Professional Doctorate*. Washington, DC: Council of Graduate Schools.

Crawford, K. (2007). Continuing professional development in higher education: Debating the academic perspective. *The International Journal for Academic Development, 13*(2), 141–146. doi:10.1080/13601440802076657

Crawford, K. (2009). Continuing Professional Development in Higher Education – tensions and debates in a changing environment. In L. Bell, H. Stevenson, & M. Neary (Eds.), *The future of higher education: Policy, pedagogy and the student experience* (pp. 69–83). London: Continuum.

Cross, J. (2014). *Introduction to Kahoot for your classroom assessments*. Retrieved from https://www.youtube.com/watch?v=PYfoRRtLXys

Cude, W. (2001). *The Ph.D. trap revisited*. Toronto, Canada: The Dundurn Group.

Cummins, J. (1994). The acquisition of English as a Second Language. In K. Spangenberg-Urbschat & R. Pritchard (Eds.), *Reading Instruction for ESL Students*. International Reading Association.

Cuoco, A., Goldenberg, P., & Mark, J. (1996). Habits of mind: An organizing principle for mathematics curricula. *The Journal of Mathematical Behavior, 15*(4), 375–402. doi:10.1016/S0732-3123(96)90023-1

Dabbagh, N. (2002). Scaffolding: An Important Teacher Competency in Online Learning. *TechTrends, 47(*2), 39-44.

Dahlstrom, E., Walker, J. D., & Dziuban, C. (2013). ECAR study of undergraduate students and information technology. Academic Press.

Darling-Hammond, L., Chung, R., & Frelow, F. (2002). Variation in teacher preparation: How well do different pathways prepare teacherst o teach? *Journal of Teacher Education, 53*(4), 286–302. doi:10.1177/0022487102053004002

Davey, T. (2011). *A guide to computer adaptive testing systems*. Educational Testing Service. Retrieved from https://files.eric.ed.gov/fulltext/ED543317.pdf

Davies, R. S., Dean, D. L., & Ball, N. (2013). Flipping the classroom and instructional technology integration in a college-level information systems spreadsheet course. *Educational Technology Research and Development, 61*(4), 563–580. doi:10.100711423-013-9305-6

Davis, C. (2013). Flipped or inverted learning: Strategies for course design. In E. G. Smyth & J. X. Volker (Eds.), *Enhancing instruction with visual media: Utilizing video and lecture capture* (pp. 241–261). Hershey, PA: Information Science. doi:10.4018/978-1-4666-3962-1.ch017

Dawson, K., & Kumar, S. (2016). Guiding principles for quality professional practice dissertations. In *Contemporary approaches to dissertation development and research methods*. Hershey, PA: IGI Global.

Day, C. (1999). *Developing teachers: The challenges of lifelong learning*. London: Falmer Press.

Day, C., Harris, A., Hadfield, M., Tolley, H., & Beresford, J. (2000). *Leading schools in times of change*. Buckingham, UK: Open University Press.

Day, C., & Sachs, J. (2004). *International handbook on the continuing professional development of teachers*. Berkshire, UK: Open University Press.

Day, L. (2013). A snapshot of the use of ICT in primary mathematics classrooms in Western Australia. *Australian Primary Mathematics Classroom*, *18*(1), 16–24.

de Freitas, A. A., & de Freitas, M. M. (2013). Classroom Live: A Software-assisted Gamification Tool. *Computer Science Education*, *23*(2), 186–206. doi:10.1080/08993408.2013.780449

Dee Fink, L. (2013). *Creating significant learning experiences: An integrated approach to designing college courses, revised and updated.* San Francisco, CA: Jossey-Bass.

Delacruz, S. (2013). Online reading response using Edmodo. *The Florida Reading Journal*, *49*(2), 9–12.

Demirer & Sahin. (2013). *Effect of blended learning environment on transfer of learning: an experimental study.* doi:10.1111/jcal.12009

Deng, L., & Yuen, A. H. (2011). Towards a framework for educational affordances of blogs. *Computers & Education*, *56*(2), 441–451. doi:10.1016/j.compedu.2010.09.005

Denner, J., Werner, L., Campe, S., & Ortiz, E. (2014). Pair programming: Under what conditions is it advantageous for middle school students? *Journal of Research on Technology in Education*, *46*(3), 277–296. doi:10.1080/15391523.2014.888272

Department for Education and Employment (DfEE). (2001). *Good value CPD: A code of practice for providers of professional development for teachers.* London: DfEE.

Department for Education and Skills (DfES). (2001). *Helping you develop: guidance on producing a professional development record.* London: DfES.

Department for Education and Skills (DfES). (2003). *The future of higher education.* London: The Stationery Office.

Deterding, S., Dixon, D., Khaled, R., & Nacke, L. (2011). From game design elements to gamefulness: defining "gamification". In *Proceedings of the 15th International Academic MindTrek Conference: Envisioning future media environments* (pp. 9–15). Tampere, Finland: ACM. 10.1145/2181037.2181040

Dewey, J. (1916). *Democracy and Education. An introduction to the philosophy of education* (1966 ed.). New York: Free Press.

Dewey, J. (1916). *Democracy and education: an introduction to the philosophy of education.* New York, NY: MacMillan Company.

Dewey, J. (1916). *The Child & The Curriculum: The School & Society.* Chicago: University of Chicago Press.

Dewey, J. (1933). *How We Think.* Boston: Heath.

Dewey, J. (1934). *Art as Experience.* New York: Capricorn Books.

Dewey, J. (1938). *Experience & Education.* New York: Macmillan Publishing Co. Inc.

Dewey, J. (1968). *Democracy & Education.* New York: Macmillan.

Dexter, B. (2007). Continuous professional development for dual professionals: Introducing CPD in a university setting. *The International Journal of Interdisciplinary Social Sciences: Annual Review*, *2*(4), 21–327. doi:10.18848/1833-1882/CGP/v02i04/52363

Dichev, C., & Dicheva, D. (2017). Gamifying education: What is known, what is believed and what remains uncertain: A critical review. *International Journal of Educational Technoilogy in Higher Education*, *14*(9), 1–36.

Dick, B. (2013). Crafting learner-centered processes using action research and action learning. In S. Hase & C. Kenyon (Eds.), *Self-determined Learning: Heutagogy in Action*. London: Bloomsbury Academic.

Dickenson, P. (2014). Flipping the Classroom in a Teacher Education Course. In J. Keengwe, G. Onchwari & J. Oigara (Eds.), Promoting Active Learning through the Flipped Classroom Model (pp. 145-162). Hershey, PA: Information Science Reference. doi:10.4018/978-1-4666-4987-3.ch008

Dinkins, C. S., & Sorrell, J. M. (2014). *Our dissertations, ourselves: Shared stories of women's dissertation journeys*. New York, NY: Palgrave Macmillan.

Donald, J. L. (2003). *Wireless Messaging Demystified: SMS, EMS, MMS, IM, and Others*. McGraw-Hill Professional.

Drake, P., & Heath, L. (2011). *Practitioner research at doctoral level*. London, UK: Routledge.

Dreiseitl, S., & Ohno-Machado, L. (2002). Logistic regression and artificial neural network classification models: A methodology review. *Journal of Biomedical Informatics*, *35*(5-6), 352–359. doi:10.1016/S1532-0464(03)00034-0 PMID:12968784

Droste, M. (2015). *Bauhaus*. Cologne, Germany: Taschen.

Drucker, P. (2014). *Innovation and entrepreneurship*. New York: Routledge. doi:10.4324/9781315747453

Du, H., Xu, H., Rosson, M. B., & Carroll, J. M. (2013). Effects of Fear Appeals and Point of Reference on the Persuasiveness of IT Security Communications. In IEEE Intelligence and Security Informatics 2013. IEEE Digital Library. doi:10.1109/ISI.2013.6578791

DuFour, R., DuFour, R., & Eaker, R. (2008). *Revisiting professional learning communities at work. New insights for improving schools*. Bloomington, IN: Solution Tree.

DuFour, R., DuFour, R., Eaker, R., & Karhanek, G. (2004). *Whatever It Takes: How Professional LearningCommunities Respond When Kids Don't Learn*. Bloomington, IN: National Educational Service.

DuFour, R., DuFour, R., Eaker, R., & Many, T. (2006). *Learning by doing: A handbook for professional learning communities at work* (2nd ed.). Bloomington, IN: Solution Tree.

Dujin, A., Geissler, C., & Horstkötter, D. (2014). *Industry 4.0*. The new industrial revolution-How europe will succeed. Munich: Roland Berger Strategy Consultants.

Duke, N. K., & Beck, S. W. (1999). Education should consider alternative formats for the dissertation. *Educational Researcher*, *28*(3), 31–36.

Dunlop, J. C., & Lowenthal, P. R. (2009). Horton Hears a Tweet. *EDUCAUSE Quarterly Magazine*, *32*(4). Retrieved 6th July 2018, from http://goo.gl/P9Wig

Ebner, M., Lienhardt, C., Rohs, M., & Meyer, I. (2010). Microblogs in Higher Education–A chance to facilitate informal and process-oriented learning? *Computers & Education*, *55*(1), 92–100. doi:10.1016/j.compedu.2009.12.006

Edmodo. (2013, March 5). Edmodo acquires education technology start-up root-1 [Web log post]. Retrieved from https://blog.edmodo.com

Edmodo. (2014). *20 ways to use Edmodo (teacher)*. Retrieved from https://support.edmodo.com/home#entries/21720784-20-ways-to-use-edmodo teacher

Edmodo. (2017). Platform overview. *Edmodo Inc.* Retrieved from https://developers.edmodo.com/resources/overview/

Edmodo. (2018). *Get more from your classroom*. Retrieved from https://go.edmodo.com/

EDUCAUSE Learning Initiative. (2012). *7 things you should know about flipped classrooms*. Retrieved from http://net.educause.edu/ir/library/pdf/ELI7081.pdf

Eierman, M. A., & Iversen, J. (2018). Comparing Test-Driven Development and Pair Programming to Improve the Learning of Programming Languages. *Journal of the Midwest Association for Information Systems, 1*(1), 23.

Ekici, D. I. (2017). The use of Edmodo in creating an online learning community of practice for learning to teach science. *Malaysian Online Journal of Educational Sciences 2017, 5*(2). Retrieved from https://eric.ed.gov/?id=EJ1142512

Ellison, N., & Wu, Y. (2008). Blogging in the classroom: A preliminary exploration of student attitudes and impact on comprehension. *Journal of Educational Multimedia and Hypermedia, 17*(1), 99–122.

Embi, M. A. (2010). e-Learning in Malaysian Higher Education Institutions. Klang: Ministry of Higher Education, Malaysia.

Emmer, M. J., & Brunhoeffer, G. C. F. (2015). Knowledge and attributes of forecasting index: Self-assessment for graduating Construction Management students. *Proceedings of the 2015 Associated Schools of Construction Annual International Conference.*

Ennis, R. H. (1996). *Critical thinking*. Upper Saddle River, NJ: Prentice Hall.

Eraut, M. (1994). *Developing Professional Knowledge and Competence*. London: The Falmer Press.

Eristi, S. D., Kurt, A. A., & Dindar, M. (2012). Teachers ' views about effective use of technology in classrooms. *Research on Computing in Education, 33*(5), 1–27.

Essa, A., & Ayad, H. (2012, April). *Student success system: Risk analytics and data visualization using ensembles of predictive models.* Paper presented at the 2nd International Conference on Learning Analytics and Knowledge, Vancouver, Canada. 10.1145/2330601.2330641

Eunmi, Y. U. (2017). Influence of Flipped Classroom Model on Korean Translation Teaching. *2017 International Conference on Education Innovation and Economic Management (EIEM 2017).*

Facione, P. A. (2013). *Critical thinking: What it is and why it counts*. Retrieved from http://www.insightassessment.com

Falk, C. (2008). All Pedagogy is Military. In T. Kvernbekk, H. Simpson, & M. A. Peters (Eds.), *Military Pedagogies: And Why They Matter*. Rotterdam: Sense Publishers.

Farrell, G. A., Shafiei, T., & Salmon, P. (2010). Facing up to 'challenging behaviour': A model for training in staff–client interaction. *Journal of Advanced Nursing, 66*(7), 1422–1434. doi:10.1111/j.1365-2648.2010.05340.x PMID:20497270

Felder, R. M., & Silverman, L. K. (1988). Learning and teaching styles in engineering education. Engr. *Education, 78*(7), 674–681.

Fell, A. (2016). *Generation Z defined; The five characteristics of today's students*. Retrieved on https://www.youtube.com/watch?v=QvCwcQmnyfE

Fenwick, T., Nerland, M., & Jensen, K. (2014). *Professional Learning in Changing Contexts*. Oxon, UK: Routledge.

Fernández-Montes, A., Ortega, J. A., Alvarez, J. A., & Gonzalez-Abril, L. (2009). Smart environment software reference architecture. In *2009 Fifth International Joint Conference on INC, IMS and IDC* (pp. 397-403). IEEE. 10.1109/NCM.2009.115

Fidalgo-Blanco, Á., Sein-Echaluce, M. L., & García-Peñalvo, F. J. (2013). MOOC cooperativo. Una integración entre cMOOC y xMOOC. *Proceedings from II Congreso Internacional sobre Aprendizaje, Innovación y Competitividad, CINAIC 2013.*

Fidge, C., Hogan, J., & Lister, R. (2013). What vs. how: comparing students' testing and coding skills. *Proceedings of the Fifteenth Australasian Computing Education Conference, 136*, 97-106.

Fiell, C., & Fiell, P. (2006). *Industrial design A-Z*. Cologne, Germany: Taschen.

Figueroa, J. F. (2015). Using gamification to enhance second language learning. *Digital Education Review*, *27*, 32–54.

Figueroa, J. F. (2016). Gamification and Game-based Learning: Two Strategies for the 21st Century Learner. *World Journal of Education Reseacrh*, *3*(2), 507–522. doi:10.22158/wjer.v3n2p507

Finkel, E. (2012). Flipping the script in k12. *District Administration*, *48*(10), 28–30.

Finlay, J., Pears, R., & Connor, A. M. (2014). Data stream mining for predicting software build outcomes using source code metrics. *Information and Software Technology*, *56*(2), 183–198. doi:10.1016/j.infsof.2013.09.001

Firmin, M. W., & Genesi, D. J. (2013). History and implementation of classroom technology. *Procedia: Social and Behavioral Sciences*, *93*(2), 1603–1617. doi:10.1016/j.sbspro.2013.10.089

Flavell, J. H. (1976). Metacognitive aspects of problem solving. *The Nature of Intelligence, 12*, 231-235.

Fleck, J. (2012). Alternative Models Blended learning and communities: Opportunities and challenges. *Journal of Management Development*, *31*(4), 398–441. doi:10.1108/02621711211219059

Flipped Learning Network (FLN). (2012). *7 things you should know about flipped classrooms*. Retrieved from https://library.educause.edu/resources/2012/2/7-things-you-should-know-about-flipped-classrooms

Flipped Learning Network (FLN). (2014). *The four pillars of F-L-I-P*. Retrieved from https://flippedlearning.org/definition-of-flipped-learning/

Foord, K. A., & Haar, J. M. (2013). *Professional learning communities: An implementation guide and toolkit*. New York: Routledge.

Force, M. T. (2014). *Report of the MLA Task Force on Doctoral Study in Modern Language and Literature*. Modern Language Association.

Forsey, M., Low, M., & Glance, D. (2013). Flipping the sociology classroom: Towards a practice of online pedagogy. *Journal of Sociology (Melbourne, Vic.)*, *49*(4), 471–485. doi:10.1177/1440783313504059

Forty, A. (1992). *Objects of desire: Design and society since 1750*. London, UK: Thames & Hudson.

Foust, T. (2012). Special guest article: A tip of the hat to the flip of the class. *Illinois Music Educator*, *73*(2), 100.

Freeman, D. (1991, December). Three views to teachers' knowledge. *IATEFL Teacher Development Newsletter*, 1-4.

Freeman, D. (1993). Renaming experience/reconstructing practice: Developing new understandings of teaching. *Teaching and Teacher Education*, *94*(5-6), 485–497. doi:10.1016/0742-051X(93)90032-C

Freeman, D., & Johnson, K. E. (1998). Reconceptualizing the knowledge-base of language teacher education. *TESOL Quarterly*, *32*(3), 397–418. doi:10.2307/3588114

Frey, N., Fisher, D., & Allen, A. (2009). Productive group work in middle and high school classrooms. In S. Parris, D. Fisher, & K. Headley (Eds.), *Adolescent literacy: Effective solutions for every classroom* (pp. 70–81). Newark, DE: International Reading Association. doi:10.1598/0695.06

Friedman, A., Davis, K., & Phillips, M. (2001). *Continuing professional development in the UK: Attitudes and experiences of practitioners*. Bristol: PARN.

Friedman, A., Durkin, C., & Phillips, M. (2000). CPD: What are the true costs of continuing professional development? *Continuing Professional Development, 3*(3), 78–87.

Friedman, A., & Phillips, M. (2004). Continuing professional development: Developing a vision. *Journal of Education and Work, 17*(3), 361–376. doi:10.1080/1363908042000267432

Friedman, A., Williams, C., Hopkins, S., & Jackson, L. (2008). *Linking professional associations with higher education institutions in relation to the provision of continuing professional development*. Bristol: Professional Associations Research Network.

Fuad-Luke, A. (2009). *Design activism: Beautiful strangeness for a sustainable world*. London, UK: Routledge.

Fukuda, Mariz, & Mesquita. (2017). Impactos da Indústria 4.0 na gestão de operações. In XXXVII Encontro Nacional de Engenharia de Produção, 2017. Joinville/SC: Enegep.

Fullan, M. (1992). *Successful school improvement: The implementation perspective and beyond*. Birmingham, UK: Open University Press.

Fullan, M. (1996). Turning Systemic Thinking on its Head. *Phi Delta Kappan, 77*(6), 420–423.

Fullan, M. (2001). *Leading in a Culture of Change: Being Effective in Complex Times*. San Francisco: Jossey-Bass Wiley.

Fullan, M., & Hargreaves, A. (Eds.). (2013). *Teacher development and educational change*. London: Routledge.

Furberg, A., Kluge, A., & Ludvigsen, S. (2013). Student sense making with science diagrams in a computer-based setting. *International Journal of Computer-Supported Collaborative Learning, 8*(1), 41–64. doi:10.100711412-013-9165-4

Furze, G., & Pearcey, P. (1999). Continuing education in nursing: A review of the literature. *Journal of Advanced Nursing, 29*(2), 355–363. doi:10.1046/j.1365-2648.1999.00896.x PMID:10197935

Gagné, R. M., Briggs, L. J., & Wager, W. W. (1992). *Principles of instructional design* (4th ed.). Fort Worth, TX: Harcourt Brace Jovanovich College Publishers.

Gallardo-Virgen, J. A., & DeVillar, R. A. (2011). Sharing, talking, and learning in the elementary school science classroom: Benefits of innovative design and collaborative learning in computer-integrated settings. *Computers in the Schools, 28*(4), 278–290. doi:10.1080/07380569.2011.621803

Galloway, A. (2018). Engineering at home. In The future starts here. London, UK: V&A Publishing.

Garcia, D. H., Leles, A. D., & Romano, R. R. (2017). Program Entrepreneurship and Innovation: Education as the Core of Innovation. In *Advances in The Human Side of Service Engineering* (pp. 235–244). Springer International Publishing. doi:10.1007/978-3-319-41947-3_22

Gardner, M., & Elliott, J. (2014). The Immersive Education Laboratory: understanding affordances, structuring experiences, and creating constructivist, collaborative processes, in mixed-reality smart environments. *EAI Endorsed Transactions on Future Intelligent Educational Environments, 14*(1).

Garrison, D. R., Anderson, T., & Archer, W. (2003). A theory of critical inquiry in online distance education. Handbook of distance education, 1, 113-127.

Garrison, D. R. (1985). Three generations of technological innovations in distance education. *Distance Education, 6*(2), 235–241. doi:10.1080/0158791850060208

Garrison, D. R., Anderson, T., & Archer, W. (2001). Critical thinking, cognitive presence, and computer conferencing in distance education. *American Journal of Distance Education, 15*(1), 7–23. doi:10.1080/08923640109527071

Garrison, D. R., & Archer, W. (2000). *A transactional perspective on teaching-learning: A framework for adult and higher education.* Oxford, UK: Pergamon.

Garrison, D. R., & Aykol, Z. (2009). Role of instructional technology in the transformation of higher education. *Journal of Computing in Higher Education, 21*(1), 19–30. doi:10.100712528-009-9014-7

Garrison, D. R., & Baynton, M. (1987). Beyond independence in distance education: The concept of control. *American Journal of Distance Education, 1*(3), 3–15. doi:10.1080/08923648709526593

Garrison, D. R., & Kanuka, H. (2004). Blended learning: Uncovering its transformative potential in higher education. *The Internet and Higher Education, 7*(2), 95–105. doi:10.1016/j.iheduc.2004.02.001

Garrison, D. R., & Vaughan, N. (2008). *Blended Learning in Higher Education: Framework, Principles, and Guidelines.* Jossey-Bass.

Garrison, D. R., & Vaughan, N. D. (2008). *Blended learning in higher education: Framework, principles, and guidelines.* John Wiley & Sons.

Garrison, R. (2000). Theoretical challenges for distance education in the 21st century: A shift from structural to transactional issues. *International Review of Research in Open and Distance Learning, 1*(1), 1–17. doi:10.19173/irrodl.v1i1.2

Gavriel, J. (2015). The flipped classroom. *Education for Primary Care, 26*(6), 424–425. doi:10.1080/14739879.2015.1109809 PMID:26808941

Gehringer, E. F., & Peddycord, B. W. III. (2013). Experience with online and open-web exams. *Journal of Institutional Research, 2*, 10–18. doi:10.9743/JIR.2013.2.12

George, A. L., & Bennett, A. (2005). *Case Studies and Theory Development in the Social Sciences.* MIT Press.

Gershenfeld, N., Krikorian, R., & Cohen, D. (2004). The internet of things. *Scientific American, 291*(4), 76–81. doi:10.1038cientificamerican1004-76 PMID:15487673

Gerstein, J. (2016). *Maker Education: Pedagogy, Andragogy and Heutagogy.* Retrieved from https://usergeneratededucation.wordpress.com/2016/06/14/maker-education-pedagogy-andragogy-heutagogy/

Gewertz, C. (2011). *Computer-adaptive testing poses challenges, expert warns.* Retrieved from http://blogs.edweek.org/edweek/curriculum/2011/05/computer-adaptive_testing_pose.html?qs=Reckase

Ghazal, S., Samsudin, Z., & Aldowah, H. (2015). Students' perception of synchronous courses using skype-based video conferencing. *Indian Journal of Science and Technology, 8*(30). doi:10.17485/ijst/2015/v8i1/84021

Gibb, A. A. (1993). Enterprise culture and education understanding enterprise education and its links with small business, entrepreneurship and wider educational goals. *International Small Business Journal, 11*(3), 11–34. doi:10.1177/026624269301100301

Gibson, D., Ostashewski, N., Flintoff, K., Grant, S., & Knight, E. (2013). *Digital Badges in Education. Education and Information Technology.* New York: Springer.

Gibson, I., Rosen, D., & Stucker, B. (2014). *Additive manufacturing technologies: Rapid prototyping to direct digital manufacturing* (2nd ed.). New York: Springer.

Glassman, M. (2016). *Educational psychology and the internet.* Cambridge University Press. doi:10.1017/CBO9781316155134

Gleason, N. W. (2018). *Higher education in the era of the fourth industrial revolution*. Retrieved from https://link.springer.com/content/pdf/bfm%3A978-981-13-0194-0%2F1.pdf

Gleason, B. L., Peeters, M. J., Resman-Targoff, B. H., Karr, S., McBane, S., Kelley, K., ... Denetclaw, T. H. (2011). An active-learning strategies primer for achieving ability-based educational outcomes. *American Journal of Pharmaceutical Education*, *75*(9), 1–12. doi:10.5688/ajpe759186 PMID:22171114

Goh, R., & Fang, Y. (2017). Improving English language teaching through lesson study: Case study of teacher learning in a Singapore primary school grade level team. *International Journal for Lesson and Learning Studies*, *6*(2), 135–150. doi:10.1108/IJLLS-11-2015-0037

Gojak, L. (2012). *To flip or not to flip: That is not the question*. National Council of Teachers of Mathematics. Retrieved from http://www.nctm.org/about/content.aspx?id=34585

Gokhberg, L., Shmatko, N., & Auriol, L. (Eds.). (2016). *The science and technology labor force: The value of doctorate holders and development of professional careers*. Springer International. doi:10.1007/978-3-319-27210-8

Golde, C. M., & Dore, T. M. (2001). *At Cross Purposes: What the experiences of doctoral students reveal about doctoral education*. Philadelphia, PA: A Report Prepared for The Pew Charitable Trusts.

Golde, C. M., & Walker, G. E. (2006). *Envision the future of doctoral education*. San Francisco, CA: Jossey-Bass.

Gomes, A., & Mendes, A. J. (2007). An environment to improve programming education. *Proceedings of the 2007 International Conference on Computer Systems and Technologies*, *88*, 1-6.

Goodwin, B., & Miller, K. (2013). Research says evidence on flipped classrooms is still coming in. *Educational Journal of Information Systems Education*, *25*(1).

Gore, A. (2013). *The future*. New York: Random House.

Goyal, M., & Vohra, R. (2012). Applications of data mining in higher education. *International Journal of Computer Science Issues*, *9*(2), 113–120.

Grabe, M., & Grabe, C. (2004). *Integrating Technology for Meaningful Learning* (4th ed.). Boston: Houghton Mifflin Company.

Graham, C. R., Woodfeld, W., & Harrison, J. B. (2013). A framework for institutional adoption and implementation of blended learning in higher education. *The Internet and Higher Education*, *18*, 4–14. doi:10.1016/j.iheduc.2012.09.003

Graube, G., & Mammes, I. (2017). Industry Involvement in Technology Education. In M. J. Vries (Ed.), *Handbook of Technology Education* (pp. 843–856). New York, NY: Springer. doi:10.1007/978-3-319-38889-2_60-1

Gray, A. (2016). *The 10 skills you need to thrive in the Fourth Industrial Revolution*. New York, NY: World Economic Forum. Retrieved from https://www.weforum.org/agenda/2016/01/the-10-skills-you-need-to-thrive-in-the-fourth-industrial-revolution/

Greason N. W. (2018). *Higher education in the era of the fourth industrial revolution*. PELLGRAVE.

Greenfield, A. (2017). *Radical technologies: The design of everyday life*. London, UK: Verso.

Griffin, P., McGaw, B., & Care, E. (2012). *Assessment and teaching of 21st century skills*. Springer. doi:10.1007/978-94-007-2324-5

Gubbi, J., Buyya, R., Marusic, S., & Palaniswami, M. (2013). Internet of Things (IoT): A vision, architectural elements, and future directions. *Future Generation Computer Systems*, *29*(7), 1645–1660. doi:10.1016/j.future.2013.01.010

Gul, S., Asif, M., Ahmad, S., Yasir, M., Majid, M., & Arshad, M. S. (2017). A survey on role of internet of things in education. *IJCSNS*, *17*(5), 159.

Gummesson, E. (2000). *Qualitative methods in management* (2nd ed.). Thousand Oaks, CA: Sage.

Gunn, A. (2018). Metrics and methodologies for measuring teaching quality in higher education: Developing the Teaching Excellence Framework (TEF). *Educational Review*, *70*(2), 129–148. doi:10.1080/00131911.2017.1410106

Hacker, D. J. (1998). Definitions and empirical foundations. In D. J. Hacker, J. Dunlosky, & A. C. Graesser (Eds.), *Metacognition in educational theory and practice* (pp. 1–23). Hillsdale, NJ: Lawrence Erlbaum Associates.

Halasek, K., McCorkle, B., Selfe, C. L., DeWitt, S. L., Delagrange, S., Michaels, J., & Clinnin, K. (2014). A MOOC with a view: How MOOCs encourage us to reexamine pedagogical doxa. *Invasion of the MOOCs,* 156. Retrieved from http://www.parlorpress.com/pdf/invasion_of_the_moocs.pdf

Hamari, J., Koivisto, J., & Sarsa, H. (2014). Does gamification work?-A literature review of empirical studies on gamification. In *Proceedings of the 47th Hawaii International Conference on System Sciences (HICSS)* (pp. 3025–3034). Waikoloa. 10.1109/HICSS.2014.377

Hamdan, N., McKnight, P. E., McKnight, K., & Artfstrom, K. M. (2013a). *The flipped learning model: A white paper based on the literature review.* Retrieved from http://flippedlearning.org/cms

Hamdan, N., McKnight, P. E., McKnight, K., & Artfstrom, K. M. (2013b). A review of flipped learning. *Flipped Learning Network.* Retrieved from https://flippedlearning.org/wp-content/uploads/2016/07/WhitePaper_FlippedLearning.pdf

Hamon, K. (2014). *The benefits of flipped classes on student workloads.* Retrieved from: https://www.fractuslearning.com/2014/07/22/flipped-classes-student-workloads/

Hanapi, Z., & Nordin, M. S. (2014). Unemployment among Malaysia Graduates: Graduates' attributes, lecturers' competency and quality of education. *Procedia: Social and Behavioral Sciences*, *112*, 1056–1063. doi:10.1016/j.sbspro.2014.01.1269

Hanapi, Z., Nordin, M. S., & Khamis, A. (2015). Challenges faced by engineering lecturers in integrating technical and employability skills in the curriculum: A case study in Community College, Malaysia. *International Journal of Social Science and Humanity*, *5*(5), 483. doi:10.7763/IJSSH.2015.V5.504

Harmer, J. (2003). *The practice of English language teaching* (3rd ed.). Harlow, UK: Pearson.

Harris, P., Connolly, J., & Feeney, L. (2009). Blended learning: Overview and recommendations for successful implementation. *Industrial and Commercial Training*, *41*(3), 155–163. doi:10.1108/00197850910950961

Harris, R. N., Dwyer, W. O., & Leeming, F. C. (2003). Are learning styles relevant in Web-based instruction? *Journal of Educational Computing Research*, *29*(1), 10–27. doi:10.2190/YHL4-UP7P-K0GD-N5LJ

Harwell, S. H. (2003). *Teacher professional development: It's not an event, it's a process.* Waco, TX: Cord.

Haseeb, A. M. (2018, Jan 10). Higher education in the era of IR 4.0. *New Straits Times.* Retrieved from https://www.nst.com.my

Hativa, N. (2001). *Teaching for effective learning in higher education.* Kluwer Academic Publishers.

Hattie, J. (2016). *On Formative Assessment: Readings from Educational Leadership.* Alexandria, VA: ASCD.

Hawken, P. (2010). The ecology of commerce revised edition: A declaration of sustainability. New York: Harper Business.

Henri, F. (1992). Computer conferencing and content analysis. In *Collaborative learning through computer conferencing* (pp. 117–136). Springer. doi:10.1007/978-3-642-77684-7_8

Herbart, J. F. (1906). *A Study in Pedagogics (A. M. Williams, Trans.)*. Kessinger Publishing.

Herreid, C., & Schiller, N. (2013). Case studies and the flipped classroom. *Journal of College Science Teaching*, *42*, 62–66.

Herreid, C., & Schiller, N. (2013). Caswe Studies and The Flipped Classroom. *Journal of College Science Teaching*, *42*, 62–66.

Herz, C. J. (2001). *Gaming the system: What higher education can learn from multiplayer online worlds*. The Internet and the University, Educause Forum on the Future of Higher Education. Retrieved from http://www.educause.edu/ir/library/pdf/ffpiu019.pdf

Heyberi, E. (2013). *Integrating Technology in the Curriculum for Enhanced Learning : A Comparative Study in England and North Cyprus* (Doctoral dissertation). University of Birmingham.

Hickman, L. A. (1990). *John Dewey's Pragmatic Technology*. Bloomington, IN: Indiana University Press.

Hickman, L. A. (2001). *Philosophical Tools for Technological Culture*. Bloomington, IN: Indiana University Press.

Higher Education Academy. (2012). *A Marked improvement: Transforming assessment in higher education*. Retrieved from https://www.cardiff.ac.uk/__data/assets/pdf_file/0007/583990/A-Marked-Improvement-transforming-assessment-in-higher-education-HEA.pdf

Hodell, C. (2011). *ISD from the ground up: A no-nonsense approach to instructional design* (3rd ed.). Alexandria, VA: ASTD Press.

Holley, K. (2017). Interdisciplinary Curriculum and Learning in Higher Education. *Oxford Research Encyclopedia of Education*. Retrieved from http://education.oxfordre.com/view/10.1093/acrefore/9780190264093.001.0001/acrefore-9780190264093-e-138

Holmberg, B. (1983). Guided didactic conversation in distance education. In Distance education: International perspectives (pp. 114-122). London: Croom Helm.

Holmberg, B. (1995). *The sphere of distance education: Theory revisited*. Retrieved from www.eric.ed.gov/ERICDocs/data/ericdocs2sql/content_storage_01/0000019b/80/14/25/f8.pdf

Holton, D., & Symons, D. (2015). *Creative Activities in Mathematics: Book 2. Problem-Based Maths Investigations for Upper Primary*. ACER.

Hong, C. L., & Seyal, A. H. (2014). Problem-based learning: An analysis of its application to the teaching of programming. *International Proceedings of Economics Development and Research*, 68–75.

Hook, L. J., & Eckerdal, A. (2015). On the bimodality in an introductory programming course: An analysis of student performance factors. *International Conference on Learning and Teaching in Computing and Engineering*, 79-86. 10.1109/LaTiCE.2015.25

Horton, B. J. (1940). *The graduate school*. New York, NY: New York University.

Huang, H. Y. W., & Soman, D. (2013). A Practitioner's Guide to Gamification of Education. Research Report Series: Behavioral Economics in Action. University of Toronto –Rotman School of Management.

Huckabee, I., & Bissette, T. (2014). *Learning made fun. Training Industry Magazine*. Retrieved from http://www.nxtbook.com/nxtbooks/trainingindustry/tiq_2014spring/

Hu, H., & Garimella, U. (2014). iPads for STEM teachers: A case study on perceived usefulness, perceived proficiency, intention to adopt, and integration in K-12 instruction. *Journal of Educational Technology Development and Exchange*, *7*(1), 49–66. doi:10.18785/jetde.0701.04

Hung, W., Jonassen, D. H., & Liu, R. (2008). Problem-based learning. Handbook of research on educational communications and technology, 3, 485-506.

Hurst, B., Wallace, R., & Nixon, S. B. (2013). The impact of social interaction on student learning. *Reading Horizons*, *52*(4). Retrieved from https://scholarworks.wmich.edu/reading_horizons/vol52/iss4/5

Hu, S. J. (2013). Evolving Paradigms of Manufacturing: From Mass Production to Mass Customization and Personalization. *Procedia CIRP*, *7*, 3–8. doi:10.1016/j.procir.2013.05.002

Husnjak, S., Perakovic, D., & Jovovic, I. (2014). Possibilities of using speech recognition systems of smart terminal devices in traffic environment. *Procedia Engineering*, *69*, 778–787. doi:10.1016/j.proeng.2014.03.054

Hyde, R., & Pestana, M. (Eds.). (2018). The Future Starts Here. London, UK: V&A Publishing.

Hyland, K. (2004). *Genre and second language writing*. University of Michigan Press.

Iaremenko, N. V. (2017). Enhancing English language learners' motivation through online games. *Information Technologies and Learning Tools*, *59*(3), 126–132.

Icard, S. B. (2014). Educational technology best practices. *International Journal of Instructional Technology and Distance Learning.*, *11*(3), 37–41.

IEA. (2015). *25 Years > Washington Accord 1989-2014 Celebrating International Engineering Education Standards and Recognition*. International Engineering Alliance.

Industry 4.0: The Fourth Industrial Revolution – Guide to Industrie 4.0. (2016). Retrieved from https://www.i-scoop.eu/industry-4-0/

Iqbal, S., & Harsh, O. K. (2013). A self-review and external review model for teaching and assessing novice programmers. *International Journal of Information and Education Technology (IJIET)*, *2*(3), 120–123. doi:10.7763/IJIET.2013.V3.247

Ismail, B., Ayob, J., Alias, M., & Kaprawi, N. (2004). Penjuruteraan Semula Pendidikan Teknikal dan Vokasional: Satu Cetusan Pemikiran dan Tinjauan. *Jurnal Pendidikan Teknikal.*, *3*, 24–35.

Ismail, M. A.-A., & Mohammad, J. A.-M. (2017). Kahoot: A promising tool for formative assessment in medical education. *Education in Medicine Journal*, *9*(2), 19–26.

Ismail, R., Yusoff, I., & Sieng, L. W. (2011). Employer Perceptions on Graduates in Malaysian Services Sector. *International Business Management*, *5*(3), 184–193. doi:10.3923/ibm.2011.184.193

Ismail, W., Abedlazeez, N., & Hussin, Z. (2011). Epistemological beliefs of students at high schools: A survey study in Malaysia. *International Journal of Sustainable Development*, *2*(08), 39–46.

Isong, B. (2014). A Methodology for Teaching Computer Programming: First year students' perspective. *International Journal of Modern Education and Computer Science*, *6*(9), 15–21. doi:10.5815/ijmecs.2014.09.03

Issham, I., Rozhan, M.I., & Thenmolli, G. (2010). *Motivation, psychology and language effect on mobile learning in University Sains Malaysia*. Academic Press.

Issham, I., Rozhan, M. I., & Thenmolli, G. (2010). Motivation, psychology and language effect on mobile learning in University Sains Malaysia. *International Journal of Interactive Mobile Technologies*, *4*(4), 31–36. doi:10.3991.ijim.v4i4.1408

Ivanitskaya, L., Clark, D., Montgomery, G., & Primeau, R. (2002). Interdisciplinary learning: Process and outcomes. *Innovative Higher Education*, *27*(2), 95–111. doi:10.1023/A:1021105309984

Jackson, D., & Chapman, E. (2012). Non-technical skill gaps in Australian Business graduates. *Education + Training*, *54*(2/3), 95–113. http://dio.org/10.1108/00400911211210224. doi:10.1108/00400911211210224

Järvelä, S., Volet, S., & Järvenoja, H. (2010). Research on motivation in collaborative learning: Moving beyond the cognitive-situative divide and combining individual and social processes. *Educational Psychologist*, *45*(1), 15–27.

Jazby, D., & Symons, D. (2015). Mathematical Problem Solving Online: Opportunities for Participation and Assesment. In *The 39th Conference for the Psychology of Mathematics Education* (Vol. 3, pp. 105-112). PME.

Jenkins, A., Breen, R., Lindsay, R., & Brew, A. (2003). *Reshaping teaching in higher education: Linking teaching with research*. London: Kogan Page. doi:10.4324/9780203416884

Jenkins, T. (2002). On the difficulty of learning to program. *Proceedings of the 3rd Annual Conference of the LTSN Centre for Information and Computer Sciences, 4*, 53-58.

Jensen, J. L., Kummar, T. A., & Godoy, P. D. (2015). Improvements from a flipped classroom may simply be the fruits of active learning. *CBE Life Sciences Education*, *14*(1), 1–12. doi:10.1187/cbe.14-08-0129 PMID:25699543

Johns Hopkins University. (1878). *Third Annual Report*. Author.

Johnson, G. B. (2013). *Student perceptions of the flipped classroom* (Unpublished Master's Thesis). University of British Columbia. Retrieved from: https://open.library.ubc.ca/cIRcle/collections/24/items/1.0073641

Johnson, L., Adams, S., Cummins, M., Estrada, V., Freeman, A., & Ludgate, H. (2013). *The NMC Horizon Report: 2013 higher education addition*. Austin, TX: The New Media Consortium. Retrieved from http://www.nmc.org/pdf/2013-horizon-report-HE.pdf

Johnson, L., Adams, S., Cummins, M., Estrada, V., Freeman, A., & Ludgate, H. (2013). *The NMC Horizon Report: 2013 Higher Education Edition*. Austin, TX: The New Media Consortium. Retrieved from http://www.nmc.org/pdf/2013-horizon-report-HE.pdf

Jonassen, D. H. (2000). *Computers As Mindtools For Schools: Engaging Critical Thinking* (2nd ed.). Merrill Prentice Hall.

Jones, B. (2017). *Is WhatsApp using too much space on your phone?* Retrieved from https://www.psafe.com/en/blog/whatsapp-using-much-space-phone/

Jones, C., & Jones, K. (2007). Language of Professional Learning. *Teaching Expertise*. Available from: www.teaching-expertise.com/articles/language-of-professional-learning-1791

Jones, M., & Gott, R. (1998). Cognitive acceleration through science education: Alternative perspectives. *International Journal of Science Education*, *20*(7), 755–768. doi:10.1080/0950069980200701

Joyce, B. R., & Showers, B. (1995). *Student achievement through staff development* (2nd ed.). White Plains, NY: Longman.

Junco, R. (2014). *Engaging students through social media: Evidence-based practices for use in student affairs*. John Wiley & Sons.

Junior, Leme, & Santos. (2018). Indústria 4.0: fundamentos, perspectivas e aplicações. São Paulo: Érica.

Jusoh. (2018). *Annual Speech by the Minister of Higher Education.* Retrieved from https://www.mohe.gov.my/speech-text/483-teks-ucapan-amanat-tahun-baharu-2018-yb-menteri-pendidikan-tinggi

Kahoot! Game-based blended learning & classroom response system. (2014). Retrieved from https://getkahoot.com

Kanter, M. R., & Litow, S. S. (2009). *Informed and Interconnected: A Manifesto for Smarter Cities.* Harvard Business School Working Paper 09-141. Boston, MA: Harvard Business School.

Kanuka, H., & Conrad, D. (2003). The name of the game: Why "distance education" says it all. *Quarterly Review of Distance Education, 4*(4), 285–394.

Kapp, K. M. (2012). *The gamification of learning and instruction: game-based methods and strategies for training and education.* New York, NY: John Wiley & Sons.

Karseth, B., & Nerland, M. (2007). Building professionalism in a knowledge society: Examining discourses of knowledge in four professional associations. *Journal of Education and Work, 20*(4), 335–355. doi:10.1080/13639080701650172

Kasten, W. C. (1997). Learning is noisy; the myth of silence in the reading-writing classroom. In J. R. Paratore & R. L. McCormack (Eds.), *Peer talk in the classroom: Learning from research* (pp. 88–101). Newark, DE: International Reading Association.

Keane, L. (2014). *The 21st century skills of the flipped Classroom.* Retrieved from https://www.linkedin.com/pulse/20141117162612-162064004-the-21st-century-skills-of-the-flipped-classroom

Keefe, K., Sheard, J., & Dick, M. (2006). Adopting XP practices for teaching object oriented programming. *Proceedings of the 8th Australasian Conference on Computing Education, 52,* 91-100.

Keegan, D. (1993). Reintegration of the teaching acts. In Theoretical principles of distance education (pp. 113-134). London: Routledge.

Keegan, D. (1986). *The foundations of distance education.* Londres: Croom Helm.

Keengwe, J., Onchwari, G., & Oigara, J. (2014). *Promoting Active Learning through the Flipped Classroom Model.* Hershey, PA: Information Science Reference. doi:10.4018/978-1-4666-4987-3

Kehm, B. M. (2010). Quality in higher European education: The influence of the Bologna Process. *Change, 42*(May/June), 40–46. doi:10.1080/00091381003704677

Kehm, B. M., & Teichler, U. (2016). Doctoral education and labor market: Policy questions and data needs. In L. Gokhberg, N. Shmatko, & K. Auriol (Eds.), *The Science and Technology Labor Force* (pp. 11–29). Springer International; doi:10.1007/978-3-319-27210-8_2

Kelly, K. (2016). *The inevitable: Understanding the 12 technological forces that will shape our future.* New York: Penguin Books.

Kember, D., & Wong, A. (2000). Implications for evaluation from a study of students' perceptions of good and poor teaching. *Higher Education Journal., 40*(1), 69–97. doi:10.1023/A:1004068500314

Kemp, A. T., Preston, J., Page, C. S., Harper, R., Dillard, B., Flynn, J., & Yamaguchi, M. (2014). Technology and teaching :A conversation among faculty regarding the pros and cons of technology. *Qualitative Report, 19*(3), 1–23.

Kennedy, E. (2014). The flipside of blended learning. *Journal of Lifelong Learning in Europe*, 4.

Kennedy, A. (2005). Models of continuing professional development: A framework for analysis. *Journal of In-service Education, 31*(2), 235–250. doi:10.1080/13674580500200277

Kenyon, C., & Hase, S. (2013). Heutagogy Fundamentals. In S. Hase & C. Kenyon (Eds.), *Self-determined Learning: Heutagogy in Action*. London: Bloomsbury Academic.

Khaddage, F., Lattemann, C., & Acosta-Díaz, R. (2014). Mobile gamification in education engage, educate and entertain via gamified mobile apps. In *Proceedings of Society for Information Technology & Teacher Education International Conference* (pp. 1654–1660). Jacksonville, FL: Association for the Advancement of Computing in Education (AACE).

Khan Academy. (2012). *Watch. practice. learn almost anything for free*. Retrieved from http://www.khanacademy.org/about

Khan, S. (2011). *Let's use video to reinvent education, Speech presented at TED 2011*. Retrieved from https://www.ted.com/talks/salman_khan_let_s_use_video_to_reinvent_education

Khan, B. H. (Ed.). (2005). *Managing e-learning: Design, delivery, implementation, and evaluation*. IGI Global. doi:10.4018/978-1-59140-634-1

Kheswa, J. G. (2014). Action Research: A practice to improve academic performance among undergraduate psychology students at the South African university campus. *Mediterranean Journal of Social Sciences*, *5*(2), 1043–1049.

Kim, J. (2013). *On bubbles, online education, and confused reporting*. Retrieved from https://www.insidehighered. com/blogs/technology-and-learning/bubbles-online-education-and-confused-reporting

Kim, M. K., Kim, S. M., Khera, O., & Getman, J. (2014). The experience of three flipped classrooms in an urban university: an exploration of design principles. *The Internet and Higher Education*, *22*, 37-50.

Kimura, T. (2017). The impacts of computer adaptive testing from a variety of perspectives. *Journal of Educational Evaluation for Health Professions*, *14*(12). doi:10.3352/jeehp.2017.14.12 PMID:28552899

King, A. (1993). From sage on the stage to guide on the side. *College Teaching*, *41*(1), 30–35. doi:10.1080/87567555.1993.9926781

King, J. (2007). *Educating Engineers for the 21st Century*. London: Royal Academy of Engineering.

Kingsley, T., & Grabner-Hagen, M. (2015). Gamification questing to integrate context, knowledge, literacy, and 21st century learning. *Journal of Adolescent & Adult Literacy*, *59*(1), 51–61. doi:10.1002/jaal.426

Kiryakova, G., Yordanova, L., & Angelova, N. (2017). Can we make Schools and Universities smarter with the Internet of Things? *TEM Journal*, *6*(1), 80.

Kissock, C., & Iyortsuun, P. T. (1982). *A guide to questioning: Classroom procedures for teachers*. London: Macmillan.

Klandt, H. 1993. Methods of teaching: what is useful for entrepreneurship education? *Internationalizing Entrepreneurship Education and training. Proceedings of the Int. Ent. 93 conference Vienna*, 31-41.

Klaus, S. (2016). The fourth industrial revolution. In *World Economic Forum* (p. 11). Academic Press.

Klein, J. T., & Newell, W. H. (1997). Advancing interdisciplinary studies. Handbook of the undergraduate curriculum: A comprehensive guide to purposes, structures, practices, and change, 393-415.

Klüsener, M., & Fortenbacher, A. (2015, September). *Predicting students' success based on forum activities in MOOCs*. Paper presented at IEEE 8th International Conference, Warsaw, Poland. 10.1109/IDAACS.2015.7341439

Kongchan, C. (2013). How Edmodo and Google Docs can change traditional classrooms. *Proceedings of The European Conference on Language Learning 2013*, 629-637.

Kortuem, G., Bandara, A. K., Smith, N., Richards, M., & Petre, M. (2013). Educating the Internet-of-Things generation. *Computer*, *46*(2), 53–61. doi:10.1109/MC.2012.390

Köse, U. (2010). A blended learning model supported with Web 2.0 technologies. *Procedia: Social and Behavioral Sciences, 2*(2), 2794–2802. doi:10.1016/j.sbspro.2010.03.417

Kraiger, K. (2002). Decision-based evaluation. In K. Kraiger (Ed.), *Creating, implementing, and maintaining effective training and development: State-of-the-Art Lessons for Practice* (pp. 331–375). San Francisco, CA: Jossey-Bass.

Kreber, C. (2001). *Scholarship revisited: perspectives on the scholarship of teaching.* San Francisco, CA: Jossey-Bass.

Krueger, R. A. (1988). *Focus groups: A practical guide for applied research.* Newbury Park, CA: Sage Publications.

Kubiatko, M., & Haláková, Z. (2009). Slovak high school students' attitudes to ICT using in biology lesson. *Computers in Human Behavior, 25*(3), 743–748. doi:10.1016/j.chb.2009.02.002

Kuh, G. D., & Schneider, C. G. (2008). *High impact educational practices: What they are, who has access to them, and why they matter.* Washington, DC: Association of American Colleges and Universities.

Kukulska-Hulme, A., & Traxler, J. (2005). *Mobile Learning: A Handbook For Educators And Trainers.* London, UK: Routledge.

Kurt, G. (2017). Implementing the flipped classroom in teacher education: evidence from Turkey. *Educational Technology & Society*, (99), 211-221.

Lai, K. W., & Pratt, K. (2004). Information and communication technology (ICT) in secondary schools: The role of the computer coordinator. *British Journal of Educational Technology, 35*(4), 461–475. doi:10.1111/j.0007-1013.2004.00404.x

Lampert, M. (1990). When the problem is not the question and the solution is not the answer: Mathematical knowing and teaching. *American Educational Research Journal, 27*(1), 29–63. doi:10.3102/00028312027001029

Lans, T., Hulsink, W. I. M., Baert, H., & Mulder, M. (2008). Entrepreneurship education and training in a small business context: Insights from the competence-based approach. *Journal of Enterprising Culture, 16*(04), 363–383. doi:10.1142/S0218495808000193

Lasi, H., Fettke, P., Kemper, H. G., Feld, T., & Hoffmann, M. (2014). Industry 4.0. *Business & Information Systems Engineering, 6*(4), 239–242. doi:10.100712599-014-0334-4

Lattuca, L. R., & Knight, D. (2010). *In the eye of the beholder: Defining and studying interdisciplinarity in engineering education.* Paper presented at the American Society for Engineering Education, Louisville, KY.

Lattuca, L. R., Knight, D. B., & Bergom, I. M. (2012). *Developing a measure of interdisciplinary competence for engineers.* Paper presented at the American Society for Engineering Education.

Lattuca, L. R., Knight, D. B., Ro, H. K., & Novoselich, B. J. (2017). Supporting the development of engineers' interdisciplinary competence. *Journal of Engineering Education, 106*(1), 71–97. doi:10.1002/jee.20155

Lattuca, L. R., Trautvetter, L., Codd, S., & Knight, D. (2011). Working as a team: Enhancing interdisciplinarity for the engineer of 2020. *Proceedings of the 118th Annual Conference of the American Society for Engineering Education Vancouver.*

Lattuca, L. R., Voigt, L. J., & Fath, K. Q. (2004). Does interdisciplinarity promote learning? Theoretical support and researchable questions. *The Review of Higher Education, 28*(1), 23–48. doi:10.1353/rhe.2004.0028

Laurillard, D. (1993). *Rethinking university teaching: A framework for the effective use of educational technology.* London: Routledge.

Law, K. M., Lee, V. C., & Yu, Y. T. (2010). Learning motivation in e-learning facilitated computer programming courses. *Computers & Education, 55*(1), 218–228. doi:10.1016/j.compedu.2010.01.007

Lawson, C., Beer, C., Rossi, D., Moore, T., & Fleming, J. (2016). Identification of 'at risk'students using learning analytics: The ethical dilemmas of intervention strategies in a higher education institution. *Educational Technology Research and Development, 64*(5), 957–968. doi:10.100711423-016-9459-0

Lawson, M. J., & Chinnappan, M. (2015). Knowledge connectedness and the quality of student and teacher mathematical knowledge. In E. A. Silver & P. A. Kenney (Eds.), *More lessons learned from research* (pp. 272–277). Reston, VA: National Council of Teachers of Mathematics.

Lee & Danby. (Eds.). (2012). Reshaping doctoral education: International approaches and pedagogies. New York, NY: Routledge.

Lee, J. J., & Hammer, J. (2011). Gamification in education: What, how, why bother? *Academic Exchange Quarterly, 15*(2).

Lee, J., Bagheri, B., & Kao, H. A. (2015). A Cyber-Physical Systems architecture for Industry 4.0-based manufacturing systems. *Manufacturing Letters, 3*, 18–23. doi:10.1016/j.mfglet.2014.12.001

Leemkuil, H., de Jong, T., de Hoog, R., & Christoph, N. (2003). KM Quest: A Collaborative Internet Based Simulation Game. *Simulation & Gaming, 34*(1), 89–111. doi:10.1177/1046878102250605

Lee, Y., & Rofe, J. S. (2016). Paragogy and Flipped Assessment: Experience of Designing and Running a MOOC on Research Methods. *Open Learning, 31*(2), 116–129. doi:10.1080/02680513.2016.1188690

Lesage, E., Valcke, M., & Sabbe, E. (2013). Scoring Methods for Multiple Choice Assessment in Higher Education – Is it still a Matter of Number Right Scoring or Negative Marking? *Studies in Educational Evaluation, 39*(3), 188–193. doi:10.1016/j.stueduc.2013.07.001

Lesh, R., & Zawojewski, J. (2007). Problem solving and modeling. Second handbook of research on mathematics teaching and learning, 2, 763-804.

Lever-Duffy, J., & McDonald, J. B. (2011). *Teaching and learning with technology* (4th ed.). Boston, MA: Pearson.

Lewis, K. E. (2018). *How to ensure the Fourth Industrial Revolution is "Made in the USA"*. Retrieved from http://theconversation.com/how-to-ensure-the-fourth-industrial-revolution-is-made-in-the-usa-81385

Licorish, S. A., Owen, H. E., Daniel, B., & George, J. L. (2018). Students' perception of Kahoot! influence on teaching and learning. *Research and Practice in Technology Enhanced Learning, 13*(9), 1–23. PMID:30595732

Likitrattanaporn, W. (2017). The development of English language teaching skills for graduate students through the process of learning by doing. *English Language Teaching, 10*(7), 96. doi:10.5539/elt.v10n7p96

Lipman, M. (1991). *Thinking in education*. Cambridge, UK: Cambridge University Press.

Llorente, R., & Morant, M. (2011). Data mining in higher education. In *New fundamental technologies in data mining* (pp. 201-220). Available from: http://www.intechopen.com/books/new-fundamental-technologies-in-data-mining/data-mining-in-higher-education

Lomax (2014). Current Problems in Business Education. *The Journal of Business Education, 13*.

Lothridge, K., Fox, J., & Fynan, E. (2013). Blended learning: Efficient, timely and cost effective. *The Australian Journal of Forensic Sciences, 45*(4), 407–416. doi:10.1080/00450618.2013.767375

Loveder, P. (2017). Australian Apprenticeships: Trends, Challenges and Future Opportunities for Dealing with Industry 4.0. Conference Paper. *National Centre for Vocational Education Research (NCVER).*

Loveless, A., Burton, J., & Turvey, K. (2006). Developing conceptual frameworks for creativity, ICT and teacher education. *Thinking Skills and Creativity*, *1*(1), 3–13. doi:10.1016/j.tsc.2005.07.001

Loy, J. (2012). Creating confidence in an alienating educational environment. In *Proceedings International conference on Engineering and Design education*. Artesis University College.

Lucke, D., Constantinescu, C., & Westkämper, E. (2008). Smart factory-a step towards the next generation of manufacturing. In *Manufacturing systems and technologies for the new frontier* (pp. 115–118). London: Springer. doi:10.1007/978-1-84800-267-8_23

Lueddeke, G. R. (2003). Professionalizing teaching practice in higher education: A study of disciplinary variation and 'teaching scholarship'. *Studies in Higher Education*, *28*(2), 213–228. doi:10.1080/0307507032000058082

Luppicini, R. (2012). *Ethical Impact of Technological Advancements and Applications in Society*. Hershey, PA: Information Science Reference. doi:10.4018/978-1-4666-1773-5

Lykke, M., Coto, M., Jantzen, C., Mora, S., & Vandel, N. (2015). Motivating Students through Positive Learning Experiences: A Comparison of Three Learning Designs for Computer Programming Courses. *Journal of Problem Based Learning in Higher Education*, *3*(2), 80–108.

Lynam, S., & Cachia, M. (2018). Students' perceptions of the role of assessments at higher education. *Assessment & Evaluation in Higher Education*, *43*(2), 223–234. doi:10.1080/02602938.2017.1329928

MacLeod, M. (2016). What makes interdisciplinarity difficult? Some consequences of domain specificity in interdisciplinary practice. *Synthese*, 1–24. doi:10.100711229-016-1236-4

Maguire, P., Maguire, R., Hyland, P., & Marshall, P. (2014). Enhancing collaborative learning using paired-programming: Who benefits? *AISHE-J: The All Ireland Journal of Teaching and Learning in Higher Education*, *6*(2), 1411–14125.

Maki, P. L. (2009). Moving beyond a national habit in the call for accountability. *Peer Review: Emerging Trends and Key Debates in Undergraduate Education*, *11*(1), 13–14.

Makrakis, V. (2005). Training teachers for new roles in the new era: Experiences from the United Arab Emirates ICT program. *Proceedings of the 3rd Pan-Hellenic Conference on Didactics of Informatics.*

Malaysian Digest. (2016, January 19). *Lack Of Self-Initiative, Soft Skills Leads To Unemployed Fresh Graduates.* Retrieved from http://www.malaysiandigest.com/frontpage/29-4-tile/590693-lack-of-self-initiative-soft-skills-leads-to-unemployed-fresh-graduates.html

Malaysian Education Blueprint (2015-2025) Higher Education. (2018). Retrieved November 18, 2008, from http://www.mohe.gov.my

Malik, S. I. (2018). Improvements in introductory programming course: Action research insights and outcomes. *Systemic Practice and Action Research*, 1–20.

Malik, S. I., & Coldwell-Neilson, J. (2016). A model for teaching an introductory programming course using ADRI. *Education and Information Technologies*, 1–32.

Mangan, K. (2013). Inside the flipped classroom. *The Chronicle of Higher Education.* Retrieved from https://www.chronicle.com/article/Inside-the-Flipped-Classroom/141891

Mansilla, V. B., & Duraising, E. D. (2007). Targeted assessment of students' interdisciplinary work: An empirically grounded framework proposed. *The Journal of Higher Education*, *78*(2), 215–237. doi:10.1080/00221546.2007.11780874

Marcano-Cedeño, A., Quintanilla-Domínguez, J., & Andina, D. (2011). WBCD breast cancer database classification applying artificial metaplasticity neural network. *Expert Systems with Applications*, *38*(8), 9573–9579. doi:10.1016/j.eswa.2011.01.167

Margolin, V. (2010). Doctoral Education in Design: Problems and Prospects. *Design Issues*, *26*(3), 70–78. doi:10.1162/DESI_a_00031

Mark, J., Cuoco, A., Goldenberg, E. P., & Sword, S. (2010). Developing Mathematical Habits of Mind. *Mathematics Teaching in the Middle School*, *15*(9), 505–509.

Marks, D. B. (2015). Flipping the classroom: Turning an instructional methods course upside down. *Journal of College Teaching and Learning*, *12*(4), 241–247. doi:10.19030/tlc.v12i4.9461

Marlowe, C. A. (2012). *The Effect of the Flipped Classroom on Student Achievement and Stress* (Unpublished Master of Science Thesis). Montana University.

Marques, G., Garcia, N., & Pombo, N. (2017). A survey on IoT: architectures, elements, applications, QoS, platforms and security concepts. In *Advances in Mobile Cloud Computing and Big Data in the 5G Era* (pp. 115–130). Cham: Springer. doi:10.1007/978-3-319-45145-9_5

Martakis, A., & Daneva, M. (2013). Handling requirements dependencies in Agile projects: A focus group with Agile software development practitioners. *7th International Conference on Research Challenges in IS*, 1-11. 10.1109/RCIS.2013.6577679

Martin, M. H. (2003). *Factors influencing faculty adoption of Web-based courses in teacher education programs within the State University of New York* (Doctoral Dissertation).

Martin, A. J., & Lazendic, G. (2017). Computer-adaptive testing: Implications for students' achievement, motivation, engagement, and subjective test experience. *Journal of Educational Psychology*. doi:10.1037/edu0000205

Martins. (2014). Guia para elaboração de monografia e TCC em engenharia de produção. São Paulo: Atlas.

Matthews, K. E., Garratt, C., & Macdonald, D. (2018). *The higher education landscape: trends and implications*. Academic Press.

Maxwell, C. (2016). *What blended learning is - and isn't*. Retrieved from https://www.blendedlearning.org/what-blended-learning-is-and-isnt/

Maxwell, T. W., & Kupczyk-Romanczuk, G. (2009). Producing the professional doctorate: the portfolio as a legitimate alternative to the dissertation. *Innovations in Education and Teaching International*, *46*(2), 135-145.

Maziah, Y. (2005). *Kajian Kecenderungan Pemilihan Kerjaya di Kalangan Siswazah. Jurnal Pengurusan Awam. Jilid 4 bilangan 1*. Jabatan Perkhidmatan Awam Malaysia.

Mazur, E. (1997). *Peer instruction: A user's manual*. Upper Saddle River, NJ: Prentice Hall.

McAlister, A., & Crappell, C. (2014). Twenty-first century pedagogy: A whole new world again. *Clavier Companion*, *6*(1), 56.

McCallum, W. (2018). *How to give up plastic: A guide to changing the world, one plastic bottle at a time*. London, UK: Penguin Life.

McCarthy, J. F., & boyd, d. (2004). Digital Backchannels in Shared Physical Spaces: Attention, Intention and Contention. *Proc. of the 2004 ACM Conf. on Computer Supported Cooperative Work (CSCW 2004)*, 550-553. 10.1145/1031607.1031700

McDonald, S. D. (2017). Enhanced critical thinking skills through problem-solving games in secondary schools. *Interdisciplinary Journal of e-Skills and Lifelong Learning, 13*, 79-96. Retrieved from http://www.informingscience.org/Publications/3711

McDowell, C., Hanks, B., & Werner, L. (2003). Experimenting with pair programming in the classroom. *SIGCSE Conference on Innovation and Technology in Computer Science Education*, 60-64. 10.1145/961511.961531

McDowell, C., Werner, L., Bullock, H. E., & Fernald, J. (2006). Pair programming improves student retention, confidence, and program quality. *Communications of the ACM*, *49*(8), 90–95. doi:10.1145/1145287.1145293

McGee, P., & Reis, A. (2012). Blended course design: A synthesis of best practices. *Journal of Asynchronous Learning Networks*, *16*(4), 7–22.

McGraw-III, H. (2007). *Shaping tomorrow's business leader: Principles and practices for a model business ethics programme.* Business Roundtable. Institute of Corporate Ethics.

McLaughlin, J. E., Roth, M. T., Glatt, D. M., Gharkholonarehe, N., Davidson, C. A., Griffin, L. T. M., ... Mumper, R. J. (2014). The flipped classroom: A course redesign to foster learning and engagement in a health professions school. *Academic Medicine*, *89*(2), 236–243. doi:10.1097/ACM.0000000000000086 PMID:24270916

McLaughlin, M. W., & Talbert, J. (1993). *Contexts that Matter for Teaching and Learning. Stanford, CA Stanford University.* Stanford, CA: Centre for Research on the Context of Secondary School Teaching.

McLaughlin, M. W., & Talbert, J. E. (2006). *Building school-based teacher learning communities: Professional strategies to improve student achievement.* Chicago: University of Chicago Press.

MCMC. (2008). *Fact & Figures (Statistics & Record).* Retrieved from http://www.skmm.gov.my/facts_figures/stats/index.asp

McMurtry, A., Rohse, S., & Kilgour, K. N. (2016). Socio-material perspectives on interprofessional team and collaborative learning. *Medical Education*, *50*(2), 169–180. doi:10.1111/medu.12833 PMID:26812996

McSporran, M., & King, C. (2005, June). Blended is better: Choosing educational delivery methods. In *EdMedia: World Conference on Educational Media and Technology* (pp. 4932-4939). Association for the Advancement of Computing in Education (AACE).

Means, B., Toyama, Y., Murphy, R., Bakia, M., & Jones, K. (2009). *Evaluation of Evidence-based Practices in Online Learning: A Meta-analysis and Review of Online-learning Studies.* Washington, DC: U.S. Department of Education.

Medina, E. G. L., & Hurtado, C. P. R. (2017). Kahoot! A digital tool for learning vocabulary in a language classroom. *Revista Publicando*, *4*(12), 441–449.

Megginson, D., & Whitaker, V. (2007). *Continuing Professional Development* (2nd ed.). London: Chartered Institute of Personnel and Development.

Mercer, N., & Wegerif, R. (1999). Is exploratory talk productive talk? In K. Littleton & P. Light (Eds.), *Learning with computers: Analysing productive interaction* (pp. 79–101). London: Routledge.

Mertes, L. (1991). Thinking and writing. *Middle School Journal*, *22*(1), 24–25. doi:10.1080/00940771.1991.11496002

MIDF Research. (2018, May 3). *Youth unemployment hit record high in 2017: MIDF Research.* Retrieved from http://www.thesundaily.my/news/2018/05/03/youth-unemployment-hit-record-high-2017-midf-research

Miller, C., Smith, C., & Tilstone, C. (1998). Professional development by Distance Ediucation: Does distance lend enhancement? *Cambridge Journal of Education*, *28*(2), 221–230. doi:10.1080/0305764980280207

Mills, K., & Chandra, V. (2011). Microblogging as a literacy practice for educational communities. *Journal of Adolescent & Adult Literacy*, *55*(1), 35–45. doi:10.1598/JAAL.54.1.4

Mineraud, J., Mazhelis, O., Su, X., & Tarkoma, S. (2016). A gap analysis of Internet-of-Things platforms. *Computer Communications*, *89*, 5–16. doi:10.1016/j.comcom.2016.03.015

Missildine, K., Fountain, R., Summers, L., & Gosselin, K. (2013). Flipping the classroom to improve student performance and satisfaction. *The Journal of Nursing Education*, *52*(10), 597–599. doi:10.3928/01484834-20130919-03 PMID:24044386

MOE. (2013). *Education Blueprint 2013 – 2025 (Preschool to Post-secondary education)*. Putrajaya: MOE.

Mohamad, M. M., Saud, M. S., & Ahmad, A. (2011). The need in training and retraining for TVET teachers in Malaysia. *Journal of Technical Education and Training*, *1*(1), 51–57.

Mohd, F. A. M. R. (2009). *Persepsi Pelajar SPI Terhadap Amalan Kemahiran Insaniah Dalam Pembelajaran Projek Sarjana Muda UTM*. Tidak diterbitkan.

Mokhtar, F. A. (2018). Breaking barriers through Edmodo: A qualitative approach on the perceptions of University of Malaya undergraduates. *Online Learning*, *22*(1), 61–80. doi:10.24059/olj.v22i1.1026

MOOCs. (2002). *Online education for the masses!* Retrieved from http://www.distancelearningportal.com/articles/ 241/moocs-online-education-for-the-masses.htm

Moore, A. J., Gillett, M. R., & Steele, M. D. (2014). Fostering student engagement with the flip. *Mathematics Teacher*, *107*(6), 420–425. doi:10.5951/mathteacher.107.6.0420

Moore, M. (1973). Toward a theory of independent learning and teaching. *The Journal of Higher Education*, *44*(12), 661–679. doi:10.2307/1980599

Moore, M. (1989). Three types of interaction. *American Journal of Distance Education*, *3*(2), 1–6. doi:10.1080/08923648909526659

Moore, M. (1990). Distance education theory. *American Journal of Distance Education*, *5*(3), 10–15.

Moore, M. G., & Kearsley, G. (2011). *Distance education: A systems view of online learning* (3rd ed.). Belmont, CA: Wadsworth.

Moore, M., & Kearsley, G. (1996). *Distance education. A system view*. Belmont, CA: Wadsworth Publishing Company.

Morrar, R., Arman, H., & Mousa, S. (2017). The Fourth Industrial Revolution (Industry 4.0): A Social Innovation Perspective. *Technology Innovation Management Review*, *7*(11), 12–20. doi:10.22215/timreview/1117

Multon, K. D., Brown, S. D., & Lent, R. W. (1991). Relation of self-efficacy beliefs to academic outcomes: A meta-analytic investigation. *Journal of Counseling Psychology*, *38*(1), 30–38. doi:10.1037/0022-0167.38.1.30

Muresan, M. (2013). A Blended Learning System within the Cybergogy Paradigm. *Procedia: Social and Behavioral Sciences*, *89*, 193–198. doi:10.1016/j.sbspro.2013.08.833

Murphy, J., & Irwin, F. (2018). Design thinking: Can it prepare students for industry 4.0? Independence, 43(1), 46.

Murugiah, C. (2015). *Article Review Paper on Soft Skills Development in Malaysia Through Shared-value Partnerships*. Retrieved from https://www.researchgate.net/publication/283297939_Articles_Review_on_Development_Soft_Skills_in_Malaysia

Mustafa, Z. (2018, April 25). Way forward for higher education in 4IR era. *News Straits Times.* Retrieved from https://www.nst.com.my/education/2018/04/361452/way-forward-higher-education-4ir-era

NAE. (2005). *Educating the Engineer of 2020: Adapting Engineering Education to the New Century.* Washington, DC: The National Academies Press.

Nah, F., Telaprolu, V., Rallapalli, S., & Venkata, P. (2013). Gamification of education using computer games. In S. Yamamoto (Ed.), *HCI 2013, Part III* (Vol. 8018, pp. 99–107). LNCS.

Naismith, L., Lonsdale, P., Vavoula, G., & Sharples, M. (2004). *Literature Review in Mobile Technologies and Learning.* Academic Press.

National Association of Colleges and Employers. (2016*). Job Outlook 2016: Attributes Employers Want to See on New College Graduates' Resumes.* Retrieved from http://www.naceweb.org/s11182015/employers-look-for-in-new-hires.aspx

National Education Blueprint. (2006-2010). Putrajaya: Kementerian Pelajaran Malaysia.

National Education Blueprint. (2015-2025). Putrajaya: Kementerian Pelajaran Malaysia.

Nerad, M., & Evans, B. (Eds.). (2014). *Globalization and its impacts on the quality of PhD education: Forces and forms in doctoral education worldwide.* Boston, MA: Sense Publishers/Center for Innovation and Research in Graduate Education. doi:10.1007/978-94-6209-569-4

Nerland, M., & Jensen, K. (2012). Epistemic practices and object relations in professional work. *Journal of Education and Work, 25*(1), 101–120. doi:10.1080/13639080.2012.644909

Nerland, M., & Jensen, K. (2014a). Changing cultures of knowledge and professional learning. In S. Billett, C. Harteis, & H. Gruber (Eds.), *International Handbook of Research in Professional and Practice-based Learning* (pp. 611–640). London: Springer. doi:10.1007/978-94-017-8902-8_23

Nerland, M., & Jensen, K. (2014b). Learning through epistemic practices in professional work. In T. Fenwick & M. Nerland (Eds.), *Reconceptualising Professional Learning: Sociomaterial Knowledges, Practices and Responsibilities* (pp. 25–37). London: Routledge.

Nersessian, N. J. (2009). How Do Engineering Scientists Think? Model-Based Simulation in Biomedical Engineering Research Laboratories. *Topics in Cognitive Science, 1*(4), 730–757. doi:10.1111/j.1756-8765.2009.01032.x PMID:25163455

Nersessian, N. J., & Newstetter, W. C. (2014). Interdisciplinarity in engineering research and learning. In A. Johri & B. M. Olds (Eds.), *Cambridge Handbook of Engineering Education Research* (pp. 713–730). Cambridge, UK: Cambridge University Press.

Neumann, R. (2005). Doctoral Differences: Professional doctorates and PhDs compared. *Journal of Higher Education Policy and Management, 27*(2), 173–188. doi:10.1080/13600800500120027

New England Association of Schools and Colleges, Commission on Institutions of Higher Education. (2018). *Standard Four: The Academic Program.* Retrieved September 28, 2018, from: https://cihe.neasc.org/standards-policies/standards-accreditation/standards-effective-july-1-2016#standard_four

Newell, W. H., & Gagnon, P. (2013). The State of the Field: Interdisciplinary Theory. *Issues in Interdisciplinary Studies,* (31), 21.

Newell, W. H., & Klein, J. T. (1998). *Interdisciplinarity: Essays from the Literature.* College Board.

Newell, W. H., Wentworth, J., & Sebberson, D. (2001). A theory of interdisciplinary studies. *Issues in Interdisciplinary Studies, 19,* 1–25.

Newman, D. R., Webb, B., & Cochrane, C. (1995). A content analysis method to measure critical thinking in face-to-face and computer supported group learning. *Interpersonal Computing and Technology*, *3*(2), 56–77.

Newstetter, W. C. (2011). *Unpacking the interdisciplinary mind: Implications for teaching and learning The challenge.* Paper presented at the American Society for Engineering Education, Vancouver, Canada.

Ng, P. Y., Voges, K., & Goi, C. L. (2010). The importance of mobile phone applications to young consumers: An exploratory study in Malaysia. Proceedings of Australian & New Zealand Marketing Academy.

Ng, P.Y., Goi, C.L., & Voges, K. (2010). *The importance of mobile phone applications to young consumers: an exploratory study in Malaysia.* Retrieved from http://anzmac2010.org/proceedings/ pdf/anzmac10Final00103.pdf

Nicholson, S. (2015). *A recipe for meaningful gamification. In Gamification in Education and Business* (pp. 1–20). Berlin, Germany: Springer.

Niemeyer, C. M. (2017). Engineering for life sciences: A fruitful collaboration enabled by chemistry. *Angewandte Chemie International Edition*, *56*(8), 1934–1935. doi:10.1002/anie.201611865 PMID:28105666

Niemiec, M., & Otte, G. (2009). An Administrator's Guide to the Whys and Hows of Blended Learning. *Journal of Asynchronous Learning Networks*, *13*(1), 19–30.

Nilson, L. B. (2010). *Teaching at its best: A research-based resource for college instructors* (2nd ed.). San Francisco: Jossey-Bass.

Ning, H., & Hu, S. (2012). Technology classification, industry, and education for Future Internet of Things. *International Journal of Communication Systems*, *25*(9), 1230–1241. doi:10.1002/dac.2373

Nixon, J., Marks, A., Rowland, S., & Walker, M. (2001). 'Towards a New Academic Professionalism: A manifesto of hope'. *British Journal of Sociology of Education*, *22*(2), 227–244. doi:10.1080/01425690124202

Noble, K. A. (1994). *Changing doctoral degrees: An international perspective.* Buckingham, UK: Open University Press.

Norris, S., & Ennis, R. (1989). What is critical thinking. *The practitioner's guide to teaching thinking series: Evaluating critical thinking*, 1-26.

Novak, J. (2018). Self-directed learning in the age of open source, open hardware and 3D printing. In E. Ossiannilsson (Ed.), *Ubiquitous inclusive learning in a digital era* (pp. 154–178). Hershey, PA: IGI Global.

Novak, J., & Loy, J. (2017). Digital technologies and 4D customized design: Challenging conventions with responsive design. In V. C. Bryan, A. T. Musgrove, & J. R. Powers (Eds.), *Handbook of research on human development in the digital age* (pp. 403–426). Hershey, PA: IGI Global.

Ntiko, A. J., & Brookhart, S. M. (2007). *Educational assessment of students* (5th ed.). Upper Saddle River, NJ: Pearson.

Nuutila, E., Torma, S., & Malmi, L. (2005). PBL and computer programming—The seven steps method with adaptations. *Computer Science Education*, *15*(2), 123–142. doi:10.1080/08993400500150788

Oravec, J. A. (2003). Blending by blogging: Weblogs in blended learning initiatives. *Journal of Educational Media*, *28*(2-3), 225–233. doi:10.1080/1358165032000165671

Oriwoh, E., & Conrad, M. (2015). 'Things' in the Internet of Things: Towards a definition. *International Journal of Internet of Things*, *4*(1), 1–5.

Ory, C. J. (2009). *Improving Your Test Questions, Evaluation and Examination Service.* University of Iowa.

Osakwe, C., Keavey, K., Uzoka, F. M., Fedoruk, A., & Osuji, J. (2015). The relative importance of academic activities: Autonomous values from the Canadian professoriate. *Canadian Journal of Higher Education*, *45*(2), 1–22.

Pacansky-Brock, M. (2013). *Best practices for teaching with emerging technologies*. New York: Routledge.

Packer, M. J., & Goicoechea, J. (2000). Sociocultural and constructivist theories of learning: Ontology not just epistemology. *Educational Psychologist*, *35*(4), 227–242. doi:10.1207/S15326985EP3504_02

Pádua, . (2016). *Inovação é tudo*. São Paulo: Atlas.

Page, B. (2010). 12 things teachers must know about learning. *Education Digest*, *75*(8), 54–56.

Paliwal, M., & Kumar, U. A. (2009). Neural networks and statistical techniques: A review of applications. *Expert Systems with Applications*, *36*(1), 2–17. doi:10.1016/j.eswa.2007.10.005

Palmer, E. J., & Devitt, P. G. (2007). Assessment of higher order cognitive skills in undergraduate education: Modified essay or multiple choice questions? Research paper. *BMC Medical Education*, *7*(49), 1–7. doi:10.1186/1472-6920-7-49 PMID:18045500

Pandiyan, V. (2017, Sept 6). Industry 4.0: The future is here. *The Star*. Retrieved from https://thestar.com.my

Papanek, V. (2005). *Design for the real world: Human ecology and social change* (revised ed.). Chicago: Chicago Review Press.

Pape, L., Sheehan, T., & Worrell, C. (2012). How to do more lessons from online learning? *Learning and Leading with Technology*, *39*(6), 18–22.

Pappano, L. (2012). The Year of the MOOC. *The New York Times*, *2*(12).

Pappas, C. (2017). What is a learning management system? LMS basic functions and features you must know. *eLearning Industry*. Retrieved from https://elearningindustry.com/what-is-an-lms-learning-management-system-basic-functions-features

Pappas, J. P., & Jerman, J. (Eds.). (2011). *Meeting adult learner needs through the nontraditional doctoral degree*. San Francisco, CA: Jossey-Bass.

Park, C. (2005). New Variant PhD: The changing nature of the doctorate in the UK. *Journal of Higher Education Policy and Management*, *27*(2), 189–207. doi:10.1080/13600800500120068

Park, Y., Yu, J. H., & Jo, I. H. (2016). Clustering blended learning courses by online behavior data: A case study in a Korean higher education institute. *The Internet and Higher Education*, *29*, 1–11. doi:10.1016/j.iheduc.2015.11.001

Paulsen, M. P. (1995). Moderating educational computer conferences. In Z. L. Berge & M. P. Collins (Eds.), *Computer-mediated communication and the on-line classroom in distance education* (pp. 113–258). Cresskill, NJ: Hampton Press.

Peffers, K., & Bloom, S. (1999). Internet-based innovations for teaching IS courses: The state of adoption, 1998-2000. *Journal of Information Technology Theory and Application*, *1*(1), 10–41.

Pelgrum, W. J. (2001). Obstacles to the integration of ICT in education : Results from a worldwide educational assessment. *Ournal of Science Education and Technology*, *37*(2), 163–178.

Perasso. (2018). O que é a 4ª revolução industrial - e como ela deve afetar nossas vidas. *BBC News Brasil*. Disponível em: https://www.bbc.com/portuguese/geral-37658309

Perkins, C., & Murphy, E. (2006). Identifying and measuring individual engagement in critical thinking in online discussions: An exploratory case study. *Journal of Educational Technology & Society*, *9*(1).

Peters, O. (1983). Distance teaching and industrial production: A comparative interpretation in outline. In D. Sewart, D. Keegan, & B. Holmberg (Eds.), *Distance education: International perspectives*. London: Croom Helm.

Petko, D., Egger, N., Cantieni, A., & Wesoi, B. (2015). Digital media adoption in schools: Bottom-up, top-down, complementary or optional? *Computers & Education*, *84*(May), 46–61.

Phillips, A. G. (2002). John Dewey and His Religious Critics. *Religious Education (Chicago, Ill.)*, *29*(1), 31–48.

Phillips, D. K., & Carr, K. (2014). *Becoming a teacher through action research: Process, context, and self-study* (3rd ed.). New York: Routledge. doi:10.4324/9781315867496

Pierce, R., Stacey, K., & Wander, R. (2010). Examining the didactic contract when handheld technology is permitted in the mathematics classroom. *ZDM*, *42*(7), 683–695. doi:10.100711858-010-0271-8

Pinheiro. (2017). Indústria 4.0: uma análise conceitual do tema, seus impactos na economia e a colocação do profissional neste novo cenário. In XXXVII Encontro Nacional de Engenharia de Produção, 2017. Joinville/SC: Enegep.

Pintrich, P. R., Smith, D. A. F., Garcia, T., & McKeachie, W. J. (1991). *A manual for the use of the Motivated Strategies for Learning Questionnaire (MSLQ)* (Tech. Report No. 91-B-004). Boards of Regents, University of Michigan.

Piper, T. (2010). *What policy changes do experts recommend K-12 instructional leaders enact to support the implementation of online instruction and learning?* (Doctoral Dissertation). Available from ProQuest Dissertations and Theses database. (UMI No. 3430711)

Plajner & Vomlel. (2016). Student Skill Models in Adaptive Testing. *JMLR: Workshop and Conference Proceedings*, *52*, 403-414. Retrieved from http://proceedings.mlr.press/v52/plajner16.pdf

Plump, C. M., & LaRosa, J. (2017). Using Kahoot! in the classroom to create engagement and active learning: A game-based technology solution for e-learning novices. *Management Teaching Review*, *2*(2), 151–158. doi:10.1177/2379298116689783

Pombo, L., & Moreira, A. (2012). Evaluation framework for blended learning courses: A puzzle piece for the evaluation process. *Contemporary Educational Technology*, *3*(3), 201–211.

Ponti, M. (2014). Self-directed learning and guidance in non-formal open courses. *Learning, Media and Technology*, *39*(2), 154–168. doi:10.1080/17439884.2013.799073

Poon, J. (2013). Blended learning: An institutional approach for enhancing students' learning experiences. *Journal of Online Learning and Teaching*, *9*(2). Available at http://jolt.merlot.org/vol9no2/poon_0613.htm

Poore, R. (2013). Flipping the classroom. *Focus on Boise State*, *34*(1), 21-22. Retrieved from http://news.boisestate.edu/update/files /2013/08/BSU_FOCUS_FALL2013_web.pdf

Porter, J. E. (2014). Framing questions about MOOCs and writing courses. *Invasion of the MOOCs*, 14. Retrieved from http://www.parlorpress.com/pdf/invasion_of_the_moocs.pdf

Powell, A. (2011). *A case study of E-Learning initiatives in New Zealand's secondary schools*. Pepperdine University.

Preece, S. J., Goulermas, J. Y., Kenney, L. P., Howard, D., Meijer, K., & Crompton, R. (2009). Activity identification using body-mounted sensors—A review of classification techniques. *Physiological Measurement*, *30*(4), R1–R33. doi:10.1088/0967-3334/30/4/R01 PMID:19342767

Prensky, M. (2001). *Digital natives, digital immigrants*. Available at: www.marcprensky.com/writing

Prensky, M. (2001). Digital natives, digital immigrants part 1. *On the Horizon*, *9*(5), 1–6. doi:10.1108/10748120110424816

Preston, D. (2005). Pair programming as a model of collaborative learning: A review of the research. *Journal of Computing Sciences in Colleges*, *20*(4), 39–45.

Pritchard, A. (2015). *Öğrenme yolları, Sınıfta öğrenme teorileri ve öğrenme stilleri* [Learning pathways, learning theories in classroom and learning styles] (M. Çevikbaş & S. Çevikbaş, Trans. & Eds.). Ankara: Nobel Akademik Yayıncılık.

Prodoehl, D. (2015). Flipping First-Year English: Strengthening Teacher-Student Conferencing through Online Modules. In A. Abigail (Eds.), Implementation and Critical Assessment of the Flipped Classroom Experience (pp. 1-24). Hershey, PA: Information Science.

Purnawarman, P., Susilawati, S., & Sundayana, W. (2016). The use of edmodo in teaching writing in a blended learning setting. *Indonesian Journal of Applied Linguistics*, *5*(2), 242–252. doi:10.17509/ijal.v5i2.1348

Puteh, M., & Hussin, S. (2007). *The role of Malaysian government it policies & e-learning in the tertiary sector.* Paper presented at the National E-Community Seminar on Bridging the Digital Divide: Malaysia's Initiative, Kuala Lumpur, Malaysia.

Qi, A. Q., & Shen, Y. J. (2011). The Application of Internet of Things in Teaching Management System. In *Information Technology, Computer Engineering and Management Sciences (ICM), 2011 International Conference on* (Vol. 2, pp. 239-241). IEEE. 10.1109/ICM.2011.238

Quigley, A. (2018). *What is an LMS?* Retrieved from https://www.linkedin.com/learning/learning-management-systems-lms-quick-start/what-is-an-lms?autoplay=true&trk=course_tocItem&upsellOrderOrigin=trk_default_learning

Quinn, C. (2000, Fall). *mLearning: mobile, wireless, in-your-pocket learning.* Retrieved from http://linezine.com/2.1/features/cqmmwiyp.htm

Quinn, C. (2002). *MLearning: Mobile, Wireless, in Your Pocket Learning.* Academic Press.

Ramsden, P. (2003). *Learning to teach in higher education.* London: Routledge Falmer. doi:10.4324/9780203507711

Ranai, M. (2017). *World Bank praises Malaysia's educational reforms.* Retrieved from http://news.moe.gov.my/2017/10/25/malaysias-education-reforms-get-thumbs-world-bank/

Rani, M. (2018). *MANDATE: Embracing Industry 4.0.* Retrieved from http://news.moe.gov.my/2018/01/27/2018-mandate-embracing-industry-4-0/

Rathakrishnan, Ahmad, & Choi. (2017). Padlet Online Discussion in Enhancing Students' macro Critical Thinking Skills. *International Journal of Latest Research in Humanities and Social Science, 1*(3), 121-126.

Rathakrishnan, M., & Pbwiki, S. (2016). On-Line Learning: The Effectiveness of Pedagogical Role Online Teachers' On Students with Difference Learning Style In Their Critical Thinking. *Proceeding of the First International Conference on ICT for Transformation*, 98-101.

Ravenscroft, A. (2011). Dialogue and connectivism: A new approach to understanding and promoting dialogue-rich networked learning. *International Review of Research in Open and Distance Learning*, *12*(3), 3. doi:10.19173/irrodl.v12i3.934

Ravitch, S. M., & Lytle, S. (2016). Becoming practitioner-scholars: The role of practice-based inquiry dissertations in the development of educational leaders. In *Contemporary approaches to dissertation development and research methods.* Hershey, PA: IGI Global.

Reich, A., Rooney, D., Gardner, A., Willey, K., Boud, D., & Fitzgerald, T. (2015). Engineers' professional learning: A practice-theory perspective. *European Journal of Engineering Education, 40*(4), 366–379. doi:10.1080/03043797.2014.967181

Reiser, R. A. (2012). Getting an instructional design position: Lessons from a personal history. In R. A. Reiser & J. V. Dempsey (Eds.), *Trends and issues in instructional design and technology* (3rd ed.; pp. 256–262). Boston, MA: Pearson Education, Inc.

Reiser, R. A., & Dempsey, J. V. (2007). *Trends and issues i0n instructional design and technology* (2nd ed.). Pearson Education, Inc.

Reiser, R. A., & Dempsey, J. V. (Eds.). (2018). *Trends and issues in instructional design and technology* (4th ed.). Boston, MA: Pearson Education, Inc.

Renard, L. (2016, Dec 23). Top-8 online learning platforms [Web log post]. Retrieved from https://www.bookwidgets.com/blog/2016/12/top-8-online-learning-platforms

Repko, A. F. (2008). Assessing interdisciplinary learning outcomes. *Academic Exchange Quarterly, 12*(3), 171.

Riasati, J. M., Alllahyar, N., & Tan, K. A. (2012). Technology in Language Education: Benefits and Barriers. *Journal of Education and Practice, 3*(5), 25–31.

Rice, C., & Matthews, L. (2012). Designing (in) the PhD in architecture. In A. Lee & S. Danby (Eds.), *Reshaping doctoral education: International approaches and pedagogies* (pp. 99–112). New York, NY: Routledge.

Richards, J. (2008). Second language teacher education today. *RELC Journal: A Journal of Language and Teaching Research, 39*(2), 158-177.

Richards, J., & Rodgers, T. (1986). *Approaches and Methods in Language Teaching*. Cambridge, UK: Cambridge University Press.

Richter, D. M., & Paretti, M. C. (2009). Identifying barriers to and outcomes of interdisciplinarity in the engineering classroom. *European Journal of Engineering Education, 34*(1), 29–45. doi:10.1080/03043790802710185

Ridzuan. (2011). *Sedekad Akademi Tentera Malaysia (1995-2005): Dari Kontena ke Kelas Pintar*. Kuala Lumpur: Penerbit UPNM.

Rieber, L. (2018). *The Proper Way to Become an Instructional Technologist. In R. E. West, Foundations of Learning and Instructional Design Technology: The Past, Present, and Future of Learning and Instructional Design Technology*. Retrieved from http://edtechbooks.org/lidtfoundations/proper_way

Roblyer, M. D., & Hughes, J. E. (2018). *Integrating educational technology into teaching: Transforming learning across disciplines* (8th ed.). Pearson.

Rodriguez, C. O. (2012). MOOCs and the AI-Stanford like courses: Two successful and distinct course formats for massive open online courses. *European Journal of Open, Distance and E-Learning.*

Rogers, E. M. (2003). *Diffusion of innovations*. New York: Free Press.

Rogerson, C., & Scott, E. (2010). The fear factor: How it affects students learning to program in a tertiary environment. *Journal of Information Technology Education: Research, 9*, 147–171. doi:10.28945/1183

Rominger, R. (2018). Building bridges and growing wings through doctoral education: A literary meta-analysis. *International Journal of Higher Education, 7*(3), 171–182. doi:10.5430/ijhe.v7n3p171

Romiszowski, A. J. (2013). Topics for debate: What's really new about MOOCs? *Educational Technology*, *53*(4), 48–51.

Roscoe, J. (2002). Continuing professional development in higher education. *Human Resource Development International*, *5*(1), 3–9. doi:10.1080/13678860110076006

Rosling, H., Rosling, O., & Rosling Ronnlend, A. (2018). *Factfulness: Ten reasons why we're wrong about the world and why things are better than you think*. London, UK: Sceptre.

Rossett, A., Douglis, F., & Frazee, R. V. (2003). Strategies for building blended learning. *Learning Circuits*, *4*(7), 1-8.

Rothwell, A., & Arnold, J. (2005). How HR professionals rate "continuing professional development. *Human Resource Management Journal*, *15*(3), 18–32. doi:10.1111/j.1748-8583.2005.tb00151.x

Rovai, A. P., & Jordan, H. (2004). Blended learning and sense of community: A comparative analysis with traditional and fully online graduate courses. *The International Review of Research in Open and Distributed Learning*, *5*(2). doi:10.19173/irrodl.v5i2.192

Rytivaara, A. (2012). Collaborative classroom management in a co-taught primary school classroom. *International Journal of Educational Research*, *53*, 182–191. doi:10.1016/j.ijer.2012.03.008

Saavedra, A. R., & Opfer, V. D. (2012). *Teaching and learning 21st century skills: Lessons from the learning sciences*. Presented at the Joint AARE/APERA Conference, Sydney, Australia. Retrieved from https://www.aare.edu.au/data/publications/2012/ Saavedra12.pdf

Sahoo, G. (2013). Study of parametric performance evaluation of machine learning and statistical classifiers. *Int. J. Inf. Technol. Comput. Sci.*, *5*, 57–64.

Salleh, N., Mendes, E., & Grundy, J. (2011). Empirical studies of pair programming for CS/SE teaching in higher education: A systematic literature review. *IEEE Transactions on Software Engineering*, *37*(4), 509–525. doi:10.1109/TSE.2010.59

Sampson, R. (2012). The language-learning self, self-enhancement activities, and self-perceptual image. *Language Teaching Research*, *16*(3), 317–335. doi:10.1177/1362168812436898

Sanders, K. A. (1997). *Identifying research strategies for the future: Alternatives to the traditional doctoral dissertation* (Order No. 3006632). Available from ProQuest Dissertations & Theses Global. (304390109). Retrieved from https://search-proquest-com.contentproxy.phoenix.edu/docview/304390109?accountid=35812

Sani, R. (2018, January 17). Embracing Industry 4.0. *News Straits Times*. Retrieved from https://www.nst.com.my/education/2018/01/325914/embracing-industry-40

Santana-Mancilla, P. C., Echeverría, M. A. M., Santos, J. C. R., Castellanos, J. A. N., & Díaz, A. P. S. (2013). Towards smart education: Ambient intelligence in the mexican classrooms. *Procedia: Social and Behavioral Sciences*, *106*, 3141–3148. doi:10.1016/j.sbspro.2013.12.363

Santos, C., Almeida, S., Pedro, L., Aresta, M., & Koch-Grunberg, T. (2013). Students' Perspectives on Badges in Educational Social Media Platforms: The Case of SAPO Campus Tutorial Badges. *Proceedings of the IEEE 13th International Conference on Advanced Learning Technologies*, 351–353. 10.1109/ICALT.2013.108

Sarker, F., Davis, H., & Tiropanis, T. (2010). *A review of higher education challenges and data infrastructure responses*. Retrieved from http://eprints.soton.ac.uk/271695/1/A_Review_of_HE_challenges_and_data_infrastructure_responses.pdf

Sarle, W. S. (1994). Neural networks and statistical models. *Proceedings of the 19th annual SAS users group international conference*.

Saroyan, A., & dan Snell, L.D. (1997). Professional development by Distance Education: Does distance lend enhancement? *Cambridge Journal of Education*, *28*(2).

Sarros, J. C., Willis, R. J., & Hardie, T. (2004). The DBA in Australia and the Asia Pacific: Opportunities and challenges. *International Journal of Organisational Behaviour*, *7*(8), 440–455.

Sarwono, J. (2012). *Path Analysis Dengan SPSS*. Jakarta: PT Alex Media Computindo.

Savery, J. R., & Duffy, T. M. (1996). Problem based learning: An instructional model and its constructivist framework. In B. Wilson (Ed.), *Constructivist learning environments: Case studies in instructional design* (pp. 135–148). Englewood Cliffs, NJ: Educational Technology Publications.

Savill-Smith, C., & Kent, P. (2003). *The Use of Palmtop Computers for Learning: A Review of The Literature*. London, UK: Learning and Skill Development Agency.

Scannell, E. E., & Donaldson, L. (2000). *Human Resource Development*. New York: Perseus Publishing.

Scherer. (2016). *Gestão da inovação na prática: como aplicar conceitos e ferramentas para alavancar a inovação. 2.* São Paulo: Atlas.

Schneider, C. G. (2008). *Liberal education and high-impact practices: Making excellence—for once and for all—inclusive. In Forward to High impact educational practices: What they are, who has access to them, and why they matter.* Washington, DC: Association of American Colleges and Universities.

Schoenfeld, A. H. (1992). *Learning to think mathematically: Problem solving, metacognition, and sense making in mathematics*. Information Age Publishing Inc.

Schwab, K. (2017). *The fourth industrial Revolution*. New York: Crown Business.

Schwartz, L., Clark, S., Cossarin, M., & Rudolph, J. (2004). Educational wikis: Features and selection criteria. *The International Journal of Research in Open and Distance Learning*, *5*(1), 162–691.

Seale, J. (2010). Doing student voice work in higher education: An exploration of the value of participatory methods. *British Educational Research Journal*, *36*(6), 995–1015. doi:10.1080/01411920903342038

See, S., & Conry, J. M. (2014). Flip my class! A faculty development demonstration of a flipped-classroom. *Currents in Pharmacy Teaching and Learning*, *6*(4), 585–588. doi:10.1016/j.cptl.2014.03.003

Seldon, A., & Abidoye, O. (2018). *The Fourth education revolution: Will artificial intelligence liberate or infantilise humanity*. Buckingham, UK: University of Buckingham Press.

Selly, W. N. R. (2013). *Study Komparasi Hasil Belajar Pendidikan Kewarganegaraan Antara Metode Konvensional Dengan Metode Diskusi Kelas Teknik Buzz Groups Pada Siswa Kelas Viii Smp N 2 Kalasan* (Master's Thesis). Universitas Negeri Yogyakarta.

Selwyn, N. (2009). The digital native–myth and reality. *Aslib Proceedings*.

Seow, J. (2010, March). Marketing excellence via education and training. *Marketing Magazine*, 10-17.

Serrano-Cámara, L. M., Paredes-Velasco, M., Alcover, C. M., & Velazquez-Iturbide, J. Á. (2014). An evaluation of students' motivation in computer-supported collaborative learning of programming concepts. *Computers in Human Behavior*, *31*, 499–508. doi:10.1016/j.chb.2013.04.030

Shahiri, A. M., Husain, W., & Rashid, N. A. (2015). A review on predicting student's performance using data mining techniques. *Procedia Computer Science*, *72*, 414–422. doi:10.1016/j.procs.2015.12.157

Shamsutdinova, E. K., Khakimzyanova, D. F., & Melnikova, O. K. (2017). Implementing interactive methods in teaching General English for future programmers. *International Journal of Scientific Study*, *5*(5), 362–366.

Sharpe, R., Benfield, G., Roberts, G., & Francis, R. (2006). *The undergraduate experience of blended e-learning: A review of UK literature and practice*. The Higher Education Academy.

Sharples, M. (2000). The Design of Personel Mobile Technologies for Lifelong Learning. *Computers & Education*, *34*(3-4), 177–193. doi:10.1016/S0360-1315(99)00044-5

Sharples, M., Corlett, D., & Westmancott, O. (2002). The design and implementation of a mobile learning resource. *Journal of Personal and Ubiquitous Computing*, *6*(3), 220–234. doi:10.1007007790200021

Shashko, A., & Koteska, G. (2017). The virtual classroom Edmodo as a teaching tool as part of the flipped classroom method in the English language classroom. *TEACHER International Journal of Education.* 54-58.

Shea, P. (2007). Bridges and barriers to teaching online college courses: A study of experienced online faculty in thirty-six colleges. *Journal of Asynchronous Learning Networks*, *11*(2), 73–128.

Shen, Y. W., Reynolds, T. H., Bonk, C. J., & Brush, T. A. (2013). A Case Study of Applying Blended Learning in an Accelerated Post-Baccalaureate Teacher Education Program. *Journal of Educational Technology Development and Exchange*, *6*(1), 6. doi:10.18785/jetde.0601.05

Shuhidan, S. M. (2012). *Probing the minds of novice programmers through guided learning (PhD thesis).* RMIT University. retrieved from https://researchbank.rmit.edu.au/eserv/rmit:160247/MohamedShuhidan.pdf

Shulman, L. (1987). Knowledge and teaching: Foundations of the new reform. *Harvard Educational Reform*, *57*(1), 1–22. doi:10.17763/haer.57.1.j463w79r56455411

Shulman, L. S., Golde, C. M., Conklin, A., Bueschel, A. C., & Garabedian, K. J. (2006). Reclaiming education's doctorates: A critique and a proposal. *Educational Researcher*, *35*(3), 25–32. doi:10.3102/0013189X035003025

Siemens, G., & Downes, S. (2009). *Connectivism & Connective Knowledge*. Retrieved from http://ltc. umanitoba. ca/connectivism/

Siemens, G. (2005). Connectivism: A learning theory for the digital age. *International Journal of Instructional Technology and Distance Learning*, *2*(1), 3–10.

Siew-Eng, L., Ariffin, S. R. B., & Rahman, S. B. (2010). Diversity in education using blended learning in Sarawak. *US-China Education Review*, *7*(2), 83–88.

Simonson, M., Smaldino, S., Albright, M., & Zvacek, S. (2012). *Teaching and learning at a distance: Foundations of distance education* (5th ed.). Boston, MA: Pearson.

Singh, H., & Reed, C. (2001). *A White Paper: Achieving Success with Blended Learning*. Retrieved July 15,2015 from www.leerbeleving.nl/wbts/wbt2014

Sivakumar, D., Jusman, M. F. B., & Mastan, A. N. B. M. (n.d.). *A Case Study Review: Future of Internet of Things (IoT) in Malaysia*. Academic Press.

Skelton, A. (2005). *Understanding Teaching Excellence in Higher Education. Towards a Critical Approach*. Abingdon, UK: Routledge. doi:10.4324/9780203412947

Smaldino, S. E., Donaldson, J. A., & Herring, M. (2012). Professional ethics: Rules applied to practice. In R. A. Reiser & J. V. Dempsey (Eds.), *Trends and issues in instructional design and technology* (3rd ed.; pp. 342–347). Boston, MA: Pearson Education, Inc.

Smith, J. D. (2013). Student attitudes toward flipping the general chemistry classroom. *Chemistry Education Research*. Available from https://pubs.rsc.org/en/content/article landing/2013/rp/c3rp00083d#!divAbstract

Smith, P. L., & Ragan, T. J. (2005). Instructional Design (3rd ed.). Hoboken, NJ: John Wiley Sons, Inc.

Smith, J. D. (2013). Student attitudes toward flipping the general chemistry classroom. *Chemistry Education Research and Practice*, *14*(4), 607–614. doi:10.1039/C3RP00083D

Smith, K. (2008). "Who do you think you're talking to?" – the discourse of learning and teaching strategies. *Higher Education*, *56*(4), 395–406. doi:10.100710734-007-9100-6

Smith, S. (2016). *Manifesto for the humanities: Transforming doctoral education in good enough times*. Ann Arbor, MI: University of Michigan Press.

Soares, A., Fonseca, F., & Martin, N. L. (2015). Teaching introductory programming with game design and problem-based learning. *Issues in Information Systems*, *16*(3), 128–137.

Somr, M., & Hruskova, L. (2014). Herbart's Philosophy of Pedagogy and Educational Teaching (The Views and Differences of Opinion). *Studia Edukacyjne*, *33*, 413–429. doi:10.14746e.2014.33.21

Sparke, P. (2013). *An introduction to design and culture: 1900 to the present* (3rd ed.). London, UK: Routledge. doi:10.4324/9780203129999

Speck, M. (1996). Best Practice in Professional Development for Sustained Educational Change. *ERS Spectrum*, 33–41.

Spiro, R. J., Feltovich, P. J., Jacobson, M. J., & Coulson, R. L. (1991). *Cognitive Flexibility, Constructivism and Hypertext*. Retrieved July 6, 2018, from http://www.ilt.columbia.edu/ilt/papers/Spiro et al.html

Stacey, G. (2012). Current developments in cell culture technology. In M. Balls, R. Combes, & N. Bhogal (Eds.), *New Technologies for Toxicity Testing* (pp. 1–13). Springer. doi:10.1007/978-1-4614-3055-1_1

Stahl, G. (2007). *Meaning making in CSCL: Conditions and preconditions for cognitive processes by groups*. Paper presented at the international conference on Computer Support for Collaborative Learning (CSCL 2007), New Brunswick, NJ. 10.3115/1599600.1599723

Stakyer, H., & Horn, M. B. (2012). *Classifying K–12 blended learning*. Mountain View, CA: Innosight Institute, Inc.

Stein, J., & Graham, C. R. (2014). *Essentials for blended learning: A standards-based guide*. Routledge. doi:10.4324/9780203075258

Stentoft, D. (2017). From saying to doing interdisciplinary learning: Is problem-based learning the answer? *Active Learning in Higher Education*, *18*(1), 51–61. doi:10.1177/1469787417693510

Sterling, S. (2012). *The flipped classroom: A pedagogy for differentiating instruction and teaching essential skills*. Retrieved from http://www.scilearn.com/blog/flipped-classroom-pedagogy-differentiating-instruction

Sterling, S. (2015). *Sustainable Education: Revisioning learning and change (Schumacher Briefings)*. Cambridge, UK: Green Books UIT.

Stewart, D. W., & Shamdasani, P. N. (2015). *Focus groups - Theory and practice. Applied Social Science Research Series* (3rd ed.). Thousand Oaks, CA: Sage Publications, Inc.

Stewart, M. (2001). Towards a global definition of patient centred care. *British Medical Journal*, *322*(7284), 444–445. doi:10.1136/bmj.322.7284.444 PMID:11222407

Storey, V. A. (Ed.). (2016). *International perspectives on designing professional practice doctorates*. New York, NY: Palgrave Macmillan. doi:10.1057/9781137527066

Storr, R. J. (1953). *The beginnings of graduate education in America*. Chicago, IL: The University of Chicago Press.

Strayer, J. F. (2007). *The effects of the classroom flip on the learning environment: A comparison of learning activity in a traditional classroom and a flip classroom that used an intelligent tutoring system* (Unpublished doctoral dissertation). Ohio State University.

Strayer, J. (2012). How learning in an inverted classroom influences cooperation, innovation and task orientation. *Learning Environments Research*, *15*(2), 171–193. doi:10.100710984-012-9108-4

Strayer, J. F. (2009). *Inverting the classroom: A study of the learning environment when an intelligent tutoring system is used to help students learn*. Saarbrücken, Germany: VDM.

Study Malaysia. (2018, May 2). *The Fourth Industrial Revolution (IR 4.0) and what it means for students like you*. Retrieved from https://www.studymalaysia.com/education/top-stories/the-fourth-industrial-revolution-ir-4.0-and-what-it-means-for-students-like-you

Subedi, B. S. (2015). Assessing the effectiveness of teacher training programs to improve the quality of school education in Nepal. *Journal of Training and Development*, *1*(1), 9–14. doi:10.3126/jtd.v1i0.13084

Sweller, J., Ayres, P. L., & Kalyuga, S. (2011). *Cognitive load theory*. New York: Springer. doi:10.1007/978-1-4419-8126-4

Symons, D., & Pierce, R. (2015). Mathematical Language Development and Talk Types in Computer Supported Collaborative Learning Environments. In *The 38th Annual Conference of the Mathematics Education Research Group of Australasia* (pp. 579-586). MERGA.

Tajudin, N. M., & Chinnappan, M. (2016). The link between higher-order thinking skills, representation and concepts in enhancing TIMSS tasks. *International Journal of Instruction*, *9*(2), 199–214. doi:10.12973/iji.2016.9214a

Talbert, J. (2010). Professional learning communities at the crossroads: How systems hinder or engender change. In A. Hargreaves, A. Lieberman, M. Fullan, & D. Hopkins (Eds.), *Second international handbook of educational change* (pp. 555–571). New York: Springer. doi:10.1007/978-90-481-2660-6_32

Tan, D. (2015). *Engineering Technology, Engineering Education and Engineering Management: Proceedings of the 2014 International Conference on Engineering Technology, Engineering Education and Engineering Management (ETEEEM 2014), Hong Kong, 15-16 November 2014*. CRC Press.

Tan, Malini, & Kaur. (2018). Kahoot! It: Gamification in Higher Education. *Pertanika Journal of Social Science & Humanities*, *26*(1), 565–582.

Tan, P., Strinbach, M., & Kumar, V. (2013). *Introduction to data mining* (International Edition). Pearson.

Tapsir & Puteh. (2018). *Framing Malaysian Higher Education 4.0: Future-Proof Talent*. Putrajaya: Department of Higher Education, Ministry of Higher Education Malaysia.

Tapsir. (2017). 9th *UTM University Presidents Forum 2017*. Universiti Teknologi Malaysia.

Tarmudi, Muhiddin, Rossdy, & Tamsin. (2016). Fuzzy delphi method for evaluating effective teaching based on students' perspective. *E-Academic Journal UiTMT*, *5*, 1-10.

Tatum, J. (2004). The challenge of responsible design. Design Issues, 20(3), 66-80.

Taylor, P. J., O'Driscoll, M. P., & Binning, J. F. (1998). A new integrated framework for training needs analysis. *Human Resource Management Journal*, *8*(2), 29–50. doi:10.1111/j.1748-8583.1998.tb00165.x

Teaching and Teacher in Blended Learning Model. (n.d.). Retrieved from http://www.google.com

Teaching with Technology Online Workshop Series. (2012-2013). *Blended Learning Models & Examples*. Retrieved from https://teachingwithtech.lss.wisc.edu/m3w1.htm

TeachThought Staff. (2015). *The definition of blended learning*. Retrieved from https://www.teachthought.com/learning/the-definition-of-blended-learning/

TeachThought Staff. (2018). *Find the model that works for you: 12 types of blended learning*. Retrieved from https://www.teachthought.com/learning/12-types-of-blended-learning/

Teague, D. M., & Roe, P. (2009). Learning to program: from pear-shaped to pairs. *Proceedings of the First International Conference on Computer Supported Education*, *2*, 151-158.

Teague, D., & Roe, P. (2008). Collaborative learning: towards a solution for novice programmers. *Proceedings of the Tenth Conference On Australasian Computing Education*, *78*,147-153.

Tearle, P. (2003). ICT implementation: What makes the difference? *British Journal of Educational Technology*, *34*(5), 567–583. doi:10.1046/j.0007-1013.2003.00351.x

The Carnegie Project on the Education Doctorate (CPED). (n.d.). Retrieved from https://cpedinitiative.site-ym.com/

The Higher Learning Commission North Central Association of Colleges and Schools. (2017). *A Report to the Board of Trustees from the Task Force on the Professional Doctorate. Received by the Board of Trustees of The Higher Learning Commission, June 6, 2006*. The Higher Learning Commission North Central Association of Colleges and Schools. Retrieved December 18, 2017, from: http://www.hlcommission.org/Publications/publications-list.html

The Woodrow Wilson National Fellowship Foundation. (2005). *The responsive Ph.D.: Innovations in U.S. doctoral education*. Princeton, NJ: Author.

Thessin, R. A., & Starr, J. P. (2011). Supporting the GROWTH of effective professional learning communities districtwide. *Phi Delta Kappan*, *92*(6), 48–54. doi:10.1177/003172171109200611

Thomas, C. (2014). *Kahoot!* Retrieved from https://www.graphite.org/website/kahoot

Thomas, G. (2009). *How to do Your Research Project*. London: Sage.

Thomas, J. R., Nelson, J. K., & Silverman, S. J. (2011). *Research methods in physical activity*. Champaign, IL: Human Kinetics.

Thompson, B., & Stapleton, J. C. (1979). A method for validating semantic differential referents. *Journal of Experimental Education*, *48*(2), 110–113. doi:10.1080/00220973.1979.11011721

Thompson, N. A., & Prometric, T. A. (2007). Practitioner's guide for variable-length computerized classification testing. *Practical Assessment, Research & Evaluation*, *12*, 1–13.

Thompson, N. A., & Weiss, D. J. (2011). A framework for the development of computerized adaptive tests. *Practical Assessment, Research & Evaluation*, *16*(1), 1–9. Retrieved from http://pareonline.net/getvn.asp?v=16&n=1

Timmons, J. A. (1997). *The entrepreneurship mind*. Prentice-Hall, Inc.

Toiskallio, J. (2003). Ethics, Military Pedagogy and Action Competence. In E. R. Micewski (Ed.), *Civil-Military Aspects of Military Ethics*. Frankfurt, Germany: Verlag.

Tomlinson, B., & Whittaker, C. (Eds.). (2013). *Introduction. Blended in English language teaching: Course design and implementation.* London, UK: British Council.

Torres, C., Lopes, A. P., Babo, L., & Azevedo, A. M. (2009). Developing multiple-choice questions in mathematics. *Proceedings of ICERI 2009 - International Conference of Education, Research and Innovation*, 6218-6229.

Towns, M. H. (2014). Guide to developing high-quality, reliable, and valid multiple-choice assessments. *Journal of Chemical Education, 91*(9), 1426–1431. doi:10.1021/ed500076x

Triantafillou, E., Georgiadou, E., & Economides, A. A. (2006). *He Design and Evaluation of a Computeriaed Adaptive Test on Mobile Devices. In Science Direct* (pp. 1319–1330). Elsevier.

Trifonova, A., Georgieva, E., & Ronchetti, M. (2006). Determining students' readiness for mobile learning. In *Proceedings of the 5th WSEAS International Conference on E-activities* (pp. 84-89). Venice, Italy: Academic Press.

Tsai, C. F., & Lu, Y. H. (2009). Customer churn prediction by hybrid neural networks. *Expert Systems with Applications, 36*(10), 12547–12553. doi:10.1016/j.eswa.2009.05.032

Tsai, S., & Chai, C. S. (2012). The "Third " order barrier for technology-integration instruction : Implications for teacher education. *Australasian Journal of Educational Technology, 28*(6), 1057–1060. doi:10.14742/ajet.810

Tsai, W. T., Li, W., Elston, J., & Chen, Y. (2011). Collaborative learning using wiki web sites for computer science undergraduate education: A case study. *IEEE Transactions on Education, 54*(1), 114–124. doi:10.1109/TE.2010.2046491

Tseng, M. M., Jiao, R. J., & Wang, C. (2010). Design for mass personalization. *CIRP Annals - Manufacturing Technology, 59*(1), 175-178. doi:10.1016/j.cirp.2010.03.097

Tucker, B. (2012). The Flipped Classroom. *Education Next, 12*(1).

Umapathy, K., & Ritzhaupt, A. D. (2017). A Meta-Analysis of Pair-Programming in Computer Programming Courses: Implications for Educational Practice. *ACM Trans. Comput. Educ., 17*(4), 16:1–16:13. doi:10.1145/2996201

Vacca, R. T., & Vacca, J. L. (2002). *Content area reading: Literacy and learning across the curriculum* (7th ed.). Boston: Allyn and Bacon.

Van Eerde, W., Tang, K., & Talbot, G. (2008). The mediating role of training utility in the relationship between training needs assessment and organizational effectiveness. *International Journal of Human Resource Management, 19*(1), 63–73.

Vardi, M. Y. (2012). Will MOOCs destroy academia? *Communications of the ACM, 55*(11), 5. doi:10.1145/2366316.2366317

Vassileva, J. (2008). Toward social learning environment. *IEEE Transactions on Learning Technologies, 1*(4), 199–213. doi:10.1109/TLT.2009.4

Vatanartiran, S., & Karadeniz, S. (2015). A needs analysis for technology integration plan: Challenges and needs of teachers. *Contemporary Educational Technology, 6*(3), 206–220.

Veeramanickam, M. R. M., & Mohanapriya, M. (2017). IOT enabled Futurus Smart Campus with effective E-Learning: i-Campus. *GSTF Journal of Engineering Technology, 3*(4).

Volpe. (2017). Habilidades e competências do profissional para o ambiente da Indústria 4.0: uma revisão sistemática. In XXXVII Encontro Nacional de Engenharia de Produção, 2017. Joinville/SC: Enegep.

Vygotsky, L. (1978). *Mind and society.* Cambridge, MA: Harvard University Press.

Vygotsky, L. S. (1986). *Thought and language.* Cambridge, MA: MIT Press. Wainer.

Wachira, P., & Keengwe, J. (2011). Technology Integration Barriers: Urban School Mathematics Teachers Perspectives. *Journal of Science Education and Technology*, *20*(1), 17–25. doi:10.100710956-010-9230-y

Walker, G. (2005). *Critical thinking in asynchronous discussions.* Retrieved April 5, 2018, from http://www.itdl.org/Journal/Jun_05/article02.htm

Walker, G. E., Golde, C. M., Jones, L., Conklin Bueschel, A., & Hutchings, P. (2008). *The formation of scholars: Rethinking doctoral education for the twenty-first century.* San Francisco, CA: The Carnegie Foundation for the Advancement of Teaching.

Walker, M. (1993). Developing the theory and practice of action research: A South African case. *Educational Action Research*, *1*(1), 95–109. doi:10.1080/0965079930010106

Walker, S. (2006). *Sustainable by design: Explorations in theory and practice.* London, UK: Routledge.

Wan Din, W. C. R. (2001). Wanita dalam pendidikan. *Jurnal Pengurusan dan Kepimpinan Pendidikan*, *11*, 1-14.

Wang, H., & Shin, C. D. (2010). Comparability of Computerized Adaptive and Paper-Pencil Tests. *Test, Measurement & Research Services Bulletin*, 13. Retrieved from https://pdfs.semanticscholar.org/6f1e/bcee90a66fcc969c4caba1362bedb-39d505a.pdf

Wang, A. L. (2015). The wear out effect of a game-based student response system. *Computer Education*, *82*, 217–227. doi:10.1016/j.compedu.2014.11.004

Wang, A. L., Zhu, M., & Satre, R. (2016). The effect of digitizing and gamifying in classrooms. In *Proceedings of the 10th European Conference on Game Based Learning*. University of the West of Scotland.

Wang, M., & Kang, M. (2006). Cybergogy for engaged teaming: A Framework for Creating Learner Engagement through Information and Communication Technology. In D. Hung & V. I. S. Khine (Eds.), *Engaged Learning with Emerging Technologies*. Dordrecht, The Netherlands: Springer. doi:10.1007/1-4020-3669-8_11

Wang, S.-L., & Lin, S. S. J. (2007). The effects of group composition of self-efficacy and collective efficacy on computer-supported collaborative learning. *Computers in Human Behavior*, *23*(5), 2256–2268. doi:10.1016/j.chb.2006.03.005

Wang, Y. (2004). Context awareness and adaptation in mobile learning. *Proceedings of the International Workshop on Wireless and Mobile Technologies in Education*, 154-158.

Wanyama, T., Singh, I., & Centea, D. (2018). A Practical Approach to Teaching Industry 4.0 Technologies. In *Online Engineering & Internet of Things* (pp. 794–808). Springer. doi:10.1007/978-3-319-64352-6_74

Washer, P. (2001). Barriers to the use of web-based learning in nurse education. *Nurse Education Today*, *21*(6), 455–460. doi:10.1054/nedt.2001.0583 PMID:11466008

Wasserman, E. (2013). How do Lecturers Evaluate their Own Work? *Procedia: Social and Behavioral Sciences*, *89*, 144–147. doi:10.1016/j.sbspro.2013.08.824

Waters, R. (2016, January 5). Artificial Intelligence: Can Watson Save IBM? *Financial Times.* Accessed November 1, 2017: https://www.ft.com/content/dced8150-b300-11e5-83589a82b43f6b2f

Watson, D. L., & Nehls, K. (2016). Alternative dissertation formats: Preparing scholars for the academy and beyond. In *Contemporary approaches to dissertation development and research methods.* Hershey, PA: IGI Global.

Watson, J., Murin, A., Vashaw, L., Gemin, B., & Rapp, C. (2010). *Keeping pace with K-12 online learning: An annual review of policy and practice.* Academic Press.

Way, W. D., Twing, J. S., Camara, W., Sweeney, K., Lazer, S., & Maeo, J. (2010). *Some considerations related to the use of adaptive testing for the common core assessments*. Educational Testing Service.

Weber, A. S. (2013). Cloud computing in education. In *Ubiquitous and mobile learning in the digital age* (pp. 19–36). New York, NY: Springer. doi:10.1007/978-1-4614-3329-3_2

Weber, R. H. (2009). Internet of things–Need for a new legal environment? *Computer Law & Security Review*, *25*(6), 522–527. doi:10.1016/j.clsr.2009.09.002

Weber, R. H. (2011). Accountability in the Internet of Things. *Computer Law & Security Review*, *27*(2), 133–138. doi:10.1016/j.clsr.2011.01.005

Weimer, M. (2013). *Learner-centered teaching: Five key changes to practice* (2nd ed.). San Francisco, CA: Jossey Bass.

Weindling, A. M. (2001). Education and training: Continuing professional development. *Current Paediatrics*, *11*(5), 369–374. doi:10.1054/cupe.2001.0217

Weiser, M. (1991). The Computer for the 21 st Century. *Scientific American*, *265*(3), 94–105. doi:10.1038cientificamerican0991-94 PMID:1675486

Weiss, D. J., & Kingsbury, G. (1984). Application of computerized adaptive testing to educational problems. *Journal of Educational Measurement*, *21*(4), 361–375. doi:10.1111/j.1745-3984.1984.tb01040.x

Wellington, J. (2000). *Teaching and learning secondary Science: Contemporary Issues and Practical Approaches*. London: Routledge. doi:10.4324/9780203059876

Wellington, J., & Osborne, J. (2001). *Language and Literacy in Science Education*. Buckingham, UK: Open University Press.

Wenglinsky, H. (2000). *How teaching matters: Bringing the classroom back into discussions of teacher quality*. Princeton, NJ: Educational Testing Service.

Werbach, K., & Hunter, D. (2012). *For the win: How game thinking can revolutionize your business*. Philadelphia: Wharton Digital Press.

West-Burnham, J., & O'Sullivan, F. (1998). *Leadership and professional development in schools*. London: Pitman.

West, G. (2017). *Scale: The universal laws of life and death in organisms, cities and companies*. London, UK: Weidenfeld & Nicholson.

Whitaker, P. (1993). *Managing change in schools*. Buckingham, UK: Open University Press.

Whitaker, P. (1997). Changes in professional development: The personal dimension. In L. Kydd, M. Crawford, & C. Riches (Eds.), *Professional development and educational management*. Buckingham, UK: Open University Press.

Wilkinson, I. A. G., Soter, A. O., & Murphy, P. K. (2010). Developing a model of quality talk about literacy text. In M. G. McKeown & L. Kucan (Eds.), *Bringing reading research to life* (pp. 142–169). New York: The Guilford Press.

Willems, G. M., Stakenborg, J. H. J., & Veugelers, W. (2000). *Trends in Dutch Teacher Education*. Garant.

Williams, L. A., & Kessler, R. R. (2000). All I really need to know about pair programming I learned in kindergarten. *Communications of the ACM*, *43*(5), 108–114. doi:10.1145/332833.332848

Williams, L., McDowell, C., Nagappan, N., Fernald, J., & Werner, L. (2003). Building pair programming knowledge through a family of experiments. *Proceedings. International Symposium on Empirical Software Engineering*, 143-152. 10.1109/ISESE.2003.1237973

Williams, L., & Upchurch, R. L. (2001). In support of student pair-programming. *ACM SIGCSE Bulletin*, *33*(1), 327–331. doi:10.1145/366413.364614

Winn, M. R. (2012). Promote digital citizenship through school-based social networking. *Learning and Leading with Technology*, *39*(4), 10–13.

Wiske, M. S., Franz, K. R., & Breit, L. (2005). *Teaching for understanding with technology*. San Francisco: Jossey-Bass Inc.

Witherspoon, A. (2011). *Edmodo a learning management system*. Retrieved from http://www.poweredwithtechnology.com/2011/01/edmodoa-learning-management-system.html

Wlodkowski, J. R. (2011). *Enhancing adult education to learn. A comprehensive guide for teaching all adults* (3rd ed.). Hoboken, NJ: John Wiley & Sons.

Wolff, A., Zdrahal, Z., Nikolov, A., & Pantucek, M. (2013, April). Improving retention: predicting at-risk students by analysing clicking behaviour in a virtual learning environment. In *Proceedings of the Third International Conference on Learning Analytics and Knowledge* (pp. 145-149). ACM. 10.1145/2460296.2460324

Wollschlaeger, M., Sauter, T., & Jasperneite, J. (2017). The future of industrial communication: Automation networks in the era of the internet of things and industry 4.0. *IEEE Industrial Electronics Magazine*, *11*(1), 17–27. doi:10.1109/MIE.2017.2649104

Wood, K. (2003). *Introduction to Mobile Learning*. M Learning.

Woodrow, L. (2014). *Writing about quantitative research in applied linguistics*. New York, NY: Springer. doi:10.1057/9780230369955

Woodward, I. (1996). Dimensions of Continuing Professional Development. In I. Woodward (Ed.), *Continuing professional development* (pp. 1–18). London: Cassell.

Wu, X., Kumar, V., Quinlan, J. R., Ghosh, J., Yang, Q., Motoda, H., ... Zhou, Z. H. (2008). Top 10 algorithms in data mining. *Knowledge and Information Systems*, *14*(1), 1–37. doi:10.100710115-007-0114-2

Wynn, D., & Williams, C. K. (2012). Principles for conducting critical realist case study research in information systems. *Management Information Systems Quarterly*, *36*(3), 787–810. doi:10.2307/41703481

Xing, B., & Marwala, T. (2017). *Implications of the Fourth Industrial Age on Higher Education*. arXiv preprint arXiv:1703.09643

Xing, B., & Marwala, T. (2017). *Implications of the Fourth Industrial Age on Higher Education*. Retrieved from https://arxiv.org/ftp/arxiv/papers/1703/1703.09643.pdf

Yadin, A. (2011). Reducing the dropout rate in an introductory programming course. *ACM Inroads*, *2*(4), 71–76. doi:10.1145/2038876.2038894

Yagci, T. (2015) Mobile Social Media in Higher Education & Implementation of "EDMODO" in reading classes. *Proceedings of 6th International Visible Conference on Educational Studies and Applied Linguistics*, 436-442.

Yanai. (2017). O desenvolvimento da Indústria 4.0: um estudo bibliométrico. In XXXVII Encontro Nacional de Engenharia de Produção, 2017. Joinville/SC: Enegep.

Youngblood, G. M., Heierman, E. O., Holder, L. B., & Cook, D. J. (2005). Automation intelligence for the smart environment. *International Joint Conference on Artificial Intelligence*, *19*, 1513.

Young, P. M. (2010). Teaching the Ethical Values Governing Mediator Impartiality Using Short Lectures, Buzz Group Discussions, Video Clips, a Defining Features Matrix, Games, and an Exercise Based on Grievances Filed Against Florida Mediators. *Pepp. Disp. Resol. LJ*, *11*, 309.

Zarzycka-Piskorz, E. (2016). Kahoot it or not? Can games be motivating in learning grammar? *Teaching English with Technology*, *16*(3), 17–36.

Zhiqiang, H., & Junming, Z. (2011). The Application of Internet of Things in Education and Its Trend of Development. *Modern Distance Education Research, 2*, 19.

Zichermann, G., & Cunningham, C. (2011). *Gamification by design*. Sebastopol, CA: O'Reilly Media.

Zikmund, W. G., Babin, B. J., Carr, J. C., & Griffin, M. (2010). Business research methods (8th ed.). South-Western: Cengage Learning.

Zimmaro, D. M. (2016). *Writing good multiple-choice exams*. Retrieved from https://facultyinnovate.utexas.edu/sites/default/files/writing-good-multiple-choice-exams-fic-120116.pdf

Zuber-Skerritt, O. (1992). *Professional development in higher education*. London: Kogan Page.

About the Contributors

Arumugam Raman is a faculty member in the School of Education and Modern Languages, College of Arts and Sciences in Universiti Utara Malaysia. His research interests include Educational Technology, Instructional Technology, Computers in Education, Educational Research Methodology, Technology and ICT, Statistics for Educational Research and Partial Least Squares Structural Equation Modeling (PLS-SEM). He had engaged in many research projects and had received five grants including university, national (FRGS, RAGS) and college (Leads). The researcher is better known as an Educational Technologist and currently teaching ICT in Education, Research Methods and Statistics in Education. Besides, he has written International articles and books which have been published in national and international levels. He is a member of few Editorial boards for International Journals such as Journal of International Education Studies, International Journal of Education and Development using ICT, and Journal of Studies in Education. Furthermore, he is also a member of professional bodies namely International Association of Science and Technology for Development (IASTED), Theosophical Society, Association for Computing Machinery (ACM), Internet society (ISOC), and Qualitative Research Association of Malaysia. Apart from this, the researcher has in-depth knowledge of Tamil Siddha Philosophy. He too has been involved in community services through various ways.

Mohan Rathakrishnan is a research fellow of Institut of Tun Mahathir's Thought (IPDM) (2018-2022) and senior lecturer in University Utara Malaysia. Currently he is attached to School of Language, Civilization and Philosophy. He is an e-learning and instructional designer in teaching and learning program. He also teaches critical thinking and creative thinking. He has conducted many workshops on Blended Learning, Critical Thinking, General Studies and Web 2.0. He has also written many books on General Studies and e-learning. He supervises and advise students who likes to do research on online learning (social media, open learning or blended learning) based on any suitable subject discipline. Currently he is supervising few Ph.d and master students regarding collaborative learning and online learning in applied linguistic and education. He has received Top Article Writer award and excellent lecturer award in 2018. He received RAGS, University grant, sTOL and DBP external grant to do research on teaching and online learning. He has become reviewer in many international online learning conferences paper. He is one of the editor for IGI global Redesigning Higher Education Initiates for Revolution of 4.0. and a member of few Editorial boards for International Journals such as Journal of International Education Studies and Internet Society.

* * *

Azlida Ahmad has served as the Head Department of General Studies (English language and Islamic Education) of Tuanku Syed Sirajuddin Polytechnic, Perlis from 2006 till now. She is also the International Affairs Officer at the institution. She completed her bachelor degree in Modern English Language Education from Lancaster University, U. K in 1990 and her master's degree in Education (Instructional Technology) from Science University of Malaysia. She was an English teacher at Matriculation College (University Utara Malaysia) from 1990 – 1997 before joining Sultan Abdul Halim Mu'adzam Polytechnic, Kedah. Azlida has been actively involved in research and innovation related to English as well as Technical and Vocational Education and Training (TVET). She won two silver medals for the innovative projects entitled "Interview skills using Sign Language for Deaf Polytechnic Students" and "Application of Gagne's nine events of instruction into Teaching and Learning" at Malaysian Technology Expo (MTE) 2011 and 2012. She has conducted a national event where well-known speakers came and contributed their ideas at the institution. She has been active in organizing international events with Indonesian and Thailand community colleges, polytechnics and university. Besides teaching, she also collaborates with local and national key industrial players that can eventually contribute to polytechnic students significantly.

Gary A. Berg, PhD, MFA, is the author of eight books including Low-Income Students and the Perpetuation of Inequality: Higher Education in America, and numerous articles on current issues in higher education, public policy, and distance learning. This chapter was support by a research fellowship at the University of Phoenix.

Mohan Chinnappan is a K-12 mathematics teacher and teacher educator. He has national and international recognition for his research on mathematics teaching and learning with particular interest in knowledge that supports students' understandings. His research endeavours have made significant contributions to the scholarship of teaching, and to the mathematics teaching profession. These contributions have been informed and supported by internationally recognised research and consultancy work in Australia and Asia. His research program focuses on a) the elucidation of organisation of mathematical information that fosters learning/problem solving and b) pedagogical approaches that would support the construction of robust mathematical schemas by students and teachers of mathematics. His research on the development of deep understanding of mathematics/numeracy and teaching has been significant in examining current issues of national and global concern including the numeracy for disadvantaged children and the implications of NAPLAN for innovations in classroom practices.

Kathleen Dass is currently a teacher educator in the Institute of Teacher Education, Sultan Abdul Halim Campus, Kedah, Malaysia. She received her M.Ed. TESOL from University Sains Malaysia in 2006. She has nineteen years of experience teaching in the primary and secondary schools before she joined the ITE in 2007.

Dennis Foung is a Teaching Fellow at the Hong Kong Polytechnic University. He holds a doctorate in language education and has been teaching English to different tertiary students in Hong Kong. Having a keen interest in computer-assisted language learning, he coordinated the design and development of various language learning platforms. Other than CALL, his research interest includes learning analytics, classroom discourse, and teacher behaviors. He is the facilitator of his university's Community of Practice on Learning Analytics (LA), and encourages the use of LA to evaluate online learning.

Jowati Juhary is an author who is a member of the Language Centre at the National Defence University of Malaysia. She is currently heading the UPNM Press, the university's publishing centre. The author has 18 years of teaching experience, and her research interests include language education, military pedagogy and higher education.

Liew-Kee Kor is a PhD graduate in Mathematics education. Her PhD project on the use of graphics calculator had won an award in the USM Graduate Research Exposition 2004. Since then, she has been actively researching and publishing articles on technology in mathematics education. She has presented papers at national and international conferences and has written journal articles pertaining to the use of educational technology in the teaching of mathematics. She is also actively involved in reviewing papers for journals and conferences. She has authored papers published in The Encyclopedia of the Science of Learning and the ZDM Mathematics Education. She was also a regional editor and contributing author in B. Sriraman (Ed.). The first sourcebook on Asia research in mathematics education: China, Korea, Singapore, Japan, Malaysia, and India.

Jennifer Loy is Professor of Product Design & Design for Additive Manufacturing at the University of Technology Sydney and research theme co-leader for Transformative Technologies and Data Poetics for Design, Architecture and Building, working across Design and the School of Engineering. Professor Loy teaches design for additive manufacturing and Major Project in the final year of the Product Design program. Her research is focussed on the development and impact of digital technologies on education, products and environments. Her additive manufacturing interests span design for health and wellbeing, clinical and industrial product, to commercial and creative product. She has a particular interest on the impact of disruptive design on business models and product service systems and the implications for design education.

Mohd Mahmud is a university lecturer at the School of Electrical and Electronic Engineering, Universiti Sains Malaysia. He has a degree in Electronic Systems Engineering from the University of Essex, an MPhil in Technology Policy and a PhD in Engineering, both from the University of Cambridge

Sergio Mazini is a Master in Production Engineering from UNESP, has specialized in analysis, development and management of systems projects by UNILINS and Graduate Technologist in Data Processing by FATEB. Course coordinator UniToledo of Production Engineering, University Professor of Business Administration courses Business, Information Systems and Production Engineering. Consultant and analyst in project development, implementation and management of integrated management systems, ORION Solutions Management. Has business consulting expertise, business management, production engineering, developing information systems, management innovation, knowledge management and enterprise portals.

Abdul Halim Mohamed is currently serving as a senior lecturer in Educational Technology Department, School of Education and Modern Language (SEML), Universiti Utara Malaysia (UUM), Malaysia since 1996. He served as a secondary school teacher in a school and UUM Matriculation Centre, Kedah from 1991 – 1996. He completed his PhD in ICT Training in University of East Anglia, United Kingdom in 2006. He finished his master's degree in Instructional Design and Technology from the University of Iowa, Iowa City, USA in 1997 and completed his Bachelor Degree in Teaching English of Speakers of Other Languages (TESOL) in 1990. Abdul Halim used to be an educational consultant/coach and mentor for Multimedia Development Corporation (MDeC), Cyberjaya and Educational Technology Division, Ministry of Education in 2007 and 2009 where he conducted training to improve teachers in managing changes and ICT enculturation successfully in smart schools. He has been recently appointed as an educational consultant and module writer for ILHAM Educare.

Affah Mohd Apandi is an English teacher at the Center of Language Studies at the Universiti Tun Hussein Onn Malaysia. She is also a PhD candidate at the Universiti Utara Malaysia where she furthers her research on Instructional Technology. Her research and publication interests include E-learning, Mobile learning, Blended learning, Flipped Classroom, Distance learning, Online learning and MOOC.

Ruuhina Mohd Sani graduated at the School of Education and Arts at the Universiti Utara Malaysia, Malaysia and completed her Master's Degree of Instructional Technology at the same university. Ruuhina has a PhD in Education and her research focuses on developing ICT competency standard for teacher in Malaysia. Her research interests also include Education 4.0 and educational policy. She has 7 years of experience as Executive Information Officer and she has worked closely with students and teachers in schools.

Mohd Faiz Mohd Yaakob is a Senior Lecturer at the School of Education and Modern Languages, College of Arts and Sciences, Universiti Utara Malaysia. He has more than 10 years of teaching experience in school. His expert and research interest area are about Educational Planning and Policy, Educational Administration and Islamic Studies.

Malar Muthiah is a Senior Academic Lecturer in IPG Campus Sultan Abdul Halim. 27 years of teaching experience in History Subject. Involve in module writing for Education Ministry for Professional Learning Community and northern region subject matter expert for implementation of PLC in schools. Have published many research papers in conference especially in pedagogical method or approaches in teaching and learning History subject in schools.

James Novak is a postdoctoral researcher in Product Design and Design for Additive Manufacturing at the University of Technology Sydney, and has experience with 3D printing dating back to 2009. He has won international awards for his research including the prestigious Dick Aubin Distinguished Paper Award in 2015, and is most well known for his full-size 3D printed bicycle frame which has been exhibited around the world at venues including the Red Dot Design Museum in Germany. James has previously been a lecturer and course convener at Griffith University within the Industrial Design and Product Design degrees, and has been appointed an Advance Queensland Community Digital Champion by the Queensland Government for his work educating the community about 3D printing and associated digital technologies.

Abdul Hamid Nur is a senior lecturer at School of Education and Modern Languages, SEML, Universiti Utara Malaysia specialized in Instructional Technology. His research interest is in enhancing teaching and learning process in the classroom such as using different technological tools in teaching. His research is also related to TPACK. He teaches various IT subjects such Fundamentals of Instructional Technology, Web Development, IT in Education and other related subjects.

Quinie Ong is currently a teacher educator in Institute of Teacher Education, Sultan Abdul Halim Campus, Kedah, Malaysia. She received her M.Ed. TESOL from University Sains Malaysia in 2004. She was teaching in the primary and secondary schools for twenty years before she joined the ITE in 2007.

Shafinah Farvin Packeer Mohamed is a lecturer in the School of Computing, Universiti Utara Malaysia, Kedah, Malaysia. She is also a member of Human-Centered Computing Lab of this school. She finished her Ph. D in Universiti Utara Malaysia in the field of Software Engineering. In particular, her research areas are the software process quality evaluation, software process certification, and multi-criteria decision making. Currently she is teaching programming subjects which uses Java as the programming language.

Robyn Pierce is an Associate Professor in mathematics education. Her current interests are in statistical literacy for students and teachers, using technology to support learning in mathematics, algebra and the use of symbols and diagnostic assessment.

S. Kanageswari Shanmugam is a senior lecturer and Head of the Mathematics Unit at the School of Education and Modern Languages, Universiti Utara Malaysia. She holds a Bachelor's degree with double majors in Mathematics and English, master's degree in Psychometrics and Educational Measurement and doctoral degree in Psychometrics and Educational Evaluation from Universiti Sains Malaysia, Penang. She served as a teacher for 14 years before joining SEAMEO RECSAM as a Mathematics Education Specialist where she conducted training programmes and research on secondary data analyses on TIMSS. She is an active member of the National Council on Measurement in Education, the International Community of Emergent Researchers in Mathematics Education based at the University of Hamburg, Germany and the Associate Member for the Network on Education Quality Monitoring in the Asia-Pacific UNESCO Bangkok and a committee member of the Malaysian Psychometric Association. She has published several books on TIMSS, and has conducted several international and national colloquia/workshops/seminar on TIMSS, PISA, Higher-order Thinking Skills (HOTs), HOTs Item Development, Authentic Assessment, Lesson Study/Action Research and Mathematics Pedagogy. She was also a key personnel in the development of KBAT (Kemahiran Berfikir Aras Tinggi) [HOTs] Manual by the Examination Syndicate, Ministry of Education, Malaysia that is being used nationwide.

Duncan Symons is a lecturer in science and mathematics education. His primary responsibilities involve preparatory teacher education in the fields of primary mathematics and science. Duncan's research interests include inquiry, investigative and problem based approaches to mathematics education in the primary years. He is also interested in how mathematics can be embedded within the broader curriculum. The adoption and promotion of STEM (science, technology, engineering and mathematics) as a means to achieve integration has become an area of research and teaching interest, and Duncan facilitates a program for teacher candidates at the University of Melbourne with this as a focus. Duncan has many research publications in the areas of technology embedded mathematics, dialogic approaches to mathematics instruction and integrated STEM education.

Raamani Thannimalai is a Senior Teacher at Kepala Batas National Secondary School. Completed her Ph.D. at Universiti Utara Malaysia.

Byabazaire Yusuf is currently a senior faculty member at the School of Education and Modern Languages (SEML), College of Arts and Sciences (CAS), Universiti Utara Malaysia (UUM), Kedah, Malaysia. He completed his PhD in Instructional Technology at the University of Malaya, Malaysia in 2010. During his professional carrier, he has worked in various school settings and higher education institutions for many years. He teaches subjects including; Fundamentals of Instructional Technology, ICT in Education, Multimedia for Education and Training, Web Development for Instruction, Instructional Technology Assessment and Evaluation, and research methods at graduate level. Areas of research interest include; ICT integration in education and training, use of technology in special needs education, and library and information science. He has published both research and conceptual articles in National and International Journals and in conference proceedings. He is also the co-editor for- Fast forwarding Higher Education Institutions for Global Challenges: Perspectives and Approaches (2016).

Mohd Muslim Md Zalli is a Lecturer at the Faculty of Human Development, Universiti Pendidikan Sultan Idris (UPSI); and a doctoral student at the School of Education and Modern Languages, Universiti Utara Malaysia (UUM). He does research in Higher Learning Education specifically in the area of self-regulated learning, community of inquiry, and Massive Open Online Courses (MOOCs).

Index

E

F

G

H

I

K

L

M

N

O

P

S

T

W

Purchase Print, E-Book, or Print + E-Book

IGI Global books are available in three unique pricing formats:
Print Only, E-Book Only, or Print + E-Book. Shipping fees apply.

www.igi-global.com

Recommended Reference Books

ISBN: 978-1-5225-2145-7
© 2017; 408 pp.
List Price: $210

ISBN: 978-1-5225-2334-5
© 2017 ; 375 pp.
List Price: $195

ISBN: 978-1-5225-1718-4
© 2017; 457 pp.
List Price: $270

ISBN: 978-1-5225-0672-0
© 2017; 678 pp.
List Price: $295

ISBN: 978-1-5225-2630-8
© 2018; 209 pp.
List Price: $145

ISBN: 978-1-5225-1621-7
© 2017; 173 pp.
List Price: $135

Do you want to stay current on the latest research trends, product announcements, news and special offers?
Join IGI Global's mailing list today and start enjoying exclusive perks sent only to IGI Global members.
Add your name to the list at **www.igi-global.com/newsletters.**

Publisher of Peer-Reviewed, Timely, and Innovative Academic Research

IGI Global
DISSEMINATOR OF KNOWLEDGE

www.igi-global.com Sign up at www.igi-global.com/newsletters facebook.com/igiglobal twitter.com/igiglobal linkedin.com/igiglobal

Ensure Quality Research is Introduced to the Academic Community

Become an IGI Global Reviewer for Authored Book Projects

The overall success of an authored book project is dependent on quality and timely reviews.

In this competitive age of scholarly publishing, constructive and timely feedback significantly expedites the turnaround time of manuscripts from submission to acceptance, allowing the publication and discovery of forward-thinking research at a much more expeditious rate. Several IGI Global authored book projects are currently seeking highly qualified experts in the field to fill vacancies on their respective editorial review boards:

Applications may be sent to:
development@igi-global.com

Applicants must have a doctorate (or an equivalent degree) as well as publishing and reviewing experience. Reviewers are asked to write reviews in a timely, collegial, and constructive manner. All reviewers will begin their role on an ad-hoc basis for a period of one year, and upon successful completion of this term can be considered for full editorial review board status, with the potential for a subsequent promotion to Associate Editor.

If you have a colleague that may be interested in this opportunity, we encourage you to share this information with them.

IGI Global
DISSEMINATOR OF KNOWLEDGE
www.igi-global.com

Celebrating 30 Years of Scholarly Knowledge Creation & Dissemination

CELEBRATING 30 years EST. 1988 Academic Publishing Excellence

InfoSci®-Books

A Collection of 4,000+ Reference Books Containing Over 87,000 Full-Text Chapters Focusing on Emerging Research

This database is a collection of over 4,000+ IGI Global single and multi-volume reference books, handbooks of research, and encyclopedias, encompassing groundbreaking research from prominent experts worldwide. These books are highly cited and currently recognized in prestigious indices such as: Web of Science™ and Scopus®.

Librarian Features:

- No Set-Up or Maintenance Fees
- Guarantee of No More Than A 5% Annual Price Increase
- COUNTER 4 Usage Reports
- Complimentary Archival Access
- Free MARC Records

Researcher Features:

- Unlimited Simultaneous Users
- No Embargo of Content
- Full Book Download
- Full-Text Search Engine
- No DRM

To Find Out More or To Purchase This Database:

www.igi-global.com/infosci-books

eresources@igi-global.com • Toll Free: 1-866-342-6657 ext. 100 • Phone: 717-533-8845 x100

IGI Global
DISSEMINATOR OF KNOWLEDGE
www.igi-global.com

IGI Global Proudly Partners with

eContent Pro International

Expert Copy Editing

eContent Pro International copy editors, with over 70 years of combined experience, will provide complete and comprehensive care for your document by resolving all issues with spelling, punctuation, grammar, terminology, jargon, semantics, syntax, consistency, flow, and more. In addition, they will format your document to the style you specify (APA, Chicago, etc.). All edits will be performed using Microsoft Word's Track Changes feature, which allows for fast and simple review and management of edits.

Additional Services

eContent Pro International also offers fast and affordable proofreading to enhance the readability of your document, professional translation in over 100 languages, and market localization services to help businesses and organizations localize their content and grow into new markets around the globe.

IGI Global Authors Save 25% on eContent Pro International's Services!

Scan the QR Code to Receive Your 25% Discount

The 25% discount is applied directly to your eContent Pro International shopping cart when placing an order through IGI Global's referral link. Use the QR code to access this referral link. eContent Pro International has the right to end or modify any promotion at any time.

Email: customerservice@econtentpro.com

econtentpro.com

Are You Ready to Publish Your Research?

IGI Global
DISSEMINATOR OF KNOWLEDGE

IGI Global offers book authorship and editorship opportunities across 11 subject areas, including business, healthcare, computer science, engineering, and more!

Benefits of Publishing with IGI Global:

- Free one-to-one editorial and promotional support.
- Expedited publishing timelines that can take your book from start to finish in less than one (1) year.
- Choose from a variety of formats including: Edited and Authored References, Handbooks of Research, Encyclopedias, and Research Insights.
- Utilize IGI Global's eEditorial Discovery® submission system in support of conducting the submission and blind-review process.
- IGI Global maintains a strict adherence to ethical practices due in part to our full membership to the Committee on Publication Ethics (COPE).
- Indexing potential in prestigious indices such as Scopus®, Web of Science™, PsycINFO®, and ERIC – Education Resources Information Center.
- Ability to connect your ORCID iD to your IGI Global publications.
- Earn royalties on your publication as well as receive complimentary copies and exclusive discounts.

Get Started Today by Contacting the Acquisitions Department at:

acquisition@igi-global.com